DEVELOPMENTS IN
VOLCANOLOGY 9

VOLCANISM IN THE CAMPANIA PLAIN

Vesuvius, Campi Flegrei and Ignimbrites

FURTHER TITLES IN THIS SERIES

A. Freundt and M. Rosi (Editors)
From Magma to Tephra: Modelling Physical Processes of Explosive Volcanic Eruptions

B. De Vivo and R.J. Bodnar (Editors)
Melt Inclusions in Volcanic Systems: Methods, Applications and Problems

V. Zobin
Introduction to Volcanic Seismology

M. Fytikas and G.E. Vougioukalakis (Editors)
The South Aegean Active Volcanic Arc: Present Knowledge and Future Perspectives

F. Dobran (Editor)
Vesuvius 2000: Education, Security and Prosperity

Front cover illustration shows a gouache by Adriana Pignatelli: “Grande Eruzione del Vesuvio del 1767” (Vesuvius, Great Eruption of 1767) and on the back cover: “Vesuvio e Castel dell’ Ovo”, also gouache by Adriana Pignatelli.

Developments in Volcanology, 9

VOLCANISM IN THE CAMPANIA PLAIN

Vesuvius, Campi Flegrei and Ignimbrites

Edited by

B. DE VIVO
Dipartimento di Scienze della Terra
Università di Napoli Federico II
Napoli, Italy

ELSEVIER

Amsterdam - Boston - Heidelberg - London - New York - Oxford
Paris - San Diego - San Francisco - Singapore - Sydney - Tokyo

Elsevier
Radarweg 29, PO Box 211, 1000 AE Amsterdam, The Netherlands
The Boulevard, Langford Lane, Kidlington, Oxford OX5 1GB, UK

First edition 2006

Library of Congress Cataloguing in Publication Data
A catalogue record for this book is available from the Library of Congress

British Library Cataloguing in Publication Data
A catalogue record for this book is available from the British Library

ISBN-13: 978-0-444-52175-0
ISBN-10: 0-444-52175-5

Printed and bound in The Netherlands

06 07 08 09 10 10 9 8 7 6 5 4 3 2 1

Contents

Preface

The volume "Volcanism in the Campania Plain: Vesuvius, Campi Flegrei, Ignimbrites" contains selected papers presented at the "Workshop on Vesuvius and the volcanism of the Campanian Plain" held in Napoli on 4–6 October 2004. The Workshop was organized on the occasion of the 60th anniversary of the last eruption of Vesuvius, which occurred in March 1944. After this small-energy eruption, Vesuvius entered a repose period which hopefully will last for many more years. Actually, there are good scientific reasons to think that the current repose period might indeed last some centuries (Lima et al., 2003); this was amply discussed in the scientific session of the Workshop. Nevertheless, the attention and vigilance of the political authorities towards the hazards posed by Vesuvius must not be lessened. The possibility of a long repose time following the 1944 eruption only means that the politicians and public authorities should include long-term plans in their agenda, hopefully improving the bizarre emergency evacuation plan prepared by the Department of Civil Protection.

As a matter of fact, the creators of the evacuation plan, commendable for making the Vesuvius hazard a priority for the Neapolitan territory, have assumed that volcanologists will be able to provide, at least, an unequivocal two-week advance warning before an eruption. We know that this will most likely not be the case, and perhaps only a few days warning will be possible, at best. The cases of Montserrat (1995), and Saint Vincent (1979) volcanoes in Antilles Islands are good examples of this circumstance. In particular, the latter volcano erupted violently, with only one day of alert, after 77 years of quiescence. In that case, it was relatively easy to evacuate 3000 people, but in the case of Vesuvius, the people to evacuate are about 800,000! The results of inadequate warning, ill-prepared civil authorities and insufficient science recently occurred with tragic consequences to the people of southeast Asia following the December 2004 earthquake-tsunami.

The main reason to organize the Workshop, and subsequently to publish this volume with Elsevier, is that, in the last few years, the Vesuvius hazard problem has become mostly an argument and discourse for politicians, territory planners, sociologists, etc., whereas the important scientific problems of Vesuvius seem to be left in the background. To the "outside" world of the non-scientist, it may seem as if all the scientific problems concerning Vesuvius have been solved. Many Italian Earth scientists offer to Civil Authorities various models based on assumed "certainties" which in reality are far from being such. These "certainties" assume the importance of scientific dogmas, and as such are passed from the politicians to the population. Naturally, politicians are eager to find scientists who give them "certainties" to be sold to the population; likewise, some scientists are eager to find politicians who support them with generous public funds for their "certainties". This creates a situation which I consider lethal for an impartial and balanced evaluation of research results and progress (of course, this is true for all science fields with a high political profile).

The "certainties" given to the politicians are only models based on available data, but I want to emphasize that we are still unable to construct a realistic, reliable model on how

Vesuvius works. There is still much to be done in order to increase our knowledge on the numerous variables which control the dynamics of an active volcano. I think that our ignorance is certainly greater than our knowledge, especially regarding the many internal variables which control magma formation and its extrusion on the land surface; besides, we have no guarantee that even a good knowledge of such parameters would ever allow deterministic prediction of a volcano's behaviour, particularly in the long term. My personal point of view is that the scientific community should indeed tell the public exactly what we know, but also what we do not know about how a volcano works. The lack of scientific knowledge is not what blocks the public from thoughtfully considering most highly scientific issues. Far more important than facts and figures is a honest framework within which the issue can be assessed.

We know a lot about the geological and geochemical history of Vesuvius, and in these terms Vesuvius is probably the best-known volcano in the world – and this certainly is very important regarding predictions on the future behaviour of the volcano. Vesuvius is among the most studied active volcanoes on the Earth, not only for the great interest of the scientific community in the origin of silica-undersaturated alkaline rocks, but also for assessing the risk that this volcano presents to the 800,000 people inhabiting its slopes. Detailed fieldwork, historical accounts and a wealth of whole-rock geochemical data have enabled an unusually good reconstruction of its eruptive history (De Vivo et al., 2003). However, the magmatic system which feeds and drives both the plinian and non-plinian eruptions is far from being well understood. If we want to progress in the knowledge and possibly have some keys to forecast eruptions, we must investigate and develop fundamental research on the internal dynamics of the volcano.

In the last 15 years, as principal leader of my research group, in collaboration with many foreign Institutions – such as United States Geological Survey (Reston, VA, USA), American Museum of Natural History (New York, USA), Virginia Polytechnic Institute and State University (Blacksburg, VA, USA), University of California (Santa Barbara, USA), Central Washington University (Ellensburg, WA, USA), University of Bristol (UK), University College London (UK) and University of Tasmania (Hobart, Australia) – I have worked both in the direction of obtaining better and more detailed knowledge on historic and ancient eruptions of the Somma-Vesuvius system, and more recently, on research concerning the internal behaviour of the volcano, through studies of fluid and melt inclusions (MI) (small droplets of trapped melts and volatiles) in the erupted crystals, combined with solubility experiments involving complex volatile systems (H_2O, SO_2, Cl) (De Vivo et al., 2005) (see Lima et al., this volume; Webster et al., this volume).

A second important contribution to the Workshop and to this volume is the problem of the ignimbrites in the Campania Plain, which has been studied and debated since the beginning of XIX Century (see Scandone et al., this volume). The ignimbritic deposits, known locally as Tufo Grigio Campano (Campanian Gray Tuff) attracted the attention of Scacchi (1890), which attributed them to eruptions originating from different sources in the Campanian Plain. Later this view was opposed by Franco (1900), who instead attributed the Campanian Gray Tuff to a unique source in the Campi Flegrei. The latter hypothesis (which later became another dogma of the Italian volcanological community) has been favoured by recent authors (Rosi and Sbrana, 1987; Fisher et al., 1993; Orsi et al., 1996; Ort et al., 1999), who suggest that the Campanian Ignimbrite was fed by Campi Flegrei and the eruption resulted in the formation of a 12 km wide caldera, centred, in the Gulf of Pozzuoli.

This view has been challenged by De Vivo et al. (2001) and Rolandi et al. (2003), who demonstrate that different ignimbrite events (at least 6) occurred in the Campania Plain, spanning, at least the period from >315 ka to 19 ka BP. The Campanian Ignimbrite dated at 39 ka is just the largest, but not a "unique" event in the Campania Plain. According to De Vivo et al. (2001) and Rolandi et al. (2003), the ignimbrites originated from a fracture system related with the subsidence of the Campania Plain. A contribution concerning the ignimbrites in the Campania Plain is the paper by Bohrson et al. (this volume).

The third problem in the Campania Plain is the caldera unrest (bradyseism) of Campi Flegrei (see De Vivo and Lima, this volume; Scandone et al., this volume). The hypothesis about this ground deformation phenomenon is presented to the population with the view that at any bradyseismic event might correspond to an eruption, though, at calderas, a distinctive feature of such deformation episodes is that they are not followed by eruptions (Dzurisin and Newall, 1984). In the Campi Flegrei, indeed, since Roman times many of such events have occurred, but only once there was an eruption (Monte Nuovo eruption, 1538 AD). This points to the fact that between a bradyseismic event and an eruption there is no necessary cause–effect relationship.

De Vivo and Lima (this volume) propose a model suggesting that ground deformations could be generated by conductive heating of the hydrothermal fluids overlying the magmatic chamber. The authors elaborate the details of the hydrothermal model, comparing the evolution of the Campi Flegrei system through time, to the model of the porphyry systems (Henley and McNabb, 1978; Burnham, 1979; Fournier, 1999). In other words, according to the authors, the Campi Flegrei might represent a modern analogue of former intrusive-volcanic systems, now mineralized porphyry systems (Beane and Titley, 1981; Beane and Bodnar, 1995; Roedder and Bodnar, 1997; see also Rapien et al. (2003) about White Island, New Zealand). In this view, the fluids at Campi Flegrei, heated by underlying crystallizing magma, under lithostatic pressure for long periods of time, generate overpressure (volatile accumulation) in the upper, apical, part of the magma chamber (*senso lato*), that confined by impermeable rind, causes uplift of the overlying rocks (positive bradyseism). A crisis occurs when the conditions change from lithostatic to hydrostatic pressure, with consequent boiling (De Vivo et al., 1989), hydraulic fracturing, seismic tremor and then pressure release. At this point, the area experiences the maximum degree of inflation, which is then followed by pressure release and beginning of subsidence (deflation of the ground). Afterwards, the system, saturated with boiling fluids, begins to seal again due to the precipitation of newly formed minerals. The beginning of a new positive bradyseism phase will occur only after several years when the system "reloads" under new lithostatic pressure conditions.

Whatever will be the real scenario in the short- and long term for Vesuvius, Ignimbrites and Campi Flegrei in the Campania Plain, the results demonstrate once again that research progress is attained only if there is a non-dogmatic approach, which favours an impartial and balanced evaluation of the research results. This, unfortunately, has not been the case in Italy in recent years, mostly because of a too-close, unhealthy connection between politics and science. May these new research results attract and motivate new researchers from all over the world. The field is still open as many contentious issues exist and anyone capable to improve the knowledge of Vesuvius and Campania Plain volcanism should be welcome for the benefit of science and of the people living around Vesuvius and Campi Flegrei.

This volume contains 14 papers that deal with particular aspects of volcanic activity in the Campania Plain.

The paper by Scandone and co-workers is a comprehensive review of the volcanological history of the volcanoes in the Neapolitan area: Vesuvius, Campi Flegrei and the Ignimbrites.

Turco and co-workers describe the process of extension and associated magmatic activity in the Tyrrhenian margin of the Apennines chain. Their model realistically assembles in a unique kinematic framework, the first-order structures that are observed in the Apennine area and in the Tyrrhenian basin.

Milia and co-workers, present an interpretation of an exactly spaced seismic grid. This permits the reconstruction of the paleogeography of Naples Bay before the onset of volcanic activity and the paleogeographic changes that followed the emplacement of the volcanic units. The authors also question the existence of a caldera offshore Campi Flegrei.

Milia, Torrente and Giordano discuss the slope instability processes occurring on the flanks of the submerged volcanoes in Naples Bay off Campi Flegrei and consider these events as elements to be taken into account when evaluating the tsunami risk for the densely populated Naples Bay coast.

Perrotta and co-workers support the hypothesis of the location of the Campanian Ignimbrite caldera as occupying the Campi Flegrei region. According to these authors, new exposures show that proximal deposits are associated with the Campanian Ignimbrite and allow a better localization of the caldera boundary, which include part of the city of Naples.

The study of Insinga and co-workers, performed on terrestrial and marine successions, helps to better understand the late-Holocene volcanological and stratigraphical evolution of the southwestern rim of Campi Flegrei caldera, previously reported as quiescent during the last 10,000 years. These authors report new chronostratigraphic data by $^{40}Ar/^{39}Ar$ and ^{14}C dating methods.

Fedele and co-workers report the results of a study on syenite nodules from the Breccia Museo deposit in the Campi Flegrei. Such nodules record convincing evidence of a transition from a magma-dominated regime to a fluid-dominated hydrothermal phase at the margins of a magma chamber, where a magma of trachytic composition was sufficiently evolved to exsolve an aqueous fluid carrying a complex solute, containing, among other components, high amounts of REE elements.

Bellucci and co-workers present a study of the Upper Pleistocene ignimbrites of the Campania margin in the Neapolitan area performed using outcrops, cores and seismic reflection data. The authors make a physical correlation between onshore and offshore stratigraphic units and evaluate NW-SE faults as being active during ignimbrite emplacement, in agreement with a model which attributes the Upper Pleistocene ignimbrites of the Neapolitan area as being related to emission from a regional fault system.

The paper by Piochi and co-workers reviews major, trace and isotopic data (Sr, Nd, Pb, O) relative to the entire volcanic activity of Somma-Vesuvius. The data strongly suggest a major role for evolutionary processes such as fractional crystallization, contamination, crystal trapping and magma mixing, occurring after magma genesis in the mantle. Chemical and isotopic data together with fluid inclusion data points to the existence of three main levels of magma storage, the two deepest ones (at ~8 and >12 km) being probably long-lived reservoirs, and an uppermost crustal level (at ~5 km) that probably coincides with the volcanic conduit.

Cecchetti and co-workers highlight the role of magmatic volatiles and of the deep system in the explosive dynamics of the eruptions during this period of activity. The authors demonstrate that input of volatile-rich magma blobs caused the recent violent strombolian and subplinian eruptions at Vesuvius.

Webster and co-workers determined through silicate melt compositions, and new experimental volatile solubility data for the complex system – phonolite melt + H_2O + NaCl + KCl + $CaCl_2$. Their data provide a more accurate interpretation of the past explosive and passive-effusive eruptive activities of Somma-Vesuvius in terms of magma geochemistry and degassing processes. The authors also report new 200-Mpa experiments which reveal that small to modest levels of S in oxidized phonolitic melt have a substantial capacity to promote degassing by reducing Cl solubility in melt.

Lima and co-workers present compositional data of reheated silicate MI in olivine and clinopyroxene crystals from cumulate nodules ejected by 79 AD plinian and by 1944 AD interplinian eruptions. Variation diagrams of some element ratios as a function of host crystal (olivine and cpx) Mg# MI in cumulate nodules and in bulk rocks from 79 AD and 1944 AD eruptions, are interpreted to depend on hydrothermal processes active in the upper parts of the shallow magma chamber, before and during explosive plinian and interplinian eruptions.

Bohrson and co-workers highlight the impact crystal–liquid separation has on melt compositional evolution and particularly focus on trace element and Th isotope evidence for open-system processes in the magma body associated with the Campanian Ignimbrite. For their interpretation, the authors utilize thermodynamic and quantitative mass-balance modelling of major and trace element data and semi-quantitative limits on Th and Sr isotopes to evaluate the role of crystal–melt separation, magma–fluid interaction, and assimilation of wall rock on the geochemical evolution of the Campanian Ignimbrite.

De Vivo and Lima elaborate a hydrothermal model to explain the ground movements (bradyseism) at Campi Flegrei. To develop such a model, the authors use as an analogue for the evolving Campi Flegrei subvolcanic system, the model of the porphyry mineralized systems.

I am grateful to the contributors of this volume, who with their papers have made possible this publication and, whose results, I am confident, will be well received by the world scientific community.

B. De Vivo

Acknowledgements

I wish to thank the Università degli Studi di Napoli Federico II, Scafi (Società di Navigazione SpA, Napoli), Servizi Tecnici Integrati Srl and Ordine dei Geologi della Campania for the support given for the organization of the Workshop; Stefano Albanese, Domenico Cicchella, Paola Frattini and Luca Fedele for their help for the activities prior to and during the Workshop.

References

Beane, R.E., Ad Bodnar, R.J., 1995. Hydrothermal fluids and hydrothermal alteration in porphyry copper deposits. Ariz. Geol. Soc. Dig., 20, 83–93.

Beane, R.E., Titley, S.R., 1981. Porphyry copper deposits. Part II. Hydrothermal alteration and mineralization. Econ. Geol. 75th Anniversary Vol., 235–263.

Bohrson, W.A., Spera, F.J., Fowler, S.J., Belkin, H.E., De Vivo, B., Rolandi, G., this volume. Petrogenesis of the 39.3 ka Campanian Ignimbrite: implications for open-system processes from trace element and Th isotopic data.

Burnham, W.C., 1979. Magmas and hydrothermal fluids. In: Barnes, H.L. (Ed.), Geochemistry of Hydrothermal Ore Deposits. Wiley, New York, pp. 71–136.

De Vivo, B., Ayuso, R.A., Belkin, H.E., Fedele, L., Lima, A., Rolandi, G., Somma, R., Webster, J.D., 2003. Chemistry, Fluid/Melt Inclusions and Isotopic Data of Lavas, Tephra and Nodules from >25 ka to 1944 AD of the Mt. Somma-Vesuvius Volcanic Activity. Mt. Somma-Vesuvius Geochemical Archive. Dipartimento di Geofisica e Vulcanologia, Università di Napoli Federico II, Open File Report 1-2003, 143pp.

De Vivo, B., Belkin, H.E., Barbieri, M., Chelini, W., Lattanzi, P., Lima, A., Tolomeo, L., 1989. The Campi Flegrei (Italy) geothermal system: a fluid inclusion study of the Mofete and San Vito fields. J. Volcanol. Geotherm. Res. 36, 303–326.

De Vivo, B., Lima, A., this volume. An hydrothermal model to explain the ground movements (bradyseism) at Campi Flegrei.

De Vivo, B., Lima, A., Webster, J.D., 2005. Volatiles in magmatic–volcanic systems. Elements 1, 19–24.

De Vivo, B., Rolandi, G., Gans, P.B., Calvert, A., Bohrson, W.A., Spera, F.J., Belkin, H.E., 2001. New constraints on the pyroclastic eruptive history of the Campanian volcanic plain. Mineral. Petrol. 73, 47–66.

Dzurisin, D., Newhall, C.G., 1984. Recent ground deformation and seismicity at Long Valley (California), Yellowstone (Wyoming), the Phlegrean Fields (Italy) and Rabaul (Papua, new Guinea). In: Hill, D.P., Bailey, R.A., Ryall, A.S. (Eds), Proceedings of Workshop XIX; Active Tectonic and Magmatic Processes Beneath Long Valley Caldera, Eastern California. Open-File Report – U.S. Geological Survey, pp. 784–829.

Fisher, R.V., Orsi, G., Ort, M., Heiken, G., 1993. Mobility of large volume pyroclastic flow – emplacement of the Campanian Ignimbrite, Italy. J. Volcanol. Geotherm. Res. 56, 205–220.

Fournier, R.O., 1999. Hydrothermal processes related to movement of fluid from plastic into brittle rock in the magmatic-epithermal environment. Econ. Geol. 94(8), 1193–1211.

Franco, P., 1900. Il Tufo della Campania. Boll. Soc. Nat. XIV, 9–25.

Henley, R.W., McNabb, A., 1978. Magmatic vapour plumes and ground water interaction in porphyry copper emplacement. Econ. Geol. 73, 1–20.

Lima, A., Danyushevsky, L.V., De Vivo, B., Fedele, L., 2003. A model for the evolution of the Mt. Somma-Vesuvius magmatic system based on fluid and melt inclusion investigations. In: De Vivo, B., Bodnar, R.J. (Eds), Melt Inclusions in Volcanic Systems. Methods, Applications and Problems. Series: Developments in Volcanology, Vol. 5. Elsevier, Amsterdam, pp. 227–249.

Lima, A., De Vivo, B., Fedele, L., Sintoni, F., this volume. Influence of hydrothermal processes on geochemical variations between the 79 AD and 1944 AD Vesuvius eruptions.

Orsi, G., de Vita, S., Di Vito, M., 1996. The restless, resurgent Campi Flegrei nested caldera (Italy): constraints on its evolution and configuration. J. Volcanol. Geotherm. Res. 74, 179–214.

Ort, M., Rosi, M., Anderson, C.D., 1999. Correlation of deposits and vent locations of the proximal Campanian Ignimbrite deposits, Campi Flegrei, Italy, based on natural remnant magnetization and anisotropy of magnetic susceptibility characteristics, Flegrei. J. Volcanol. Geotherm. Res. 91, 167–178.

Rapien, M.H., Bodnar, R.J., Simmons, S.F., Szabo, C.S., Wood, C.P., Sutton, S.R., 2003. Melt inclusion study of the embryonic porphyry copper system at White island, New Zealand. Econ. Geol., Spec. Publ. 10, 41–59.

Roedder, E., Bodnar, R.J., 1997. Fluid inclusion studies of hydrothermal ore deposits. In: Barnes, H.L. (Ed.), Geochemistry of Hydrothermal Ore Deposits, 3rd ed. Wiley, New York, pp. 657–698.

Rolandi, G., Bellucci, F., Heizler, M.T., Belkin, H.E., De Vivo, B., 2003. Tectonic controls on the genesis of ignimbrites from the Campanian Volcanic Zone, Southern Italy. In: De Vivo, B., Scandone, R. (Eds), Ignimbrites of the Campanian Plain, Italy. Mineral. Petrol. 79, 3–31.

Rosi, M., Sbrana, A. (Eds), 1987. Phlegraean Fields, Vol. 114. CNR. Quad. Ric. Sci., Roma., 167pp.

Scacchi, A., 1890. La regione vulcanica fluorifera della Campania, II editione. Mem. Regio Com. Geol. It., Vol. IV, Firenze.

Scandone, R., Giacomelli, L., Fattori Speranza, F., this volume. The volcanological history of the volcanoes of Naples: a review.

Webster, J.D., Sintoni, M.F., De Vivo, B., this volume. The role of sulfur in promoting magmatic degassing and volcanic eruption at Mt. Somma-Vesuvius.

Volcanism in the Campania Plain:
Vesuvius, Campi Flegrei and Ignimbrites
edited by B. De Vivo

Chapter 1

The volcanological history of the volcanoes of Naples: a review

Roberto Scandone, Lisetta Giacomelli and Francesca Fattori Speranza*

Dipartimento di Fisica, Università Roma Tre, Via Vasca Navale 84, 00146, Roma, Italy

Abstract

Campi Flegrei and Vesuvius were mainly formed after the eruption of the Campanian Ignimbrite (39 kyr) along newly formed tectonic faults. The caldera of Campi Flegrei was formed after another voluminous eruption (the Neapolitan Yellow Tuff occurred between 12 and 15 kyr). The formation of the caldera favored the occurrence of the subsequent activity mostly within the collapsed structure. Mt. Vesuvius was entirely built after 25 kyr. The activity displays different styles ranging from plinian eruptions with average return period of thousands years, to mild effusive activity. Effusive activity has been predominant in the last hundred years. The subsurface structure of the volcano provides evidence of a peculiar shallow rigid central core where extensive hydrothermal processes are still active. The occurrence of magmatic reservoir at a depth below 8 km is also suggested.

1. Introduction

The Campanian plain (Fig. 1) in southern Italy is bordered by Mesozoic carbonate platforms, which subsided during the Pliocene and Pleistocene with a maximum vertical extent of 5 km (Ippolito et al., 1973). Its origin has been related to the stretching and thinning of the continental crust by a counter-clockwise rotation of the Italian peninsula and the contemporaneous opening of the Tyrrhenian sea with a consequent subsidence of the carbonate platform along most of the Tyrrhenian coast (Scandone, 1979a). Campi Flegrei, Vesuvius and the island of Procida are located to the south-east along the coast. Campi Flegrei activity spans the period from 47 kyr (age of the oldest products outcropping in Campi Flegrei) to the present (Di Girolamo et al., 1984; Rosi and Sbrana, 1987). Most of the explosive activity of Somma-Vesuvius occurred after 25 kyr (Santacroce, 1987), whereas activity on Procida occurred between >40 kyr and 18 kyr (Di Girolamo et al., 1984; De Astis et al., 2004). A widespread pyroclastic deposit called the "Campanian Ignimbrite" (Barberi et al., 1978) is found all over the plain.

The city of Naples lies in the middle of the plain and is bordered by the two active volcanoes of Campi Flegrei and Vesuvius. The high volcanic risk related with the possible renewal of activity of one of these volcanoes close to a densely inhabited area promoted an intense scientific effort to improve the knowledge on the eruptive history of

*Corresponding author. *E-mail address*: scandone@amaldi.fis.uniroma3.it (R. Scandone).

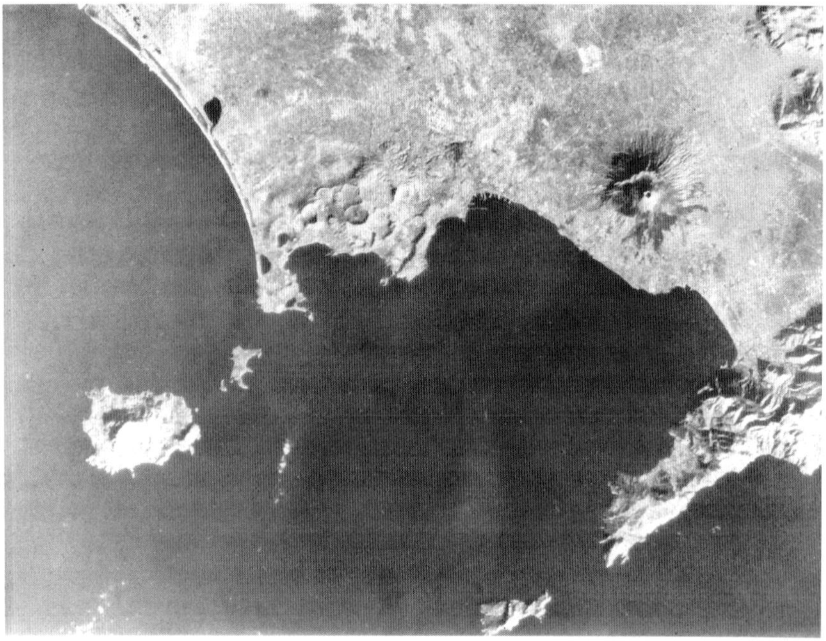

Figure 1. Satellite image of southern Campanian plain with the active volcanoes around the bay of Naples. From left to right are visible: Ischia Island, Procida Island, Campi Flegrei and Vesuvius. The southern promontory is the Sorrento Peninsula made by the outcropping carbonatic platforms.

the volcanoes as well as their style of eruption (a summary of these efforts is reported in several special issues of scientific journals) (Barberi et al., 1984; Luongo and Scandone, 1991; De Vivo et al., 1993; Orsi et al., 1999; Spera et al., 1998; De Vivo and Rolandi, 2001; De Vivo and Scandone, 2003; Civetta et al., 2004).

The aim of this paper is an attempt to summarize the volcanological history of the volcanoes of the Campanian plain with an emphasis to the known facts and the remaining problems.

2. The Campanian Ignimbrite(s)

The term Campanian Ignimbrite (CI) has been given to a unique pyroclastic-flow deposit occurring mostly in the Campanian plain and in the close valleys of the Apennine chain up to 800–900 m above sea level (Barberi et al., 1978). This deposit was first identified by Breislak (1798) and later studied by different authors who called it "Tufo Pipernoide" or "Tufo Grigio Campano" (Scacchi, 1848, 1890; De Lorenzo, 1904; Rittmann, 1950; Di Girolamo, 1968).

In the following sections, when not differently specified, the term "Campanian Ignimbrite" is referred to the huge deposit of a single volcanic eruption that occurred at 39 kyr (De Vivo et al., 2001).

2.1. The deposit

The deposit of the distal facies of the Campanian Ignimbrite is made up of pumice and black scoriae, with a different degree of flattening, embedded in an ashy matrix with subordinate lithics and crystals. Columnar jointing and fumarolic pipes are often observed. Lateral facies variation produces a change in color from a poorly welded gray deposit to a more welded yellow one.

Di Girolamo et al. (1973) identified a pumice fall deposit (Fig. 2) at the base of the Campanian Ignimbrite, in some places separated from the overlaying pyroclastic flow deposit, by a paleosol. Scandone et al. (1991), Rosi et al. (1999), and Polacci et al. (2003) found this pumice deposit in direct contact with the Ignimbrite and related it with the air fall deposition from a plinian eruptive column which eventually resulted in the collapse and subsequent deposition of an ash flow deposit. Perrotta and Scarpati (2003) provided an estimate of the partition between the pumice fall deposit at the base of the CI and the co-ignimbrite ash fall.

A discrepancy exists in the identification of the air fall deposit of the CI in the marine deposits of the Eastern Mediterranean. Keller et al. (1978) correlate a tephra layer (the Y-3 layer dated at 26 kyr), found mostly in the Ionian and Tyrrhenian seas with the fall deposit of the CI. Thunell et al. (1979), on the contrary, associate the co-ignimbrite layer of the Campanian Ignimbrite with the widespread Y-5 ash layer dated by the oxigen isotope record, at approximately 38 kyr. Munno and Petrosino (2004) associate the Y-5 layer, with the CI eruption and the Y-3 layer with another eruption of Campi Flegrei.

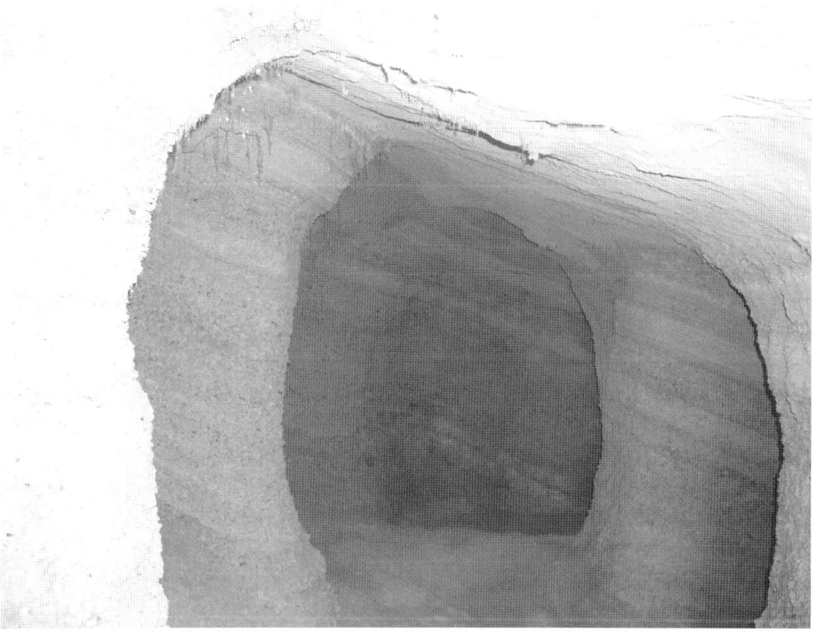

Figure 2. Air fall pumice deposit at the base of the Campanian Ignimbrite in the locality of San Martino in the city of Naples.

Rosi and Sbrana (1987) suggested that the proximal facies of the Campanian Ignimbrite were made up by the "Piperno" deposit of Campi Flegrei and the overlaying breccia (Museum Breccia of Johnston-Lewis, 1889).

The volume estimates of the deposit vary with a factor of 2. Thunnell et al. (1979) estimate the co-ignimbrite ash layer at 30–40 km^3 of dense rock equivalent, and hypothesize a similar volume for the ignimbrite for a total volume of 80 km^3. Rolandi et al. (2003) give a total estimate of 200 km^3 DRE for both the on-land distribution of the ash deposit and the distal air fall.

2.2. The source problem

De Lorenzo (1904) thought that the "Tufo Pipernoide Campano" and the pipernoid tuff of Campi Flegrei "Piperno" were similar deposits that erupted from several vents located in the proximity of the Camaldoli hill on the rim of Campi Flegrei. Rittmann (1950) suggested that the Tufo Grigio Campano and the Piperno Tuff resulted from different eruptions but their source area was proximal to Campi Flegrei. Di Girolamo (1970), Barberi et al. (1978), Di Girolamo et al. (1984) and Lirer et al. (1987) suggest that the Campanian Ignimbrite was fed through an arcuate fracture on the northern edge of Campi Flegrei. Rosi and Sbrana (1987), Fisher et al. (1993), Orsi et al. (1996) and Ort et al. (1999) suggest that the Campanian Ignimbrite was fed by Campi Flegrei, and the eruption resulted in the formation of a 12-km-wide caldera centered on the Gulf of Pozzuoli. Scandone et al. (1991) propose that the Campanian Ignimbrite was erupted through a NE-SW fracture bordering at the southern edge the Campi Flegrei and on the northern one, the so-called Acerra Graben. De Vivo et al. (2001) and Rolandi et al. (2003) suggest that the Campanian Ignimbrite was fed by a fracture system related with the sinking of the Campanian plain.

2.3. The age problem

The age of the Campanian Ignimbrite has been the object of an intense debate (see a summary in Scandone et al., 1991). Available ^{14}C dates ranged between 28 and 40 kyr; K-Ar age of 37 kyr was also provided. More recently, De Vivo et al. (2001) put more precise constraints on the age of the Campanian Ignimbrite, identifying different pyroclastic deposits spread over the plain and having different ^{40}Ar/^{39}Ar ages of 205 kyr, 184 kyr, 157 kyr, 39 kyr and 18 kyr. These authors (De Vivo et al., 2001) correlate, the most voluminous deposit, having an age of 39 kyr, with the Campanian Ignimbrite.

3. Campi Flegrei

Volcanic products younger than 1 Ma are found in several drillings all over the Campanian plain (Ippolito et al., 1973; Brocchini et al., 2001). Volcanic products younger than 200 kyr outcrop on Ischia, Procida islands, Campi Flegrei and Vesuvius in the southern part of the Campanian Region. Procida islands is separated by Campi Flegrei by a narrow strait and its activity may be considered as similar to the one occurring in Campi Flegrei (Di Girolamo et al., 1984). The products outcropping on Procida span a period between younger than 55 kyr (Pozzo Vecchio Formation of De Astis et al., 2004), and 14 kyr of Solchiaro Volcano (Pescatore and Rolandi, 1981; Di Girolamo et al., 1984; Rosi et al., 1988a,b). The activity of this sector has a sudden end immediately before the voluminous eruption of the Neapolitan Yellow Tuff (NYT) (12 kyr) in Campi Flegrei.

The oldest products outcropping in the Campi Flegrei are lava domes with K-Ar ages of approximately 47 and 37 kyr (Rosi and Sbrana, 1987). A series comprising ash-beds with pumice and scoria deposits interbedded with paleosols directly underlies the deposit of the Piperno-"Museum Breccia" formation. The Piperno is a welded ash with abundant fiamme and some pipe-structures that pass into the overlying breccia and has been considered as the proximal deposit of the Campanian Ignimbrite (39 kyr) by Rosi and Sbrana (1987). Before 12 kyr, the products of scattered eruptions were found on the eastern rim of Campi Flegrei, the so-called Whitish tuffs (16 kyr) (Di Girolamo et al., 1984).

The eruption of the NYT has been dated at ~12 kyr BP by the ^{14}C method and at 15 kyr by the $^{39}Ar/^{40}Ar$ method (Deino et al., 2004). We continue to use the ^{14}C date because of the number of concordant ages obtained by this method (Scandone et al., 1991) and in view of the relative ages with the other products of Campi Flegrei. Lirer and Munno (1975), Di Girolamo et al. (1984) and Lirer et al. (1987) proposed that all the deposits of yellow tuff outcropping outside and on the rim of Campi Flegrei were the results of a unique eruption. Scarpati (1990) and Scarpati et al. (1993) suggest that the eruption of the NYT (Fig. 3) was characterized by the deposition of a phreato-plinian deposit of alternating pumice and ashes, followed by the deposition of a huge sequence of surge and pyroclastic flows.

Figure 3. Outcrop of the Nepolitan Yellow Tuff at Cuma. The deposit in the foreground is a pyroclastic flow deposit welded because of syneruptive zeolitization. In the background, the non-welded upper part cover unconformably a lava dome of the earlier activity at 47 kyr.

Figure 4. In the foreground is an aerial view of the cone of Monte Nuovo formed during the last eruption of Campi Flegrei in 1538. In the background is seen the lake of Averno (eruption occurred 3700 a BP).

Several other eruptions occurred between 11 and 9 kyr. Among these are the eruptions of Gauro, Archiaverno and of Agnano Pumices (Monte Ruscello and probably the volcanoes along the northern margin of the caldera; Montagna Spaccata, Pisani, etc.). All these eruptions occurred near the rim of Campi Flegrei (Di Girolamo et al., 1984). Cole et al. (1994) also report the occurrence of volcanic edifices younger than the NYT within the boundaries of the city of Naples.

A new period of renewed activity occurred after 4.5 kyr in the inner part of Campi Flegrei. The most important eruptions were those of Agnano-Monte Spina (4.0 kyr), Astroni (3.7 kyr) and Averno (3.7 kyr) (Rosi and Sbrana, 1987). Di Vito et al. (1999) report at least 61 eruptions after the eruption of the NYT; it is however difficult to understand if the reported eruptions represent the different building stages of single monogenic edifices or actual individual events separated in space and time.

The last eruption in the area occurred in historical time and was that of Monte Nuovo (1538 AD) (Fig. 4).

3.1. The caldera problem

Campi Flegrei has a broadly circular symmetry bounded to the east by the hills of Posillipo and Camaldoli, and to the west by the reliefs of Monte di Procida. The center of Campi Flegrei, on land, is marked by a raised marine terrace, named "La Starza," which is presently at 40 m asl (Cinque et al., 1985). There is however evidence of at least two different levels of marine erosion separated by a step (at 40 and 50–54 m asl) (Rosi and Sbrana, 1987). Orsi et al. (1996) considered the block delimited by the La Starza terrace, a resurgent dome.

The southern half of Campi Flegrei is below sea level and defines the Gulf of Pozzuoli. An en-echelon pattern of normal faults in the northern sector of the Gulf of Pozzuoli (Colantoni et al., 1972) marks the transition between the two halves of the caldera.

The eruption of the large volume Campanian Ignimbrite posed the problem of the existence of a possible caldera related with it (Thunnel et al., 1979). The idea that Campi Flegrei is a caldera had already been proposed by Rittmann (1950), who suggested that it was formed after the eruption of the Campanian Ignimbrite, and that the eruption decapitated a stratovolcano called "Archiflegreo." Rosi et al. (1983) and Rosi and Sbrana (1987) have argued for the existence of a previous stratovolcano and proposed the existence of a caldera 12 km wide centered on the Gulf of Pozzuoli. Lirer et al. (1987) suggested that the caldera was the result of two eruptions: that of the Campanian Ignimbrite and the smaller one of the NYT which produced a central, more collapsed zone. Scandone et al. (1991) suggest that Campi Flegrei was a caldera related only with the eruption of the NYT, basing it mostly on geophysical evidence. A negative residual Bouguer anomaly defines a circular area that includes the Gulf of Pozzuoli and part of Campi Flegrei (Fig. 5). This anomaly has been interpreted as resulting from the caldera collapse (Nunziata and Rapolla, 1981; Cassano et al., 1986). The caldera defined by this anomaly has a diameter of about 8 km and is much smaller than the one proposed by Rosi and Sbrana (1987).

Orsi et al. (1996) propose that the caldera was the result of the two major eruptions, but considerably enlarged its eastern rim encompassing also the city of Naples.

Recently, new and stronger geophysical evidence puts firm constraints to the extension of the caldera. Fedi et al. (1991) and Florio et al. (1999) follow Scandone et al. (1991) and suggest that the caldera is approximately 6–8 km wide on the basis of gravity data and potential field data. The limited extent of the caldera is also proposed by Zollo et al. (2003) on the basis of new tomographic studies. Another argument in favor of the limited extent of the caldera relates to the present-day active unrest (discussed in the next section), which affects only the central part of the caldera. Overall, we believe that there are strong arguments against a large caldera, and that it was formed after the eruption of the NYT. In our view, the problem of the source of the Campanian Ignimbrite is still an open problem as well as the lack of evidence of a caldera associated with it. We suspect that this large-scale eruption, as well as those of the Roman province, may be related with the development of tectonic fissures, which tap deep-seated magma reservoir. In this respect, we believe that the most likely source area is the NE-SW fault, bordering to the east the Campi Flegrei, and passing through the city of Naples.

4. The caldera unrest (Bradyseism)

The term "Bradyseism" derives from the Greek language and means "slow movement." It is widely used in the Italian literature to identify the slow movement of the ground occurring in the area of Pozzuoli in Campi Flegrei, which has been the site of slow vertical movements of the ground since at least Roman times.

A slow regular subsidence of the floor of the caldera occurred possibly for most of the period since the last eruption in 1538. This process became evident after the excavation, in the first half of the 18th century, of the ruins of an ancient Roman market, the "Serapeum,"

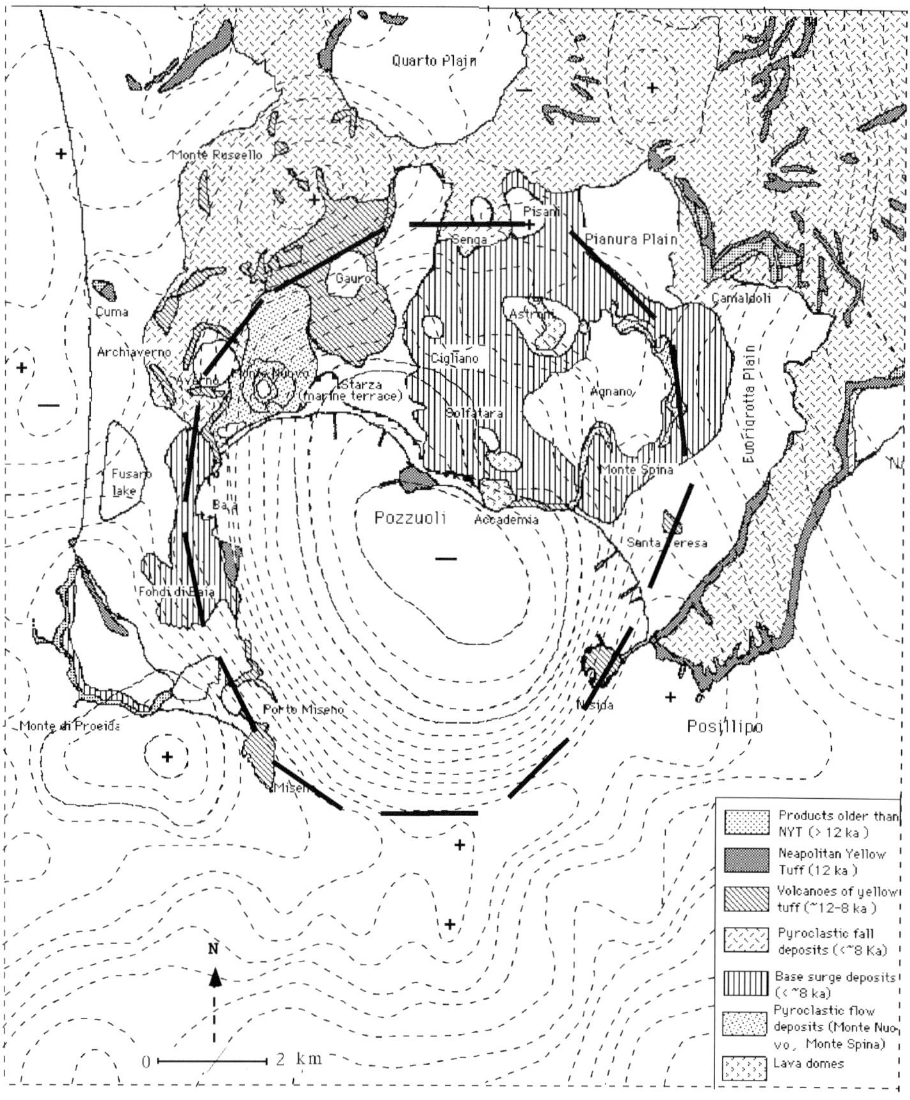

Figure 5. Bouguer anomaly of Campi Flegrei (1 milligal interval) and limits of the caldera as defined by the anomaly minimum (modified after Scandone et al., 1991).

in the city of Pozzuoli (Parascandola, 1947) (Fig. 6). The Serapeum has three high standing columns with evidence of a marine submersion provided by the holes of lithodomes, which reached a level of 10.26 m above the ground floor. Several scholars interpreted the phenomenon as an evidence of sea-level fluctuations (Nicolini as quoted by Parascandola, 1947). In 1828, Charles Lyell visited the place and in his “Principles of Geology,” first interpreted the phenomenon of Pozzuoli as owing to the submersion and subsequent emergence of the ground (Lyell, 1830).

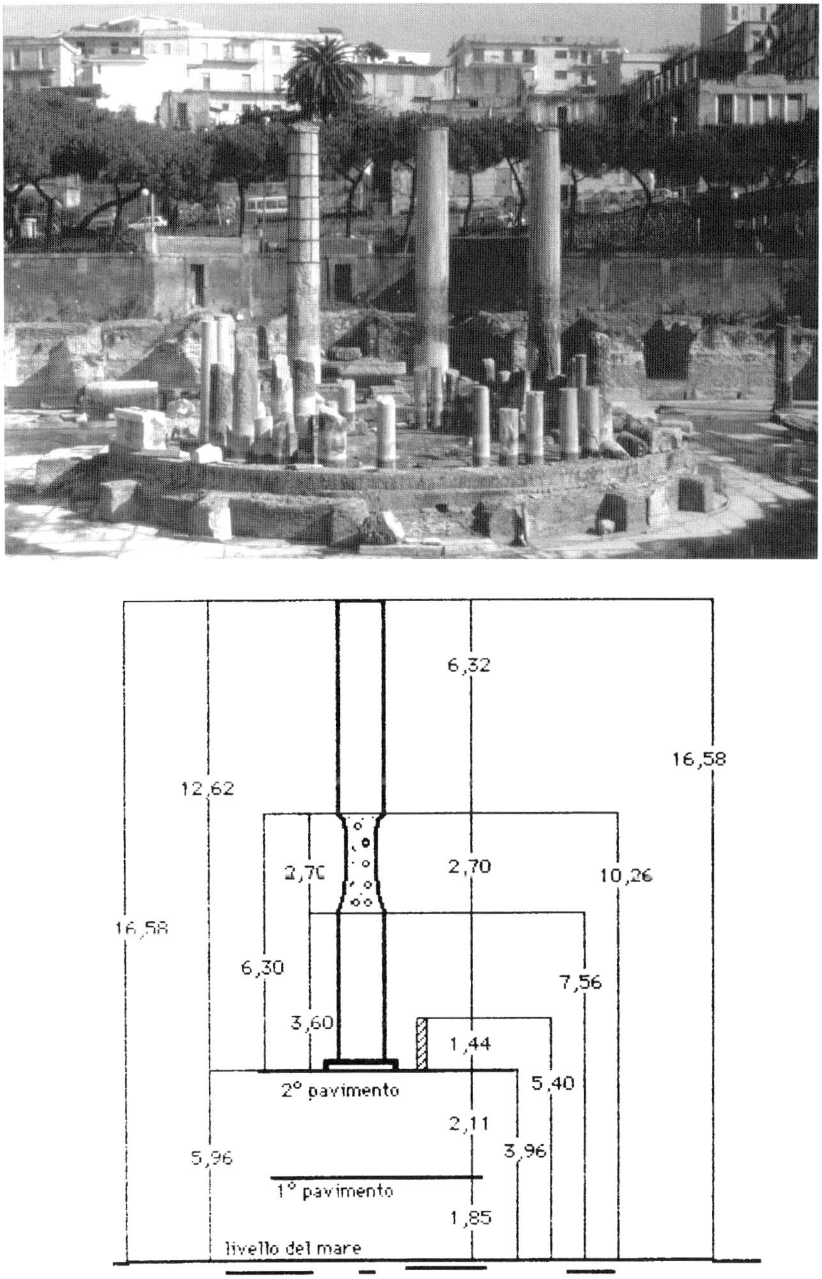

Figure 6. The Roman market known as the "Serapeum" in Pozzuoli. The columns mark the level of maximum subsidence where the signs left by lithodomes are evident. In the lower drawings are reported the height of the lithodomes with respect to the original floor (redrawn after Parascandola, 1947).

There are historical evidences of a slow subsidence of the area since at least Roman times (Parascandola, 1947). These evidences are based on several elevations, and the restoration of the original floor of the "Serapeum" made in the first century of the Christian era (Parascandola, 1947). Another proof is the repairing works made at the seaside road of Pozzuoli, because of the action of the sea during the time of Caesar (Parascandola, 1947). Several Roman ruins, including the palace of Emperor Nero (1st century AC), and the Roman port of Pozzuoli, "Portus Iulius," are now at a depth of about 10–12 m bsl (Fig. 7). There is a later reference of a submersion of the seafront of Pozzuoli as early as the 9th century AC (Fredericksen, 1977). A summary of the history of the vertical movements in the area has been compiled by Dvorak and Mastrolorenzo (1990).

The first evidences of an inversion of the movement of the ground from subsidence to uplift are dated at 1503 and 1511 AD (Parascandola, 1947; Dvorak and Gasparini, 1991) when two edicts of the viceroys of Naples assigned to the city of Pozzuoli, for the purpose of taxation, the new lands that were drying up around the town. This uplift has been taken as an evidence of a long-term precursor of the following eruption (see for example Dvorak and Gasparini, 1991), which occurred in 1538. The eruption, called Monte Nuovo, was preceded by only a few hours by a drying up of a large portion (200 paces) of the seashore in front of the eruption site. A similar phenomenon has recently occurred before the eruption of Rabaul in the September of 1994.

A slow subsidence of the floor of the caldera appears to have occurred for most of the period since 1538, and the area of the Serapeum was invaded by seawater around 1820.

Figure 7. The submerged ruins of the Roman port "Portus Iulius" in the foreground of Pozzuoli in Campi Flegrei.

Since that time, numerous measurements were made of the height of the sea level with respect to the market floor. The Serapeum floor had a depth of 1.396 m bsl as measured by precise leveling in 1905. The average subsidence rate between 1819 and 1968 has been of about 1.4 cm/year. The ground floor should have been about 25 m asl in the year 1 AD, had this same rate persisted since the beginning of the Christian era. However, we know that the slow subsidence was interrupted at least once in 1500.

The levelings made at the beginning of this century showed that the maximum measured sinking of the caldera floor was occurring in the city of Pozzuoli in the proximity of the Serapeum, and regularly decreased eastward and westward along the coast (Lirer et al., 1987). The movement marked by the Serapeum was an amplified mirror of the movement of the entire caldera floor. The slow sinking of the ground continued until 1968. In the periods 1970–1972 and 1982–1984, two important episodes of inflation occurred in the Pozzuoli area (Berrino et al., 1984). These episodes produced an uplift of 170 cm (inferred with respect to the previous leveling) and 182 cm, respectively, at the points of maximum deformation (located in Pozzuoli). The inflation geometry was the inverted mirror image of the slow sinking observed until 1968 (Lirer et al., 1987); it had a circular symmetry around Pozzuoli and regularly decreased toward the margin of the caldera. One particular feature of the movement of the ground is the constancy of the areal extent of the deformation. Repeated levelings (Berrino et al., 1984) showed that the bell-shaped form of the deformation did not change appreciably during 1982–1984: although it displayed a marked vertical variation, its horizontal extent remained the same.

The episodes of inflation were accompanied by seismic crises. The first seismic crisis occurred between 1970 and 1972, and was characterized by only a few felt earthquakes mostly occurring in the Gulf of Pozzuoli (Corrado et al., 1976). A second more intense seismic crisis began in 1983, some months after the beginning of the inflation of the ground (Barberi et al., 1984). The main features can be summarized as follows. Earthquakes occurred mostly in the coastal region around Pozzuoli, only a few, deeper events occurred within the gulf; however, they did not extend outside the border of the Campi Flegrei caldera. Hypocenters were between 1.5 and ~5 km depth. The maximum observed magnitude was 4.2 on October 4, 1983; the most important swarm occurred on April 1, 1984 with 513 earthquakes in 4 h. A distinctive areal difference was observed in the pattern of seismicity. Earthquakes with the shallowest foci occurred mostly as swarms and were located in an area west of Pozzuoli; earthquakes located in the eastern area, however, had higher magnitudes, occurred as single or double events, and generally had deeper hypocenters. The events occurring in the gulf resembled those of the eastern area, but with generally lower magnitudes. The swarms of the shallowest earthquakes had a high Gutenberg b-value, implying a highly fractured medium and smaller stress-drops. In contrast, the swarms located in the eastern region, in the proximity of Solfatara, had an anomalous increase in the frequency of earthquakes with higher magnitudes, and a smaller b-value (Vilardo et al., 1991).

The inflation which occurred in 1970–1972 was followed by a deflation of some 20 cm. Between 1985 and 2002, a deflation amounting to a total of about 70 cm has occurred. The regular pattern of deflation was interrupted in 1989, 1994 and 2000 when small episodes of inflations occurred again. Each episode was accompanied by a mild seismicity. During the episode of 1989, the maximum observed uplift was 7 cm, and there was a swarm of 316 earthquakes located in the proximity of the Solfatara crater. In March 2000, there was a new uplift of 5 cm followed in July and August by two earthquake swarms. During this crisis, earthquakes with a characteristic low frequency content (Lp events) were recorded for the first time since the monitoring of the area began (Osservatorio Vesuviano, 1995; Bianco et al., 2004).

One of the major problems, when dealing with active volcanoes, is the correlation between the present-day observed dynamics and the past geological record. The Campi Flegrei is a typical example where such a problem arises, and sometimes errors are made in the attempt to reconcile phenomena that occur at different timescales.

The explanation of the long-term deflation and shorter inflation episodes of Campi Flegrei are different. According to one school of thought, the uplift is related with the intrusion at shallow depth of new magma (Corrado et al., 1976; Barberi et al., 1984; Berrino et al., 1984). The deflation is on the contrary due to a compaction of the pore space. According to another view, the inflation episodes are due to a pressure increase in the fluids circulating in the hydrothermal system of Campi Flegrei (Casertano et al., 1976; Bonafede, 1991; Gaeta et al., 1998). De Natale et al. (1991) propose a model, which takes into account both a pressure increase in a shallow magma chamber and a pressure increase of the fluids circulating in the hydrothermal system.

Orsi et al. (1996) consider that all the recent dynamics of Campi Flegrei is related with the uplift of the so-called block resurgence delimited by La Starza Terrace.

The relevance of the discussion relies on the fact that the only historical eruption of Campi Flegrei of Monte Nuovo in 1538 was preceded by an observed uplift between 1502 and 1538.

A comprehensive model of the recent dynamics of Campi Flegrei should be able to reconcile the phenomena observed at different timescales.

In the shorter timescale, deflation and inflation episodes provide evidence of a deformation of the floor of the caldera as a unique body without apparent discontinuity. The floor deforms almost as an elastic plate bounded by a circular discontinuity. The spatial definition of the deformation is in favor of a pressure source at constant depth eventually bounded by caldera discontinuities (Lirer et al., 1987; De Natale and Pingue, 1993). Within this frame we favor the hypothesis put forward by De Natale et al. (1991) of a pressure source mostly related with the inflation of the fluids in the geothermal system.

On the other hand, the longer-term deformations provide important insights into the mechanics of the uplift of the northern sector of Campi Flegrei and its bearing with the eruption dynamics.

The relative movement between the northern and southern caldera blocks seems to have started since at least 10000–5300 years (Rosi and Sbrana, 1987). The subsided part of the Gulf of Pozzuoli is presently at a depth varying from 50 to 100 m bsl, has an approximately quadrilateral shape and an area of about 19 km^2. The uplifted part (the on land Campi Flegrei) is delimited with much difficulty because it has been partially filled by the products of younger eruptions which occurred after the main uplift.

The "Serapeum" and much of the seafront of Pozzuoli are located on the border side of the two blocks in that part where the action of the sea has eroded the "La Starza" raised cliff. Much of the ruins of the ancient Roman port "Portus Iulius" and those of the Emperor palace of Baia lay on the subsided block that has not been greatly affected by uplift (Fig. 8a).

The situation is sketched in the cartoon of Figure 8b. The normal long-term dynamics of the caldera floor is a slow subsidence, which may have a maximum rate similar to that

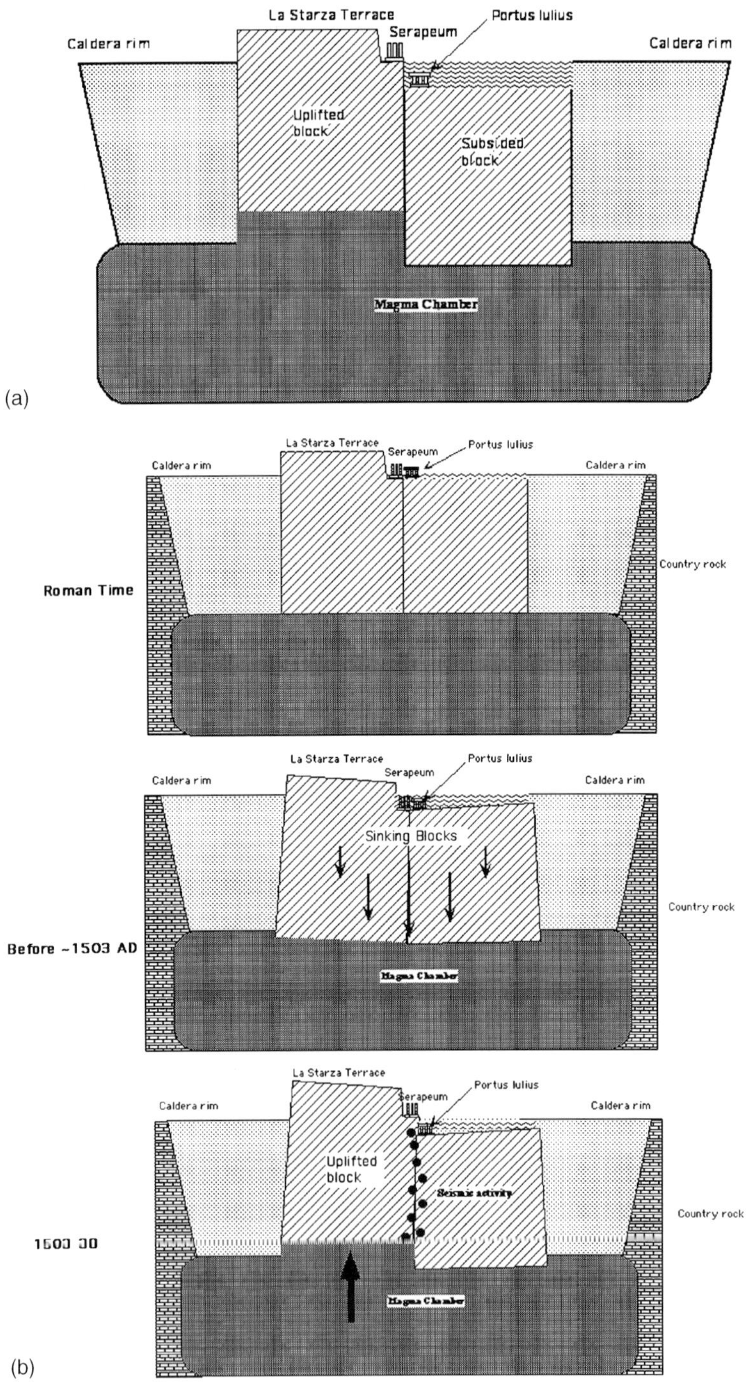

Figure 8. (a) Relative position of the Serapeum and Portus Iulius with respect to the different caldera blocks. (b) Dynamics of different caldera blocks during the subsidence and uplift before the Monte Nuovo eruption.

observed between 1819 and 1968 of about 1.4 cm/year. The subsidence appears to affect the entire caldera floor with an areal pattern similar to that of an uplift (Lirer et al., 1987). During these phases, the two blocks are locked and no seismicity accompanies the subsidence. Occasionally, the two blocks unlock producing a seismicity along the edge of the blocks and an uplift of the northern block.

Figure 8b depicts what may have occurred in the time between Roman time and the eruption of Monte Nuovo in 1538. In the period between Roman time until about 1503, the two blocks had been sinking with a maximum subsidence in the center of the Gulf of Pozzuoli. From 1503 until 1534, the two blocks suffered an uplift of several meters (to allow the drying of new land) at a relatively slow rate. The rate increased in the period between 1534 and 1538 during which a strong seismic crisis ruined most of Pozzuoli. The peak of the crisis was reached on September 28, 1538 when an inferred uplift of 7 m (Parascandola, 1947) preceded, by a few hours, the eruption of Monte Nuovo. The strong earthquake swarms associated with the days before the eruption represent the unlocking of the two blocks. As a consequence, the northern part resulted in an uplift with respect to the southern part. After that episode, the blocks have been locked again and the slow sinking has continued until 1970 when a new crisis produced between 1970 and January 1982 a cumulative uplift of 320 cm in the proximity of Serapeum. Currently, the two blocks still behave as a unique body. We infer that a new unlocking could cause a sudden depressuriztion of the hydrothermal fluids and the shallow magma, and favor the initiation of eruptive episodes.

5. Mount Vesuvius

Volcanic products of age between 0.4 and 0.3 Myr have been found in the drillhole Trecase 1 on the southern flank of Vesuvius (Brocchini et al., 2001). The age of these products are similar to that of other volcanic products found in many drillings all over the Campanian plain (Scandone et al., 1991) and give support to the idea of a diffuse volcanic activity within the plain since at least 1 Myr. The same drillhole provides evidence of a lack of activity between 0.3 Ma and the eruption of the Campanian Ignimbrite. Vesuvius volcano was mainly built after this eruption (Scandone et al., 1991; Brocchini et al., 2001). This coincidence gives a strong support to the idea that the eruption of the Campanian Ignimbrite may have caused a reactivation of tectonic lineaments along the Acerra Graben (Scandone et al., 1991) and caused the localization of activity on the western side in Campi Flegrei and on the eastern side at Vesuvius.

The early activity of the volcano is mainly effusive with numerous lava flows outcropping on the flanks of the older part of the volcano called Mt. Somma. Breislak (1798) and Johnston-Lavis (1884) first identified these products. The oldest dated explosive product is a pumice fall deposit dated at 25 kyr BP (Alessio et al., 1974) and called "Codola Pumice" after the locality where it is found.

The first attempt to a systematic dating of eruption products was made by Delibrias et al. (1979), who identified several plinian eruptions during the past 17,000 years and suggested a stepwise caldera collapse following each of these eruptions. The last largest explosive eruption was the one of 79 AD (Lirer et al., 1973; Sigurdsson et al., 1985) (Fig. 9). The volcanological history has been detailed by Santacroce (1987) and several other authors (e.g. De Vivo et al., 1993; Spera et al., 1998).

Figure 9. The pumice fall and pyroclastic flow deposit of the 79 AD eruption of Vesuvius (Villaregina, Boscotrecase). The cast of a tree was covered by the fall deposit and cut at the level of the upper pyroclastic flows of higher energy.

The most recent findings identify four major plinian events (VEI = 5–6) in the last 20,000 years of activity (Cioni et al., 1999).

Eruption Name	Age (y BP)	Volume (DRE)
Pomici di Base	18,300	4.4 km^3 (fall)
(Sarno)		? (pf)
Mercato Pumices	8000	2–3 km^3 (fall)
(Ottaviano)		0.25 km^3 (pf)
Avellino Pumices	3400–3700	0.7 km^3 (fall)
		0.5 km^3 (pf)
Pompei Pumices	79 AD	1–1.5 km^3 (fall)
		0.75 km^3 (pf)

The volume estimates generally vary of a factor 2 according to different authors and methods of estimate. Other authors define the eruptions with other names (reported in parenthesis in the table); here we follow Cioni et al. (1999).

In between these eruptions, several minor explosive eruption products have been identified and dated. Several eruptions are defined subplinian and have a VEI ≤ 4–5.

Eruption Name	Age (y BP)	Source
Greenish Pumice	16,000	Delibrias et al. (1979)
AP1	3250	Andronico and Cioni (2002)
AP2	3000	Andronico and Cioni (2002)
Pollena	472 AD	Rosi and Santacroce (1983)
1631	1631 AD	Historical reports

Six other minor explosive eruptions (VEI ≤ 3) are reported in the time period after 2700 a BP and 79 AD (Rolandi et al., 1998; Andronico and Cioni, 2002).

The eruption of 79 AD opens the historical period when more detailed information on the record of activity is available.

We have no information on the state of Vesuvius immediately after the eruption of 79 AD. The first account of continuing activity is from Galenus (c. 172 AD) who testifies that "*the matter in it* (Vesuvius) *is still burning.*" Dio Cassius in 203 AD reports a violent eruption that was heard in Capua, some 40 km from the volcano. The same eruption is reported by another source (Manuele) referred to by Gasparini and Musella (1991). Two large eruptions occurred in 472 and 512 AD (Alfano and Friedlaender, 1929). Several other eruptions are reported in 685 (Paulus Diaconus), 787 and 968 AD. Several authors report other eruptions in 991, 993 and 999 AD (see in Alfano, 1924). Leo Marsicanus refers to another eruption on January 27, 1037, which lasted for 6 days. The chronicle of the Cassino monastery records an explosive eruption between 1068 and 1078 (Gasparini and Musella, 1991). The last eruption before a long quiescent period occurred on June 1, 1139. Several sources refer to it as a strong explosive eruption (Falcone Beneventano, the Chronicle of the Monastery of Cava dei Tirreni, John of Salisbury). It lasted for eight days and ashes covered Salerno, Benevento, Capua and Naples. Figliolo and Marturano (1994) made a critical revision of the historical sources for the period between the 7th to the 12th century AC. They suggested that the eruption of 685 was merely explosive, that of 1036 had effusive and explosive character and that of 1139 was mainly explosive. According to these authors, the eruption of 787 had lava flows reaching the sea. No reliable report of volcanic activity is available until 1500, when Ambrogio di Nola reports a small explosion. From 1500 until to 1631, no eruption occurred on Vesuvius. Records are good during this period, and none mentions volcanic activity. Historical documents have been cross-checked with eruption deposits. Rosi and Santacroce (1983) identified the products of a subplinian eruption referred to as "Pollena eruption" and have referred it to the 472 AD eruption basing it on ^{14}C datings. Arnò et al. (1987), Andronico et al. (1995) and Rolandi et al. (1998) correlated, on the basis of ^{14}C datings, pyroclastic deposits with the eruptions of 512, 685, 787 and 1139. They also found other deposits of more difficult attribution. Principe et al. (2004) made a detailed archeomagnetic study of the lava flows outcropping on the southern margin of the volcano, mainly between Portici and Torre Annunziata. They recognize lava flows referred to eruptions in the 9th and 10th century AD as well as lava flows referred to the eruptions of 968, 999, 1037 and 1139.

A strong explosive eruption started in the night between December 15 and 16, 1631 and its paroxysmal stage lasted for two days (Fig. 10). The eruption started a period of persistent activity that lasted, with a few breaks, for more than three centuries until 1944.

The close observation of the activity, firstly from local scholars, and then by an increasing number of foreign students permitted to develop new theories on the character of volcanic activity. The careful descriptions of the volcano activity permitted Baratta (1897), and Alfano and Friedlaender (1929) to formulate models of activity between 1631 and 1906 suggesting recurring cycles of activity. Each cycle was characterized by the succession (Carta et al., 1981) of:

(a) a period of repose (generally not exceeding a few years) (R = repose);
(b) a phase of strombolian activity with the building of a conelet within the crater (A = permanent activity), and eventually, the emission of some lava flows (IE = intermediate eruptions) (either within the crater or outside it);
(c) a violent eruption usually with a lava flow and strong explosions followed by a new repose (FE = final eruption).

The idea of cycles of activity was also used in the most recent compilation of the Vesuvian activity (Arnò et al., 1987). Scandone et al. (1993a) cast some doubts about the real existence of cycles of activity. We suggest that the activity observed in this period shows an evolutionary trend and that the cycles are only an artifact due to the violence of a few eruptions which caused small caldera collapses.

Scandone et al. (1993a) report 99 magmatic eruptions following the one in 1631; 5 FE had a VEI of 3+ (1737, 1779, 1794, 1822, 1906), and 12 had a VEI of 3. Fifty-three

Figure 10. The eruption of Vesuvius of 1631 in a contemporary engraving.

eruptions were accompanied by (or were entirely) explosive phenomena. Arrighi et al. (2001) have shown that some explosive eruptions (not followed by a repose) had a violence comparable with that of eruptions ending a cycle. At the same time eruptions with predominantly effusive activity may have volumes as large as that of mainly explosive ones. Explosive activity was predominant until 1872 (49 events); since this date, effusive eruptions became more numerous and longer and there was a slow accumulation of lava either on the flanks of the cone (building of several lava domes between 1872 and 1899) or filling of the crater and outpouring of lava from it. Since 1872, the length of IE became longer (Carta et al., 1981). Such pattern is not uncommon on other volcanoes. For example, a similar behavior, although on a shorter timescale, has been observed at Paricutin between 1943 and 1952 (Scandone, 1979b) and St Helens between 1980 and 1986. We show in Figure 11 that, since the beginning of 1700, there is a regular decrease in the length of periods of explosive activity and a regular increase in that of effusive activity.

Scandone et al. (1986), Arrighi et al. (2001) and Marianelli et al. (2004) suggest that the explosive events during this period are due to the arrival of a gas-rich magma that cause the emission of high lava fountains (up to 1–2 km height), and an eruptive column up to 10 km height. We further suggest that also the initial effusive phase of these eruptions is characterized by a high effusion rate as suggested by the morphology of the lava field (single-channel flow) opposed to that typical of eruption with smaller effusion rate (multiple flow) such as that of 1858 or 1891–1894 and 1895–1899. Eichelberger et al. (1986) suggest that non-explosive silicic eruptions may be due to a water loss from the magma during its ascent from mid-crustal magma chambers.

Recent studies (Devine et al., 1998; Rutherford and Devine, 2003) give support to this hypothesis, and further suggest that the episodes of effusive activity during several eruptive periods (Mt. St Helens, 1980–1986; Soufriere Hills of Montserrat, 1995–) are related to magmas that ascend with a slower velocity than that of explosive episodes. We suggest

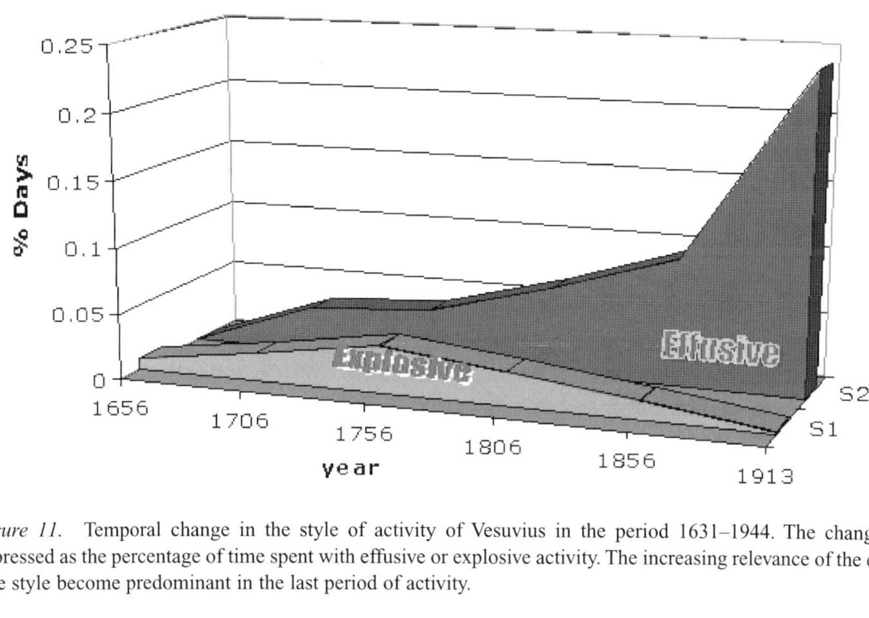

Figure 11. Temporal change in the style of activity of Vesuvius in the period 1631–1944. The change is expressed as the percentage of time spent with effusive or explosive activity. The increasing relevance of the effusive style become predominant in the last period of activity.

that this is due to the time available to degassing during the ascent and that this principle is also valid for more mafic magmas.

The ascent velocity of magmas, ascending as liquid-filled cracks, is dependent on the rheological properties of magma, buoyancy and stress field (Shaw, 1980; Ryan, 1994). The chemical composition of magmas erupted during the 1631–1944 period does not change in a regular fashion such as to affect the rheology and buoyancy. So we imply that the main cause of the regular change of activity is due to a change of the stress field inside the volcano either due to a progressive sealing of pathways after the major explosive eruption of 1631 or a general change of regional stress. The first hypothesis is suggested by the similarity with the trend of other eruptions (Paricutin, 1943–1952; Mt. St Helens, 1980–1986; Soufriere Hills of Montserrat, 1995–) and the second one is suggested by the occurrence, at the beginning of this period, of a cluster of tectonic earthquakes with a magnitude of >6 in the nearby Apennine Chain (1626, 1688, 1694, 1702, 1732) which has no equivalent in recent times (Bonasia et al., 1985).

In either case, the predominance of one type of activity with respect to the other is on a statistical base: the ascent is governed by the predominant stress field, but the actual ascent rate is governed by a number of different casual phenomena that allow, in the same period, different ascent rates. In the earlier period the ascent rate was generally higher, permitting the arrival of more gas-rich magma batches, with consequent higher chance of explosive eruptions. In the more recent period, the ascent rate was generally lower permitting the arrival of gas-depleted magma batches, with consequent higher chance of effusive eruptions. In the end, this regime shifted toward much slower ascent rates, not even permitting the arrival of magma to the surface.

Overall, the activity record identified through the historical and geological investigations provides the evidence of a complex volcanic activity alternating long periods of quiescence with major eruptions, or periods of persistent volcanic activity. Although the geological studies indicate the occurrence only of eruptions with VEI = 4–6 before the eruption of the "Pomici di Avellino" 3500–3770 years ago, we believe that this is an artifact due to the loss of information, because the deposits of smaller eruptions are easily destroyed by the following activity and completely lost to our investigation. It is likely that the older volcanological history of the volcano will never be known with sufficient detail.

6. The structure and feeding system of Vesuvius

Finetti and Morelli (1974), Scandone and Cortini (1982), Vilardo et al. (1996), Bianco et al. (1998) and Bruno et al. (1999) suggest that Mt. Vesuvius is built at the crossing of two fault systems with NE-SW and NW-SE directions, respectively, identified by seismic reflection profiles on land and at sea, by focal mechanism of local earthquakes and alignment of lateral vents of the volcano. The south-western part of the volcano is lowered by these fault systems (Bianco et al., 1998; Bruno et al., 1998), and the surface aspect of the volcano takes the resemblance of a horse-shoe-shaped aspect. This shape has been, for some time, interpreted as the result of a sector collapse of the old volcanic Somma structure (Milia et al., 1998; Bruno and Rapolla, 1999). Generally, sector collapses are more common in volcanoes erupting andesitic or dacitic lavas with steep slopes (≥30°). Mt. Somma only rarely does attain such critical slopes (only in a limited portion on the

northern flank). Further on, the identification of the landslide deposit and the hummocky surface resulting from the collapse is problematic and uncertain. Contrary to these interpretations, the SW rim of the caldera is still visible, although lowered with respect to the north-eastern side. This is not a feature observed in most sector collapse of other volcanoes. We favor the interpretation of a progressive caldera collapse due to several plinian eruptions as suggested by Santacroce (1987). The lowering of the SW rim is possibly related with regional NW-SE faults lowering the sedimentary basement toward the Gulf of Naples (Bianco et al., 1998).

Vesuvius is built mostly on alluvial and marine sediments that filled up the graben formed by the subsidence of the carbonatic platforms, which make up the basement of the Campanian plain during the Pliocene and Pleistocene with a maximum vertical extent of 2–3 km below the volcano (Ippolito et al., 1973).

The volcanic activity occurred mainly after the eruption of the Campanian Ignimbrite (39 kyr). As a consequence, the structure of the volcano is in a state of isostatic disequilibrium due to its rapid formation compared to the isostatic re-equilibration of the load.

Bouguer gravimetric anomalies of the volcano provide evidence of a shallow structure without deep roots (Cassano and La Torre, 1987). Appreciable anomalies are observed only in close connection with the central cone of the volcano. Cubellis et al. (2001) suggest the existence of a structure along the axis of the volcanic edifice, with a density of 2100 kg/m^3 down to a depth of about 2000 m and 2400 kg/m for the lowermost part.

Magnetic anomalies provide a more complex picture (Cassano and La Torre, 1987; Fedi and Rapolla, 1999). Vesuvius has a high magnetization with a maximum inside the volcanic structure. The marginal parts of Vesuvius show low magnetization. Magnetized rocks extend down to about 2000 m bsl, but the magnetization becomes weak down to 4–5 km bsl (Fedi and Rapolla, 1998) (Fig. 12).

A prominent high-density core has been identified by seismic tomography (Zollo et al., 1996) concentric with the caldera structure. The anomalous high-velocity region starts from about 400 m below the crater and extends down to at least 3000 m. The highest velocities 3.8–4.0 km/s are observed at about 1500–2000 m below the Earth's surface (Zollo et al., 1996, 1998; De Natale et al., 1998). The zone was interpreted, by the above authors, as a plexus of solidified dykes. Combined magneto-telluric investigations and time-domain electromagnetic (TDEM) soundings across Vesuvius (Di Maio et al., 1998; Manzella et al., 2004) reveal the presence of a resistive cover layer underlain by an anomalous conductive layer (c. 250–2500 m below the ground surface) interpreted as the shallow hydrothermal system of the volcano inferred also by geochemical data (Chiodini et al., 2001). The occurrence of hydro-fracturing induced seismicity at depth as high as 4–5 km bsl (Bianco et al., 1999) suggests the possible downward continuation of the hydrothermal system.

Seismic data evidenced an extended low-velocity layer at about 8–10 km depth, interpreted as the top of a magma reservoir, having a surface area of at least 400 km^2 (Auger et al., 2001).

In conclusion, geophysical studies provide evidence of a volcano made up by a shallow structure mostly above sea level and a more complex central core with high rigidity and intense hydrothermal circulation. The core traverses the low-density, alluvial terrain and is pinned into the low-standing carbonate platforms, causing intense alteration and decarbonation.

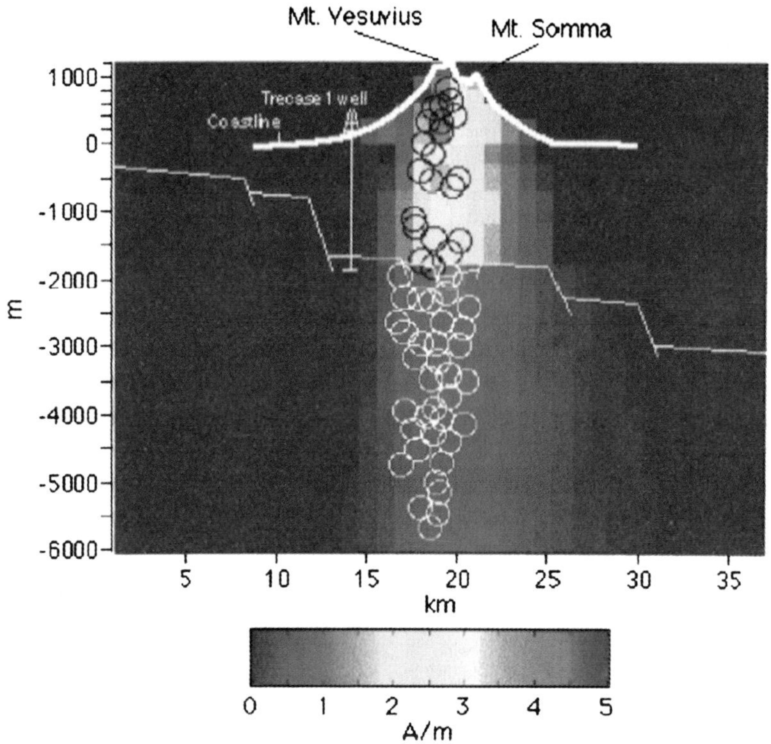

Figure 12. The source of the magnetic anomaly of Vesuvius is explained in terms of a high magnetic inner core of the volcano extending down to the carbonatic platform (after Fedi et al., 1998).

Other important information about the lower part of the volcanic system is provided by geochemical studies. The study of fluid inclusions provided evidence of a series of intermediate storage of magma at depth ranging between 12 and 2 km (see for example Belkin et al., 1985, 1993, 1998; Marianelli et al., 2004).

7. Conclusions

The study of the volcanism of the Campanian plain provides a striking example of the interplay between tectonics and volcanic activity in controlling the location of the volcanic centers and the control on the eruptive style. Both Campi Flegrei and Vesuvius resulted as a consequence of the giant eruption of the Campanian Ignimbrite at 39 kyr; they emplaced along the newly formed line of weakness, erupted magma with chemical similarities and displayed a different style of activity, which led to the building of volcanic edifices of different types.

The control on the style of activity is exerted by the local stress field, which may vary in time and permit either an easy ascent of relatively less differentiated magmas or a slower ascent and longer residence times in crustal magma chambers.

The high level of risk in the area (Scandone et al., 1993b; Alberico et al., 2002) requires a continuous effort for a better understanding of the structure of the area and the factors controlling the dynamics of volcanic eruptions.

Acknowledgements

We acknowledge financial support from GNV-INGV, Vesuvius sub-project by Protezione Civile Nazionale of Italy, and MIUR-Prin Project 2005 (Risalita dei Magmi e Dinamica delle Eruzioni).

References

Alberico, I., Lirer, L., Petrosino, P., Scandone, R., 2002. A methodology for the evaluation of long-term volcanic risk from pyroclastic flows in Campi Flegrei (Italy). J. Volcanol. Geotherm. Res. 116, 63–78.

Alessio, M., Bella, F., Improta, S., Belluomini, G., Cortesi, C., Turi, B., 1974. University of Rome Carbon-14 Dates XII. Radiocarbon 16(3), 358–367.

Alfano, G.B., 1924. Le eruzioni del Vesuvio fra il 79 ed il 1631, Valle, Pompei (Roma, Istituto Pio IX tip).

Alfano, G.B., Friedlaender, I., 1929. La storia del Vesuvio illustrata da documenti coevi, Karl Hohn Verlag, Ulm, 69 pp.

Andronico, D., Calderoni, G., Cioni, R., Sbrana, A., Sulpizio, R., Santacroce, R., 1995. Geological map of Somma-Vesuvius volcano. Per. Mineral. 64, 77–78.

Andronico, D., Cioni, R., 2002. Contrasting styles of Mount Vesuvius activity in the period between the Avellino and Pompeii Plinian eruptions, and some implications for assessment of future hazards. Bull. Volcanol. 64, 372–391.

Arnò, V., Principe, C., Rosi, M., Santacroce, R., Sbrana, A., Sheridan, M.F., 1987. Eruptive history. In: Santacroce R. (Ed), Somma-Vesuvius. CNR Quaderni della Ricerca Scientifica 114, vol. 8, pp. 53–104.

Arrighi, S., Principe, C., Rosi, M., 2001. Violent strombolian and subplinian eruptions at Vesuvius during post-1631 activity, Bull. Volcanol. 63, 126–150.

Auger, E., Gasparini, P., Virieux, J., Zollo, A., 2001. Seismic evidence of an extended magmatic sill under the Mt. Vesuvius. Science 294, 1510–1512.

Baratta, M., 1897. Il Vesuvio e le sue Eruzioni, Società Editrice Dante Alighieri, Roma, 203 pp.

Barberi, F., Corrado, G., Innocenti, F., Luongo, G., 1984. Phlegraean Fields 1982–1984: brief chronicle of a volcano emergency in a densely populated area. Bull. Volcanol. 47(2), 175–185.

Barberi, F., Innocenti, F., Lirer, L., Munno, R., Pescatore, T., Santacroce, R., 1978. The Campanian Ignimbrite: a major prehistoric eruption in the neapolitan area (Italy). Bull. Volcanol. 41(1), 1–22.

Belkin, E.H., De Vivo, B., 1993. Fluid inclusions studies of ejected nodules from plinian eruptions of Mt. Somma-Vesuvius. J. Volcanol. Geotherm. Res. 58, 89–100.

Belkin, H.E., De Vivo, B., Roedder, E., Cortini, M., 1985. Fluid inclusion geobarometry from ejected Mt. Somma-Vesuvius nodules. Am. Mineral. 70, 288–303.

Belkin, H.E., De Vivo, B., Török, K., Webster, J.D., 1998. Pre-eruptive volatile content, melt-inclusion chemistry, and microthermometry of interplinian Vesuvius lavas (pre-AD 1631). J. Volcanol. Geotherm. Res. 82, 79–95.

Berrino, G., Corrado, G., Luongo, G., Toro, B., 1984. Ground deformation and gravity changes accompanying the 1982 Pozzuoli uplift. Bull. Volcanol. 47, 187–200.

Bianco, F., Castellano, M., Milano, G., Ventura, G., Vilardo, G., 1998. The Somma-Vesuvius stress field: seismological and mesostructural data. J. Volcanol. Geotherm. Res. 82, 199–218.

Bianco, F., Castellano, M., Milano, G., Ventura, G., Vilardo, G., Ferrucci, F., Gresta, S., 1999. The seismic crises at Vesuvius during 1995 and 1996. Phys. Chem. Earth 24, 977–983.

Bianco, F., Del Pezzo, E., Saccorotti, G., Ventura, G., 2004. The role of hydrothermal fluids in triggering the July–August 2000 seismic swarm at Campi Flegrei, Italy evidence from seismological and mesostructural data. J. Volcanol. Geotherm. Res. 133, 229–246.

Bonafede, M., 1991. Hot fluid migration: an efficient source of ground deformation. Application to the 1982–85 Crisis at Campi Flegrei-Italy. J. Volcanol. Geotherm. Res. 48, 187–198.

Bonasia, V., Del Pezzo, E., Pingue, F., Scandone, R., Scarpa, R., 1985. Eruptive history, seismic activity, and ground deformations at Mt. Vesuvius, Italy, Ann. Geoph. 3(3), 395–406.

Breislak, S., 1798. Topografia Fisica della Campania. Brazzini, Firenze, 368 pp.

Brocchini, D., Principe, C., Castradori, D., Laurenzi, M.A., Gorla, L., 2001. Quaternary evolution of the southern sector of the Campanian Plain and early Somma Vesuvius activity: insights from the Trecase 1 well. Mineral. Petrol. 73, 67–92.

Bruno, P.P., Cippitelli, G., Rapolla, A., 1998. Seismic study of the Mesozoic carbonate basement around Mt. Somma–Vesuvius, Italy. J. Volcanol. Geotherm. Res. 84, 311–322.

Bruno, P.P.G., Rapolla, A., 1999. Study of the sub-surface structure of Somma-Vesuvius (Italy) by seismic reflection data. J. Volcanol. Geotherm. Res. 92(3–4), 373–387.

Carta, S., Figari, R., Sartoris, G., Sassi, E., Scandone, R., 1981. A statistical model of Vesuvius and its volcanological implication. Bull. Volcanol. 44(2), 129–151.

Casertano, L., Oliveri del Castillo, A., Quagliariello, M.T., 1976. Hydrodynamics and geodynamics in the Phlegraean Fields area of Italy. Nature 264, 161–164.

Cassano, E., Guglielminetti, M., Verdiani, G., 1986. Phlegrean Fields area, first results of geothermal research. Paper presented at International Symposium on Earthquake Engineering, Bari, Italy.

Cassano, E., La Torre, P., 1987. Geophysics. In: Santacroce, R. (Ed.), Somma Vesuvius. Quaderni de 'La Ricerca Scientifica' 114/8, CNR, Rome, pp. 175–196.

Chiodini, G., Marini, L., Russo, M., 2001. Geochemical evidence for the existence of high-temperature hydrothermal brines at Vesuvio volcano, Italy. Geochim. Cosmochim. Acta 65(13), 2129–2147.

Cinque, A., Rolandi, G., Zamparelli, V., 1985. L'Estensione dei Depositi Marini Olocenici nei Campi Flegrei in Relazione alla Vulcano-Tettonica. Boll. Soc. Geol. It. 104, 327–348.

Cioni, R., Santacroce, R., Sbrana, A., 1999. Pyroclastic deposits as a guide for reconstructing the multi-stage evolution of the Somma-Vesuvius caldera. Bull. Volcanol. 60, 207–222.

Civetta, L., Orsi, G., Patella, D., 2004. Special issue on the Neapolitan volcanoes. J. Volcanol. Geotherm. Res. 133, 1–4.

Colantoni, P., Del Monte, M., Fabbri, A., Gallignani, P., Selli, R., Tomadi, L., 1972. Ricerche geologiche nel golfo di Pozzuoli, CNR, Quaderni Ric. Sci. 83, 26–76.

Cole, P.D., Perrotta, A., Scarpati, C., 1994. The volcanic history of the southwestern part of the city of Naples. Geol. Mag. 131, 785–799.

Corrado, G., Guerra, I., Lo Bascio, A., Luongo, G., Rampoldi, F., 1976. Inflation and microearthquake activity of Phlegraean Fields, Italy. Bull. Volcanol. 40(3), 169–188.

Cubellis, E., Ferri, M., Luongo, G., Obrizzo, F., 2001. The roots of Mt. Vesuvius deduced from gravity anomalies. Mineral. Petrol. 73, 23–38.

De Astis, G., Pappalardo, L., Piochi, M., 2004. Procida volcanic history: new insights into the evolution of the Phlegraean Volcanic District (Campania region, Italy). Bull. Volcanol, 66, 622–641.

Deino, A., Orsi, G., de Vita, S., Piochi, M., 2004. The age of the Neapolitan Yellow Tuff caldera forming eruption (Campi Flegrei Caldera, Italy) assessed by $^{40}Ar/^{39}Ar$ dating method. J. Volcanol. Geotherm. Res. 133, 157–170.

Delibrias, G., Di Paola, G.M., Rosi, M., Santacroce, R., 1979. La storia del complesso eruttivo Somma-Vesuvio ricostruita dalle successioni piroclastiche del Monte Somma. Rend. Soc. It. Mineral. Petrol. 35, 411–438.

De Lorenzo, G., 1904. L'Attivita' Vulcanica dei Campi Flegrei. Rend. Acc. Sci. Fis. Mat. Napoli 10, 203–221.

De Natale, G., Capuano, P., Troise, C., Zollo, A., 1998. Seismicity at Somma-Vesuvius and its implications for tomography of the volcano. J. Volcanol. Geotherm. Res. 82, 175–197.

De Natale, G., Pingue, F., 1993. Ground deformations in collapsed caldera structures. J. Volcanol. Geotherm. Res. 57, 19–38.

De Natale, G., Pingue, F., Allard, P., Zollo, A., 1991. Geophysical and geochemical modeling of the Campi Flegrei caldera. J. Volcanol. Geotherm. Res. 48, 199–222.

Devine, J.D., Rutherford, M.J., Gardner, J.E., 1998. Petrologic determination of ascent rates for the 1995–1997 Soufriere Hills Volcano andesitic magma. Geophys. Res. Lett. 25(19), 3673–3676.

De Vivo, B., Rolandi, G. (Eds), 2001. Special issue: Mt. Somma-Vesuvius and volcanism of the Campanian Plain. Mineral. Petrol. 73, 1–232.

De Vivo, B., Rolandi, G., Gans, P.B., Calvert, A., Bohrson, W.A., Spera, F.J., Belkin, H.E., 2001. New constraints on the pyroclastic eruptive history of the Campanian volcanic plain. Mineral. Petrol. 73, 47–66.

De Vivo, B., Scandone, R. (Eds), 2003. Special issue: Ignimbrites of the Campanian Plain. Mineral. Petrol. 79, 1–125.

De Vivo, B., Scandone, R., Trigila R.R. (Eds), 1993. Mount Vesuvius. J. Volcanol. Geotherm. Res. 58, 1–381.

Di Girolamo, P., 1968. Petrografia dei Tufi Campani: Il Processo di Pipernizzazione (tufo→tufo pipernoide→piperno). Rend. Acc. Sci. Fis. Mat. Napoli 4(25), 5.

Di Girolamo, P., 1970. Differenziazione Gravitativa e Curve Isochimiche nell' Ignimbrite Campana. Rend. Soc. It. Mineral. Petrol. 26(2), 547–588.

Di Girolamo, P., Ghiara, M.R., Lirer, L., Munno, R., Rolandi, G., Stanzione, D., 1984. Vulcanologia e Petrologia dei Campi Flegrei. Boll. Soc. Geol. 103, 349–413.

Di Girolamo, P., Rolandi, G., Stanzione, D., 1973. L'Eruzione di Pomici a Letto dell'Ignimbrite Campana (Tufo Grigio Camapano Auct). Period. Mineral. 42, 439–468.

Di Maio, R., Mauriello, P., Patella, D., Petrillo, Z., Piscitelli, S., Siniscalchi, A., 1998. Electric and electromagnetic outline of the Mount Somma–Vesuvius structural setting. J. Volcanol. Geotherm. Res. 82, 219–238.

Di Vito, M.A., Isaia, R., Orsi, G., Southon, J., de Vita, S., D'Antonio, M., Pappalardo, L., Piochi, M., 1999. Volcanic and deformational history of the Campi Flegrei caldera in the past 12 ka. J. Volcanol. Geotherm. Res. 91, 221–246.

Dvorak, J.J., Gasparini, P., 1991. History of earthquakes and vertical ground movement in Campi Flegrei caldera, Southern Italy: comparison of precursory events to the AD 1538 eruption of Monte Nuovo and of activity since 1968. J. Volcanol. Geotherm. Res. 48, 77–92.

Dvorak, J.J., Mastrolorenzo, G., 1990. History of vertical movement in Pozzuoli Bay, Southern Italy: the result of regional extension related to evolution of the Tyrrhenian Sea and of local volcanic activity. Am. Spec. Pap. 263, 47.

Eichelberger, J.C., Carrigan, C.R., Westrich, H.R., Price, R.H., 1986. Non explosive silici volcanism. Nature 323, 598–602.

Fedi, M., Florio, G., Rapolla, A., 1998. 2.5D modelling of Somma-Vesuvius structure by aeromagnetic data. J. Volcanol. Geotherm. Res. 82, 239–247.

Fedi, M., Nunziata, C., Rapolla, A., 1991. The Campania–Campi Flegrei area: a contribution to discern the best structural model from gravity interpretation. J. Volcanol. Geotherm. Res. 48, 51–59.

Figliolo, B., Marturano, A., 1994. The eruptions of Vesuvius from the 7th to the 12th centuries. In: Morello, N. (Ed.), Volcanoes and History, Genova, Brigati, pp. 133–156.

Finetti, I., Morelli, C., 1974. Esplorazione sismica a riflessione nei golfi di Napoli e Pozzuoli. Boll. Geof. Teor. Appl. 16, 175–222.

Fisher, R.V., Orsi, G., Ort, M., Heiken, G., 1993. Mobility of large volume pyroclastic flow – emplacement of the Campanian Ignimbrite, Italy. J. Volcanol. Geotherm. Res. 56, 205–220.

Florio, G., Fedi, M., Cella, F., Rapolla, A., 1999. The Campanian Plain and Phlegrean Fields: structural setting from potential field data. J. Volcanol. Geotherm. Res. 91, 361–379.

Fredericksen, M., 1977. Una fonte trascurata sul bradisimo puteolano, Atti Convegni Lincei 33, 117–130.

Gaeta, F.S., De Natale, G., Peluso, F., Mastrolorenzo, G., Castagnolo, D., Troise, C., Pingue, F., Mita, D.G., Rossano, S., 1998. Genesis and evolution of unrest episodes at Campi Flegrei caldera: the role of thermal fluiddynamical processes in the geothermal system, J. Geophys. Res. 103, 921– 933.

Gasparini, P., Musella, S., 1991. Un Viaggio al Vesuvio, Liguori Editore, 307 pp.

Ippolito, F., Ortolani, F., Russo, M., 1973. Struttura Marginale dell'Appennino Campano: Reinterpretazione dei Dati di Antiche Ricerche di Idrocarburi. Mem. Società Geologica Italiana XII, 227–250.

Johnston-Lavis H.F., 1884. The geology of Mt Somma and Vesuvius, being a study of volcanology. Q. J. Geol. Soc. London 40, 35–149.

Johnston-Lavis, H.F., 1889. Report of the committee appointed for the investigation of the volcanic phenomena of Vesuvius and its neighbour. Royal Society, London.

Keller, J., Ryan, W.B.F., Ninkovich, D., Altherr, R., 1978. Explosive volcanic activity in the Mediterranean over the past 200000 yr as recorded in deep-sea sediments. Geol. Soc. Am. Bull. 89, 591–604.

Lirer, L., Luongo, G., Scandone, R., 1987. On the volcanological evolution of Campi Flegrei. EOS, Trans. Am. Geophys. Union 68, 226–234.

Lirer, L., Munno R., 1975. Il Tufo Giallo Napoletano (Campi Flegrei). Period. Mineral. 44, 103–118.

Lirer, L., Pescatore, T., Booth, B., Walker, G.P.L., 1973. Two plinian fall deposits from Somma-Vesuvius. Geol. Soc. Am. Bull. 84, 759–772.

Luongo, G., Scandone, R. (Ed.), 1991. Special issue on Campi Flegrei. J. Volcanol. Geotherm. Res. 48, 1–2.

Lyell, C., 1830. Principles of Geology, John Murray, London, 810 pp.

Manzella, A., Volpi, G., Zaja, A., Meju, M., 2004. Combined TEM-MT investigation of shallow-depth resistivity structure of Mt Somma-Vesuvius. J. Volcanol. Geotherm. Res. 131, 19–32.

Marianelli, P., Sbrana, A., Metrich, N., Cecchetti, A., 2004. The deep feeding system of Vesuvius involved in recent violent strombolian eruptions. Geophys. Res. Lett. 32, L02306, doi: 10.1029/2004GL021667.

Milia, A., Mirabile, L., Torrente, M.M., Dvorak, J.J., 1998. Volcanism offshore of Vesuvius volcano in Naples Bay. Bull. Volcanol. 59, 404–413.

Munno, R., Petrosino, P., 2004. New constraints on the occurrence of Y-3 Upper Pleistocene tephra marker layer in the Thyrrenian sea. Il Quaternario 17(1), 11–20.

Nunziata, C., Rapolla, A., 1981. Interpretation of gravity and magnetic data in the Phlegrean Fields geothermal area, Naples, Italy. J. Volcanol. Geotherm. Res. 10, 209–226.

Orsi, G., Civetta, L., Valentine, G.A., 1999. Special issue on Campi Flegrei. J. Volcanol. Geotherm. Res. 91, 1–4.

Orsi, G., de Vita, S., Di Vito, M., 1996. The restless, resurgent Campi Flegrei nested caldera (Italy): constraints on its evolution and configuration. J. Volcanol. Geotherm. Res. 74, 179– 214.

Ort, M., Rosi, M., Anderson, C.D., 1999. Correlation of deposits and vent locations of the proximal Campanian Ignimbrite deposits, Campi Flegrei, Italy, based on natural remanent magnetization and anisotropy of magnetic susceptibility characteristics, Flegrei. J. Volcanol. Geotherm. Res. 91, 167–178.

Osservatorio Vesuviano, 1995. La sorveglianza delle aree vulcaniche napoletane, Open File Report, Teti, Napoli.

Parascandola, A., 1947. I fenomeni bradisismici del Serapeo di Pozzuoli. Genovese, Naples, Italy, 156 pp.

Perrotta, A., Scarpati, C., 2001. Volume partition between the plinian and co-ignimbrite air fall deposit of the Campanian Ignimbrite eruption. Mineral. Petrol. 79, 67–78.

Pescatore, T., Rolandi, G., 1981. Preliminary observations on stratigraphy of volcanoclastic deposits of the SW sector of Campi Flegrei (in Italian). Boll. Soc. Geol. It. 100, 233–254.

Polacci, M., Pioli, L., Rosi, M., 2003. The Plinian phase of the Campanian Ignimbrite eruption (Phlegrean Fields, Italy): evidence from density measurements and textural characterization of pumice. Bull. Volcanol. doi: 10.1007/s00445-002-0268-4.

Principe, C., Tanguy, J.C., Arrighi, S., Paiotti, A., Le Goff, M., Zoppi, U., 2004. Chronology of Vesuvius activity from AD 79 to 1631 based on archeomagnetism of lavas and historical sources. Bull. Volcanol. doi: 10.1007/s00445-004-0348-8.

Rittmann, A., 1950. Sintesi Geologica dei Campi Flegrei. Boll. Soc. Geol. It. LXIX-II, 117–128.

Rolandi, G., Bellucci, F., Heizler, M.T., Belkin, H.E., De Vivo, B., 2003. Tectonic controls on the genesis of ignimbrites from the campanian volcanic zone. Mineral. Petrol. 79, 2–31.

Rolandi, G., Petrosino, P., McGeehin, I., 1998. The interplinian activity at Somma-Vesuvius in the last 3500 years. J. Volcanol. Geotherm. Res. 82, 19–52.

Rosi, M., Santacroce, R., 1983. The AD 472 "Pollena Eruption": volcanological and petrological data for this poorly known, plinian-type event at Vesuvius. J. Volcanol. Geotherm. Res. 17, 249–272.

Rosi, M., Sbrana, A., Principe, C., 1983. The Phlegrean Fields: structural evolution, volcanic history, and eruptive mechanisms. J. Volcanol. Geotherm. Res. 17, 273–288.

Rosi, M., Sbrana, A. (Eds), 1987. Phlegraean Fields, Vol. 114. CNR. Quad. Ric. Sci., Roma. 167 pp.

Rosi, M., Sbrana, A., Vezzoli, L., 1988a. Tephrostratigraphy of Ischia, Procida and Campi Flegrei volcanic products (In Italian). Mem. Soc. Geol. It. 41, 1015–1027.

Rosi, M., Sbrana, A., Vezzoli, L., 1988b. Stratigraphy of Procida and Vivara islands (In Italian). Boll. GNV 4, 500–525.

Rosi, M., Vezzoli, L., Castelmenzano, A., Grieco, G., 1999. Plinian pumice fall deposit of the Campanian Ignimbrite eruption (Phlegrean Phields, Italy). J. Volcanol. Geotherm. Res. 91, 179–198.

Rutherford, M.J., Devine, J.D., 2003. Magmatic conditions and magma ascent as indicated by hornblende phase equilibria and reactions in the 1995–2002 Soufriere Hills magma. J. Petrol. 44(8), 1433–1454.

Ryan, M.P., 1994. Neutral-buoyancy controlled magma transport and storage in mid-ocean ridge magma reservoirs and their sheeted-dike complex: a summary of basic relationships. In: Ryan, M.P. (Ed.), Magmatic Systems, Academic Press, New York, pp. 97–138.

Santacroce, R., 1987. Somma-Vesuvius. CNR, Quaderni, 114, Roma, 251 pp.

Scacchi, A., 1848. Memorie Geologiche della Campania. Rend. Acc. Sci. Napoli, 41, 8–43.

Scacchi, A., 1890. La Regione Vulcanica Fluorifera della Campania, 2nd ed. Mem. Regio Com. Geol. d'It. Vol IV, Firenze.

Scandone, P., 1979a. Origin of the Thyrrenian Sea and Calabrian Arc. Boll. Soc. Geol. It. 98, 27–34.

Scandone, R., 1979b. Effusion rate and energy balance of Paricutin eruption (1943–1952), Michoacan, Mexico. J. Volcanol. Geotherm. Res. 6, 49–59.

Scandone, R., Arganese, G., Galdi, F., 1993b. The evaluation of volcanic risk in the Vesuvian area. J. Volcanol. Geotherm. Res. 58, 263–271.

Scandone, R., Bellucci, F., Lirer, L., Rolandi, G., 1991. The structure of the Campanian Plain and the activity of the Neapolitan volcanoes Italy. J. Volcanol. Geotherm. Res. 48, 1–31.

Scandone, R., Cortini, M., 1982. Il Vesuvio: un vulcano ad alto rischio. Le Scienze. 163, 92–105.

Scandone, R., Giacomelli, L., Gasparini, P., 1993a. Mount Vesuvius: 2000 years of volcanological observations. J. Volcanol. Geotherm. Res. 58, 5–25.

Scandone, R., Iannone, F., Mastrolorenzo, G., 1986. Stima dei Parametri Dinamici dell'Eruzione del 1944 del Vesuvio. Boll. Gruppo Nazionale di Vulcanol. 2, 487–512.

Scarpati, C., 1990. Stratigrafia, Geochimica e Dinamica Eruttiva del Tufo Giallo Napoletano. Ph.D. thesis, University of Naples, 159 pp.

Scarpati, C., Cole, P., Perrotta, A., 1993. The Neapolitan YellowTuff – a large volume m multiphase eruption from Campi Flegrei, Southern Italy. Bull. Volcanol. 55, 343–356.

Shaw, H.R., 1980. The fracture mechanism of magma transport from the mantle to the surface. In: Hargraves, H. (Ed.), The Physics of Magmatic Processes, pp. 201–264.

Sigurdsson, H., Carey, S., Cornell, W., Pescatore, T., 1985. The eruption of Vesuvius in AD 79. Nat. Geograph. Res. 1(3), 332–387.

Spera, F.J., De Vivo, B., Ayuso, R.A., Belkin, H.E. (Eds), 1998. Special issue: Vesuvius. J. Volcanol. Geotherm. Res. 82, 1–247.

Thunell, R., Federman, A., Sparks, S., Williams, D., 1979. The age, origin and volcanological significance of the Y-5 ash layer in the Mediterranean. Quaternary Res. 12, 241–252.

Vilardo, G., Alessio, G., Luongo, G., 1991. Analysis of the magnitude-frequency distribution for the 1983–1984 earthquake activity of Campi Flegrei, Italy. J. Volcanol. Geotherm. Res. 48, 115–126.

Vilardo, G., De Natale, G., Milano, G., Coppa, U., 1996. The seismicity of Mt. Vesuvius. Tectonophysics 261, 127–138.

Zollo, A., Gasparini, P., Virieux, J., Biella, G., Boschi, E., Capuano, P., de Franco, R., Dell'Aversana, P., de Matteis, R., De Natale, G., Iannaccone, G., Guerra, I., Le Meur, H., Mirabile, L., 1998. An image of Mount Vesuvius obtained by 2D seismic tomography. J. Volcanol. Geotherm. Res. 82, 161–174.

Zollo, A., Gasparini, P., Virieux, J., Le Meur, H., de Natale, G., Biella, G., Boschi, E., Capuano, P., de Franco, R., dell'Aversana, P., de Matteis, R., Guerra, I., Iannaccone, G., Mirabile, L., Vilardo, G., 1996. Seismic evidence for a low-velocity zone in the upper crust beneath Mount Vesuvius. Science 274, 592–594.

Zollo, A., Judenherc, S., Auger, E., D'Auria, L., Virieux, J., Capuano, P., Chiarabba, C., de Franco, R., Makris, J., Michelini, A., Musacchio, G., 2003. Evidence for the buried rim of Campi Flegrei caldera from 3-d active seismic imaging. Geoph. Res. Lett. 30, doi: 10.1029/2003GL018173.

Volcanism in the Campania Plain:
Vesuvius, Campi Flegrei and Ignimbrites
edited by B. De Vivo

Chapter 2

The Pleistocene extension of the Campania Plain in the framework of the southern Tyrrhenian tectonic evolution: morphotectonic analysis, kinematic model and implications for volcanism

*E. Turco**, *A. Schettino, P.P. Pierantoni and G. Santarelli*

Dipartimento di Scienze della Terra, Università di Camerino, via Gentile III da Varano, 62032 Camerino (MC), Italy

ABSTRACT

The Tyrrhenian margin of the Apennine chain (TMAC) experienced widespread extensional tectonics characterized by volcanism and the formation of several marine and intermontane troughs and basins in Pleistocene times. The Campania Plain is part of this extensional system, which encompasses an area from southern Tuscany to the northern margin of Calabria. Extensional tectonics affecting these continental areas is likely to be related with the final stages of the opening of the southern Tyrrhenian Sea, which developed since Middle Tortonian times. This work presents a quantitative kinematic model explaining the relationships between extension in the Tyrrhenian Sea, basin formation in the TMAC, migration of the Apenninic arcs and geotectonic setting of the volcanism. A synthesis of the volcanic, structural and geophysical data available in the literature, coupled with a detailed morphotectonic analysis of the study areas were used in computer-aided reconstruction techniques based on interactive modelling of rigid block rotations to realistically assemble in a unique kinematic framework the first-order structures that are observed in the Apennines area and in the Tyrrhenian basin.

Once established, the extension directions in the various sectors of the Apennine chain, by comparing the results of the morpho-structural analysis with data collected from the abundant geological literature, we identified two distinct kinematic elements characterizing the Apennine chain that, from Plio-Pleistocene times, moved independently with respect to the Eurasian reference plate: the Northern Apennines Arc (NAA) and the Southern Apennines Arc (SAA). On the basis of the first-order geological and geophysical constraints, as well as on trial and error experiments, we identified two distinct rotation stages for the Apennine chain. During the first stage, from 3.5 to 0.78 Ma, the NAA and the SAA migrated independently. In the second stage, from 0.78 Ma to the present, the NAA stopped migrating, while the SAA continued migrating towards SE. Thus, N-S extension in the Campania Plain is the result of the relative motion of the NAA with respect to the SAA during the first stage only, whereas the present-day NW-SE extension in this area, which is characterized by intense volcanism (e.g. Ignimbrites, Somma-Vesuvio, Ischia, Campi Flegrei), is related to the migration towards the SE of the SAA with respect to the NAA.

The simplifying assumption of rigidity of the two arcs does not substantially affect the model presented, which only aims at describing the process of extension and associated magmatic activity in the TMAC. Furthermore, the model presented above could not take into account many aspects of the complex tectonic evolution of the TMCA. Nevertheless, it realistically assembles in a unique kinematic framework the first-order structures that are observed in the Apennine area and in the Tyrrhenian basin.

*Corresponding author. *E-mail address:* eturco@camserv.unicam.it (E. Turco).

1. Introduction

Large-scale extensional tectonics coupled with orogenic processes is a typical feature of the Miocene to the recent peri-Tyrrhenian orogenic belt of Italy and Sicily. In central-southern Italy, while the thrust belt-foredeep system of the Apennine chain continued migrating towards the present-day Adriatic–Ionian foreland (Patacca et al., 1990), the Tyrrhenian margin of the Apennine chain (TMAC) experienced widespread extensional tectonics characterized by volcanism and the formation of several marine and intermontane troughs and basins in Pleistocene times. The Campania Plain, an E-W elongated basin infilled by up to 3000 m of Pleistocene volcaniclastic and alluvial sediments (Milia and Torrente, 1999), is part of this extensional system, which encompasses an area extending from southern Tuscany to the northern margin of Calabria (Fig. 1).

Extensional tectonics affecting these continental areas is likely to be related with the final stages of opening of the southern Tyrrhenian Sea (STS). The Tyrrhenian Sea, which developed since Middle Tortonian times, is the youngest basin of the western Mediterranean (Sartori et al., 2004) and, since the 1960, it has been subject to several geological and geophysical explorations and surveys. In spite of the huge amount of available data, the geodynamic evolution of the Tyrrhenian basin and surrounding regions are yet to be coherently described and have been subject to controversial interpretations (Biju-Duval et al., 1977; Dercourt et al., 1986; Malinverno and Ryan, 1986; Dewey et al., 1989; Boccaletti et al., 1990; Carmignani et al., 1995; Lavecchia et al., 1995; Faccenna et al., 1996; Ferranti et al., 1996; Turco and Zuppetta, 1998; Jolivet and Faccenna, 2000; Faccenna et al., 2001; Rosenbaum et al., 2002; Lavecchia et al., 2003; Peccerillo and Turco, 2004). In particular, the kinematic relationships between extension in the Tyrrhenian Sea, basin formation in the TMAC, migration of the Apenninic arcs and geotectonic setting of volcanism still remain to be determined.

In order to reconstruct the tectonic evolution of the Campania Plain during the Pleistocene, in the framework of the southern Tyrrhenian tectonic history, we tried to outline the relationship between extensional tectonics and volcanism that characterized the TMAC during the last 3.5 Myrs. We used volcanic, structural, geophysical and morphological data available in the literature, as well as computer-aided reconstruction techniques based on interactive modelling of rigid block movement.

2. Geology of the Tyrrhenian–Apennines region

In this section, we briefly discuss the main structural and geological features of the Apennine chain and the Tyrrhenian basin.

2.1. The Apennine chain

The Apennine–Maghreb chain is a Neogene thrust belt which comprises Mesozoic to Palaeogene sedimentary rocks, derived from different basins and shelf paleogeographic domains located in the Adria continental margin of the African plate (Patacca et al., 1990).

The formation of the thrust belt started with the collision between the European Corsica–Sardinia block and the Adriatic–African margin, an event that in Oligocene times led to the closure of a tract of the Neo-Tethyan ocean (Dewey et al., 1989). In fact, in Liguria, Toscana and Calabria, Mesozoic to Cenozoic metasedimentary and ophiolitic rocks, the remnants of an ancient accretionary wedge (Knott, 1994), overrode Apenninic Mesozoic carbonate rocks that belonged to the Adriatic domain. In Calabria and NE Sicily, Palaeozoic igneous and metamorphic rocks with the overlying Mesozoic to Cenozoic sedimentary cover, which are considered to be a fragment of the European margin of the Neo-Tethys (Kastens and Mascle, 1990; Knott, 1994 and references therein), overrode the ophiolitic complex.

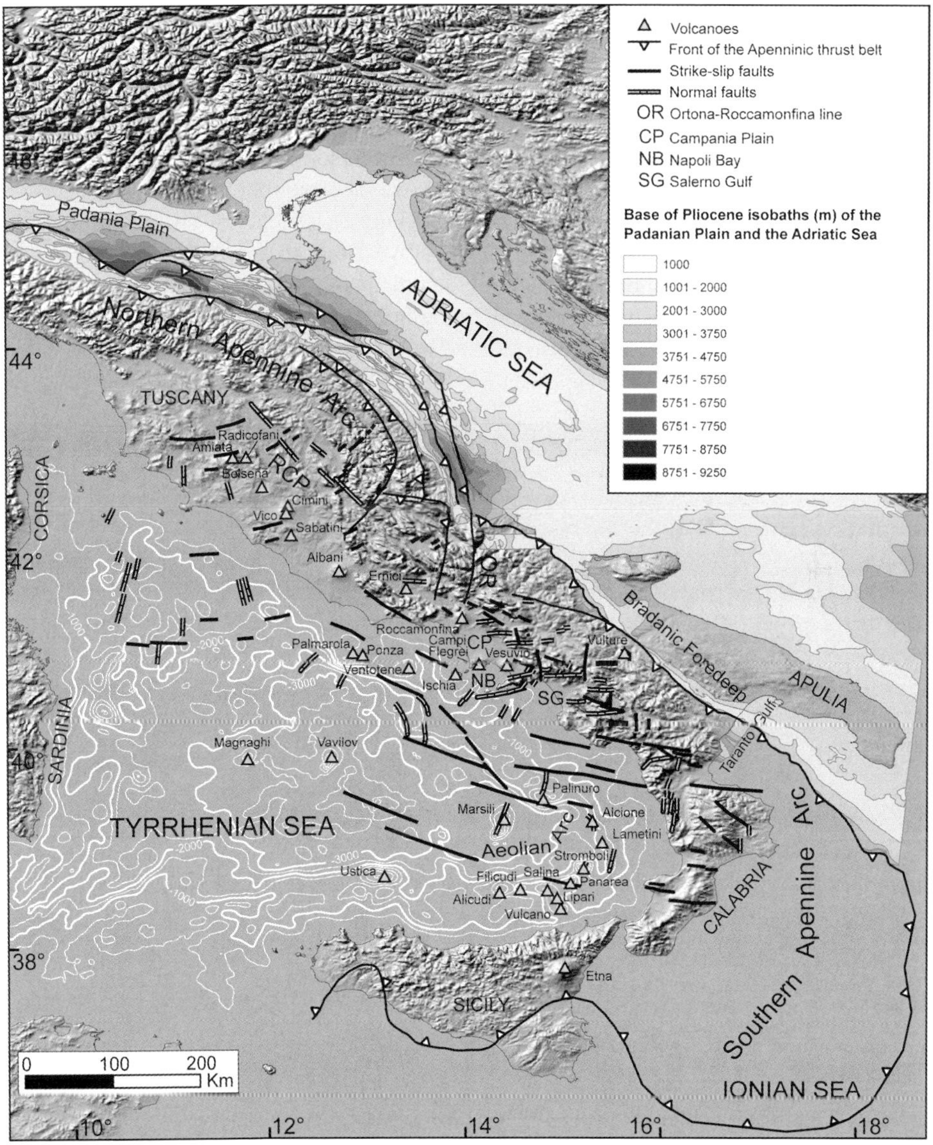

Figure 1. Structural sketch of the Tyrrhenian Sea and the Apennines.

Therefore, since the Early Miocene a collisional belt separated the Corsica–Sardinia European (Eurasian) block from the undeformed Adria domains (Patacca et al., 1990).

From a structural point of view, Patacca et al. (1990) distinguished two major arcs in the Apennine chain: the NE-verging Northern Apennines Arc (NAA), which extends from Monferrato to Molise, and the E and SE-verging Southern Apennines Arc (SAA), which extends through the Calabrian arc from Molise to Sicily (Fig. 1). The two arcs merge along a transversal lineament known as the "Ortona–Roccamonfina line" (Locardi, 1982; Patacca et al., 1990). According to these authors this lineament represents a Late Pliocene dextral strike-slip fault. A third minor arc is located between the two major arcs in the Molise area, but its origin is still unclear. The foreland of these arcs is represented by the Ionian–Adriatic domain.

The Adriatic foreland flexure is regionally drawn by the SW deepening base of Pliocene isobaths (Royden et al., 1987; Bigi et al., 1990), a feature that is particularly evident in the NAA (Fig. 1). In the STS, deep earthquakes foci (Anderson and Jackson, 1987) draw the Ionian lithosphere subducted under the Calabrian Arc. The external portions of the arcs are marked by negative Bouguer gravity anomalies (Fig. 2), except for an area around the Vulture volcano where the Apulia foreland positive Bouguer anomalies cut across the Bradanic foredeep to join the positive gravity anomalies in the STS. In this area, the positive gravity anomaly corresponds to a low topographic relief of the Apenninic thrust belt (Fig. 2). These crustal features are also marked by a high-velocity crustal body at shallow

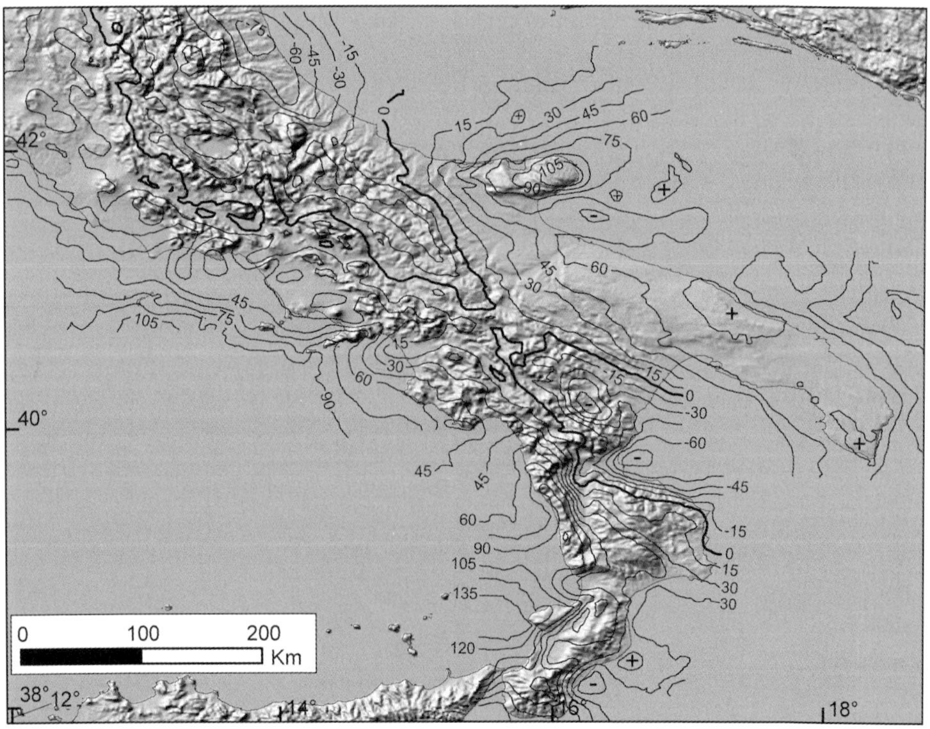

Figure 2. Gravity map of central-southern Apennines. Bouguer isoanomalies in mGal.

depth as shown by 3D crustal P-wave tomography (Alessandrini et al., 1995). Westward, the Apenninic domain boundary is represented by the Tyrrhenian margin. This boundary is marked by a volcanic alignment that span from southern Tuscany to the Aeolian Islands arc. Along the northern sector of this margin (from southern Tuscany to Campania), high-K volcanoes that have been dated from the Late Pleistocene to the present (Serri et al., 2001 and references therein) occur. The apparatuses of these volcanoes follow significant structural alignments. In particular, in the Latium–Tuscany area they are NW-SE aligned, whereas in Campania they follow an E-W trend. The southern tract of the boundary (from Bay of Naples to the Eolian Islands) comprises calc-alkaline volcanoes (Peccerillo and Turco, 2004 and references therein).

2.2. The Tyrrhenian basin

Starting from Late Tortonian times, severe extensional processes took place along the western side of the Apennine chain, with extensive rifting and rapid tectonic subsidence (Kastens et al., 1988 and references therein). Extension in the Tyrrhenian region and compression in the Apennine chain coexisted with a progressive migration of the rift-thrust belt-foredeep system towards the present-day Po Plain–Adriatic–Ionian foreland (Ricci Lucchi, 1986; Patacca et al., 1990; Cipollari and Cosentino, 1992). Marine conditions were reached in the western part of the Tyrrhenian basin in early Messinian times (Sartori et al., 2004). From Early Pliocene times, a significant volcanic activity was associated with rifting processes, leading to the onset of high-K magmatism in the Tyrrhenian continental margin of the Italian peninsula (Beccaluva et al., 1990, Peccerillo and Turco, 2004).

Trincardi and Zitellini (1987) pointed out the strong asymmetry of conjugate rifted margins in the STS, represented respectively by the eastern Sardinia continental margin and by the central southern Italy. According to these authors, in the Tyrrhenian margin of Campania the asymmetric rifting process could have been controlled by an east-dipping low-angle crustal detachment fault.

In this view, the lower plate of the detachment system is represented by the Sardinian passive margin, while the upper-plate counterpart is the Campanian margin.

The existence of oceanic crust in the STS is likely to be restricted in the Vavilov basin and in the Marsili basin (Marani, 2004; Sartori et al., 2004). Nevertheless, there is evidence of a large area encompassing the Magnaghi and Vavilov seamounts that shows an oceanic-like Moho depth of about 10 km (Carrara, 2002) (Fig. 3). An E-W seismic section across the Magnaghi and Vavilov seamounts reveals the absence of lower crust and the presence of tilted upper crust blocks (Sartori et al., 2004). Furthermore, seismic velocities recorded along the same section suggest the occurrence of a wide continent–ocean transition characterized by sub-continental serpentinized mantle (DSDP Leg 107, site 651; Kastens et al., 1988).

3. Morpho-structural analysis

In order to analyse the major structural and tectonic features characterizing the TMAC, we applied a technique based on the interpretation and synthesis of different types of

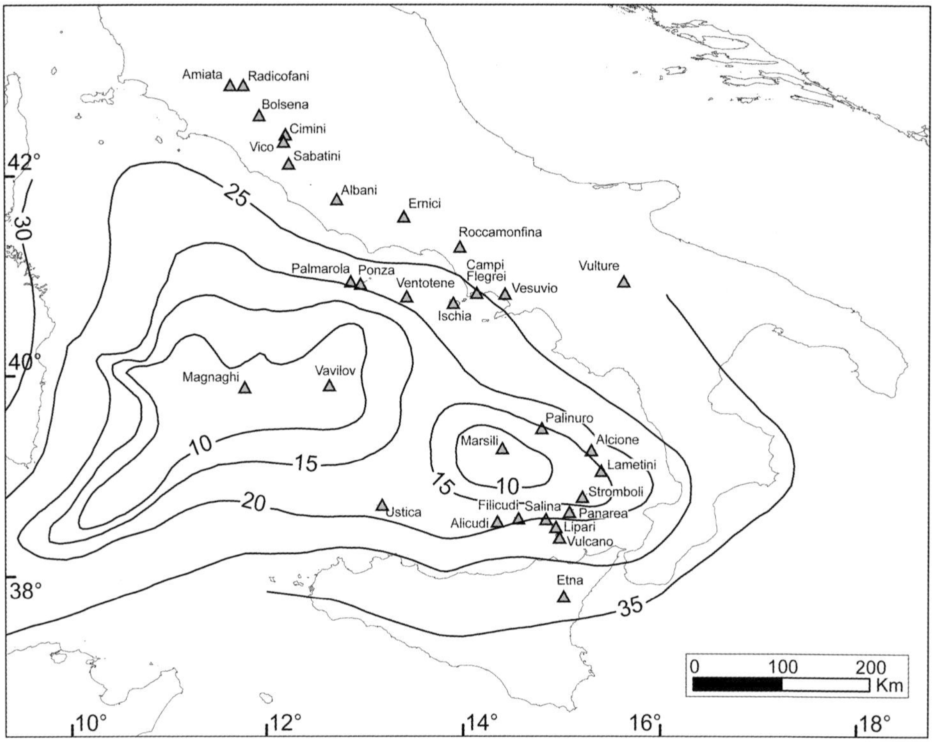

Figure 3. Schematic map of the Moho depths (in km) in the Tyrrhenian-Apennines system (after Carrara, 2002).

remote sensing data in the light of the geological and geophysical data published in the literature. Landsat ETM 7 and Shuttle Radar Topography Mission (SRTM) elevation data (ftp://edcsgs9.cr.usgs.gov/pub/data/srtm/) constituted the remote sensing data set. The Landsat ETM 7 imagery has a 30 m pixel resolution and contains seven spectral bands. Bands 1–5 and 7 contain spectral information, while band 6 contains thermal information. In this study, we choose the 7:4:2 band combination, which satisfactorily highlights the geological information. These imageries were combined with an SRTM imagery covering the same area. This SRTM image has been processed into a Digital Elevation Model (DEM) with a resolution of 90 m. The Geologic Map of Italy from Servizio Geologico d'Italia (scale 1/1,250,000) (Compagnoni and Galluzzo, 2004) and the Structural Model of Italy (scale 1/500,000) (Bigi et al., 1990) were used to insert field geological data and guide the remote sensing interpretation. The Gravity Map of Italy (Carozzo et al., 1992) and CROP-MARE (Scrocca et al., 2003) seismic data were used to further constrain the interpretations. Recently, the Institute of Marine Science (ISMAR) of the National Research Council (CNR) carried out a high-resolution bathymetric survey (Marani and Gamberi, 2004) that we combined with land topographic data to produce a high-resolution image of the in-land and sea-bottom morphology of the Italian peninsula and the Tyrrhenian Sea (Figs. 4–6). This combined map allowed us to perform the identification of tectonic features and lineaments on a large region that encompasses both on-land and marine areas.

3.1. The morpho-structures of the Apennine arcs

In the NAA, the main structural lineaments clustered into principal sets striking NW-SE and NE-SW (Fig. 4). The NW-SE trending set – that characterize the Umbria region, southern Tuscany and part of the Lazio region – shows many morphological markers, with evidence of structural highs and lows. Many authors (e.g. Deiana and Pialli, 1994; Calamita and Deiana, 1995) interpreted these structures as dip-slip normal faults, known as "faglie appenniniche" (i.e. Apennine-trending faults), related to block-faulting (Sani et al., 1998) controlling the formation of Mio-Pliocene basins.

The NE-SW striking structures, known as "faglie anti-appenniniche" (i.e. anti-Apennine-trending faults), have been interpreted as transfer faults related to the NW-SE trending extensional faults (Bartole, 1995). The evidence of small pull-apart basins formed along these lineaments supports this hypothesis (Bonini, 1997).

The structural pattern of the SAA is much more complex and at least five sets of lineaments can be identified (Fig. 5). The first set strikes N110-N120 and propagates throughout the southern Apennines from the Salerno Gulf to the Taranto Gulf and the Ionian foredeep (Fig. 5). These lineaments seem to be superimposed over a second set of N140-N150 striking lineaments. According to previous interpretations (Turco et al., 1990; Knott and Turco, 1991) both these sets of lineaments are interpreted as left-lateral strike-slip faults. Sharp escarpments mark a third set of NE-SW anti-Apenninic-striking structures (Fig. 5), which have been interpreted as normal faults (Knott and Turco, 1991; Milia and

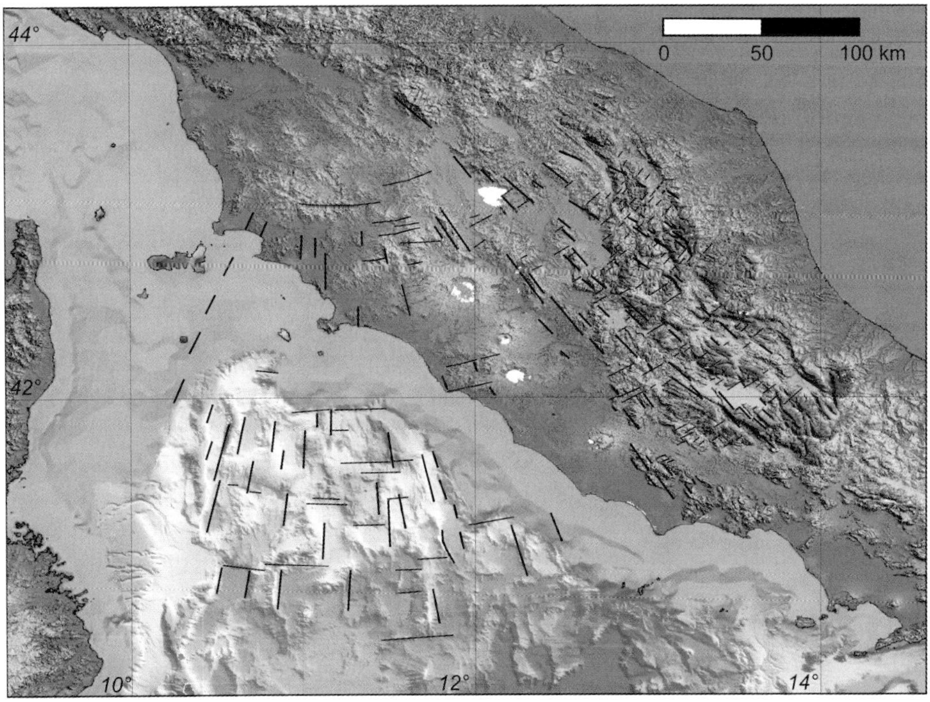

Figure 4. Tectonic lineaments in the northern Apennine Arc and in the central-northern Tyrrhenian Sea detected from remote sensing data and bathymetric data interpretation.

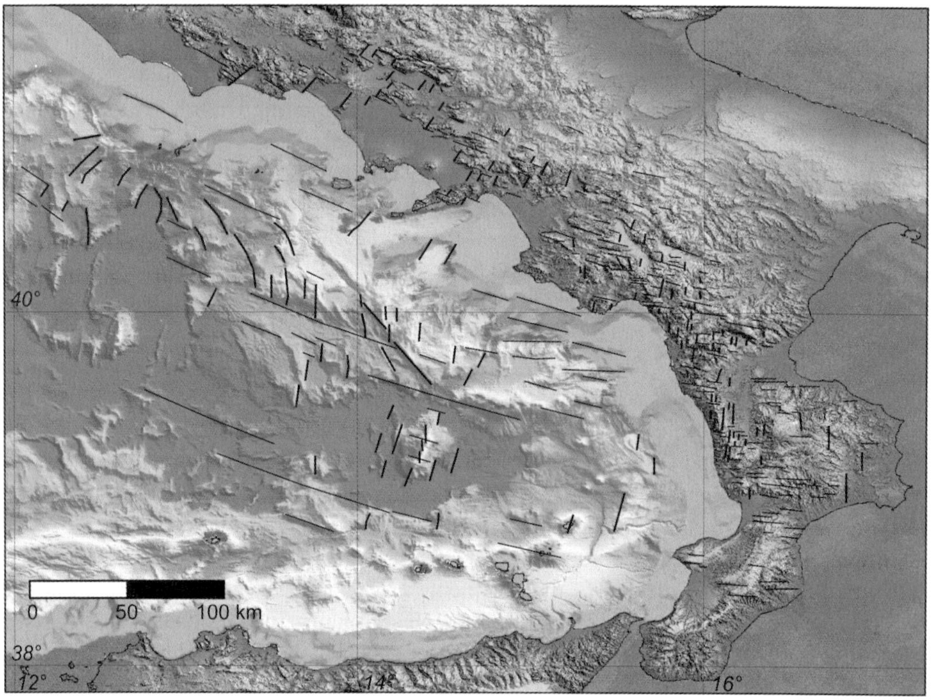

Figure 5. Tectonic lineaments in the southern Apennine Arc and in the southern Tyrrhenian Sea detected from remote sensing data and bathymetric data interpretation.

Torrente, 1999, 2000). It should be noted that in the NAA the anti-Apenninic structures are mainly strike-slip faults, whereas in the SAA they generally have an extensional kinematics. This supports a first-order sub-division of the Apennine chain into at least two main arcs on the basis of homogeneous structural patterns.

Finally, a double set of E-W and N-S trending lineaments characterizes the Irpinia area, from southern Latium and Abruzzi regions to the Monte Vulture volcanic apparatus (Fig. 6). The E-W trending lineaments are in some cases associated with basins infilled by lacustrine sediments. Examples of such basins are the Isernia and Boiano basins (Bosi et al., 2004) and the Matese lake. Lacustrine sediments also occur in the Volturno and Calore valleys (Bonardi et al., 1988). In the Picentini Mountains, which represent the SW border of the Irpinia area, Ferranti et al. (1996) suggested the existence of ENE-WSW trending low-angle normal faults that were active during the uppermost Pliocene. These features suggest that the E-W trending lineaments in the Irpinia area could be related to the same extensional event, while N-S trending lineaments, which do not always show significant morphological evidences, would represent transfer faults related to the E-W trending normal faults system.

3.2. The morpho-structures of the southern Tyrrhenian Sea

The morphotectonic analysis of the STS (Fig. 5) was mainly focused on its south-eastern margin, where the structural pattern is most likely associated with the extensional phases that affected the Marsili basin and the Campania Plain.

Figure 6. Tectonic lineaments of the Irpinia area detected from remote sensing data interpretation.

The Marsili basin, with its homonymous N15-N20 elongated seamount, is the younger of the two oceanic sub-basins that form the Tyrrhenian Sea (Marani, 2004 and references therein) (Fig. 1). Two NNE-SSW trending sets of faults, parallel to the 50 km elongated Marsili volcano, developed symmetrically in the basin floor. These features have been interpreted as horst and graben pairs at both edges of the Marsili volcano (Marani, 2004).

The SE margin of the STS is characterized by three en-echelon, N110 trending lineaments, represented by escarpments dipping towards the SW (Fig. 5). The southernmost of these lineaments connects the Palinuro seamount with the Poseidone ridge and corresponds to the northern margin of the Marsili basin. The intermediate and the northern lineaments correspond respectively to the SW-dipping escarpment of the Tacito seamount and to the Pontine Islands escarpment. Minor N-S and NNW-SSE trending lineaments connect these escarpments. Finally, the Sartori escarpment, composed of three N150 trending dextral en-echelon segments connected by NE-SW lineaments, is a further important lineament that characterizes the STS–TMAC transition (Marani, 2004).

4. Structural associations and determination of extension directions

The extension directions in the various sectors of the Apennine chain were determined by comparing the results of the morpho-structural analysis with data collected from the abundant geological literature.

In the NAA, the NW-SE trending regional normal faults and the associated NE-SW trending transfer faults indicate an NW–SE trending direction of extension (Fig. 4). Crustal extension in the internal (western) domain of the NAA took place, while the external (eastern) domain was subjected to thrusting and eastward migration of the thrust system (Patacca et al., 1990; Carmignani et al., 1995; Ferranti et al., 1996; Jolivet et al., 1998; Brunet et al., 2000; Rosenbaum et al., 2002). The internal sedimentary basins formed following the migration of the arc and become younger towards the east (Tavarnelli et al., 1998). Evidences of inactivity of thrusting and related folding in the external domain of the NAA in the last 800 ky (Di Bucci and Mazzoli, 2002) suggest that the ENE-directed migration of the NAA stopped in early Pleistocene times.

The structural complexity of the SAA (Fig. 5) is probably due to the superposition of two recent extensional phases. The difficulties in the identification of a coherent system of tectonic structures in this area led us to focus our attention to the more consistent morphotectonic features of the STS margin, which are directly linked to the relative motion between the SAA and the Western Tyrrhenian block and that were not subject to secondary deformation processes. Marani (2004) interpreted the Marsili seamount (0.78–0.1 Ma) as a N20-oriented spreading ridge. In this perspective we interpret the three en-echelon escarpments, represented by the Palinuro seamount–Poseidone ridge, the Tacito escarpment and the Pontine Islands escarpment, as part of a dextral N110 transform fault system, which transfers the extension throughout the eastern Tyrrhenian margin from the Marsili basin to the Campanian Plain area. Hence, in analogy with the NAA, if we associate the extension in the internal domain of the SAA with thrusting in the external domains, the resulting direction of arc migration is N110. Both the Sartori escarpment and the N140-N150 lineaments are incompatible with this kinematic framework, hence we suggest that they could be related with the formation of the older Vavilov basin (3.5 Ma) (Kastens et al., 1988).

The third extensional system is represented by E-W trending normal faults in the Irpinia area and related N-S transfer faults (Fig. 6). As stated above, this system is associated to a N-S direction of extension and is responsible for the moving apart of the two main arcs.

5. Chronology of the extensional phases

In order to determine the temporal sequence of the extensional tectonic events that affected the TMAC, we used the stratigraphic record from the Campania Plain, the Sele Plain, the Marsili basin and the Irpinia area. Further temporal constraints were derived from the volcanic events ages.

5.1. Extension in the Campania Plain – Bay of Naples basin

The Campania Plain is located in the merging area between the NAA and the SAA. It extends, from NW to SE, from the Aurunci Mountains and Roccamonfina volcano to the Sorrento Peninsula (Figs. 1 and 6). The Caserta Mountains, a NW-SE trending elongated relief, represent its NE limit, while its SW prosecution is open to the Tyrrhenian Sea. The basin is filled up with 3000 m Pleistocene sediments and volcanic rocks (Ippolito et al., 1973). Milia and Torrente (1999), based on chronostratigraphic data, indicated that NE–SW trending extensional faults in the Campania margin started to be active from 0.73 Ma. This extensional tectonics also affected the Bay of Naples, where seismic data show the presence of a NE-SW trending normal faults system of the same age (Milia et al., 1998; Milia and Torrente, 1999).

5.2. Extension in the Bay of Salerno – Sele Plain basin

The Sele Plain represents the on-shore prosecution of the Salerno Bay basin and is filled up with a thick succession of Quaternary sediments. A Pleistocene conglomeratic succession, known as the Eboli Conglomerates crop out on its northern margin and shows a well-developed system of conjugate N110 and N50 trending oblique faults (Cello et al., 1981). In the Salerno Bay, NW—SE trending extensional tectonics is testified by seismic data. In particular, the CROP-MARE M36 deep seismic reflection line (Scrocca et al., 2003) shows NE–SW trending normal faults and tilted blocks (Fig. 7). This confirms the connection between the on- and off-shore structures that we observed during the morpho-structural analysis. Furthermore, in the Salerno Bay, the Mina well (AGIP, 1977) showed a Plio-Pleistocene 2000–m-thick sedimentary deposit, and in particular 1000 m of Pleistocene sediments that suggest a strong tectonic subsidence affecting the basin during the Pleistocene.

5.3. Extension in the Marsili basin

The Marsili basin is a rectangular-shaped basin of roughly 80 × 50 km. It reaches a depth of more than 3000 m, and the Marsili seamount is located in its central part. The DSDP Leg 107 well 650 investigated the basin, drilling about 600 m of sediments laying above a basaltic basement (Fig. 8a). Kastens et al. (1988), on the basis of biostratigraphic and magnetostratigraphic constraints, suggest that inception of spreading in the Marsili basin took place between 1.87 and 1.67 Ma. Savelli and Schreider (1991) and Faggion et al. (1995) confirm this spreading inception age on the basis of the regional magnetic anomaly field (Fig. 8b).

5.4. Extension in the Irpinia area

The N-S extension that affected the Irpinia area is likely to be related with the formation of several Early Pleistocene lacustrine basins, for instance the Isernia and Boiano Basins (Bosi et al., 2004), and the Volturno and Calore valleys (Bonardi et al., 1988). As a matter of fact, these faults were reactivated in Middle Pleistocene times (Corrado et al., 2000; Calabrò et al., 2003), albeit with controversial kinematic interpretations.

5.5. Age of volcanic apparatuses

The oldest magmatic activity related with extensional tectonics in the Southern Apennines–Southern Tyrrhenian region is represented by the oceanic-spreading magmatism in the Marsili basin (1.8 Ma) (Kastens et al., 1988). This magmatism was followed by the

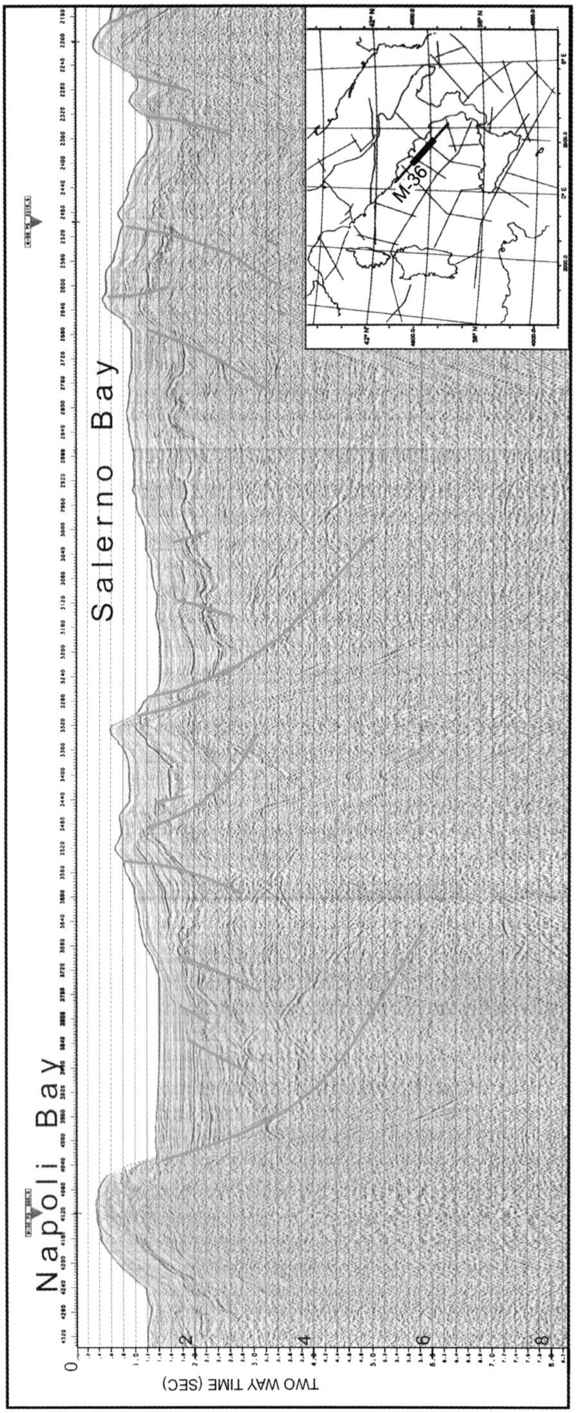

Figure 7. Seismic section and schematic line drawing from CROP-MARE M36 deep seismic line. The inserted box indicates the location of the line.

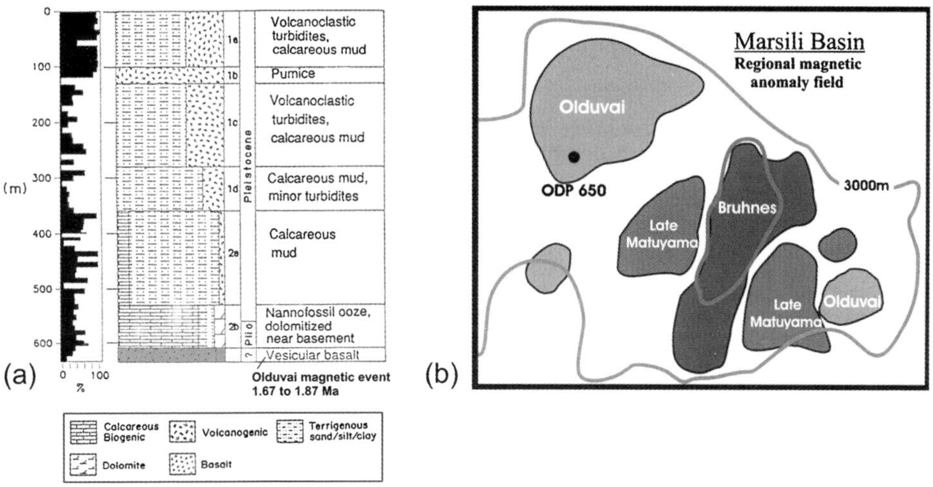

Figure 8. (a) Core log indicating the litostratigraphic units recovered at Site 650 (modified from Kastens et al., 1988); (b) Sketch map of the regional magnetic anomaly field in the Marsili basin. The closed line in correspondence of the Brhunes anomaly outlines the Marsili volcano (from Marani, 2004).

volcanic activity of the Palmarola Island at 1.6 Ma (Codeaux et al., 2004) and prosecuted northward in the Cimini Mts. (1.4 Ma) and Radicofani (1.3 Ma) (Serri et al., 2001). A second volcanic activity phase (0.8–0.1 Ma) led to the formation of Bolsena-Vico, Sabatini, Albani Hills, Ernici Mts., Ventotene Island, Marsili seamount, Roccamonfina and Vulture volcanoes (Serri et al., 2001). In the Campania Plain, the volcanic activity took place with the formation of the Ignimbrites (from 0.205 to 0.018 Ma) (De Vivo et al., 2001; Rolandi et al., 2003), which was followed by the Ischia, Campi Flegrei and Vesuvius activity (0.15–0.0 Ma) (Serri et al., 2001). In the STS, further volcanic apparatuses of the same age are: the Palinuro Smt, Alcione e Lamentini Smts and the Eolie Island (Serri et al., 2001).

6. Methods and constrains for the elaboration of the kinematic model

In order to describe quantitatively the geologic evolution of the Tyrrhenian Basin, we used a new software tool for the modelling of instantaneous motions of tectonic plates designed by one of us. Plate reconstructions were made using PCME, a computer program designed by Schettino (1998). The construction of a plate tectonic model for the geologic evolution of the Apenninic–Tyrrhenian Basin system required the following steps: (1) identification of the tectonic elements, that is, lithospheric blocks that were subject to independent motion during the considered time interval; (2) determination of the Euler poles describing relative movement between pairs of plates; (3) comparison between the predicted and observed structural patterns in order to confirm poles consistency; (4) compilation of a rotation model, which includes finite rotation parameters for pairs of plates. The tectonic elements were identified on the basis of first-order structures recognized by means of both the morpho-structural analysis and the spatial distribution of the volcanic apparatuses, whereas the rotation model was compiled based on both timing of activity of the first-order structures and regional first-approximation finite strain evaluation (e.g., either total shortening for the arcs or total extension for the basins).

6.1. Identification of the tectonic elements

The Apennine chain has been divided on the basis of homogenous structural patterns into two kinematic elements that, from Plio-Pleistocene times, moved independently with respect to the Eurasian reference plate. The two blocks are the Northern and the Southern Arcs (Fig. 9). A third element, the Western Tyrrhenian block, is considered as fixed with respect to Eurasia. The boundaries between these three kinematic elements are illustrated hereafter (Fig. 9). (1) The NW-SE trending volcanic lineaments of the Roman Comagmatic Province in the Latium region represents the boundary between the Northern Arc and the Western Tyrrhenian block. (2) The boundary between the Southern Arc and the Western Tyrrhenian is composed of three segments: the first segment runs from the Gaeta Basin to the Gortani Basin; the second segment corresponds to the N20 elongated Marsili Smt. The two segments are linked by a third lineament encompassing the Tacito and Palinuro-Poseidone escarpements. (3) The E-W

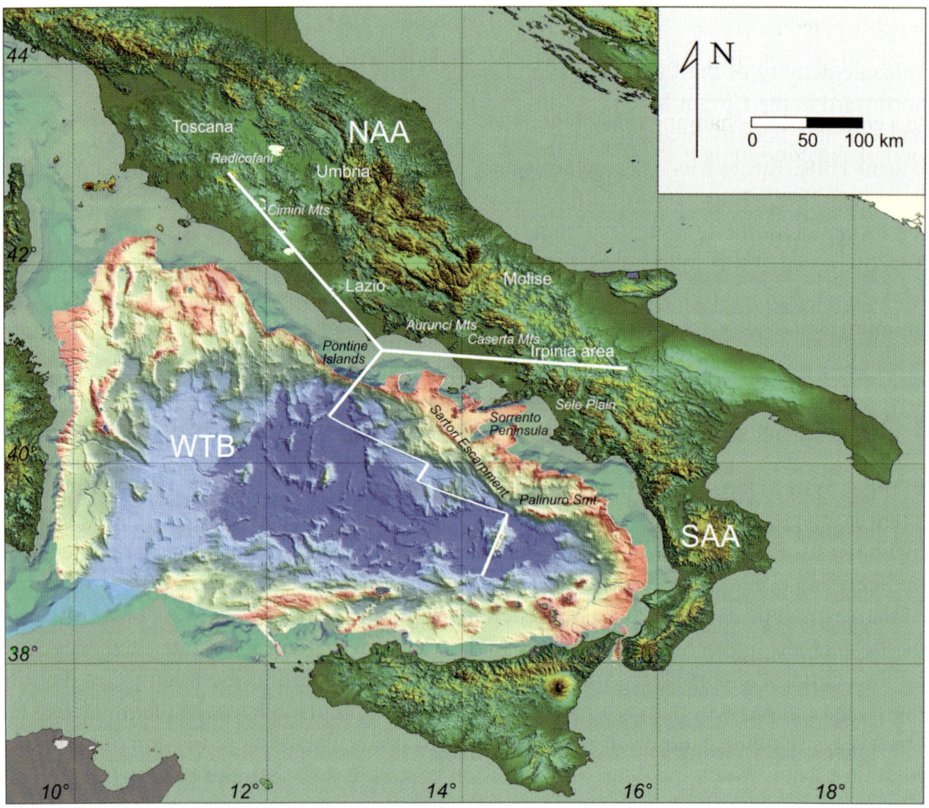

Figure 9. The three tectonic elements used to model the kinematic evolution of the Tyrrhenian margin of the Apennine chain. NAA, Northern Apennines Arc; SAA, Southern Apennines Arc; WTB, Western Tyrrhenian basin.

trending structural depression of the Irpinia area is the limit between the Northern and the Southern Arc. The individuation of this latter limit is also supported by gravimetric data (Fig. 2), which shows an E-W trending positive Bouguer magnetic anomaly in correspondence of the Irpinia area, indicating the existence of high-density body most likely related to a stretched lithosphere.

6.2. Determination of the Euler poles

The NE-SW trending system of strike-slip faults recognized by morpho-structural analysis of the central Apennines is consistent with a single rotation pole. This Euler pole, $\boldsymbol{e}_1$, determines the instantaneous rotation of the Northern Arc with respect to Eurasia from the Uppermost Pliocene to the Lower Pleistocene and is located at (44.00°N, 11.20°E). Similarly, an analysis of the three N110-trending en-echelon escarpments of the Palinuro Smt-Poseidone ridge, Tacito Smt and Pontine Island escarpment led us to identify a unique stage of rotation of the Southern Arc about a pole $\boldsymbol{e}_2$ located at (45.17°N, 17.51°E). However, this stage encompasses the whole time interval from the Uppermost Pliocene to the present.

Therefore, the Northern Arc, the Southern Arc and the Western Tyrrhenian block can be approximated as a three-plates system that can be described with the methods of instantaneous plate tectonics (McKenzie and Parker, 1974; Dewey, 1975). In this instance, the instantaneous pole of rotation of the Northern Arc with respect to the Southern Arc must be a continuously changing instantaneous Euler pole associated with structures that change their strike continuously.

6.3. Validation of the Euler poles

Our modelling software allowed us to generate grids of parallels and meridians for the Euler poles determined above. The reliability of the Euler poles was then assessed by comparing the pole grids with the actual structural lineaments recognized by the morpho-structural analysis. In fact, parallels and meridians of an Euler pole grid represent respectively strike-slip trends and normal faults. The results of such a comparison are illustrated in Figure10a and b. Good correspondence is evident between the main normal faults and the Euler poles meridians both in the Northern and Southern Arcs. In the Southern Arc, there is also a good match of the N110 and N20-N40 trending lineaments in the on-land areas with the $\boldsymbol{e}_2$ parallels and meridians.

The validation method described above is also useful to discriminate the structural associations that are likely to be related with pre-Quaternary tectonic phases. For example, in the Southern Arc the N140-N150 trending lineament represented by the Sartori escarpment mismatched the $\boldsymbol{e}_2$ grid. We interpret this first-order structure as the result of a previous tectonic phase, most likely related to the opening of the Vavilov basin (3.5 Ma). This hypothesis is supported by the fact that the Sartori escarpment shows variable morpho-structural features along its length (Fig. 5). In its NW segment the lineament is represented by a sharp ridge, probably related with strike-slip tectonics, while the SE segment is represented by an escarpment separating two portions of sea floor at different depths, thus indicative of normal faulting. These features suggests that the SE segment of the Sartori

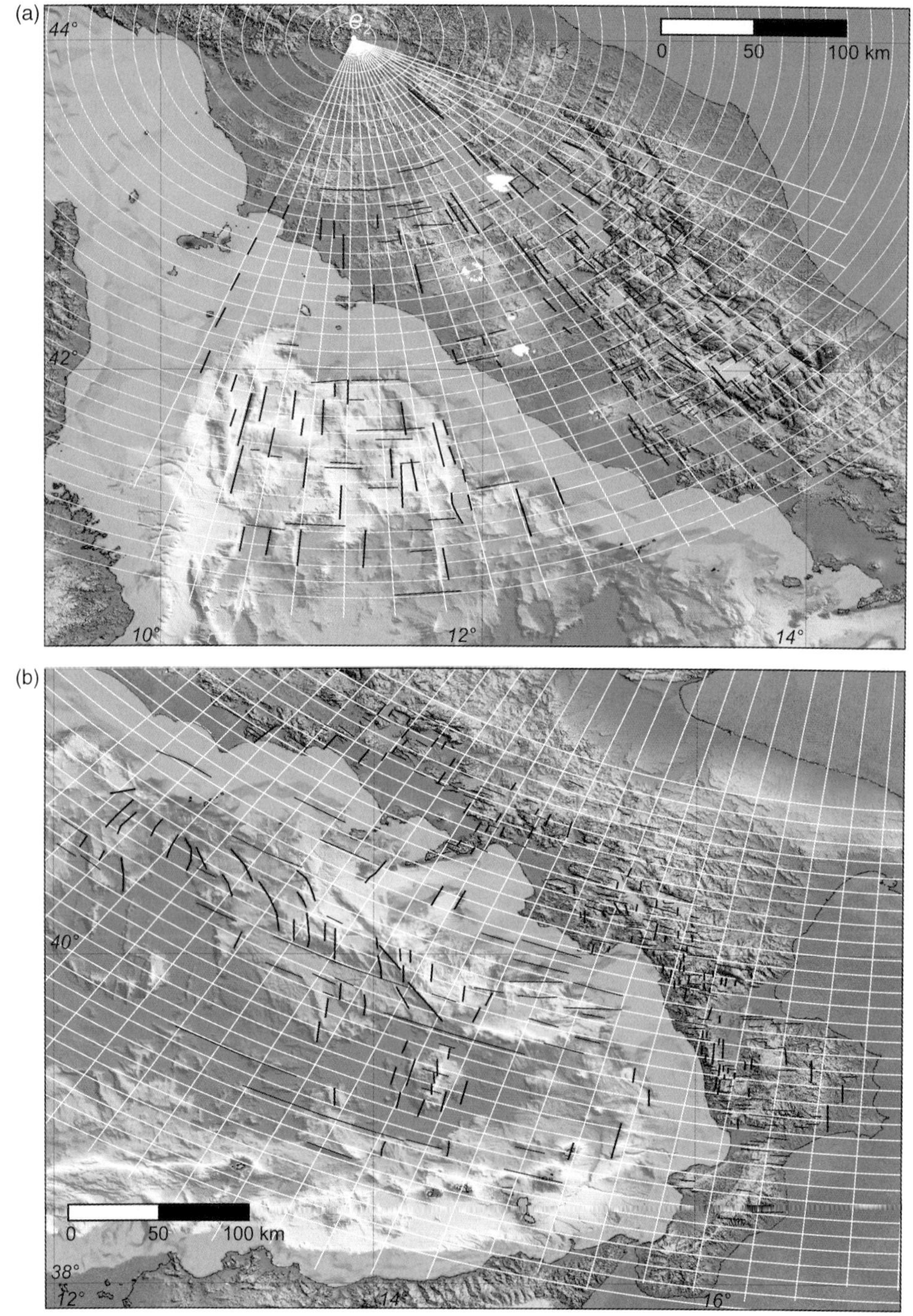

Figure 10. Comparison between the structural lineaments recognized by the morpho-structural analysis and the Euler pole grids for the rotations of the (a) Northern Apennine Arc and (b) the Southern Apennine Arc. Note the good correspondence between the main normal faults and the Euler poles meridians both in the Northern and in the Southern Arcs. In the Southern Arc, there is also a good match of the N110 and N20-40 trending lineaments in the on-land areas with the e_2 parallels and meridians.

line, which was originally related to strike-slip faulting, was reactivated as a normal fault during the activity of the superimposed N110 right-lateral strike-slip faults system widespread in the SAA. Further evidences of pre-Quaternary N140-N150 trending strike-slip tectonics, most likely related with the same tectonic phase of the Sartori line, have been recognized in Calabria (Van Dijk, 2000).

6.4. Finite strains, strain rates and finite angular rotations of the Southern and Northern Arcs

Euler poles alone do not allow a complete representation of the tectonic evolution of a region. They only constrain local directions of extension, strike-slip or convergence between two plates. In order to quantitatively describe the total deformation, a determination of the angles of rotation about these poles is needed. A technique for determining the angle of rotation of the Southern Arc with respect to the Western Tyrrhenian block is to move back the arc by the angle that removes the whole oceanic crust formed during the spreading episode of the Marsili basin.

We estimate the width of Marsili oceanic crust to be ~80 km on the basis of the magnetic anomaly field and the Moho depth. Hence, the total angle of rotation about the SAA pole $\boldsymbol{e}_2$ for the closure of the Marsili basin results to be 6.93°. This stage started during the Olduvai polarity chron (~1.87 Ma) and lasted till about 0.78 Ma (Matuyama–Bruhnes transition), when spreading ceased in the Marsili basin and extension jumped south-eastward in the Aeolian Island Arc. The corresponding spreading rate and direction result to be 77.65 mm/yr, N110E at 39.3°N, 14.4°E.

During the second stage, from ~0.78 Ma (Matuyama–Bruhnes transition) to the present time, the migration of the SAA occurred about the same Euler pole $\boldsymbol{e}_2$. If we assume that the extension rate remained the same as the previous stage, we obtain an angle of rotation of 5.00° about the SAA pole $\boldsymbol{e}_2$ from 0.78 Ma up to the present. In the Marsili basin, still continuing up-welling of magma at the (now extinct) spreading centre contributed to the edification of the Marsili Seamount. In fact, the youngest volcanic rocks in this region are ~0.1 Ma in age (Selli et al., 1979).

The angle of rotation of the Northern Arc was calculated indirectly on the basis of the kinematic parameters of the three-plates system and the observed structural pattern in the Irpinia area. As already mentioned, this pattern cannot be described by a single pole of instantaneous rotation, because it is the characteristic of a continuously migrating Euler pole. However, an average N-S direction of extension can be identified (Fig. 11). Several different patterns can be predicted by varying the rate of the angular velocities of the two arcs. Let Ω_N and Ω_S be the angular velocities of the Northern and the Southern Arcs, respectively. Using specific software, we noted that in order to obtain a mean N–S direction of extension it was necessary that the following identity was satisfied:

$$\Omega_N \cong 1.5\Omega_S \tag{1}$$

The parameter Ω_S is determined by the total angle of rotation (6.93°). Hence, the application of Equation 1 allowed us to estimate the total angle of rotation of the Northern Arc as 10.40°. Figure 11 illustrates the predicted pattern of relative linear velocities between the Northern and the Southern Arcs.

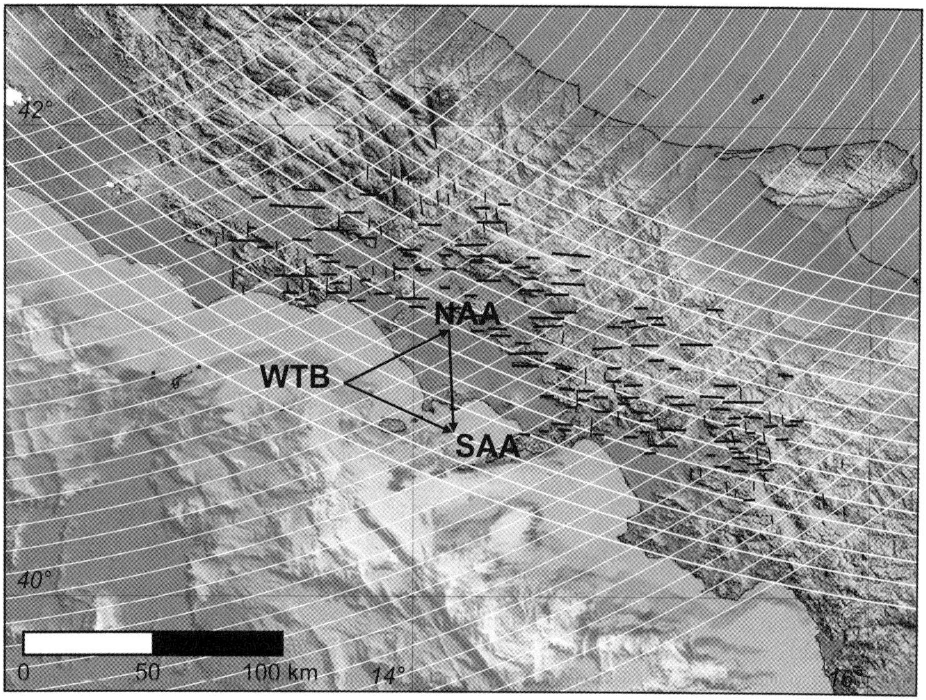

Figure 11. Predicted pattern of the relative linear velocities between the Northern and the Southern Arcs. It was calculated on the basis of the kinematic parameters of the three-plates system and the observed structural pattern in the Irpinia area. For further explanations see the text.

The simultaneous rotation of the Northern and Southern Arcs ceased at about 0.78 Ma, when the Northern Apennines chain stopped its migration. Starting from this time, only the Southern Arc continued its ESE motion. As already mentioned, the angle of rotation for this additional stage (~5°) was estimated on the basis of the assumption that the angular velocity of the Southern Arc remained approximately constant.

7. Discussion

In this section, we discuss the geologic consequences of our kinematic model as well as unsolved problems. Although the kinematic model described above was built on the basis of estimated expansion rates in the Marsili basin since 1.87 Ma, the process initiated some time before, perhaps at the same time of the cessation of spreading in the Vavilov basin (3.5 Ma, Kastens et al., 1988). In this hypothesis, the STS would be subject to a continuous process of rifting-spreading since the Early Pliocene through a series of ridge jumps.

During the first stage, between 3.5 and 0.78 Ma (Fig. 12a), the anticlockwise rotation of the Northern Arc generated both the NW-SE trending normal faults and the NE-SW trending strike-slip faults in the central-northern Apennines. The SE-directed migration of the Southern Arc was associated to extension and spreading in the Marsili basin as well as transcurrent tectonics along N110 strike-slip faults in the SAA and in correspondence of the Tyrrhenian escarpments. The NE-SW and N-S trending normal faults in the Tyrrhenian margin are interpreted as releasing step-over associated to the N110 trending strike-slip faults. The relative motion between the two arcs produced the E-W trending normal faults in the Irpinia and Campania Plain area, together with the associated N-S trending strike-slip faults. Finally, incipient volcanism took place in the Palmarola volcanic apparatus (1.6 Ma) and in the Cimini Mts (1.4 Ma), while magmatic intrusions occurred in the Radicofani area.

In the second stage (0.8 Ma–present time) (Fig. 12b), the N-E directed migration of the Northern Arc either considerably slowed down or even ceased at all, while the Southern Arc continued migrating towards the ESE. The northern limit of the Southern Arc, represented in the previous stage by the Irpinia area, is now characterized in the Campania Plain by a new generation of NE-SW trending normal faults and NW–SE trending strike-slip faults, while in the Irpinia area both the previous E-W trending

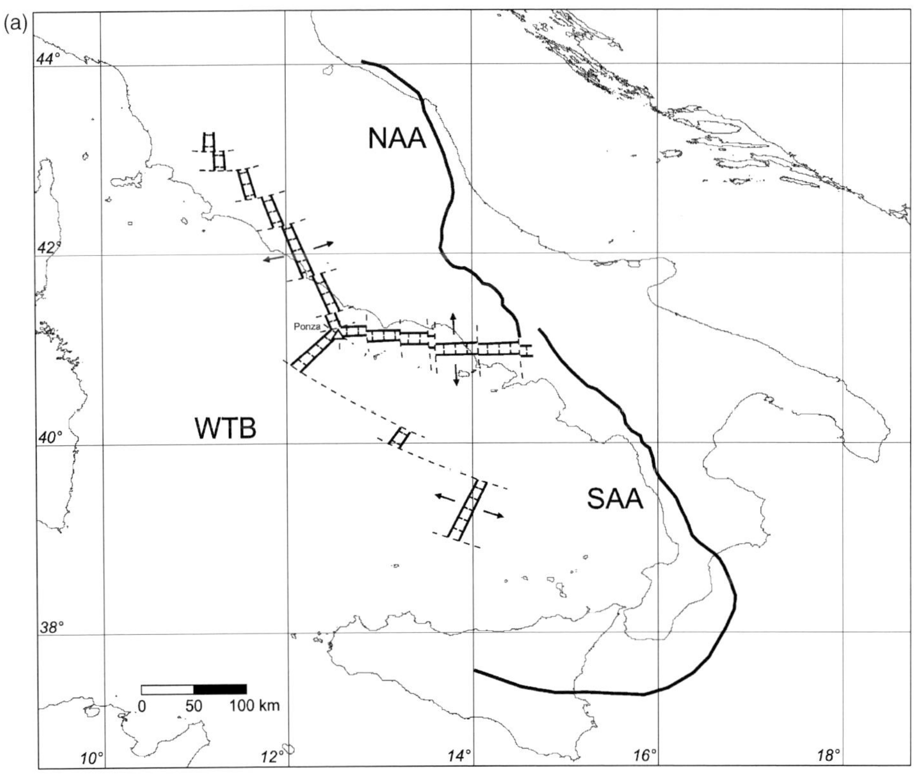

Figure 12. Kinematic model of the Tyrrhenian–Apennines system showing the correlation between the extension and volcanism in the back of the two main arcs. (a) Initial configuration of the three tectonic elements (Middle Pleistocene times); (b) first stage, between 3.5 and 0.78 Ma; (c) second stage, from 0.78 Ma to present time. See text for further discussion.

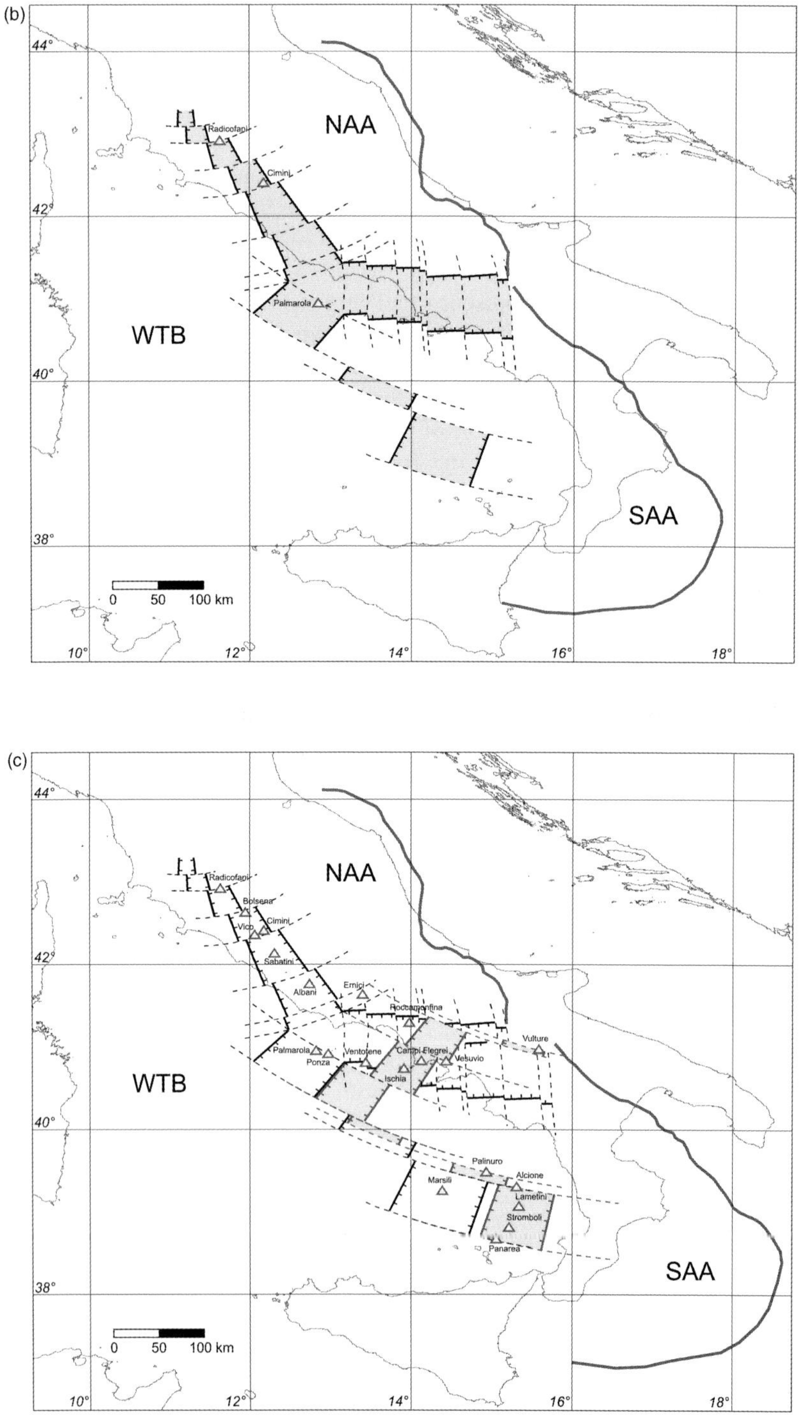

Figure 12. (Continued)

normal faults and N-S strike-slip faults are reactivated respectively as left- and right-lateral transtensional faults transferring the motion of the Southern Arc into the Bradanic foredeep. We put forward the idea that emplacement of the Vulture Mountain volcano is related to this latter transfer fault system. The extensional tectonics in the Campania Plain is transferred into the Southern Tyrrhenian through the previous formed N110-trending strike slip-faults.

Spreading in the Marsili basin stopped and the spreading centre jumped to its present position along the Aeolian Islands (0.5 Ma), Alcione (0.35 Ma) and Lametini (0.35 Ma) volcanic lineament. In a short time span between the end of the first stage and the beginning of the second one, a massive volcanic activity took place contemporaneously in correspondence of the three extensional axes (Fig. 12b). The most recent volcanic activity in the TMAC – Ignimbrites of Campania Plain (from 0.205 to 0.018 Ma), Ischia (0.15 Ma), Campi Flegrei (0.03 Ma) and Vesuvius volcano (0.03 Ma) – is limited to areas affected by extensional tectonics related to the second stage of the tectonic evolution.

In this reconstruction of the tectonic evolution of the Campania Plain, the location in present-day coordinates of the extinct triple junction between the Northern Arc, the Southern Arc and the Western Tyrrhenian block is not easy to determine. In fact, the prevalence of diffuse deformation (rifting) makes it difficult to determine a unique point of conjunction of three distinct "plate boundaries". Furthermore, in areas of incipient rifting the migration of the lithosphere extensional axes can follow different trajectories depending on the symmetry of the rift system. In other words, extension axes migration follows the same rules of oceanic ridges when extension is symmetric (i.e. when they follow the McKenzie rifting model, 1978), while they go behind the motion of the upper plate when extension is asymmetric (i.e. in the Wernicke rifting model, 1985) (Fig. 13a,b). In the TMAC, there is not enough data to constrain the symmetry of the rifting phases. Therefore, it is not possible to determine accurately the position of the extensional axes and the associated migration of the triple junction. Nevertheless, a qualitative composition of the vectors for the relative motion of the Northern Arc with respect to the Southern Arc suggests that the triple junction migrated towards east.

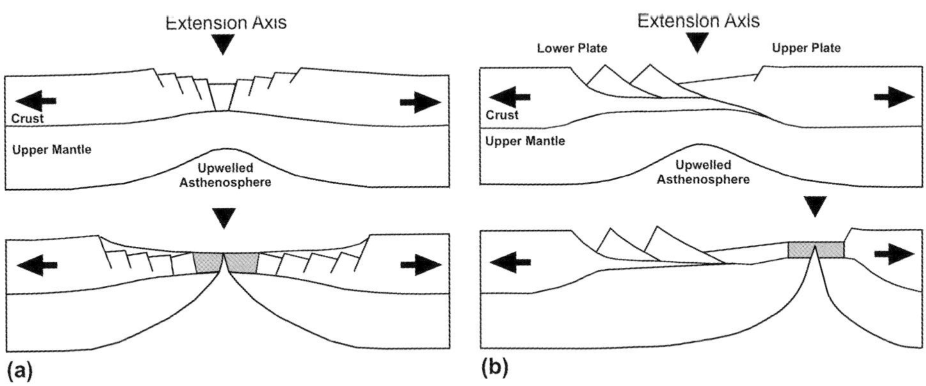

Figure 13. Two end-member model for continental extension. (a) In the symmetrical extension model (i.e. McKenzie pure-shear model, 1978) the extension axe remains fixed in the middle of the two conjugate margins. (b) In the asymmetrical extension model (i.e. Wernicke simple-shear model, 1985), after the inception of oceanic spreading the extension axe remains fix with the upper-plate margin.

8. Conclusions

The relative motion of the northern and southern Apennines chains was reconstructed on the basis of the geologic and kinematic constraints described above. When those constraints are either lacking or insufficient, tectonic motions were established by both trial-and-error tests or indirect methods based on vector calculation. The first preliminary result of this technique was the identification of two distinct rotation stages for the Apennine chain. During the first stage, from 3.5 to 0.78 Ma, the Northern and the Southern Arcs migrated independently with respect to the chosen reference system represented by the Tyrrhenian Sea–Sardinia–Corsica–Eurasia blocks. In the second stage, from 0.78 Ma to present, the Northern Arc stopped migrating, as suggested by cessation of thrusting and related folding in the external domains of the northern Apennines (Di Bucci and Mazzoli, 2002). Conversely, the Southern Arc continued migrating towards the SE. Therefore, the N-S extension in the Campania Plain is the result of the relative motion of the NAA with respect to the SAA during the first stage only, whereas the present-day NW-SE extension in this area, which is characterized by intense volcanism (e.g., Ignimbrites, Ischia, Campi Flegrei, Somma–Vesuvius), is related to the migration of the SAA with respect to a NAA block that is now fixed to the Western Tyrrhenian block. This migration is kinematically linked, through a system of right-lateral en-echelon transfer faults, with the extension centre of the STS located near the Alcione–Lametini–Aeolian Island Arc volcanic lineament.

This model of migration of the Apennine chain is based upon the assumption that the whole mountain range can be considered as a system of only two rigid arcs. This approximation is valid if we consider as negligible the internal deformation of the Southern Arc along the N110 sinistral strike-slip fault, which separates the southern block in at least two distinct tectonic elements (Dewey et al., 1989; Knott and Turco, 1991). Such separation determined a diachronism in the foredeep activity. In fact, tectonic activity in the Bradano trough ceased 0.65 Ma (Patacca and Scandone, 2001), whereas the Ionian foredeep can be still considered as active on the basis of the deep seismic activity related with the Tyrrhenian slab subduction. Conversely the Northern Arc is clearly rigid or quasi-rigid during the considered time interval, except for its southernmost end (Molise). In conclusion, the simplifying assumption of rigidity of the two arcs does not affect the model presented in this paper, which only aims at describing the process of extension and associated magmatic activity in the Tyrrhenian margin of the Apennines chain. Finally, although the model presented above does not take into account many aspects of the complex tectonic evolution of the TMCA, it realistically assembles in a unique kinematic framework the first-order structures that are observed in the Apennine area and in the Tyrrhenian basin, in order to explain the relationships existing between the main structural features of this region.

Acknowledgements

This paper was supported by University of Camerino research grants to Eugenio Turco and Pietro Paolo Pierantoni. We are grateful to reviewers Stefano Mazzoli and Benedetto De Vivo for their thorough and constructive reviews. Comments and suggestions by Mike Carroll and Giovanni Deiana are also gratefully acknowledged.

References

AGIP, 1977. Temperature Sotterranee. Segrate, Milano, 1390 pp.

Alessandrini, B., Beranzoli, L., Mele, F.M., 1995. 3-D crustal P-wave velocity tomography of the Italian region using local and regional seismicity data. Ann. Geofis. 38, 189–211.

Anderson, H., Jackson, J., 1987. The deep seismicity of the Tyrrhenian sea. Geophys. J. R. Astron. Soc. 91, 613–637.

Bartole, R., 1995. The North Tyrrhenian–Northern Apennines post-collisional system; constraints for a geodynamic model. Terra Nova 7(1), 7–30.

Beccaluva, L., Bonatti, E., Dupuy, C., Ferrara, G., Innocenti, F., Lucchini, F., Macera, P., Petrini, R., Rossi, P.L., Serri, G., Seyler, M., Siena, F., 1990. Geochemistry and mineralogy of volcanic rocks from ODP sites 650, 651, 655 and 654, in the Tyrrhenian Sea. Proc. ODP Sci. Results 107, 49–74.

Bigi, G., Cosentino, D., Parotto, M., Sartori, R., Scandone, P., 1990. Structural model of Italy, Scale 1:500.000. Quaderni de "La Ricerca Scientifica" 114 (3), CNR.

Biju-Duval, B., Dercourt, J., Le Pichon, X., 1977. From the Tethys Ocean to the Mediterranean Seas: a plate tectonic model of the evolution of the Western Alpine system. In: Biju-Duval, B., Montadert, L. (Eds), International Symposium on the Structural History of the Mediterranean Basins (Split, 1976). Editions Technip, Paris, pp. 143–164.

Boccaletti, M., Ciranafi, N., Cosentino, D., Deiana, G., Gelati, R., Lentini, F., Massari, F., Moratti, G., Pescatore, T., Lucchi, F.R., Tortorici, L., 1990. Palinspastic restoration and paleogeographic reconstrution of the peri-thyrrenian area during the Neogene. Palaeo., Palaeo., Palaeo., 77, 41–50.

Bonardi, G., D'Argenio, B., Perrone, V., 1988. Geological Map of Southern Apennines, Scale 1:250000. C.N.R. 74° Congresso della Società Geologica Italiana, S.E.L.C.A.

Bonini, M., 1997. Evoluzione tettonica plio-pleistocenica ed analisi strutturale del settore centro-meridionale del bacino Tiberino e del bacino di Rieti (Appennino Umbro-Sabino). Boll. Soc. Geol. It. 116(2), 279–318.

Bosi, C., ED., IGAG, 2004. Quaternary. In: Crescenti, U., D'Offizi, S., Merlino, S., Sacchi, L. (Eds), Geology of Italy. Special Volume of the Italian Geological Society for IGC-32, Florence.

Brunet, C., Monié, P., Jolivet, L., Cadet, J.P., 2000. Migration of compression and extension in the Thyrrenian Sea insights from 40Ar/39Ar Ages on micas along a transect from Corsica to Tuscany. Tectonophysics 321, 127–155.

Calabrò, R.A., Corrado, S., Di Bucci, D., Robustini, P., Tornaghi, M., 2003. Thin-skinned vs. thick-skinned tectonics in the Matese Massif, Central-Southern Apennines (Italy). Tectonophysics 377, 269–297.

Calamita, F., Deiana, G., 1995. Correlazione tra gli eventi deformativi neogenico-quaternari del settore tosco-umbro-marchigiano. In: Cello, G., Deiana, G., Pierantoni, P.P. (Eds), Geodinamica e tettonica attiva del sistema Tirreno-Appennino. Studi Geologici Camerti, Volume Speciale 1, 137–152.

Carmignani, L., Decandia, F.A., Disperati, L., Fantozzi, P. L., Lazzarotto, A., Liotta, D., Oggiano, G., 1995. Relationships between the tertiary structural evolution of the Sardinia-Corsica-Provencal Domain and the Northern Apennines. Terra Nova 7(2), 128–137.

Carozzo, M.T., Luzio, D., Margiotta, C., Quarta, T., 1992. Gravity Map of Italy, Scale 1:500,000. Quaderni de "La Ricerca Scientifica" 114(3), CNR.

Carrara, G., 2002. Evoluzione cinematica neogenica del margine occidentale del bacino Tirrenico. PhD thesis, Parma University, Italy, p. 160.

Cello, G., Tortorici, L., Turco, E., 1981. Analisi mesostrutturale dei depositi conglomeratici della bassa valle del F. Selle (Salerno). Atti del convegno sul tema: il terremoto del 23 novembre 1980. Rome, Italy, Societa Geologica Italiana 4(2), 113–116.

Cipollari, P., Cosentino, D., 1992. Evolution of late Miocene orogenic system in central Italy. 29th International Geological Congress. Abstracts 29, 287.

Codeaux, A., Pinti, D., Gillot, P.Y., 2004. Early Pleistocene fast magmatic transition at Pontine Islands (Central Tyrrhenian Sea, Italy): the waning stages of the subduction-related magmatism? 32nd International Geological Congress, Abstract. Florence, Italy. Abstract.

Compagnoni, B., Galluzzo, F., 2004. Geological Map of Italy, Scale 1: 1,250,000.

Corrado, S., Di Bucci, D., Naso, G., Valensise, G., 2000. The role of pre-existing structures in Quaternary extensional tectonics of the Southern Apennines, Italy; the Boiano Basin case history. J. Czech. Geol. Soc. 45(3/4), 217.

De Vivo, B., Rolandi, G., Gans, P.B., Calvert, A., Bohrson, W.A., Spera, F.J., Belkin, H.E., 2001. New constraints on the pyroclastic eruptive history of the Campanian volcanic Plain (Italy). Mineral. Petrol. 73, 47–65.

Deiana, G., Pialli, G., 1994. The structural provinces of the Umbro-Marchean Apennines. Mem. Soc. Geol. It. 48(2), 473–484.

Dercourt, J., Zonenshain, L.P., Ricou, L.E., Kazmin, V.G., Le Pichon, X., Knipper, A.L., Grandjacquet, C., Sbortshikov, I.M., Geyssant, J., Lepvrier, C., Pechersky, D.H., Boulin, J., Sibuet, J.-C., Savostin, L.A., Sorokhtin, O., Westphal, M., Bazhenov, M.L., Lauer, J.P., Biju-Duval, B., 1986. Geological evolution of the Tethys belt from the Atlantic to the Pamirs since the Lias. Tectonophysics 123, 241–315.

Dewey, J.F., 1975. Finite plate implications: some implications for the evolution of rock masses at plate margins. Am. J. Sci. 275(A), 260–284.

Dewey, J.F., Helman, M.L., Turco, E., Hutton, D.H.W., Knott, S.D., 1989. Kinematics of the Western Mediterranean. In: Coward, M.P., Dietrich, D., Park, R.G. (Eds), Alpine Tectonics. Geol. Soc. Spec. Publ. 45, London, 1989, pp. 265–283.

Di Bucci, D., Mazzoli, S., 2002. Active tectonics of the Northern Apennines and Adria geodynamics; new data and a discussion. J. Geodynamics 34, 687–707.

Faccenna, C., Becker, T.W., Lucente, F.P., Jolivet, L., Rossetti, F., 2001. History of subduction and back-arc extension in the Central Mediterranean. Geophys. J. Int. 145, 809–820.

Faccenna, C., Davy, P., Brun, J.-P., Funiciello, R., Giardini, D., Mattei, M., Nalpas, T., 1996. The dynamics of back-arc extensions: an experimental approach to the opening of the Tyrrhenian Sea. Geophys. J. Int. 126, 781–795.

Faggion, O., Pinna, E., Savelli, C., Schreider, A.A., 1995. Geomagnetism and age study of Tyrrhenian seamounts. Geophys. J. Int. 123, 915–930.

Ferranti, L., Oldow, J.S., Sacchi, M., 1996. Pre-Quaternary orogen-parallel extension in the Southern Apennine Belt, Italy. Tectonophysics 260, 325–347.

Ippolito, F., Ortolani, F., Russo, M., 1973. Struttura marginale tirrenica dell'Appennino Campano; reinterpretazione di dati di antiche ricerche di idrocarburi. Mem. Soc. Geol. It. 12, 227–250.

Jolivet, L., Faccenna, C., 2000. Mediterranean extension and the Africa-Eurasia collision. Tectonics 19, 1095–1107.

Jolivet, L., Faccenna, C., Goffe, B., Mattei, M., Rossetti, F., Brunet, C., Storti, F., Funiciello, R., Cadet, J.P., D'Agostino, N., Parra, T., 1998. Midcrustal shear zones in postorogenic extension: example from the Tyrrhenian Sea. J. Geophys. Res. 103, 12123–12160.

Kastens, K.A., Mascle, J., 1990. The geological evolution of the Tyrrhenian Sea: an introduction to the scientific results of the ODP leg 107. Proc. ODP Sci. Results 107, 3–26.

Kastens, K.A., Mascle, J., Auroux, C., Bonatti, E., Broglia, C., Channel, J., Curzi, C., Emeis, K.C., Glacon, G., Hasegava, S., Hiecke, W., Mascle, G., MacCoy, F., McKenzie, J., Mandelson, J., Muller, J., Rehault, J.P., Robertson, A., Sartori, R., Sprovieri, R., Torii, M., 1988. ODP leg 107 in the Tyrrhenian Sea: insights into passive margin and back-arc basin evolution. Geol. Soc. Am. Bull. 100, 1140–1156.

Knott, S., 1994. Structure, kinematics and metamorphism in Liguride Complex, Southern Apennines, Italy. J. Struct. Geol. 16, 1107–1120.

Knott, S., Turco, E., 1991. Late Cenozoic kinematics of the Calabrian arc, Southern Italy. Tectonics 10, 1164–1172.

Lavecchia, G., Boncio, P., Creati, N., Brozzetti, F., 2003. Some aspects of the Italian Geology not fittine with a subduction scenario. J. Virtual Explorer 10, 1–14.

Lavecchia, G., Federico, C., Stoppa, F., Karner, G., 1995. La distensione tosco-tirrenica come possibile motore della compressione appenninica. Studi Geol. Camerti, Vol. Spec. 1995, 489–497.

Locardi, E., 1982. Individuazione di strutture sismogenetiche dall'esame dell'evoluzione vulcano-tettonica dell'Appennino e del Tirreno. Mem. Soc. Geol. It. 34, 569–596.

Malinverno, A., Ryan, W.B.F., 1986. Extension in the Tyrrhenian Sea and shortening in the Apennines as result of arc migration driven by sinking of the lithosphere. Tectonics 5, 227–245.

Marani, M.P., 2004. Super-inflation of a spreading ridge through vertical accretion. Mem. Descr. Carta Geol. D'It. XLIV, 185–194.

Marani, M.P., Gamberi, F., 2004. Structural framework of the Tyrrhenian Sea unveiled by seafloor morphology. Mem. Descr. Carta Geol. D'It. XLIV, 97–108.

McKenzie, D., 1978. Some remarks on the development of sedimentary basins. Earth Planet. Sci. Lett. 40, 25–32.

McKenzie, D., Parker, R.L., 1974. Plate tectonics in omega pace. Geol. Soc. Lond. Spec. Publ. 45, 265–283.

Milia, A., Giordano, F., Nardi, G., 1998. Stratigraphic and structural evolution of Naples Harbour over the last 12 Ka. Giornale di Geologia 60(3A), 41–52.

Milia, A., Torrente, M.M., 1999. Tectonics and stratigraphic architecture of a peri-Tyrrhenian half-graben (Bay of Naples, Italy). Tectonophysics 315, 301–318.

Milia, A., Torrente, M.M., 2000. Fold uplift and synkinematic strata architectures in a region of active transtensional tectonics and volcanism, eastern Tyrrhenian Sea. Geol. Soc. Am. Bull. 112, 1531–1542.

Patacca, E., Sartori, R., Scandone, P., 1990. Tyrrhenian basin and Apenninic arcs: kinematic relations since late Tortonian times. Mem. Soc. Geol. It. 45, 425–451.

Patacca, E., Scandone, P., 2001. Late thrust propagation and sedimentary response in the thrust-belt-foredeep system of the Southern Apennines (Pliocene-Pleistocene). In: Vai, G.B., Martini, I.P. (Eds), Anatomy of an Orogen: The Apennines and Adjacent Mediterranean Basins, Kluwar Academic Publishers, Dordrecht, The Netherlands, pp. 401–440.

Peccerillo, A., Turco, E., 2004. Petrological and geochemical variations of Plio-Quaternary volcanism in the Tyrhhenian Sea area: regional distribution of magma types, petrogenesis and geodynamic implications. Per. Mineral 73, 231–251.

Ricci Lucchi, F., 1986. The Oligocene to Recent foreland basins of the Northern Apennines. In: Allen, P.A., Homewoof, P. (Eds), Foreland Basins. Int. Assoc. Sedimentol. Spec. Publ., Blackwell, Scientific, Vol. 8, pp. 105–139.

Rolandi, G., Bellucci, F., Heizler, M.T., Belkin, H.E., De Vivo, B., 2003. Tectonic controls on the genesis of ignimbrite from the Campanian Volcanic Zone, southern Italy. Mineral. Petrol. 79, 3–31.

Rosenbaum, G., Lister, G.S., Duboz, C., 2002. Reconstruction of the tectonic evolution of the western Mediterranean since the Oligocene. J. Virtual Explorer 8, 107–126.

Royden, L., Patacca, E., Scandone, P., 1987. Segmentation and configuration of subducted litosphere in Italy: an important control on thrust belt and foredeep basin evolution. Geology 15, 714–717.

Sani, F., Moratti, G., Bovini, M., 1998. The geodynamic evolution of the Northern Apennines; insights from the Neogene-Quaternary basins. Annales Tectonicae 12(1–2), 145–161.

Sartori, R., Torelli, L., Zitellini, N., Carrara, G., Magaldi, M., Mussoni, P., 2004. Crustal features along a W–E Tyrrhenian transect from Sardinia to Campania margins (Central Mediterranean). Tectonophysics 383, 171–192.

Savelli, C., Schreider, A.A., 1991. The opening processes in the deep Tyrrhenian basins of Marsili and Vavilov, as deduced from magnetic and chronological evidence of their igneous crust. Tectonophysics 189, 1–13.

Schettino, A., 1998. Computer-aided paleogeographic reconstructions. Comput. Geosci. 24, 259–267.

Scrocca, D., Doglioni, C., Innocenti, F., Manetti, P., Mazzotti, A., Bertelli, L., Burbi, L., D'Offizi, S. (Eds), 2003. CROP Atlas: seismic reflection profiles of the Italian crust. Mem. Descr. Carta Geol. D'It. 62, 194.

Selli, R., Lucchini, F., Rossi, P. L., Savelli, C., and Del-Monte, M., 1979. Geology and petrochemistry of the central Tyrrhenian volcanoes. In: Cousteau, J.-Y., (Ed.). Symposium de geology et geophysique marines. Commission Internationale pour l'Exploration Scientifique de la Mer Mediterranee, Paris, France, Vol. 25–26(2a), pp. 61–62.

Serri, G., Innocenti, F., Manetti, P., 2001. Magmatism from Mesozoic to present: petrogenesis, time-space distribution and geodynamic implication. In: Vai, G.B., Martini, I.P. (Eds), Anatomy of an Orogen: The Apennines and the Adjacent Mediterraean Basins. Kluwer, Dordrecht, The Netherlands, pp. 77–104.

Tavarnelli, E., Decandia, F.A., Alberti, M., 1998. The transition from extension to compression in the Messinian Laga Basin and its significance on the evolution of the Apennine belt-foredeep-foreland system. In: Suc, J.P. (Ed.), Neogene Basins of the Mediterranean Region; Controls and Correlation in Space and Time. Il Sedicesimo, 12(1–2), 133–144.

Trincardi, F., Zitellini, N., 1987. The rifting of the Tyrrhenian basin. Geo. Mar. Lett. 7, 1–6.

Turco, E., Zuppetta, A., 1998. A kinematic model for the Plio-Quaternary evolution of the Tyrrhenian–Apenninic system; implications for rifting processes and volcanism. J. Volcanol. Geoth. Res. 82, 1–18.

Turco, E., Maresca, R., Cappadonna, P., 1990. La tettonica pliopleistocenica del confine calabro–lucano: modello cinematico. Mem. Soc. Geol. It. 45, 519–529.

Van Dijk, J.P., Bello, M., Brancaleoni, G.P., Cantarella, G., Costa, V., Frixa, A., Golfetto, F., Merlini, S., Riva, M., Torricelli, S., Toscano, C., Zerilli, A., 2000. A regional structural model for the northern sector of the Calabrian Arc (southern Italy). Tectonophysics 324, 267–320.

Wernicke, B., 1985. Uniform sense simple shear of the continental lithosphere. Can. J. Earth. Sci. 22, 108–125.

Volcanism in the Campania Plain:
Vesuvius, Campi Flegrei and Ignimbrites
edited by B. De Vivo

Chapter 3

Rapid changes of the accommodation space in the Late Quaternary succession of Naples Bay, Italy: the influence of volcanism and tectonics

A. Milia[a,*], *M.M. Torrente*[b], *F. Giordano*[c] *and L. Mirabile*[c]

[a]*IAMC, CNR, Calata Porta di Massa, Porto di Napoli, I-80100, Naples, Italy*
[b]*DSGA, University of Sannio, Via Portarsa 11, I-82100 Benevento, Italy*
[c]*University Parthenope, Via Acton 38, I-80133 Naples, Italy*

Abstract

Naples Bay is an extensional basin that experienced an important reactivation of regional faults associated with the emplacement of thick volcanic units during the Late Quaternary. This paper is based on the interpretation of a strictly spaced seismic grid that permitted the reconstruction of the paleogeography of Naples Bay before the onset of volcanic activity and the paleogeographic changes following the voluminous volcanic unit's emplacement. Using the estimated paleo-water depth we calculated curves of space filled with volcanics and sediments, subsidence and accommodation space in order to understand the relationship between volcanic activity and tectonic subsidence at three selected sites (Penta Palummo, Pozzuoli Bay and offshore Vesuvius) of the Naples Bay basin.

Repeated volcanic events and rapid basin infill were documented for the Penta Palummo area that underwent a dramatic physiographic change, changing from a slope-basin (in the Middle Pleistocene) to a shelf (during Late Quaternary). A more gradual physiographic change from a slope-basin to a shelf occurred in Pozzuoli Bay where the basin infill was caused by clastic vertical aggradation, later followed by a tectonic subsidence related to post-15 ka faulting and folding. Finally, the area offshore Vesuvius remained a shelf and did not experience any physiographic change as a rapid increase of accommodation space was balanced by the contemporaneous filling of the space due to the emplacement of Upper Pleistocene ignimbrites. These findings question the occurrence of a caldera offshore Campi Flegrei as reported by previous workers.

1. Introduction

The accommodation space is the space that is made available within a basin for the deposition of sediment (e.g. Posamentier et al., 1988). Marine accommodation space increases during a sea-level rise and/or a tectonic subsidence, but decreases during a sea-level fall, tectonic uplift, sediment or volcanic discharge.

A large volcanic eruption can rapidly fill the accommodation space producing an instantaneous and dramatic change in the physiography of the region. By contrast, when a large-volume ignimbrite eruption is associated with caldera collapse, a circular depression forms and there is an instantaneous increase in the accommodation space equal to the algebraic sum of the volcano tectonic subsidence and the vertical volcanic aggradation.

Recent volcanological work has documented the existence of numerous ignimbrites that were emplaced over the entire Campanian Plain in the last 300 ka (e.g. Rolandi et al., 2003). Many authors (e.g. Rosi and Sbrana, 1987; Orsi et al., 1996) believe that a caldera

*Corresponding author. *E-mail address*: alfonsa.milia@iamc.cnr.it (A. Milia).

formed in the Campi Flegrei and northern Naples Bay during the eruptions of the Campania Ignimbrite (CI) (39 ka-old, De Vivo et al., 2001) and the Neapolitan Yellow Tuff (NYT) (15 ka-old, Deino et al., 2004).

The geologic evolution of the Naples Bay half graben and its stratigraphic architecture were reconstructed by Milia (1999a) and Milia and Torrente (1999). In addition, many thick volcanic deposits and a complex pattern of regional faults were recognized in Naples Bay using high-resolution and low-penetration seismic reflection profiles as reported by Milia and Torrente (2000, 2003); however, the relationship between these upper Pleistocene ignimbrite eruptions and the structure of the Campania continental margin is still a matter of debate. In order to contribute something new to this argument the authors acquired, over the last 20 years, more than 3500 km of multichannel and monochannel seismic lines in Naples Bay. Their purpose was to create a step-by-step reconstruction of the evolution of the paleogeography during the volcanic activity and to evaluate the changes in the accommodation space in order to correctly evaluate the role of tectonics and/or volcano tectonics in the emplacement of the volcanic units.

2. *Geological setting*

The Tyrrhenian Sea corresponds to a region of lithospheric stretching that started along its western margin in upper Miocene times (e.g. Patacca et al., 1990) and progressively migrated eastward reaching the Campania margin in Quaternary times. The Campania continental margin displays the typical features of a back-arc extensional domain: numerous normal faults, a very shallow Moho (Ferrucci et al., 1989), high heat flow values (Della Vedova et al., 2001) and large-volume ignimbrite eruptions. The large-scale structure of the margin, reconstructed by means of geological and geomorphological works, well, gravimetric and seismic reflection data (Ippolito et al., 1973; Nunziata and Rapolla, 1981; Mariani and Prato, 1988; Brancaccio et al., 1991; Milia and Torrente, 1999; Milia et al., 2003), corresponds to a series of structural highs and lows of the Mesozoic carbonate substrate covered by Quaternary clastic sediments and volcanic rocks.

The regional structure of the Campania continental margin is characterized by upper Miocene carbonatic nappes of the Apennine chain overprinted by Lower Pleistocene NW-SE normal faults, followed by post-700 ka NE-SW normal faults (Milia and Torrente, 1999; Turco et al., this volume) (Fig. 1). The NE-trending normal faults form an asymmetrical system featuring a half graben in the Naples Bay and Campi Flegrei region (Figs. 2 and 3). They produced a severe crustal thinning (a value of elongation $e = 0.25$ calculated by Milia et al., 2003) and accommodated much of the crustal extension of the Campania margin strictly controlling the site of sedimentation and Quaternary volcanism (Ischia, Procida and Campi Flegrei). This asymmetrical style of deformation produced lateral changes in the subsidence of the basin.

In Naples Bay tectonics controls the depositional environment, sediment distribution and rate of sediment supply. The activity of the NE-trending normal faults produced an increase in the accommodation space characterized by lateral changes in the basin and a physiographic change from shallow water to deep basin. In particular NE-SW trending hanging wall blocks and fault-bounded basins formed (Figs. 2 and 3). Consequently, in the present basin shallow water deposits were covered by debris flows and deep marine

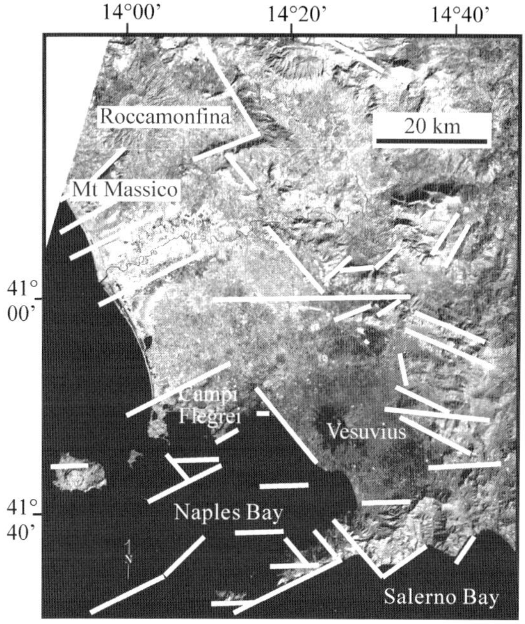

Figure 1. Tectonic sketch of the Campania margin characterized by E-W, NW-SE and NE-SW normal faults.

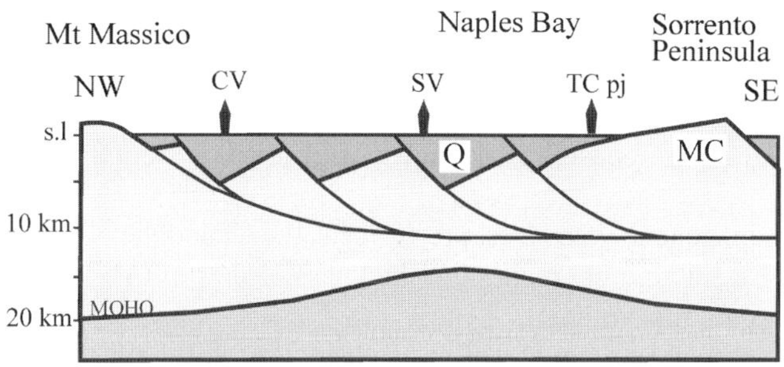

Figure 2. Balanced section across the NE-trending normal faults of the Campania margin. Q, Quaternary fill; MC, Meso-Cenozoic rocks; CV, Castelvolturno 1 well; SV, S.Vito 1 well; TC, Trecase well (modified from Milia et al., 2003).

sediments. The sedimentation along the hanging wall consists of sediments that prograde from shallow water to deep basin producing a wide shelf in the southern and middle part of Naples Bay (Figs. 3 and 4) (Milia, 1999a; Milia and Torrente, 1999). The stratigraphic succession features a Trangressive–Regressive Cycle; in detail a transgressive depositional sequence set B (formed by the three depositional sequences B1, B2 and B3) was deposited between 700 and 400 ka, whereas a regressive depositional sequence set C (formed by the three depositional sequences C1, C2 and C3) was laid down between 400 and 100 ka (Fig. 3); both sequence sets are made up of fourth order (with a 100 ka frequency) depositional sequences (e.g. Mitchum and Van Wagoner, 1991). Sequence set B

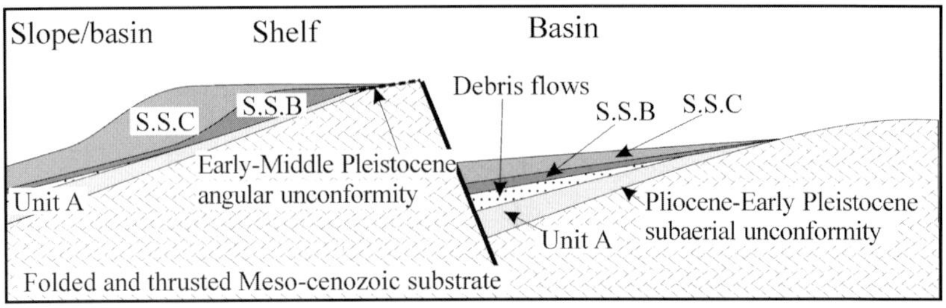

Figure 3. Schematic section along Salerno Bay-Naples Bay displaying the relation between tectonics, sedimentation and physiography. Unit A, shallow water deposits; S.S.B, sequence set B, from shallow water to deep basin sediments forming the transgressive cycle; S.S.C, sequence set C, from shallow water to deep basin sediments forming the regressive cycle. The debris flow in the basin and the angular unconformity on the shelf mark the onset of the basin formation (modified from Milia and Torrente, 1999).

displays a transgressive trend and aggradational stacking pattern in the shelf-slope due to a high basin subsidence related to fault activity (half-graben formation), whereas a period of tectonic stability and a consequent absence of basin subsidence is witnessed in the oblique-progradational stacking pattern of sequence set C (Milia, 1999a; Milia and Torrente, 1999).

The continental shelf, corresponding to the topset and/or toplap surface of the prograding wedges deposited during the Middle Pleistocene, extends northwestward to a slope. The latter corresponds to the foresets of the sedimentary units (Figs. 3 and 4a). The last depositional sequence was subdivided into: A Basin Floor Fan deposited at the mouth of the Dohrn Canyon; a Lowstand Prograding Wedge deposited at the margin of the northern continental shelf; a Transgressive System Tract deposited close to the volcanic reliefs and in the coastal area; and a Highstand System Tract deposited in the coastal area.

Numerous monogenetic volcanoes, pre-CI tuffs, the CI and the NYT cover the northern part of Naples Bay attaining a maximum thickness of 300 m (Milia and Torrente, 2000, 2003; Bellucci et al., this volume).

The Late Quaternary structural setting is characterized by NE-trending normal faults downthrowing southeastward, E-W left lateral faults and NW-SE oblique faults (Milia and Torrente, 2003) in addition to WNW-ESE synsedimentary tectonic folds enucleated during the Holocene South of Campi Flegrei (Milia and Torrente, 2000). An anticline characterized by a half wavelength of ca 1 km culminates in the Pozzuoli area with a syncline occurring in Pozzuoli Bay.

At present, the physiography of Naples Bay (Fig. 4b) is characterized by a wide continental shelf that extends to water depths of 100–180 m (Milia, 1999b). The shelf width varies from a maximum of approximately 20 km in the central bay to approximately 2.5 km off the islands of Capri and Procida. The northern area displays an irregular continental shelf, which forms a part of an extensive system of volcanic banks. An intraslope basin is bounded by Naples Bay continental slopes and, toward the southwest, by a NE-trending structural high known as Banco di Fuori (Milia, 2000). The continental slope is cut by two canyons, the Magnaghi and Dohrn Canyons.

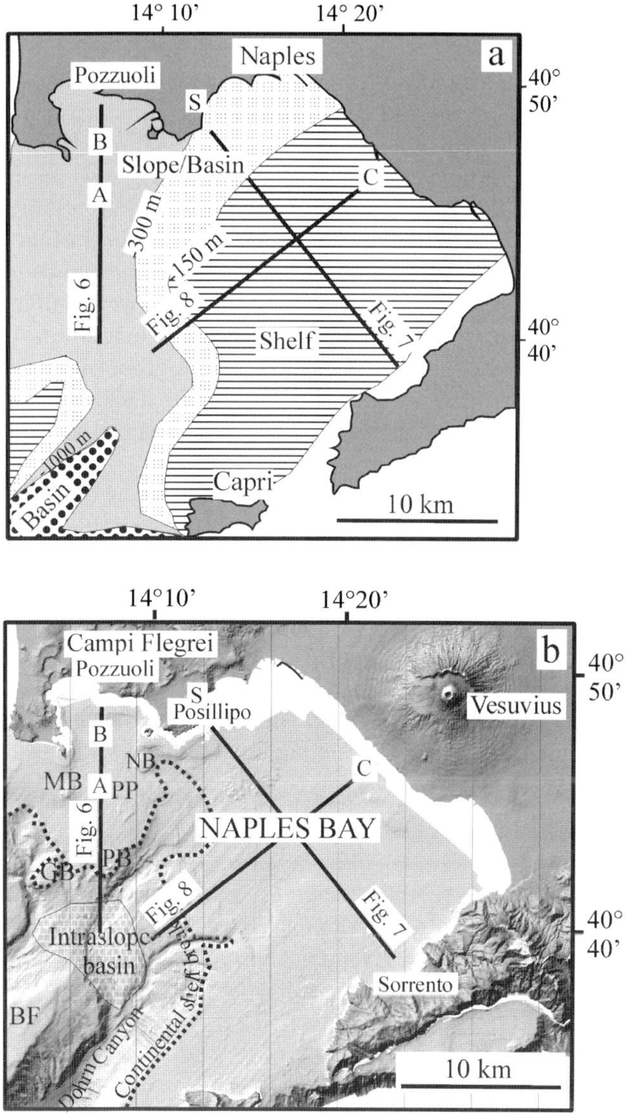

Figure 4. (a) Past physiography of Naples Bay after the deposition of the Middle Pleistocene succession. Note the existence of a wide continental shelf in the southeastern part of the bay and the continental slope to basin toward the northwest. S, Piazza Sannazzaro well. A, B and C are Naples Bay sites where the accommodation curves were calculated. (b) Present physiography of Naples Bay showing the shelf, the slope, an intraslope basin and the submarine volcanoes of Gaia Bank (GB), Pia Bank (PB), Penta Palummo Bank (PP), Nisida Bank (NB) and Miseno Bank (MB). BF, Banco di Fuori high; S, Piazza Sannazzaro well. A, B and C are Naples Bay sites where the accommodation curves were calculated. Seafloor bathymetry of Naples Bay after D'Argenio et al. (in press) and geomorphologic features after Milia (1999b).

3. Materials and methods

Naples Bay was investigated by means of closely spaced multichannel and single-channel seismic reflection profiles (Fig. 5). The latter were acquired using a 16 kJ Multispot Extended Array Sparker (MEAS) system and a 1 kJ Surfboom system. All seismic sections were recorded graphically on continuous paper sheets with a vertical recording scale of 0.25 and 0.5 s for Surfboom, and 1.0 and 2.0 s for MEAS. Ship positioning was determined by: LORAN C for MEAS and Micro-Fix Racall for Surfboom (with a position accuracy of 1 m). The best vertical resolution was approximately 6 m for the MEAS data and 1 m for the Surfboom. The multichannel profiles were acquired using a double water gun and a 24-trace streamer. The maximum recorded length was 4.5 s of two-way travel time and the fold coverage was 12 or 24; the data processing sequence included deconvolution, velocity analysis, normal moveout (NMO) stacking and time migration. These multichannel profiles were characterized by a high signal–noise ratio and a best vertical resolution of 10 m.

During the Late Quaternary, the eustatic sea-level curve was characterized by a slow sea-level fall followed by a rapid sea-level rise (Martinson et al., 1987). With reference to these distinct phases of sea-level fluctuations, sedimentary deposits on continental margins can readily be subdivided on the basis of their internal geometry and stacking pattern (Mitchum et al., 1977; Posamentier et al., 1988; Thorne and Swift, 1991). Seismic units are groups of seismic reflections, the parameters of which (configuration, amplitude, continuity and frequency) differ from those of adjacent groups. Volcanic and sedimentary units were delineated on the basis of contact relations and internal and external configurations. Sequence stratigraphy permits the interpretation of the environmental settings from

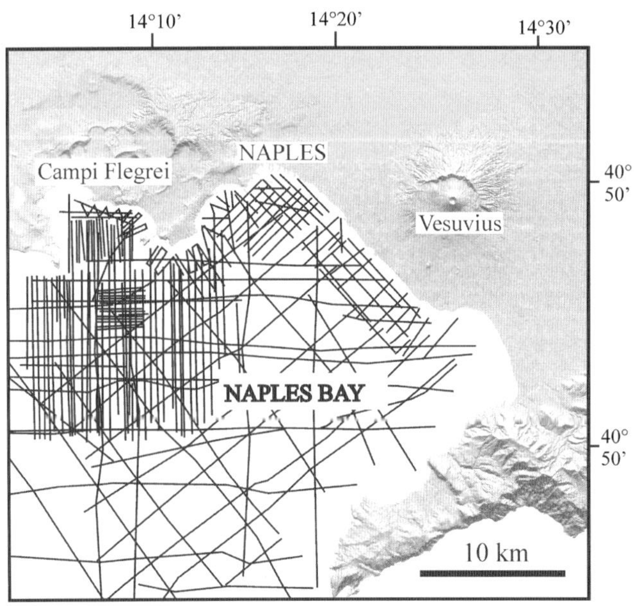

Figure 5. Index map of seismic reflection profiles.

seismic data. Using seismic characteristics, the stratigraphic relations between various units and the physical continuity of coastal outcrops we attempted to relate each seismic unit to a geologic one.

4. Results

In order to reconstruct the subsidence associated with fault activity and its relationship with volcanism, the authors shall describe the stratigraphy of Naples Bay and its relationship with the Middle Pleistocene physiography through the analysis of three sections (Fig. 4a): the first section is taken from across the Middle Pleistocene slope/basin area, the second from the Middle Pleistocene slope/basin to shelf areas and the third from across the Middle Pleistocene shelf.

4.1. Section from the Middle Pleistocene slope/basin area

The stratigraphy of the northern part of Naples Bay, across the Middle Pleistocene slope/basin area, is well illustrated by the interpretation of a multichannel seismic profile collected from the intraslope basin to Pozzuoli Bay (Fig. 6). The stratigraphic units of the shallower part of the seismic section (upper 0.2 s of two-way travel time, corresponding to Late Quaternary depositional sequence) were analyzed by means of single-channel profiles and detected in the intraslope basin, Penta Palummo shelf and Pozzuoli Bay. Using the multichannel profiles, by contrast, we were able to recognize in the Penta Palummo and Pozzuoli Bay areas the deeper stratigraphic units up to the Middle Pleistocene deposits.

From the intraslope basin to Pozzuoli Bay three areas characterized by different stratigraphic architecture were recognized (Fig. 6): the intraslope basin area, the Penta Palummo shelf area and the Pozzuoli Bay area.

The intraslope basin features parallel reflectors characterized by high- to low-frequency variable amplitude and good continuity that cover a strong reflector (SSB) with high amplitude, low frequency and good continuity; these seismic units display the typical attributes of clastic sediments. Based on the stratigraphic reconstruction made up by Milia (1999a) and Milia and Torrente (1999), we interpret the strong reflector SSB as a condensed section of the depositional sequence set B deposited during the transgressive cycle; whereas the overlying parallel reflectors, subdivided on the basis of seismic facies and angular unconformity, are associated with the fourth-order depositional sequences C1, C2 and C3. The Penta Palummo area is made up of the parallel reflectors of sequence set B and the C1 and C2 depositional sequences that display a lateral continuity with the succession of the intraslope basin, thus suggesting a similar environment and paleogeography for these adjacent areas. However, this horizontally lying sedimentary succession is covered, in the northern margin, by almost three mounded units (V5, V4, V3) interlayered with marine sediments that prograde north-ward and southward. The mounds are interpreted as volcanoes based on the internal and external seismic configuration and the volcanic nature of the area. The south margin of the Penta Palummo area is bounded by the Pia Bank volcano. A normal fault occurs in the southern margin of the Penta Palummo area in correspondence to the Pia volcano. This fault downthrows the southern block by approximately 75 m. The continental slope is formed by the southern margin of the Pia volcano up to the

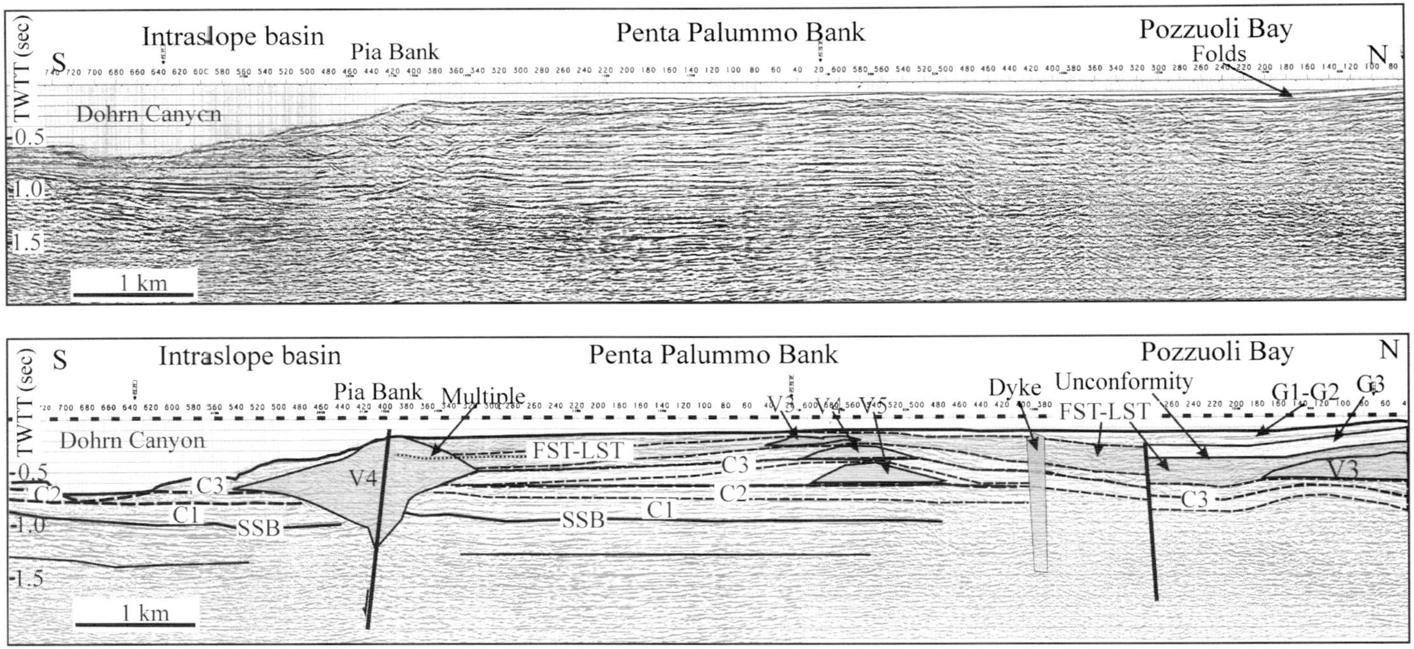

Figure 6. Migrated multichannel seismic profile extending from the intraslope Naples Bay basin toward Pozzuoli displaying a complex stratigraphic architecture due to the interplay between clastic (SSB, C1, C2, C3, FST–LST, G3, G2, G1) and volcanic (V5, V4, V3) units. The Middle Pleistocene seismic units (Sequence set B, C1, C2 and C3 units) were cross-correlated with those identified by Milia (1999a) and Milia and Torrente (1999), whereas the post-15 ka units G1, G2 and G3 with those recognized by Milia and Torrente (2000). FST–LST=Forced Regression System Tract–Lowstand System Tract. For seismic section location see Figure 4.

intraslope basin. At the base, the volcanic edifice is interlayered within the sediment of the C3 depositional sequence.

In the Pozzuoli Bay area, the parallel reflectors of sequence set B and of the C1 and C2 depositional sequences, constituting the intraslope basin and the base of the Penta Palummo shelf, are not imaged probably due to the presence of gas and high temperatures typical of the Campi Flegrei region. The prograding unit present at the northern margin of the Penta Palummo area formed a slope toward Pozzuoli Bay with their bottomsets extending toward the north in the adjacent basin with a uniform thickness. A thick chaotic unit (V3), with an external wedge form is present in the northern part of Pozzuoli Bay thinning toward the south. The bay is successively filled by a sedimentary unit (FST-LST) that onlaps the slope of Penta Palummo in the southern part of the bay and the chaotic wedge toward the north. The upper boundary of this unit corresponds to an unconformity that lies at a depth of 130–150 m in the northern part and at a depth of approximately 300 m in the southern part. This sedimentary unit (FST-LST) corresponds to the Forced Regression System Tract–Lowstand System Tract of the last depositional sequence deposited in the 100–18 ka time span. Unit V3 features a wedge geometry and corresponds to a pyroclastic wedge recognized and mapped in Naples Bay and physically correlated onshore in Naples city to pre-CI tuffs (Milia and Torrente, 2003; Bellucci et al., this volume). The youngest units detected in Pozzuoli Bay present seismic facies and architecture typical of sedimentary deposits. An isopach unit (G3) characterized by a parallel seismic reflector is covered in onlap by two (G2, G1) wedge-shaped units (Milia and Torrente, 2000).

It is possible to recognize a dyke intrusion in the southern margin of the Bay. This dyke, mapped by means of high-resolution seismic profiles, is formed by two coalescent NW-SE and NE-SW trending bodies. NW-trending normal faults downthrow to the north with a total displacement of approximately 75 m. These faults are post-NYT as reported by Milia and Torrente (2000). Tectonic gentle folds are also displayed on the northern part of the multichannel profile (Fig. 6). They affect Pozzuoli Bay since 8 ky producing a subsidence of the central part of the bay and an uplift of the Pozzuoli city area (Milia, 1998; Milia and Torrente, 2000).

4.2. Section from Middle Pleistocene slope/basin to shelf areas

A seismic line from the Middle Pleistocene slope/basin to the shelf areas shows a complex stratigraphic architecture (Fig. 7) made up of Meso-Cenozoic carbonate (MC), outcropping in the Sorrento Peninsula along the southern margin of Naples Bay and overlain by the middle Pleistocene sedimentary succession (MP) that progrades northwestward. The latter is covered by an old wedge (Pre-CI) with a chaotic seismic facies and a maximum thickness in the northwest part that thins toward the southwest. It follows a thin seismic unit with parallel reflectors that is in turn covered by an intermediate chaotic seismic facies unit (CI) forming two wedges (the first at the northwest end of the profile and the second at the southeast end). These wedges are covered by a thin seismic unit with parallel reflectors. A younger wedge (NYT) characterized by a chaotic seismic facies is present in the northwest part of the seismic profile. The physical correlation between these seismic units and the stratigraphic units drilled on the Naples city coast (Bellucci et al., this volume) and outcropping at the Sorrento Peninsula permitted the authors to interpret the oldest wedge

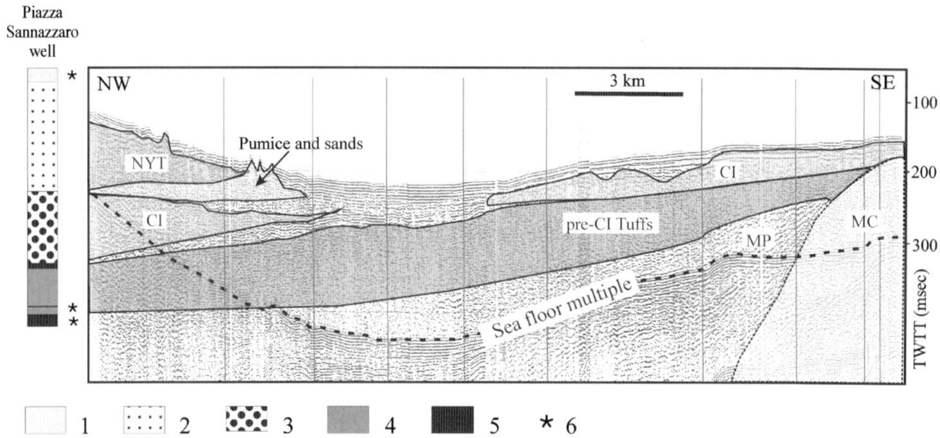

Figure 7. Sparker seismic profile showing the stratigraphic relationships between the volcanic and sedimentary units from slope to shelf areas. MC, Mesocenozoic rocks; MP, Middle Pleistocene sediments; CI, Campania Ignimbrite; NYT, Neapolitan Yellow Tuff. The seismic units are correlated to the stratigraphy of the Piazza Sannazzaro well located along the Naples coast at the northwest end of the seismic line. 1, Reworked pyroclastics and marine strata; 2, NYT; 3, Pumice deposits; 4, Tuff deposits; 5, Marine deposits; 6, Fossils (modified from Milia et al., 1998). For seismic section and well location see Figure 4.

as the pre-CI tuffs interlayered with marine sediments; the intermediate chaotic wedge as the CI pyroclastic unit outcropping at the Sorrento cliffs and Naples (Fig. 4); and the youngest wedge as the NYT that outcrops at Posillipo hill (Fig. 4).

4.3. *Section across the Middle Pleistocene shelf*

A NE-trending seismic line crossing the continental shelf reveals seaward-prograding Middle Pleistocene deposits and a landward-dipping toplap surface (Fig. 8). This toplap surface is overlain by a thick chaotic seismic unit that has a wedge geometry, thins seaward and corresponds to the pre-CI tuffs and to CI. The landward-dipping nature of the unconformity U2 is due to the platform fault block tilting in accordance with a major SW-dipping normal fault (Milia, 1999a, 2000; Bellucci et al., this volume). A younger sedimentary unit characterized by horizontal reflectors onlaps the CI.

5. Discussion

The analysis of a basin is performed by considering sea-level changes, sediment and volcanic accumulation and subsidence analysis in order to reconstruct accommodation space changes.

Changes in sea level can lead to errors in calculating the basin subsidence history. For example, in the case of sea-level rises, the stratigraphic record will show a deepening of water depth (increase in the accommodation space) that could be interpreted as being the result of an increase of the basin subsidence and vice versa.

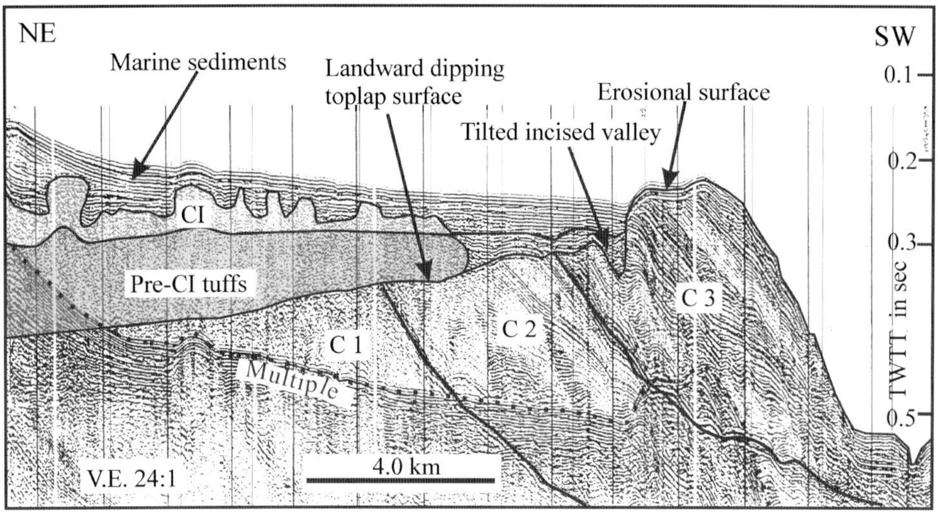

Figure 8. Sparker seismic profile extending from the slope to the continental shelf and displaying the Middle Pleistocene marine succession (C1, C2, C3) tilted and overlain by volcanic units of the pre-CI tuffs and CI, in turn covered by marine sediments (modified from Milia, 1999). For seismic section location see Figure 4.

An extensional basin is characterized by a syn-rift and a post-rift subsidence. Syn-rift subsidence is essentially controlled by fault geometry and flexural isostatic rebound both characterized by a lateral change in the subsidence along the basin, whereas post-rift subsidence is associated with the thermal effects associated with the changing density structure of the lithosphere. The amount of subsidence in a basin is very well constrained by a detailed analysis of the stratigraphic record. In the case of Naples Bay syn-rift subsidence was documented during the deposition of sequence set B (700–400 ka), followed by a period of tectonic stability (400–100 ka) when sequence set C was laid down (Milia, 1999a). In addition, we can exclude a post-rift subsidence because rifting is still in progress as witnessed by volcanism, high geothermal flux and fault activity.

Paleobathymetry is an important factor in the estimation of the subsidence and accommodation space. Paleobathymetry in a datum point of the basin changes with time because of global sea-level fluctuations. These effects can be evaluated by utilizing the available eustatic sea-level curves as the paleobathymetry changes in different points of a basin in relation to the paleogeography.

In order to understand the relationship between volcanic activity and tectonic subsidence during the Late Quaternary, the authors reconstructed the accommodation curve for three specific points of the Naples Bay basin (A, B, C, respectively, located in Penta Palummo, Pozzuoli Bay and offshore Vesuvius; Fig. 9). The accommodation space curve is the result of the algebraic sum of the eustatic curve (Martinson et al., 1987), the curve of space filled by volcanics and sediments (calculated assuming in the depth conversion a velocity of 1600–1800 m/s compatible with very shallow mainly medium-grained sediments and pyroclastic rocks) and the subsidence curve (associated with faulting and folding). A basin analysis was performed for the last 200 ka and the paleobathymetry was estimated by reconstructing the stratigraphic architecture and depositional environments at

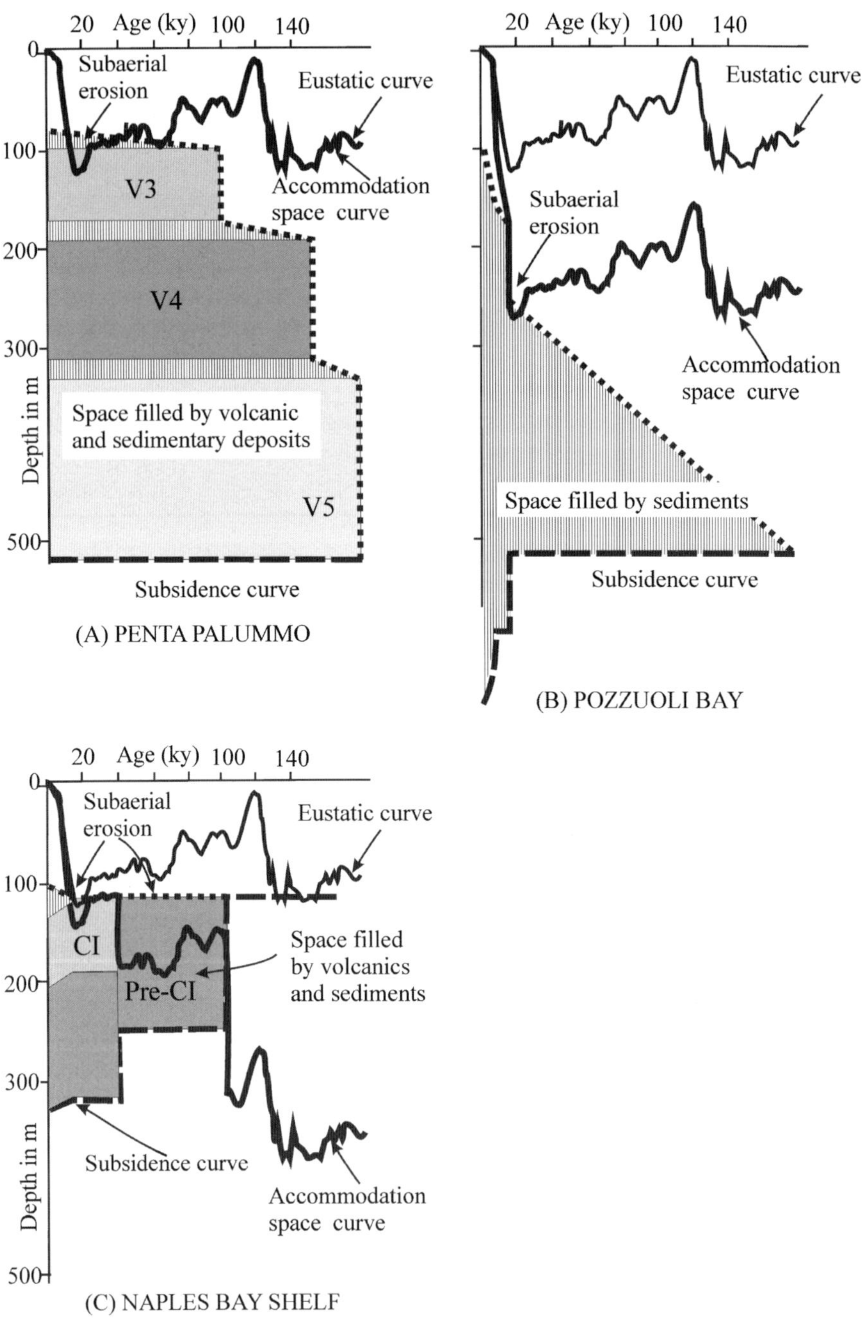

Figure 9. Accommodation curves calculated in three selected sites of Naples Bay basin. The accommodation curves are the result of the interplay between eustatic and subsidence curve. For site location see Figure 4.

the time of sequence set B deposition (SSB in Fig. 6). Any curve starts at the time of sequence set C deposition, when no regional subsidence affected the Naples Bay half-graben (Milia, 1999a; Milia and Torrente, 1999). Sediment accumulation is plotted through time. Even if an absolute age of the marine succession is not available, the age of the marine packages of the Naples Bay succession was calibrated using a sequence stratigraphic approach integrated with the ages of the thick volcanic units. Indeed, many units of the Late Quaternary succession were physically correlated to volcanic units dated onshore (Bellucci et al., this volume).

In the Penta Palummo area there was no tectonic subsidence and the rate of the accommodation space changes according to the eustatic curve minus the space filled by volcanics and sediments (Fig. 9A). Here Middle Pleistocene sedimentary strata extend horizontally from the intraslope basin northward (Fig. 6), thus supporting a paleogeographic scenario preceding volcanism. The latter is characterized by a paleo-water depth of approximately 520 m (top of C2 sequence in Fig. 6). The space between the paleo-sea-floor and the sea level was filled by isolated monogenetic volcanoes (V5, V4, V3) and clastic marine sediments until the present water depth of 75 m. Sedimentary units prograde north- and southward from these volcanoes, indicating that the latter became the source areas for the sediments during the intervals of volcanic standstill. The vertical aggradation of both volcanic and clastic deposits gave rise to the emersion of this area during the last glacial maximum and to subaerial erosion when the curve displaying the vertical aggradation of the volcanic and sedimentary deposits intersected the accommodation curve (Fig. 9A).

The paleogeography of Pozzuoli Bay before the onset of volcanism is that of the adjacent intraslope basin and Penta Palummo area (Figs. 4a and 6) characterized by a paleo-water depth of approximately 520 m. The subsidence curve shows a negligible value until 15 ka. Afterward an instantaneous increase of subsidence due to faulting and the Holocene enucleation of the syncline (Fig. 9B) can be seen. The accommodation curve presents a rapid increase over the last 15 ka corresponding to basin subsidence. The space filled by sediments increased linearly producing a gradual decrease in the water depth until this curve intersected the accommodation curve during the last glacial maximum producing an emersion of the area and subaerial erosion (see unconformity at the top of the Forced regression system tract–Lowstand system tract, FST-LST in Fig. 6). The mean rate of sedimentary supply increased over the last 15 ka as indicated by the thick Holocene succession (G3, G2, G1).

The area offshore Vesuvius displays a subsidence curve characterized by two vertical steps associated to very rapid tectonic events (Fig. 9C) characterized by throws in the order of hundreds of meters (Bellucci et al., this volume). Because the Lowstand system tracts were deposited during the glacial maximum periods, the occurrence of the toplap surface of the Middle Pleistocene sedimentary units, mainly made up of Lowstand system tracts (Milia, 1999a), suggests a paleo-water depth of approximately 120 m. The tilting of the toplap surface (Milia, 2000) created the space for the deposition of the pre-CI tuffs and of the CI. The accommodation curve (Fig. 9C) displays two rapid steps that produced dramatic increases in the space contemporaneously filled by eruption products. The two-stage evolution of this fault is also reported by Bellucci et al. (this volume). In addition, an emersion of this area occurred when the accommodation space curve was over filled by volcanic deposits approximately between 100 and 18 ky.

6. Concluding remarks

This research documents the capability of accommodation space curves, calculated at different sites throughout Naples Bay, to discern the influence of volcanism, sedimentation and tectonic subsidence.

Repeated volcanic events and a rapid infill of the basin were documented for the Penta Palummo area that experienced a dramatic physiographic change, from a slope-basin (in the Middle Pleistocene) to a shelf (during Late Quaternary) due to volcanic vertical aggradation. A more gradual physiographic change from a slope-basin to a shelf occurred in Pozzuoli Bay where the basin infill, caused by clastic vertical aggradation, was later followed by a localized tectonic subsidence. Finally, the area offshore Vesuvius remained a shelf and did not experience any physiographic change as the increase of accommodation space, associated with repeated events of volcano tectonic subsidence, was balanced by the contemporaneous filling of the accommodation space due to the emplacement of Upper Pleistocene ignimbrites.

In conclusion, this research documents a physiographic change for the area offshore Campi Flegrei (Penta Palummo–Pozzuoli Bay). The area under study went from a slope-basin to a shelf and experienced a decrease in the accommodation space due to the emplacement of volcanoes and clastic sediments. By contrast, offshore Vesuvius did not experience any physiographic change but experienced strong episodes of volcano tectonic subsidence contemporaneous to ignimbrite emplacement. These findings question the occurrence of a caldera offshore Campi Flegrei as purported by previous workers (e.g. Rosi and Sbrana, 1987; Orsi et al., 1996).

The application of these basin analysis techniques provides a relatively new and very powerful tool for understanding the role of tectonics and sea-level changes in the evolution of a sedimentary basin and for evaluating the volcano tectonic subsidence associated with the emplacement of large-volume ignimbrites.

Acknowledgments

F. Giordano and L. Mirabile have acquired the seismic data set. A. Milia and M.M. Torrente performed the geologic interpretation of the seismic data and are responsible for the results and discussion paragraphs. This manuscript has benefited from the constructive review given by E. Turco and E. Tavarnelli. Financial support was given by the Italian "Ministero dell'Università e della Ricerca Scientifica e Tecnologica" (FAR 2003, 2004, M. Torrente).

References

Bellucci, F., Milia, A., Rolandi, G., Torrente, M.M., this volume. Structural control on the Upper Pleistocene ignimbrite eruptions in the Neapolitan area (Italy): volcano tectonic faults versus caldera faults.

Brancaccio, L., Cinque, A., Romano, P., Russo, F., Santangelo, N., Santo, A., 1991. Geomorphology and neotectonic evolution of a sector of the Tyrrhenian flank of the southern Apennines (Region of Naples, Italy). Z. Geomorph. N. F. 82, 47–58.

D'Argenio, B., Aiello, G., de Alteriis, G., Milia, A., Sacchi, M., Tonielli, R., Angelino, A., Budillon, F., Chiocci, F., Conforti, A., De Lauro, M., Di Martino, G., d'Isanto, C., Esposito, E., Ferraro, L., Innangi, S., Insinga, D., Iorio, M., Marsella, E., Molisso, F., Morra, V., Passaro, S., Pelosi, N., Porfido, S., Raspini, A., Ruggirei, S., Sarnacchiaro, G., Terranova, C., Vilardo, G., Violante, C., in press. Digital elevation model of the Naples Bay and adjacent areas, eastern Tyrrhenian Sea. In: Pasquarè, G., Venturini, C. (Eds), Mapping Geology in Italy. APAT Dipartimento Difesa del Suolo-Servizio Geologico d'Italia, pp. 21–28.

Deino, A.L., Orsi, G., de Vita, S., Piochi, M., 2004. The age of the Neapolitan Yellow Tuff caldera-forming eruption (Campi Flegrei caldera-Italy) assessed by $^{39}Ar/^{40}Ar$ dating method. J. Volcanol. Geotherm. Res. 133, 157–170.

Della Vedova, B., Bellini, S., Pellis, G., Squarci, P., 2001. Deep temperatures and surface heat flow distribution. In: Vai, G.B., Martini, I.P. (Eds), Anatomy of an Orogen: the Apennines and Adjacent Mediterranean Basin. Kluwer Academic Publishers, Dordrecht, the Netherlands, pp. 65–67.

De Vivo, B., Rolandi, G., Gans, P.B., Calvert, A., Bohrson, W.A., Spera, F.J., Belkin, H.E., 2001. New constraints on the pyroclastic eruptive history of the Campanian volcanic plain (Italy). Mineral. Petrol. 73, 47–65.

Ferrucci, F., Gaudiosi, G., Pino, N.A., Luongo, G., Hirn, A., Mirabile, L., 1989. Seismic detection of a major Moho upheaval beneath the Campania volcanic area. Geophys. Res. Lett. 16, 1317–1320.

Ippolito, F., Ortolani, F., Russo, M., 1973. Struttura marginale tirrenica dell'Appennino campano: reinterpretazioni di dati di antiche ricerche di idrocarburi. Mem. Soc. Geol. It. 12, 227–250.

Mariani, M., Prato, R., 1988. I bacini neogenici costieri del margine tirrenico: approccio sismo-stratigrafico. Mem. Soc. Geol. It. 41, 519–531.

Martinson, D.G., Pisias, N.G., Hays, J.D., Imbrie, J., Moore, T.C., Schackleton, N.J., 1987. Age dating and orbital theory of the Ice Ages: development of a high-resolution 0 to 300000 year chronostratigraphy. Quat. Res. 27, 1–29.

Milia, A., 1998. Stratigrafia, strutture deformative e considerazioni sull'origine delle unità deposizionali oloceniche del Golfo di Pozzuoli. Boll. Soc. Geol. It. 117, 777–787.

Milia, A., 1999a. Aggrading and prograding infill of a pery-tyrrhenian basin (Naples Bay, Italy). Geo-Mar. Lett. 19, 237–244.

Milia, A., 1999b. The geomorphology of Naples Bay continental shelf (Italy). Geogr. Fis. Dinam. Quat. 22, 73–78.

Milia, A., 2000. The Dohrn Canyon formation: a response to the eustatic fall and tectonic uplift of the outer shelf (Eastern Tyrrhenian Sea margin, Italy). Geo-Mar. Lett. 20, 101–108.

Milia, A., Mirabile, L., Torrente, M.M., Dvorak, J.J., 1998. Volcanism offshore of Vesuvius volcano in Naples Bay. Bull. Volcanol. 59, 404–413.

Milia, A., Torrente, M.M., 1999. Tectonics and stratigraphic architecture of a pery-Tyrrhenian half-graben (Bay of Naples, Italy). Tectonophysics 315, 297–314.

Milia, A., Torrente, M.M., 2000. Fold uplift and syn-kinematic stratal architectures in a region of active transtensional tectonics and volcanism, Eastern Tyrrhenian Sea. Geol. Soc. Am. Bull. 112, 1531–1542.

Milia, A., Torrente, M.M., 2003. Late Quaternary Volcanism and transtensional tectonics in the Bay of Naples, Campanian continental margin, Italy. Mineral. Petrol. 79, 49–65.

Milia, A., Torrente, M.M., Russo, M., Zuppetta, A., 2003. Tectonics and crustal structure of the Campania continental margin: relationships with volcanism. Mineral. Petrol. 79, 33–47.

Mitchum, R.M., Vail, P., Sangree, J.B., 1977. Seismic stratigraphy and global changes of sea level, part 6: stratigraphic interpretation of seismic reflection patterns in depositional sequences. In: Payton, C.E. (Ed.), Seismic stratigraphy – Application to Hydrocarbon Exploration. Am. Assoc. Petrol. Geol., Memoir 26, 205–212.

Mitchum, R.M., Van Wagoner, J.C., 1991. High-frequency sequences and their stacking patterns: sequence-stratigraphic evidence of high-frequency eustatic cycles. Sediment. Geol. 70, 131–160.

Nunziata, C., Rapolla, A., 1981. Interpretation of gravity and magnetic data in the Phlegrean Fields geothermal area, Naples, Italy. J. Volcanol. Geotherm. Res. 9, 209–225.

Orsi, G., de Vita, S., Di Vito, M., 1996. The restless resurgent Campi Flegrei nested Caldera (Italy): constraints on its evolution and configuration. J. Volcanol. Geotherm. Res. 17, 273–288.

Patacca, E., Sartori, R., Scandone, P., 1990. Tyrrhenian Basin and Apennine Arcs: kinematic relations since late Tortonian times. Mem. Soc. Geol. It. 45, 425–451.

Posamentier, H.W., Jervey, M.T., Vail, P.R., 1988. Eustatic control on clastic deposition. I-Conceptual framework. In: Wilgus, C.K., Hastings, B.S., Posamentier, H.W., Van Wagoner, J., Ross, C.A., Kendall, C.G.C. (Eds), Sea Level Changes – An Integrated Approach. Soc. Econ. Paleontol. Mineral. Spec. Publ., Vol. 42, pp. 109–124.

Rolandi, G., Bellucci, F., Heizler, M.T., Belkin, H.E., De Vivo, B. 2003. Tectonic controls on the genesis of ignimbrites from the Campanian Volcanic Zone, southern Italy. Mineral. Petrol. 79, 3–31.

Rosi, M., Sbrana, A. (Eds), 1987. Phlegrean Fields. CNR Quad. Ric. Sci., 114, pp. 1–175.

Thorne, J.A., Swift, D.J.P., 1991. Sedimentation on continental margins, VI: a regime model for depositional sequences, their components systems tracts, and bounding surfaces. In: Swift, D.J.P., Oertel, G.F., Tillman, R.W., Thorne, J.A. (Eds), Shelf Sand and Sandstone Bodies: Geometry, Facies and Sequence Stratigraphy. Int. Assoc. Sediment. Spec. Publ., Vol. 14, pp. 189–255.

Turco, E., Schettino, A., Pierantoni, P.P., Santarelli, G., this volume. The Pleistocene extension of the Campania Plain in the framework of the southern Tyrrhenian tectonic evolution: morphotectonic analysis, kinematic model and implications for volcanism.

Volcanism in the Campania Plain:
Vesuvius, Campi Flegrei and Ignimbrites
edited by B. De Vivo

Chapter 4

Gravitational instability of submarine volcanoes offshore Campi Flegrei (Naples Bay, Italy)

A. Milia[a,*], *M.M. Torrente*[b] *and F. Giordano*[c]

[a]*IAMC,CNR, Calata Porta di Massa, Porto di Napoli, I-80100, Naples, Italy*
[b]*DSGA, University of Sannio, Via Portarsa 11, I-82100 Benevento, Italy*
[c]*University Parthenope, Via Acton 38, I-80133 Naples, Italy*

Abstract

Slope instability processes have sculpted numerous morphological features on the flanks of the submerged volcanoes in Naples Bay off Campi Flegrei. Geophysical data were used to define the time and spatial evolution of sediment failures. Four types of volcanic slopes were recognized: (1) highly inclined slopes with low-relief morphologic features resulting from shallow translational slump complexes; (2) highly inclined slopes with high-relief morphologic features associated with deep, rotational slump complexes; (3) highly inclined slopes corresponding to scars and slump deposition at the base on a subhorizontal surface; and (4) gently inclined slopes with a staircase morphology associated with shallow rotational slumps. All slumps occur immediately after the emplacement of the volcanoes with some of them remaining active for a long time after their formation. The volcanic landslides under study were characterized by different concomitant triggering factors (angle of slope margin, seismic activity, basement architecture, rapid sea-level changes, sea currents and high pore-fluid pressure) and featured volumes of up to 200 million m^3. The identification of recurrent events of flank instability affecting submarine volcanoes should be considered when evaluating the potential risk that tsunamis pose to the densely populated Naples Bay coast.

1. Introduction

It is only in the last two decades that subaerial and submarine volcanoes were recognized as having evolved over long periods of construction punctuated by short and sometimes violent destructive events. During such events, major segments of volcanic edifices may collapse catastrophically giving rise to slumps and debris avalanches (e.g. Siebert, 2002). These phenomena are largely recognized in the oceanic volcanoes, where it is commonly believed that destructive processes affect the subaerial part and terminate when the subaerial island is completely eroded to the point where it may sink below the wave base, whereas in the submarine part sediments accumulate with time at the expense of the volcano height.

The evacuation zone is commonly located in the emerged part of coastal or oceanic volcanoes, while the depositional zone occurs in the submarine part. An example of such a collapsing volcano is Stromboli (Eolian Islands), which over the past 13 ka has experienced four huge (in the order of 1 km^3) sector collapses, alternating with periods of growth, on its NW flank (Tibaldi, 2001).

*Corresponding author. *E-mail address:* alfonsa.milia@iamc.cnr.it (A. Milia).

Slumps and debris avalanches have been recognized off the flanks of oceanic volcanoes such as the Hawaiian islands. Slumps are of a relatively short duration and move episodically with a speed of approximately 10 m/y. They are generally associated with large earthquakes and are characterized by steep fronts, show terraces and move on slopes greater than 3°. Debris avalanches, the second type of mass transport, can flow, also upslope, for several hundred meters at gradients of less than 3–1.5° and leave long and hummocky deposits. They were recently recognized in Naples Bay offshore Vesuvius volcano (Milia et al., 2003).

The instability of coastal and oceanic volcanoes can also induce tsunamis. One such event was witnessed on Stromboli volcano on 30 December 2002. A subaerial landslide, that also involved the underwater flank of the volcanic edifice, occurred for a total volume of about 16 million m^3 (Chiocci et al., 2003). This instability event generated tsunami waves of up to 10 m in height and caused severe damage along the Stromboli coast, reaching the northern coast of Sicily about 50 km to the south.

Naples Bay is a pery-Tyrrhenian basin corresponding to a Quaternary half graben that bounds two important volcanic districts: Vesuvius and Campi Flegrei, the latter being made up of several monogenic volcanoes. The physiography of Naples Bay includes a relatively wide continental shelf with much of the materials erupted from the emerged and submerged volcanoes being stored in shallow water (Fig. 1).

Most research work on the growth and destabilization of submarine volcanoes concerns oceanic island volcanoes (e.g. Hawaiian and Canary Islands), whereas very few geomorphologic studies have dealt with volcanoes formed in a continental shelf environment. In order to contribute something new to this argument, we made a seismo-stratigraphic and geomorphologic analysis of the Late Quaternary succession of Naples Bay

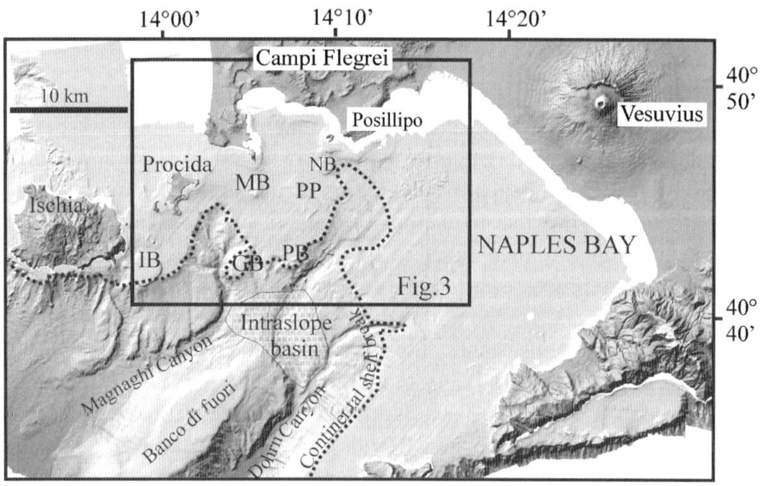

Figure 1. Physiographic map of the Campania margin displaying the volcanic districts of Campi Flegrei and Vesuvius and the submarine volcanoes of Ischia Bank (IB), Gaia Bank (GB), Pia Bank (PB), Penta Palummo Bank (PP), Nisida Bank (NB) and Miseno Bank (MB). The highlighted box shows the location of Figure 3. Seafloor bathymetry of Naples Bay after D'Argenio et al. (in press) and geomorphologic features after Milia (1999).

offshore Campi Flegrei and documented recurrent events of the flank instability of submarine volcanoes that should be considered when evaluating the risks of the densely populated Naples Bay coast.

2. Geological setting

The Campania continental margin, located between 40°N and 41°N latitude on the eastern Tyrrhenian Sea, is affected by intense volcanic activity. Late Quaternary eruptions gave rise to Campi Flegrei (e.g. Rosi and Sbrana, 1987), Somma-Vesuvius (e.g. Santacroce, 1987) and several volcanoes in Naples Bay (Milia and Torrente, 2003) (Fig. 1).

The oldest volcanoes recognized in Naples Bay (Unit V4 in Milia, 1996) have been correlated to the oldest volcanic products on Ischia, dated at 150 ka (Vezzoli, 1988) and to coeval volcanic deposits of the Campanian Plain (Rolandi et al., 2003). The Campania Ignimbrite eruption at 39 ka produced the most widespread (about 6000 km^2) and the largest (200 km^3 Dense Rock Equivalent) volcanic unit in the Campanian margin (Rolandi et al., 2003) that is present in the city of Naples and on the Sorrento coastal slope. In addition, the Campanian Ignimbrite was recognized on the continental shelf of the Naples Bay of Naples (e.g. Milia et al., 1998).

The Neapolitan Yellow Tuff (hereinafter NYT), dated at 15 ka (Deino et al., 2004), is one of the largest volcanic eruptions of the Campanian margin with an estimated total dense-rock equivalent volume of 49.3 km^3 and an inferred source in the eastern part of Campi Flegrei (Scarpati et al., 1993). The NYT forms a thick and widespread pyroclastic unit on the periphery of Campi Flegrei and reaches a thickness of approximately 150 m at Posillipo Hill (Guadagno, 1928). The stratigraphic analysis of deep holes located in the city of Naples documents that the NYT pyroclastic deposits are characterized by two superposed facies: a lower green facies, attributed to a submarine eruption, and an upper yellow one emplaced in a subaerial environment (D'Erasmo, 1931); thinning towards the east, the NYT overlies older pyroclastic deposits, including the Campania Ignimbrite and marine sediments. The NYT was also documented offshore Campi Flegrei, where it forms a wedge that thickens towards Posillipo Hill (Milia, 1998; Milia et al., 1998).

The continental shelf of Naples Bay extends to water depths of 180 m below sea level (Fig. 1). The northern sector is made up of an extensive system of banks of volcanic origin interlayered with volcaniclastic and marine sediments (Milia and Torrente, 2003). These banks, referred to as the Ischia Bank, Gaia Bank, Nisida Bank, Pia Bank, Penta Palummo Bank and Miseno Bank, form a continental shelf with an edge at a depth of about 140 m and an irregular steep slope dipping towards the south-southeast (Fig. 1). In the central sector, the continental shelf is 20 km wide, the upper slope has an average gradient of 3° and dips towards the west-northwest. An intraslope basin is present at the base of these slopes and is bounded to the southwest by the Banco di Fuori structural high. The continental slope of Naples Bay is cut by two canyons (Magnaghi Canyon and Dohrn Canyon) that cross the slope and terminate in the Tyrrhenian basin (Milia, 2000).

The Late Quaternary stratigraphic succession of Naples Bay is characterized by a complex architecture due to the presence of volcanic units and interlayered marine deposits. Depositional sequences and their systems tracts (Fall, Lowstand (LST), Transgressive (TST) and Highstand (HST)) were distinguished (Milia and Torrente, 2000, 2003).

Many Late Quaternary fault systems affect submarine volcanoes located in Naples Bay. NE-SW trending normal faults downthrow Gaia Bank and Pia Bank volcanoes towards the sea (Milia and Torrente, 2003). NW-SE and E-W trending faults, by contrast, affect Miseno Bank. The age of the latter faults is post-15 ka as these structures offset the prograding wedge overlying the NYT.

3. Materials

Naples Bay was investigated by means of multichannel seismic reflection profiles and single-channel seismic reflection profiles (Fig. 2). The single-channel profiles were acquired using: a 16 kJ Multispot Extended Array Sparker (MEAS) system in April 1989 and May 1990; a 1 kJ Surfboom system in September 1986; and a 0.2 kJ multi-electrode Sparker system in September 2000. All seismic sections were recorded graphically on continuous paper sheets with a vertical recording scale of 0.25 s for the multi-electrode Sparker, 0.25 and 0.5 s for the Surfboom and 1.0 and 2.0 s for the MEAS.

Ship positioning was determined using LORAN C for the MEAS, Micro-Fix Racall for the Surfboom (with a position accuracy of 1 m) and a differential GPS system (with a position accuracy of 1 m) for the multi-electrode Sparker. The best vertical resolution was approximately 6 m for the MEAS data and 1 m for the Surfboom and multi-electrode Sparker data. The multichannel profiles were acquired in 1988 using a double water gun and a 24-trace streamer. The maximum recorded length was 4.5 s of two-way travel time and the fold coverage was 12 and 24; the data processing sequence included a pre-stack deconvolution, velocity analysis, normal moveout (NMO) stacking and post-stack time migration. The multichannel profiles were characterized by a high signal–noise ratio and a best vertical resolution of 10 m. The geometry of the acquisition and the processing used for the restitution of the seismic reflection profiles are described by Mirabile et al. (1989).

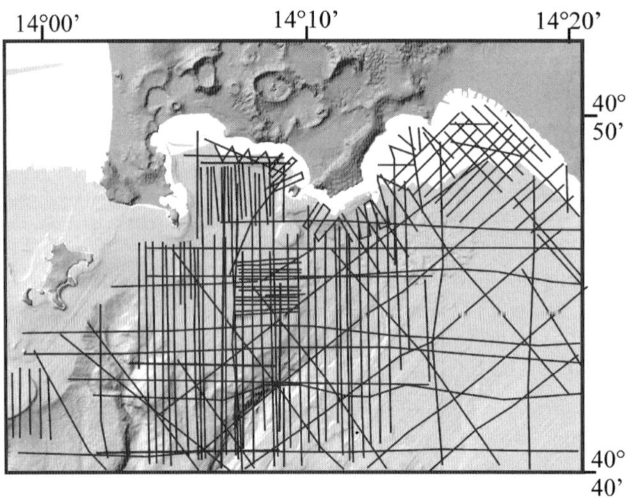

Figure 2. Index map of seismic reflection profiles.

4. Results

In the marine environment, slumping is commonly observed along the continental slopes dominated by terrigenous sediment fluxes and along the slope of oceanic volcanoes, where slump mass characteristics are observed to be different from the nature of the surrounding sediment (Dingle, 1977). Slump detection was based on a change of sediment reflection pattern from smooth parallel in the undisturbed area to chaotic and uneven topography in the slide areas.

Gravitational instability of submarine volcano flanks was detected in Naples Bay. In detail, slump scars, back tilted rotated blocks, mass flow and slump deposits were recognized (Fig. 3). A detailed description, from southern to northern volcanoes, of these instability features follows.

Gaia and Pia Bank volcanoes lie at the margin of the northern continental shelf of Naples Bay (Fig. 1). Gaia Bank was seen to be an isolated volcano with an elliptical shape and an approximately 3.0-km-long axis. The volcano top reaches a depth of 125 m and the southern flank (10–18°) of the volcano is very steep (Fig. 4). A seismic reflection profile reveals chaotic units characterized by numerous diffractions in the lower part and at the base of the slope. They are interpreted as being a slump accumulation zone that is overlain by a unit displaying low-amplitude and high-frequency reflectors corresponding to younger stratified sedimentary deposits. These mass transport deposits correspond to a rugged morphology in

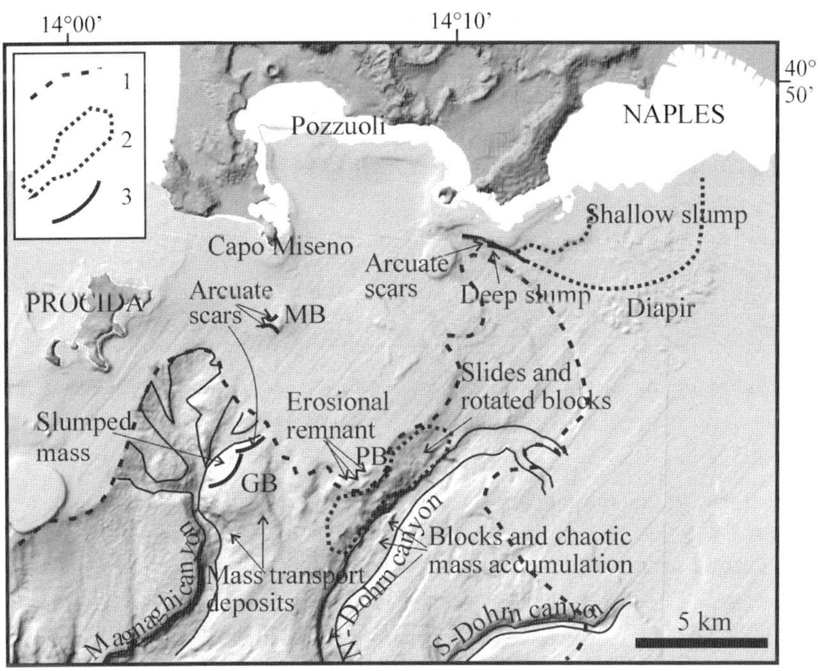

Figure 3. Bathymetric map displaying the main gravitational instabilities of the submerged volcanoes with morphologic expression (1) continental shelf break; (2) slide and slump area; and (3) slump scar. Seafloor bathymetry of Naples Bay after D'argenio et al. (in press).

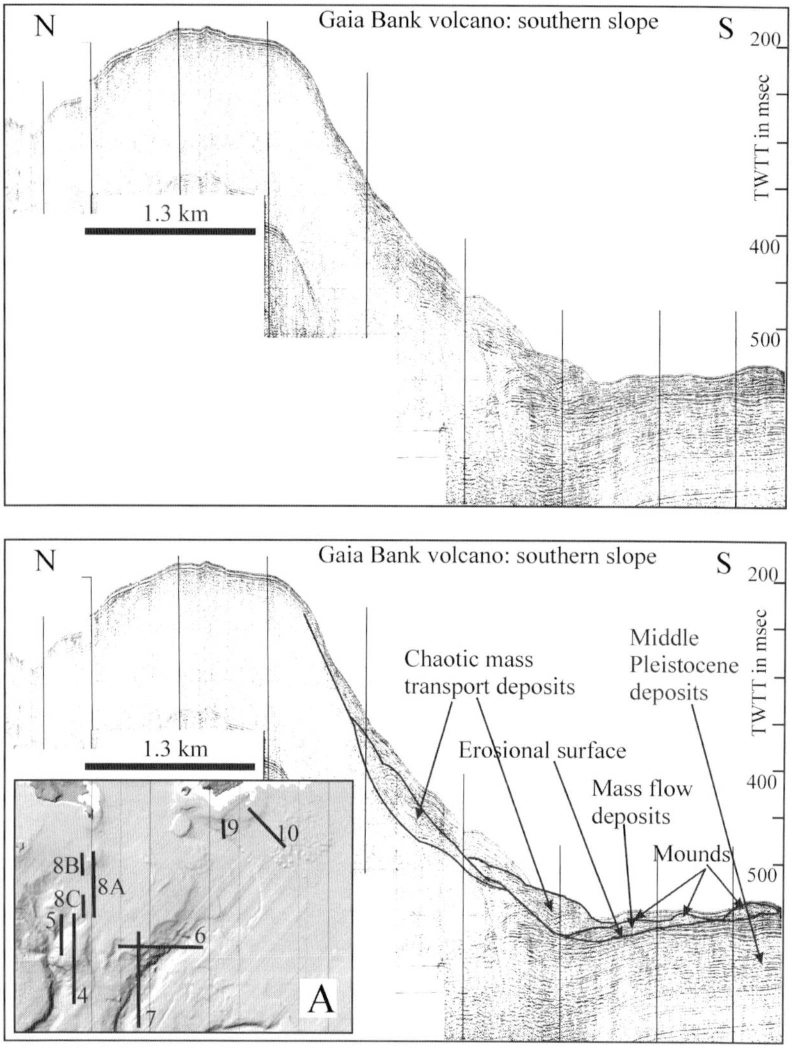

Figure 4. Surfboom seismic line across Gaia Bank volcano displaying a slumped mass along the southern slope. Inset 4A displays the location of the seismic lines presented in this paper and the corresponding figure numbers.

the southern part of Gaia Bank (Fig. 3). The northern flank of the volcano is also affected by a steep scarp (9°) located on the continental slope between depths of 200 and 300 m and features a chaotic seismic facies block unit, with a volume of approximately 3.8 million m^3, overlying a concave decollement surface (Fig. 5). This slump unit is draped by a 15-m-thick stratified unit. The seafloor map reveals arcuate scarps and an irregular morphology associated with a slumped mass northwestward of Gaia volcano (Fig. 3). Pia Bank volcano is near the continental shelf edge eastward of Gaia Bank (Fig. 1). It has a nearly circular shape and a diameter of approximately 2.3 km. The southern flank of this volcano corresponds to the continental slope with a gradient of 7° up to depths of 525 m. We imaged Pia volcano

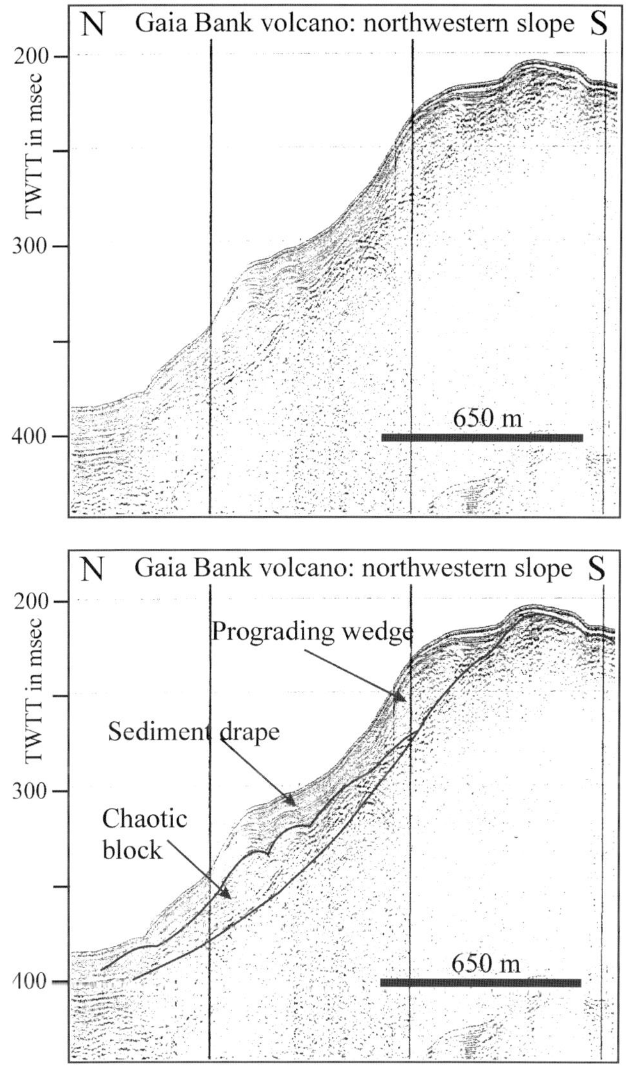

Figure 5. Surfboom seismic line across Gaia Bank volcano displaying a slump on the northwestern slope. See Figure 4A for location.

using two different seismic lines: a Sparker seismic profile characterized by a vertical exaggeration of 10:1 and a multichannel seismic profile displaying a vertical exaggeration of 2:1. Both seismic profiles (Figs. 6 and 7) clearly show evacuation zones in the upper part of the volcano and mounds and terraces bounded by steep slopes in the middle and lower part of the volcano. These mounds feature a morphologic relief of up to 20 m, an internal reflection-free seismic facies and a landward dipping side (towards the volcano). The overall features of these mounds can be associated with rotational slides and blocks. These slump complexes extend significantly into surrounding areas and develop characteristic amphitheatre-shaped terraces on the upper volcano flanks. Mass transport deposits with a

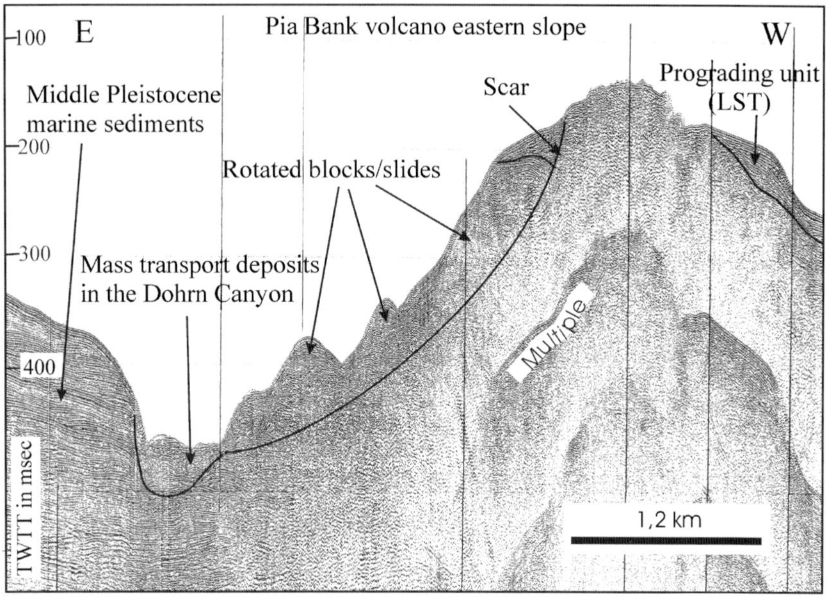

Figure 6. Sparker seismic line across Pia Bank volcano displaying the slumped mass along the eastern slope. See Figure 4A for location.

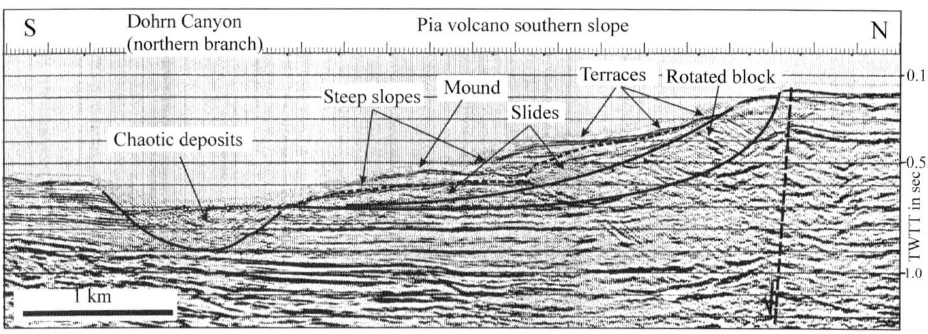

Figure 7. Multichannel seismic line across Pia Bank volcano displaying a staircase morphology associated with slumping on the southern slope. See Figure 4A for location.

chaotic seismic facies at the base of the slope partially fill the Dohrn Canyon (Fig. 6) and the diffractions at the top of these deposits indicate the occurrence of scattered blocks. The volume of these mass transport deposits is approximately 100 million m^3. The seafloor map displays (Fig. 3) three isolated remnants of the Pia volcanic cone reaching depths of 100 m: an area characterized by steep flanks with shallow failures (rotated blocks and slides); a small number of rotational slumps that give rise to elongated low-relief depressions; and scattered blocks within the Dohrn Canyon.

The Miseno Bank volcano is located on the continental shelf between Gaia Bank and Capo Miseno (Fig. 3). It has a nearly subcircular shape and a diameter of approximately 2 km. The volcano top lies at a depth of 60 m, whereas its base is located at a depth of 220 m (Fig. 8). This volcano is characterized by the presence of both a north- and south-facing arcuate scar that are partially filled by the sedimentary deposits of the last depositional sequence (Milia, 1997, 2000). The southern scar (Fig. 8B) features a dip of approximately 12° and is associated with a 40-m-thick chaotic seismic unit with a volume of approximately 200 million m^3 that accumulated far from the volcano edifice on a gently dipping surface (Fig. 8). This unit can be interpreted as being a mass movement that travelled up to 3.8 km. Both the scar and chaotic mass are buried by the last depositional sequence. In detail, the prograding wedge of the LST (Fig. 8C) lies at a depth of 130 m and covers the distal part of the volcano, whereas the TST, made up of a progradational unit arranged in backstepping with a topset at 60–75 m depth, covers both the volcanic unit and the LST (Fig. 8).

In the northern part of Naples Bay, the NYT wedge covers the continental shelf and thins gradually with a mean slope of 4.2° until a depth of −120 m and lower than 0.8° in the distal part. In the Southwest, by contrast, the NYT wedge and the underlying pyroclastic deposits exceed the shelf break and terminate on the continental slope with a mean dip of about 6°.

The NYT pyroclastic wedge displays a complex morphology due to slumps in the western and southern shelf off Posillipo Hill (Figs. 9–11).

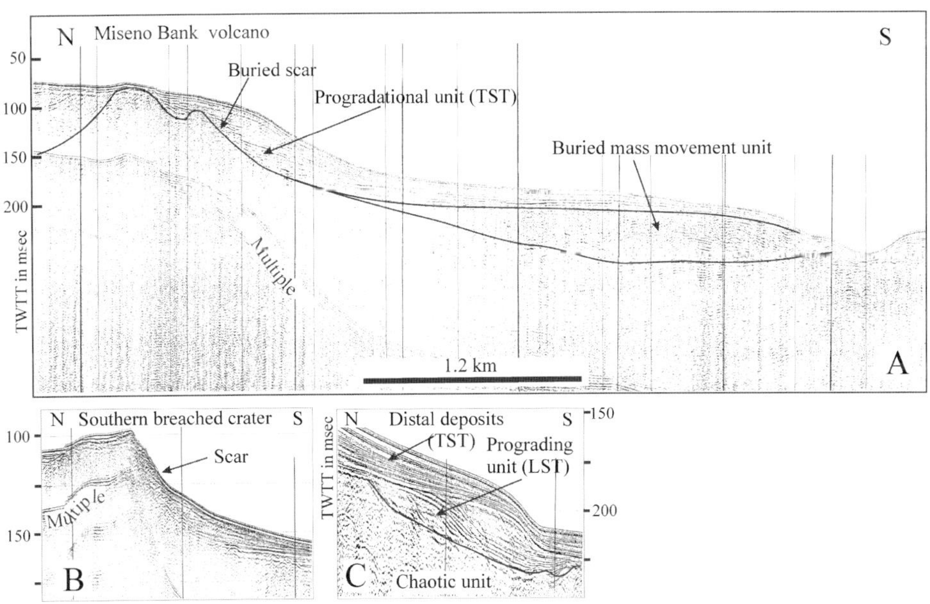

Figure 8. A surfboom seismic lines across Miseno Bank volcano displaying a slumped mass along its southern slope. Inset 8B shows the southern scar of a breached crater. Inset 8C shows the chaotic unit overlain by the Lowstand Systems Tract (LST) and the Transgressive System Tract (TST). See Figure 4A for location.

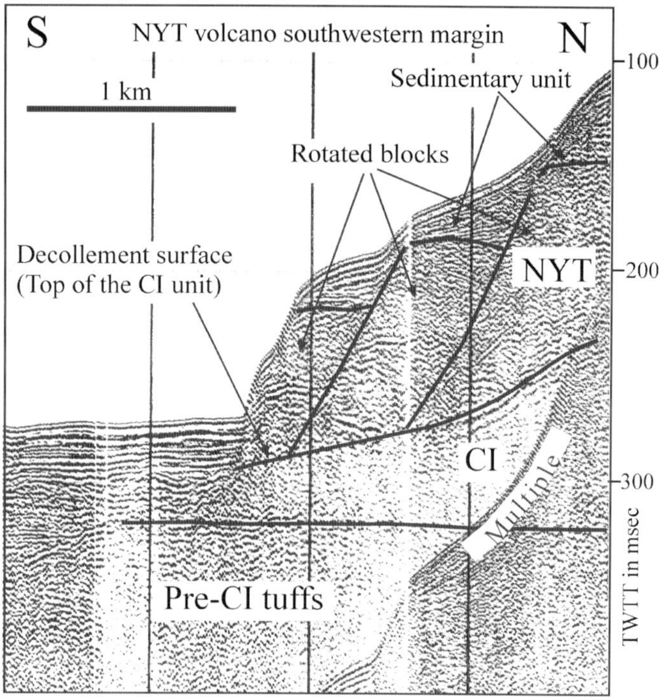

Figure 9. Sparker seismic line across the margin of the NYT volcano displaying a deep slump along the southwestern slope. Scars merge at the basal decollement surface. See Figure 4A for location.

In the western margin of the pyroclastic wedge, a thick slump was recognized on the basis of three subparallel listric scars, rotated blocks and an uneven morphology of the NYT top (Fig. 9). This slump is characterized by a basal decollement surface, extending between approximately 130 and 225 m of depth. It corresponds to the top of the Campania Ignimbrite unit that dips in the same direction as the slope. The associated slump scars are characterized by dips of up to 15° and an arcuate shape as can be seen on the map of Figure 11. This slump is overlain by a prograding seismic unit characterized by a toplap surface at a depth of 90 m. This gravitational feature corresponds to a thick slump characterized by a mean thickness of 80 m and a main scar position below −130 m.

In the southern margin of the pyroclastic wedge, by contrast, a thin slump characterized by a mean thickness of 18 m and a main scar position at −90 m occurs. Indeed, the NYT top presents an irregular surface terminating with a flat surface at approximately 90 m below the present sea level (Fig. 10). This flat surface is bounded by a morphologic step with a concave surface and rotated blocks with a reflection-free seismic facies along the slope. The top of the rotated blocks which overlie the main decollement surface is characterized by a mean dip of 11° in the upper part and is locally imaged by a strong reflector (Fig. 10). The slump interpretation is confirmed by the arched shape of the slump scars as seen in the map of Figure 11. The morphologic step, corresponding to the slump scar, is overlain by a prograding wedge characterized by a toplap surface at approximately −90 m of water depth. In particular, an angular unconformity is present above the distal sediments

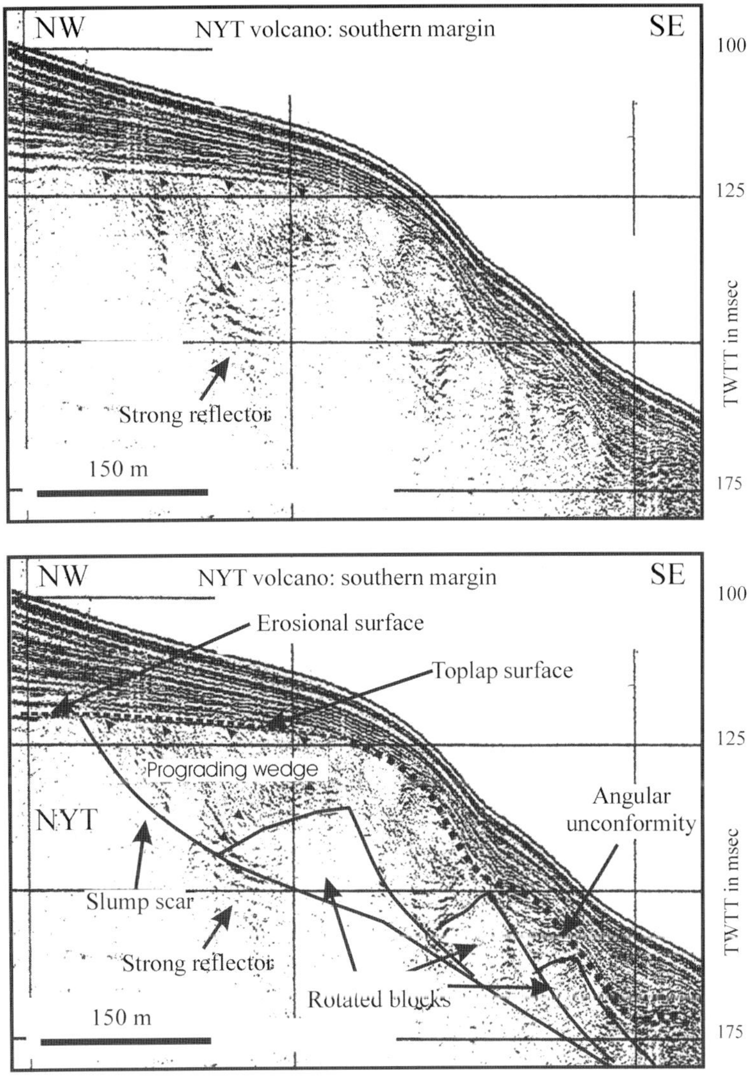

Figure 10. Surfboom seismic line and interpretation displaying a thin slump characterized by a scar and rotated blocks of NYT pyroclastic deposits covered by a prograding wedge. Note the angular unconformity at the top of the distal sediment of the prograding wedge. For seismic line location see Figure 4A.

of the prograding wedge that in turn covers the slump; this unconformity may reflect subsequent episodes of small-scale sliding of the blocks following the main event that caused the erosion of the topmost sediments that were successively covered in discordance.

Thin and thick slumps are covered by the prograding wedge corresponding to the oldest marine sediment deposited immediately after the emplacement of the NYT pyroclastic wedge (Figs. 9 and 10). In fact the position of the toplap surface (–90 m) of the prograding wedge corresponds to that of the sea level of 15 ka and consequently this unit can be attributed to the TST. According to this interpretation: the flat erosional surface is due to

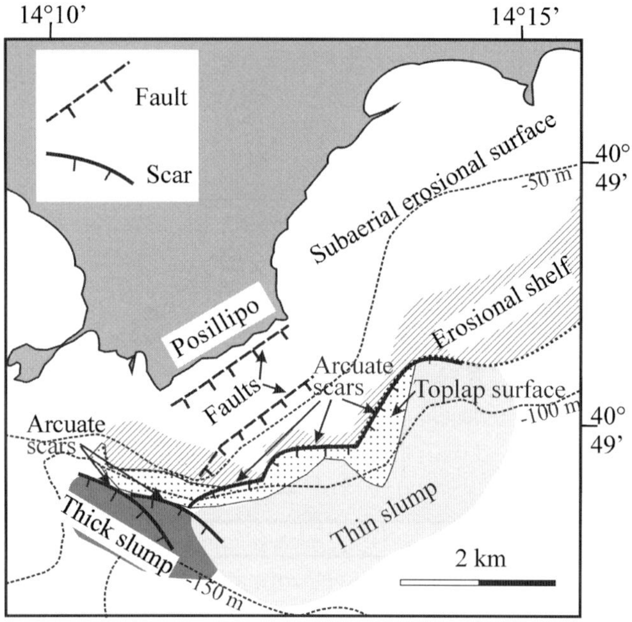

Figure 11. Paleogeographic map of the shelf off Posillipo Hill immediately after the emplacement of the NYT pyroclastic wedge.

the wave erosion along the newly created shoreface coastal zone; the morphologic step created by the slump promoted the deposition of eroded sediment below the sea level with the formation of a prograding wedge.

The resulting paleogeography of the region at the time of the NYT emplacement was characterized by three main areas off the coast of Posillipo Hill (Fig. 11): a proximal area affected by subaerial erosion; an intermediate area, corresponding to the shoreface, characterized by erosion and lateral deposition; and a distal area featuring a thick slump in the west and a thin slump in the south.

5. Discussion

The climatic cycles during the Quaternary affected all the continental margins with a stepwise fall in the eustatic sea level that culminated in the glacial maximum of oxygen isotope stages 2 and 6, and a minimum sea-level depth at 130 m followed by a much faster sea level rise during isotope stages 1 and 5 (Martinson et al., 1987).

As we mentioned earlier, according to Milia (1996), Gaia and Pia volcanoes correspond to 150 ka-old V4 unit. These volcanoes formed during isotopic stage 6 and instantaneously filled the accommodation space of a deep Pleistocene marine basin until the emersion of small volcanic islands up to 30 m high (Milia et al., this volume). In fact, the flat surface at the top of Gaia Bank and the erosional remnants at the top of Pia Bank can be explained by the marine erosion that affected the volcanoes during the glacial maximum. The Miseno Bank volcano, the top of which lies at a depth of 60 m, was mainly emerged and formed a

70-m-high volcanic island during the sea-level lowstand. This reconstruction is confirmed by the position of the LST that covers the volcano base.

Because the pyroclastic wedge of the NYT was emplaced on a continental shelf, it filled the accommodation space until its emersion formed a new coastal area. According to the eustatic curve reconstructed by Bard et al. (1990), the position of the sea level during the NYT eruption 15 ka was approximately 90 m below the present sea level. Therefore, immediately after this eruption the position of the sea level corresponded to both the flat erosional surface and the adjacent top of the prograding wedge (Figs. 10 and 11).

On the basis of the geomorphologic setting and slope angle the gravitational instabilities documented on the flanks of submerged volcanoes in Naples Bay can be subdivided into four types. The first type of instability corresponds to a submerged volcano flank that forms a very steep (7–18°) continental slope (e.g. Gaia and Pia volcanoes; Figs. 4–7). In this case, the slumped mass accumulated mainly in the middle and lower part of the slope. The second type of instability occurs on a mainly emerged volcano and is characterized by deep scars, an evacuation zone bounded by a concave surface and a slumped mass that travelled a few kilometres over a subhorizontal surface (e.g. Miseno Bank volcano; Fig. 8). The third and fourth types of instabilities affect a volcano that is partly emerged and partly submerged with slumps forming in the submarine environment (Fig. 11). In particular, the third type of instability involves the whole volcano flank and features subparallel deep scars bounding rotated blocks and a lateral displacement measuring up to a few hundred metres (e.g. NYT thick slump; Fig. 9), whereas the fourth type is characterized by a very shallow slump (e.g. NYT thin slump; Fig. 10).

Parameters that generally favour or trigger slope instabilities include: seismic activity, angle of slope margin, basement architecture, sea current and high pore-fluid pressure (Vendeville and Gaullier, 2003). The preferential development of the four types of volcano slope instability suggests different triggering processes.

Seismic activity is a parameter that could have favoured slope instability in the case under study. Indeed, recent works (Milia and Torrente, 2000, 2003) document late Quaternary activity of NE-SW, E-W and NW-SE faults that affected submarine volcanoes of Naples Bay.

In Naples Bay, large volumes of deposits associated with submarine volcanic eruptions were instantaneously emplaced and produced a rapid physiographic change (Milia et al., this volume). In particular, the flanks of both Gaia and Pia volcanoes form steep continental slopes that can be considered an important triggering factor of slope instability.

The third type of gravitational instabilities is influenced by basement architecture. Indeed, the steep scars affecting the whole wedge of the NYT are linked to the basement top that acted as a decollement surface. As a matter of fact, this pyroclastic wedge reposes on an inclined basement dipping in the same direction as the slope. This geometry allows the gravitational forces to act on the pyroclastic wedge favouring instability which leads to the formation of the thick slump.

The formation of the thin slump on the NYT pyroclastic wedge is probably triggered by sea current and high pore-fluid pressure. In fact, the main scars of the slump occur parallel to the coastal zone (Fig. 11) and is mainly influenced by such hydrodynamic conditions as current direction and strong wave action. The latter caused pore pressure changes induced by the cyclic and residual stress of storm waves.

6. Concluding remarks

We recognized, for the first time, Late Quaternary recurrent events of submarine volcano instability in both the shelf and slope of Naples Bay off the volcanic district of Campi Flegrei. These volcanic landslides were characterized by different concomitant triggering factors (angle of slope margin, fault activity, basement architecture, sea-level fluctuation, sea current and high pore-fluid pressure) and featured volumes of up to 200 million m^3. The occurrence of numerous faults displacing these submarine volcanoes suggests that faulting, probably associated with the deformation of the seafloor and seismic activity, can be considered as the main triggering factor in slump formation. In addition, slump formation was influenced by rapid sea-level changes: the building of the Gaia Bank and Pia Bank volcanoes occurred during the glacial maximum of the isotopic stage 6 and these volcanoes were successively affected by a rapid sea-level rise, during the isotopic stages 6 and 5, and a sea-level fall during the isotopic stages 5 and 2; the NYT was emplaced 15 ka BP and successively affected by the rapid sea-level rise of the last sea-level transgression that occurred between 15 ka and 6 ka BP.

In the case of any future large-volume landslide affecting a submarine volcano flank, a potential landslide-induced tsunami should be considered in the risk evaluation of the densely populated coastal area of Naples Bay.

Acknowledgements

F. Giordano made available the seismic data set. A. Milia and M.M. Torrente performed the geologic interpretation of seismic data and are responsible for the results and discussion paragraphs. Financial support was given by the Italian “Ministero dell’Università e della Ricerca Scientifica e Tecnologica” (FAR 2003, 2004, M. Torrente).

References

Bard, E., Hamelin, B., Fairbanks, R.G., Zindler, A., 1990. Calibration of the 14 C time scale over the past 30,000 years using mass spectrometric U-Th ages from Barbados corals. Nature 345, 405–410.

Chiocci, F.L., Bosman, A., Romagnoli, C., Tommasi, P., De Alteris, G., 2003. The december 2002 Sciaradel Fuoco (Stromboli island) submarine landslide: a first characterization. Geophys. Res. Abs. 5, 12069.

D’Argenio, B., Aiello, G., de Alteriis, G., Milia, A., Sacchi, M., Tonielli, R., Angelino, A., Budillon, F., Chiocci, F., Conforti, A., De Lauro, M., Di Martino, G., d’Isanto, C., Esposito, E., Ferraro, L., Innangi, S., Insinga, D., Iorio, M., Marsella, E., Molisso, F., Morra, V., Passaro, S., Pelosi, N., Porfido, S., Raspini, A., Ruggieri, S., Sarnacchiaro, G., Terranova, C., Vilardo, G., Violante, C., in press. Digital elevation model of the Naples Bay and adjacent areas, Eastern Tyrrhenian Sea. In: Pasquarè, G., Venturini, C. (Eds), Mapping Geology in Italy, APAT Dipartimento Difesa del Suolo-Servizio Geologico d’Italia, pp. 21–28.

Deino, A.L., Orsi, G., de Vita, S., Piochi, M., 2004. The age of the Neapolitan Yellow Tuff caldera-forming eruption (Campi Flegrei caldera-Italy) assessed by $^{39}Ar/^{40}Ar$ dating method. J. Volcanol. Geotherm. Res. 133, 157–170.

D’Erasmo, G., 1931. Studio geologico dei pozzi profondi della Campania. Boll. Soc. Nat. 43, 15–130.

Dingle, R.V., 1977. The anatomy of large submarine slump on sheared continental margin (SE Africa). J. Geol. Soc. Lond. 134, 293–310.

Guadagno, M., 1928. Il tufo giallo trachitico nel sottosuolo della città di Napoli. Atti Reale Istituto d’Incoraggiamento, Napoli, pp. 3–36.

Martinson, D.G., Pisias, N.G., Hays, J.D., Imbrie, J., Moore, T.C., Schackleton, N.J., 1987. Age dating and orbital theory of the Ice Ages: development of a high-resolution 0 to 300000 year chronostratigraphy. Quat. Res. 27, 1–29.

Milia, A., 1996. Evoluzione tettono-stratigrafica di un bacino peritirrenico. Ph.D. Thesis, University of Naples "Federico II" 184 p.

Milia, A., 1997. Attività di una faglia e variazioni laterali delle geometrie di un *system tract* nel Golfo di Napoli. Il Quat. 10, 461–464.

Milia, A., 1998. Le unità piroclastiche tardo-quaternarie nel Golfo di Napoli. Geogr. Fis. Dinam. Quat. 21, 147–153.

Milia, A., 1999. The geomorphology of Naples Bay continental shelf (Italy). Geogr. Fis. Dinam. Quat. 22, 73–78.

Milia, A., 2000. The Dohrn Canyon formation: a response to the eustatic fall and tectonic uplift of the outer shelf (Eastern Tyrrhenian Sea margin, Italy). Geo-Mar. Lett. 20, 101–108.

Milia, A., Mirabile, L., Torrente, M.M., Dvorak, J.J., 1998. Volcanism offshore of Vesuvius volcano in Naples Bay. Bull. Volcanol. 59, 404–413.

Milia, A., Torrente, M.M., 2000. Fold uplift and syn-kinematic stratal architectures in a region of active transtensional tectonics and volcanism, Eastern Tyrrhenian Sea. Geol. Soc. Am. Bull. 112, 1531–1542.

Milia, A., Torrente, M.M., 2003. Late Quaternary volcanism and transtensional tectonics in the Bay of Naples, Campanian continental margin, Italy. Mineral. Petrol. 79, 49–65.

Milia, A., Torrente, M.M., Giordano, F., Mirabile, L., this volume. Rapid changes of the accommodation space in the Late Quaternary succession of Naples Bay, Italy: the influence of volcanism and tectonics.

Milia, A., Torrente, M.M., Zuppetta, A., 2003. Offshore debris avalanches at Somma-Vesuvius volcano (Italy): implications for hazard evaluation. J. Geol. Soc. Lond. 160, 309–317.

Mirabile, L., Nicolich, R., Piermattei, R., Ranieri, G., 1989. Identificazione delle strutture tettono-vulcaniche dell'area flegrea: sismica multicanale nel Golfo di Pozzuoli. Atti dell' VIII Convegno GNGTS, pp. 829–838.

Rolandi, G., Bellucci, F., Heizler, M.T., Belkin, H.E., De Vivo, B., 2003. Tectonic controls on the genesis of ignimbrites from the Campanian Volcanic Zone, southern Italy. Mineral. Petrol. 79, 3–31.

Rosi, M., Sbrana, A. (Eds), 1987. Phlegrean Fields. CNR Quad. Ric. Sci. 114, 1–175.

Santacroce, R. (Ed.), 1987. Somma-Vesuvius. CNR Quad. Ric. Sci. 114, 1–251.

Scarpati, C., Cole, P., Perrotta, A., 1993. The Neapolitan Yellow Tuff – a large volume multiphase eruption from Campi Flegrei, Southern Italy. Bull. Volcanol. 55, 343–356.

Siebert, L., 2002. Landslides resulting from structural failure of volcanoes. In: Evans, S.G., DeGraff, J.W. (Eds), Catastrophic Landslides: Effects, Occurrence and Mechanisms. Geol. Soc. Am. Rev. Eng. Geol. 15, 209–235.

Tibaldi, A., 2001. Multiple sector collapses at Stromboli volcano, Italy: how they work? Bull. Volcanol. 63, 112–125.

Vendeville, B.C., Gaullier, V., 2003. Role of pore-fluid pressures and slope angle in triggering submarine mass movements: natural examples and pilot experimental models. In: Locat, J., Mienert, J. (Eds), Submarine Mass Movements and their Consequences. Kluwer, Dordrecht, pp. 137–144.

Vezzoli, L. (Ed.), 1988. Island of Ischia. CNR Quad. Ric. Sci. 114, 1–133

Volcanism in the Campania Plain:
Vesuvius, Campi Flegrei and Ignimbrites
edited by B. De Vivo

Chapter 5

The Campi Flegrei caldera boundary in the city of Naples

Annamaria Perrotta[a], *Claudio Scarpati*[a,*], *Giuseppe Luongo*[a] *and Vincenzo Morra*[b]

[a]*Dipartimento di Scienze della Terra, Università degli Studi di Napoli Federico II, Largo San Marcellino, 10, 80138, Napoli, Italy*
[b]*Dipartimento di Scienze della Terra, Università degli Studi Federico II, via Mezzocannone 8, 80138-Napoli, Italy*

Abstract

The Campanian Ignimbrite caldera occupies the Campi Flegrei region and part of the city of Naples. The previous caldera boundary throughout the northern periphery of Naples was merely inferred due to the lack of outcrops of proximal deposits associated with the Campanian Ignimbrite. The exact location of this important structural feature within the city of Naples is fundamental for the reconstruction of the volcanic evolution and hazard implications. New exposures and subsurface constraints reveal thick welded and lithic-rich successions overlying several monogenetic volcanoes. These proximal deposits are associated with the Campanian Ignimbrite and allow a better localization of the caldera boundary well inside the city of Naples, 2 km south from the previous limit. The caldera rim in this sector partially coincides with a vent alignment that represents a structurally weak zone through which the caldera collapse occurred. The minor displacement (few tens of metres) of the top of the sedimentary succession, beneath the volcanic sequence near the caldera rim compared with 3 km displacement of the top of the sedimentary succession in the central part of the caldera suggests the presence of a complex geometry of the caldera floor, which shows a piecemeal-like structure characterized by deeper blocks at the centre and shallower blocks to the sides.

1. Introduction

The Campi Flegrei caldera was first proposed by Rittmann (1950), who related this structure to the emplacement of the Grey Tuff (later named Campanian Ignimbrite). Rittmann suggested that the Campi Flegrei volcanic field was formed as a result of the collapse of an old stratovolcano, the Archiphlegrean volcano, largely sunk during the Grey Tuff eruption. The remnants of this old volcanic edifice were never recognized and, on the contrary, geological evidence shows that the pre-caldera activity was dominated by numerous explosive and effusive monogenetic centres (Rosi and Sbrana, 1987; Perrotta and Scarpati, 1994; Orsi et al., 1996). Cole et al. (1994) suggested the existence, prior to Campanian Ignimbrite eruption, of an ancient volcanic field larger than the present day Campi Flegrei that encompassed the city of Naples. Rittmann's boundaries of the Campi Flegrei caldera were re-proposed by Rosi and Sbrana (1987) on a new geological map of the Campi Flegrei area. Following the Druitt and

*Corresponding author. Fax: +39-081-5527631. *E-mail address:* claudio.scarpati@unina.it (C. Scarpati).

Sparks model (1982) related to the co-ignimbrite breccia, they identified the Piperno-Breccia Museo as the coarse and welded proximal facies of the Campanian Ignimbrite exposed along the caldera rim. Owing to the lack of lithic breccia deposits inside the city of Naples, they proposed that the eastern limit of the caldera lay in the Montesanto area (western Naples) on the basis of a welded ash layer described in an old excavation by Johnston-Lavis (1888). Geophysical investigations of the Campi Flegrei allowed Lirer et al. (1987) and Scandone et al. (1991) to re-interpret the caldera rim as the product of a younger explosive event that occurred 15 ka (Deino et al., 2004; Insinga et al., 2004), the Neapolitan Yellow Tuff eruption; while Barberi et al. (1991) suggested that the presence of three nested calderas related respectively with the Campanian Ignimbrite, the Neapolitan Yellow Tuffs and the emplacement of recent vents. Scarpati et al. (1993) illustrated that the caldera rim proposed by Rittmann (1950) cannot be related with the Neapolitan Yellow Tuff because pyroclastic sequences occurring beneath this formation overlay this structure. These authors identify an inner caldera rim related to the Neapolitan Yellow Tuff, largely buried under the products of younger eruptions. Orsi et al. (1996) have also recognized the presence of a nested structure resulting from two main collapses, the older and outer related to the Campanian Ignimbrite eruption. They included all the city of Naples in this larger caldera considering the Camaldoli-Poggioreale alignment, a scarp formed by a NE-SW trending fault related to the caldera collapse. Finally, De Vivo et al. (2001) and Rolandi et al. (2003) claim that the Campanian Ignimbrite eruption could be related to fissures activated along neotectonic Appennine faults.

Therefore, volcanological, geophysical and drill-hole data show a still controversial configuration of the Campi Flegrei caldera, the precise knowledge of which is fundamental for the reconstruction of the volcanic evolution and consequently for the volcanic hazard assessment of a highly populated urban area. The aim of this paper is to better define the caldera geometry inside the city of Naples on the basis of new field observations and a significantly revised stratigraphy.

2. Stratigraphy

In order to unravel the geology inside a large city such as Naples, it is necessary to understand the relationship between stratigraphy and structural features. This was reviewed by Cole et al. (1994), but later studies require a more updated analysis. We retain here some descriptions (Parco Margherita, Parco Grifeo and Funicolare di Chiaia volcanoes) made by Cole et al. (1994), while most of the presented stratigraphy is based on new outcrops and boreholes (Figs. 1 and 2). Finally, we address here only those details necessary for the purposes of this paper.

2.1. Pre-caldera deposits

The base of the volcanic sequence in the city of Naples crops out in few discrete places, along the Vomero and Capodimonte hills seaward sides, possibly in consequence of the Holocene denudation of these sides (Fig. 1a,b). The oldest volcanic sequence is composed of both pyroclastic deposits and lavas separated by paleosols. A scoriaceous lava flow was exposed during a building excavation in the Chiaia area (Scherillo, 1957) and represents the older volcanic product outcropping in Naples. Overlying this lava is a

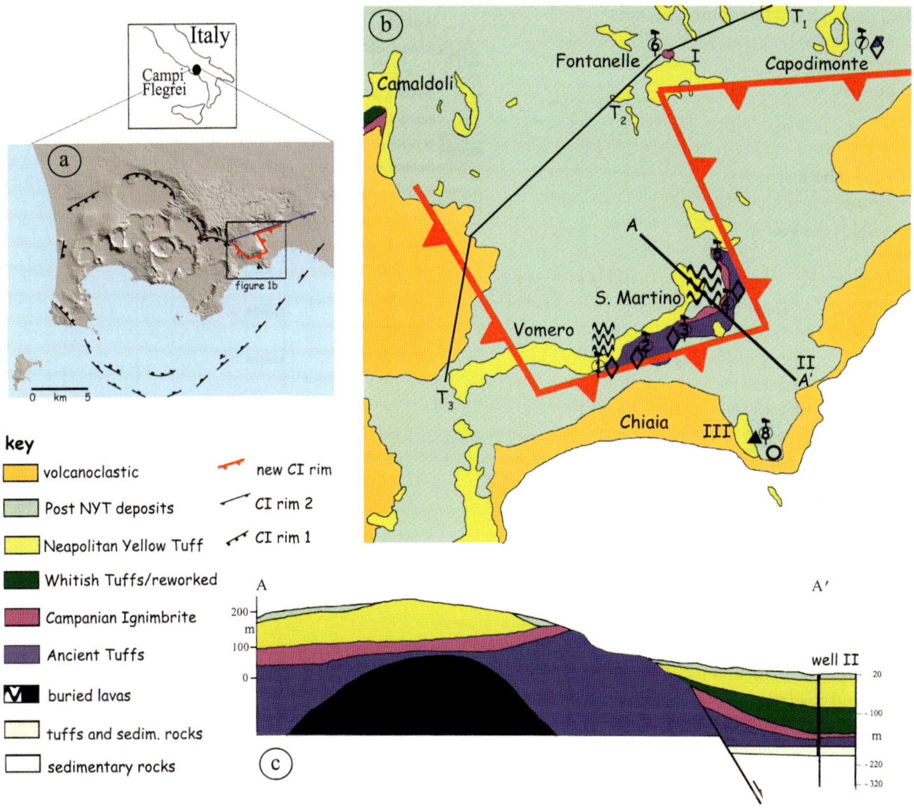

Figure 1. (a) Shaded relief of the Neapolitan region showing the postulated rim of the Campanian Ignimbrite caldera. Rim 1 was proposed by Rosi and Sbrana (1987) and Barberi et al. (1991); rim 2 was proposed by Orsi et al. (1996) who traced the Camaldoli-Poggioreale alignment as northeastern boundary of the caldera (blue rim). Box highlights the new caldera boundary in the area enlarged in Figure 1b. (b) Geological map of the study area with the inferred boundary of the Campanian Ignimbrite caldera within the city of Naples. Hammers represent the location of the stratigraphic sections reported in Figure 2a. Roman numbers: drill hole locations; diamonds: vent locations older than Campanian Ignimbrite; triangle: vent location older than Neapolitan Yellow Tuff; circle: vent location younger than Neapolitan Yellow Tuff. Thin black line shows the trace of the water gallery: from T_1 to T_2 welded grey tuff and lithic breccia, from T_2 to T_3 Neapolitan Yellow Tuff; (c) Geological cross-section through the study area based on surface and subsurface geological data (the location of the section and borehole II are reported in Fig. 1b).

sequence of coarse and ballistic-rich, lithified pyroclastic deposits that represent the remnants of monogenetic volcanoes. Parco Margherita tuff cone, a thinly bedded pyroclastic sequence of ash layers intercalated with coarser, poorly sorted ash and lapilli layers, more than 6 m thick, rests on this lava flow (Scherillo, 1957; Cole et al., 1994). A very close source to the southeast was proposed by Cole et al. (1994), who observed impact sags produced by large ballistic lithic blocks. Parco Margherita products are overlain by the Parco Grifeo volcano deposit, a yellow stratified tuff showing syn-depositional erosional surfaces filled with coarse pumice. Large lithic blocks up to 1.8 m in size occur in massive beds, while planar and sand wave bedding form fine-grained beds. The abundance

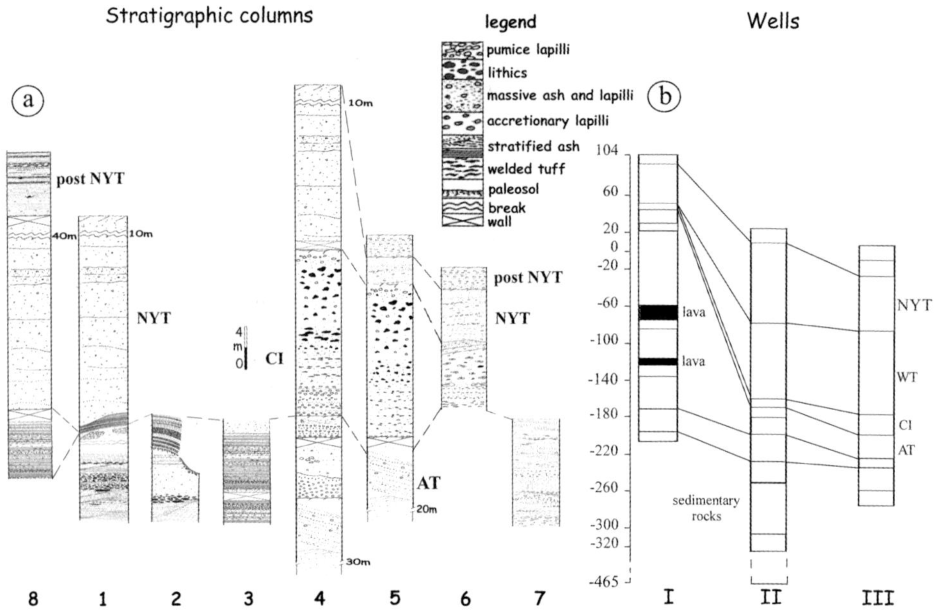

Figure 2. (a) Measured stratigraphic sections showing the volcanic deposits outcropping in the city of Naples. (b) Stratigraphic constraints of boreholes I, II and III constructed from lithological data reported in D'Erasmo (1931) and Società dell'Acquedotto di Napoli (unpublished). NYT: Neapolitan Yellow Tuff, WT: Whitish Tuffs; CI: Campanian Ignimbrite, AT: Ancient Tuffs. Numbers refer to locations shown in Figure 1b.

of coarse lithic blocks suggests that this tuff is possibly the remnants of the wall of a tuff cone with a vent to the south (Cole et al., 1994). The products of the Funicolare di Chiaia volcano rest on a strong erosive unconformity with a well-developed paleosol upon the Parco Grifeo volcano. They consist of stratified ash layers with accretionary lapilli and coarser ash and lapilli beds that retain their thickness laterally. The stratified tuff of S. Sepolcro volcano is east of Parco Grifeo volcano. A steep exposure, more then 30 m thick, shows a yellow stratified tuff dipping west. The deposit is composed of thin parallel beds, with rare cross-stratification, of fine ash with scattered rounded lithic fragments and accretionary lapilli. No overlap is seen between this tuff and the Parco Grifeo volcano; nevertheless, the temporal progression from west to east for the other three cones suggests that the S. Sepolcro edifice is the youngest of this WSW-ENE alignment. Two kilometres northeast of S. Sepolcro volcano a small remnant of a fifth edifice, the Capodimonte volcano crops out. This volcanic centre is composed of a stratified tuff dipping NNW; the lower part of the outcropping succession is made up by undulating thin ash and fine lapilli layers with a basal, 50 cm thick, coarse pomiceous blocks bed. This whole sequence is covered by a 5 m thick part showing dunes whose wavelength and amplitude are 2 and 0.3 m, respectively; they are formed of alternating layers of fine ash with accretionary lapilli and coarse lithic lens in an ash matrix. Numerous coarse juvenile and lithic bombs deform the succession at different stratigraphic heights suggesting a very close vent (Fig. 3a).

In the S. Martino area the basal monogenetic volcanoes are draped by three coarse stratified, well sorted, pumice lapilli beds. Paleosols and reworked materials separate these

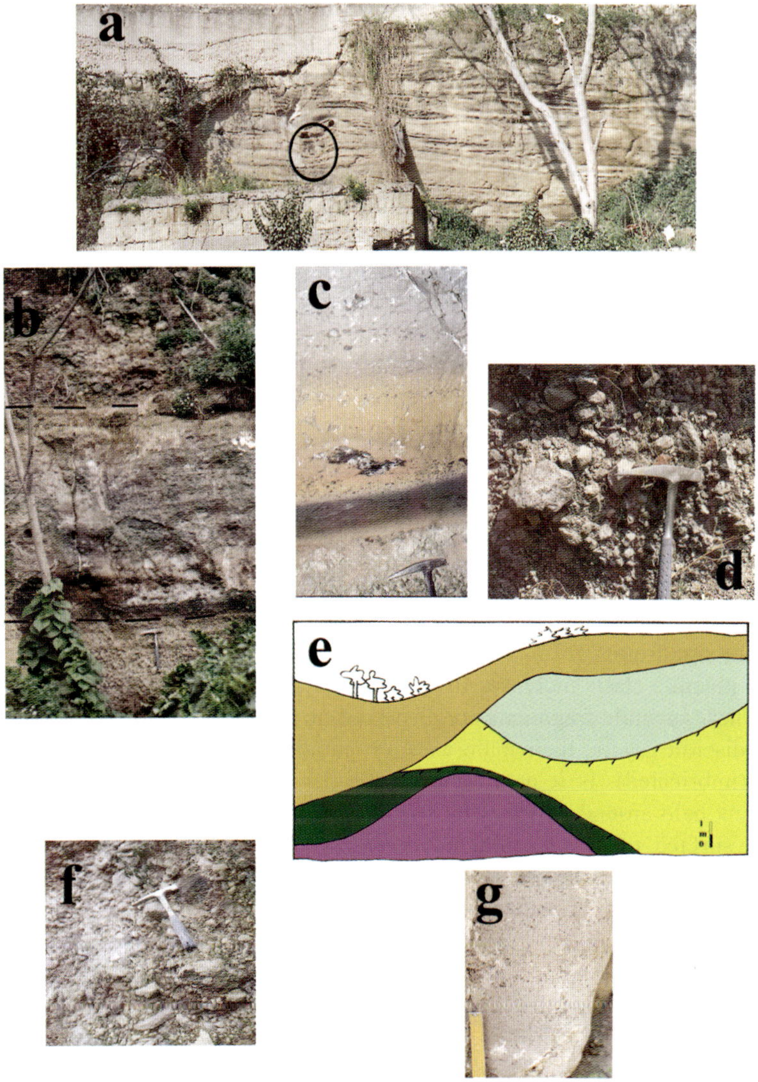

Figure 3. (a) Panorama view of the Capodimonte tuff. Large clasts have deformed into the underlying finer-grained beds on impact; (b) proximal Campanian Ignimbrite deposits at S. Martino. From the base: pumice lapilli fall deposit, welded ash deposit (piperno) and coarse lithic breccia; (c) locally, between the basal lapilli pumice deposit and piperno is present a stratified and incoherent ash deposit that changes in colour upwards; (d) closer view of the clast-supported lithic breccia deposit at S. Martino; (e) schematic illustration of the unconformities between Campanian Ignimbrite proximal deposits and the main post-caldera products at Montesanto. Colours legend as in Figure 1c; (f) closer view of the lithic-rich breccia at Montesanto; (g) grey welded tuff at Fontanelle. Locations are shown in Figure 1b.

beds. Thick ash beds with coarse, rounded pumice clasts rest on erosional surfaces in both the lower and the upper pumice lapilli beds. The name "Ancient Tuffs" is retained here for this sequence. The stratigraphic position of these tuffs is below the Campanian Ignimbrite deposits and not above them as considered by Orsi et al. (1996).

2.2. Caldera-forming deposits

The products of the Campanian Ignimbrite caldera-forming eruption crop out in three localities in the central part of the city of Naples: S. Martino hill, Montesanto and Fontanelle. The most complete sequence is that of S. Martino hill (previously studied by Rolandi et al., 2003), where a plinian pumiceous fall deposit is overlain by a stratified sequence representing the proximal facies of the Campanian Ignimbrite (Fig. 3b). The basal coarse pumice lapilli bed is 1 m thick and is eroded by the overlying welded ignimbrite. Pumice clasts are light grey in colour, well vesiculated and show aphyric to slightly porphyritic textures. Based on internal structures, textures and components five units are identified throughout the ignimbritic sequence. The lowermost unit, up to 40 cm thick, is a stratified and incoherent ash deposit that changes in colour upwards from pink to brown, to dark grey (Fig. 3c). Single layers range in thickness from 3 to 26 cm and are laterally discontinuous. Variable amounts of rounded grey pumice lapilli are dispersed within these layers. A matrix-supported lens of rounded scoriaceous fragments occurs locally. The overlying welded unit (Piperno), 2 m thick, consists of a fine-grained matrix with dispersed flattened juvenile fiamme (Fig. 3b). It is stratified by change in colour from yellowish at the base, to grey-purple to dark grey that grade into each other (Fig. 3c). Welding is more pronounced in the central part, decreasing towards base and top. This unit possesses an eutaxitic fabric, the height/width ratio of deformed juvenile pyroclasts range from a 1:3–1:5 at base to 1:6–1:7 in the central part. The mean diameter of the juvenile clasts increases from few millimetres to several centimetres towards the top. These juvenile fragments are dispersed throughout the unit and locally concentrated in discrete layers; their main axes are parallel to the stratification but some are inclined (imbricated). Is it noteworthy that a large fiamma, 28 cm large, shows an intense pink halo around it, 5 cm thick (Fig. 3c). Above this unit, separated by a sharp or erosive surface, there is a lithic, incoherent, breccia deposit 5 m thick (Fig. 3d). This clast-supported deposit is massive or, locally, inversely graded. The lithic clasts range in shape from rounded to angular and have a variety of compositions (e.g. trachytic and leucite lavas, tuff fragments, obsidians, sedimentary clasts). Johnston-Lavis (1888, 1889) named this deposit “museum breccia” to describe the great variety of rock types. A grey deposit consisting mostly of coarse and sintered spatter clasts is locally interlayered in the lower part of the lithic breccia. The spatter unit, up to 3 m thick is laterally discontinuous and seems to fill narrow channels. In most localities the spatter unit is absent and the uppermost breccia unit grades directly into the incoherent upper part of the welded unit. The deposit consists of coarse spatter clasts and a scarce fine-grained matrix. Spatter clasts, up to 40 cm in diameter, are welded and deformed. The uppermost unit, >1.5 m thick, is a weakly lithified deposit with an ash to coarse-ash reddish matrix containing a large fraction of juvenile material. The juvenile content consists of abundant rounded grey scoria clasts, obsidians and rounded pumice clasts, these latter forming discontinuous lenses confined towards the top of the unit. Lithic fragments are scattered throughout the bed. The lithified unit is capped by a thick and reddish paleosol. A few tens of metres from the main outcrop of S. Martino, in a wine-cellar along the Pedemontina alley, we have found a grey welded tuff, 5 m thick, with reverse graded, black scoriae, embedded in an ashy matrix with subordinate lithics and crystals. The contacts at the base and top are not visible. The scoriae are slightly flattened towards the base and equant at the top (maximum diameter 20 cm), where the matrix shows a reddish colour.

At Montesanto, near the tunnel excavation described by Johnston-Lavis (1888), in a cellar and in the overhanging parking-lot a grey tuff crops out, >6 m thick, rich in reverse-graded scoria clasts up to 20 cm in diameter. The overlying, incoherent, clast-supported breccia is 3 m thick (Fig. 3e). The breccia is made up of lithic blocks rounded to subangular shapes and up to 50 cm in diameter (Fig. 3f).

A similar succession crops out in the Fontanelle area (Fig. 1b), where a grey welded tuff (Fig. 3g), >3 m thick, is overlain by a lithic breccia capped by a thick paleosol. The grey tuff is crudely stratified due to variation in concentration of scoria fragments. Towards the base, flattened and imbricated fiamme, up to 17 cm in diameter, are present. In the upper part of this deposit, large, rounded, lithic clasts up to 75 cm in diameter occur. Above there is a 3 m thick, incoherent lithic breccia. The deposit is fines-poor and the matrix is reddish in colour. Rare lapilli to block scoria clasts, up to 20 cm in diameter, are dispersed in the matrix; the shape of the lithic clasts range from rounded to subangular.

2.3. Post-caldera deposits

Above the caldera-forming deposits lies, with strong unconformity, the products of the Neapolitan Yellow Tuff (Fig. 3e), up to 50 m thick, dated 15 ka (Deino et al., 2004; Insinga et al., 2004). The Neapolitan Yellow Tuff eruption resulted in the formation of a caldera, 10 km in diameter, which is now largely buried by the products of more recent activity. In this formation, two members have been distinguished (A and B from bottom to top; Scarpati et al., 1993). Member A is made up of stratified ash and pumice lapilli layers; the thinly stratified basal ash fall (unit A1) is a marker horizon. Member B is coarser and thicker than Member A. Several different lithofacies have been identified within this member: a massive valley-ponded facies, inverse-graded facies, regressive sand wave facies, stratified facies, particle aggregate facies, and vesicular ash facies (Cole and Scarpati, 1993). The Neapolitan Yellow Tuff occurs as lithified and non-lithified facies (de'Gennaro et al., 2000), the first has a yellow colour whereas the latter is grey. The lithified facies is closer to the vent (located in the western part of the city of Naples; Scarpati et al., 1993) than the unlithified. East of Chiaia, in a water reservoir drilled in the Roman time, is exposed a stratified tuff completely buried by the Neapolitan Yellow Tuff. This deposit, more than 6 m thick, dips 15°. The sequence is made up of thinly bedded ash layers intercalated with thicker, poorly sorted ash and lapilli layers. Many large rounded juvenile blocks, up to 20 cm across, are dispersed in the thicker layers. Some coarse pumice clasts, greater than 30 cm in size are ballistically emplaced. In the upper part of this deposit there are fractures filled with fragments of tuff. These angular fragments, up to 1 m in diameter, form a 4 m thick bed above the stratified tuff. This proximal sequence represents the remnants of a volcanic centre, the Chiatamone volcano, overlain by a dislodged mass of tuff coherently slid downslope. Locally, angular pumice lapilli beds, interbedded with ash layers or a poorly sorted, massive ash deposit with grey pumice lapilli, 40 cm thick, are found beneath the Neapolitan Yellow Tuff. Above the Neapolitan Yellow Tuff a 1–3 m thick stratified deposit composed of pumice lapilli beds interbedded with thin ash layers is exposed throughout the study area (Fig. 3e). It represents the product as one of the largest eruption of the Campi Flegrei (Pomici Principali), which occurred 11 ka.

2.4. Boreholes

Subsurface constraints on the structure of the Campanian Ignimbrite caldera are provided by three deep boreholes (D'Erasmo, 1931; Società dell'Acquedotto di Napoli, unpublished) located in the Fontanelle and Chiaia areas (Fig. 1b). A 5 km tunnel beneath Capodimonte and Vomero hills provides additional constraints (Fig. 1b). The drillings were performed for hydrological scope and their lithological description is presented here together with a review of the stratigraphy. Borehole I (Fig. 2b) is 310 m deep at 104 m altitude and encounters different lithologies. Near the surface are reworked materials that cover a 40 m thick Neapolitan Yellow Tuff sequence. Below the Neapolitan Yellow Tuff is present a grey tuff, which can be associated to the Campanian Ignimbrite eruption and then a yellow tuff overlying a 200 m thick sequence of loose pyroclastic deposits with minor lava horizons possibly related to the Ancient Tuffs. The Ancient Tuffs cover a tephra deposit interbedded with sandstone layers. The lowermost materials are of sedimentary nature and described as clay with fossils. Boreholes II and III were drilled by order of the king of Naples, Ferdinando II in the 1859; the successions were later examined and described by De Lorenzo (1904) and D'Erasmo (1931). They are located in the royal palace (II) and in a nearby square (III) at an altitude of 20 m and 4 m asl and a depth of 465 and 280 m, respectively. They exhibit the same lithologies with only minor variation in thickness of some stratigraphic horizons. The lowermost materials are clay, sandstone and marl, more than 100 m thick. The top of this sedimentary basement ranges between 330 and 340 m. Above this is a tuff interbedded with clay. Overlying are yellow to reddish tuffs possibly related to the Ancient Tuffs, less than 30 m thick, and then a grey tuff associated to the Campanian Ignimbrite. A 100 m thick sequence of unlithified ash with pumice is present above the Campanian Ignimbrite. This succession is thicker than the stratigraphically equivalent Whitish Tuffs, vented in the Camaldoli area, and consequently we suggest that it represents the accumulation of remobilized pyroclasts from the neighbouring high ground (see below for discussion). This succession is covered by 60–80 m of Neapolitan Yellow Tuff. The topmost products are loose pyroclasts and reworked material.

Pyroclastic products have been drilled for a water gallery, 1 m deep, at an altitude of 90.4 m asl (Fig. 1b). The gallery extends for 4867 m mainly through the Neapolitan Yellow Tuff; in the Fontanelle area the gallery cuts a breccia and grey welded tuff succession similar to that outcropping in our Section 6 (Fig. 2a).

3. Caldera geometry in the city of Naples

The Vomero-S.Martino area is a topographic height, raising 100–200 m above the southern and eastern terrains. The older pyroclastic strata (Ancient Tuffs and Campanian Ignimbrite) dip consistently *outward* on both the southern and eastern sides of the hill. Instead, the uppermost succession (e.g. Neapolitan Yellow Tuff) drapes over a strong unconformity dipping

20° towards the topographic lows (*inward-dipping*), while covering conformably the older succession at the top of the hill (see cross-section on Fig. 1c). On the west side of the hill, the Campanian Ignimbrite is not exposed because the pre-Neapolitan Yellow Tuff deposits are completely buried by the thick Neapolitan Yellow Tuff. To understand whether the studied scarps are fault-controlled, we have to investigate if part of the outcropping sequence is displaced in the underlying plain. The Campanian Ignimbrite and the Ancient Tuffs are almost 150 m lower in the wells II and III than in the well I and along the Vomero-S.Martino scarps (Fig. 1c). To ascertain that this difference is a structural displacement and is not due to the geometry of the Campanian Ignimbrite that drapes over the articulate, pre-existing topography, we have evaluated, in the same wells, the height of the top of sedimentary succession. This shows a difference in heights of almost 40 m. This may likely be interpreted as the result of down-faulting that occurred during the Campanian Ignimbrite eruption because the younger terrains overlay the unconformity. The displacement of the Campanian Ignimbrite proximal deposits allow a better definition of the caldera boundary inside the city of Naples, 2 km south from the previous limit (see Fig. 1a and Orsi et al., 1996).

To better constrain the structure of the Campanian Ignimbrite caldera, we must consider that the top of the sedimentary basement is at almost 3 km depth (below sea level) in the central part of Campi Flegrei (Rosi and Sbrana, 1987; Barberi et al., 1991) and at only 350 m depth near the caldera rim at Naples (wells I, II and III in Figs. 1b and 2b). These different depths are partially due to the effect of the younger Neapolitan Yellow Tuff caldera collapse, restricted to the central part of the Campi Flegrei, of not less than 600 m (Scarpati et al., 1993). We suggest that the different depths of the floor of the Campi Flegrei caldera at its centre and in the Chiaia area suggest that the caldera has a piecemeal-like geometry at depth, as documented for other large calderas: Aira (Aramaki, 1984), Aso (Ono and Watanabe, 1983), Grizzly Peak (Fridrich et al., 1991), and Scafell (Branney and Kokelaar, 1994).

Above the Campanian Ignimbrite the younger deposits plaster the structural relief burying it. Later, an intense erosive action, on the seaward side of the Vomero-S.Martino hill, has exhumed the Campanian Ignimbrite scarp cutting pre-caldera deposits. We suggest that the thick incoherent pyroclastic succession accumulated above the Campanian Ignimbrite only on the caldera floor (wells II in Fig. 1c and III in Fig. 2b) is largely represented by slumped and remobilized pyroclasts. Isolated patches of Neapolitan Yellow Tuff are also preserved adhering to the old caldera surface (e.g. Parco Grifeo, S. Sepolcro and Fontanelle sites).

4. Volcanism in the central part of Naples and the Campanian Ignimbrite caldera collapse

The autochthonous volcanism in the central part of the city of Naples lies on sedimentary rocks. This ancient activity is recorded in few boreholes which cut 200 m of loose pyroclastic deposits with minor lava horizons. The main lava body was a lava dome identified during the excavation of various tunnels beneath S. Martino (Cole et al., 1994 and references therein). The subsequent activity was exclusively explosive producing the monogenetic vents of Parco Margherita, Parco Grifeo, Funicolare di Chiaia, S. Sepolcro and Capodimonte. Where exposed the contacts between the remnants of the cones show a west to east trend of this pre-caldera activity. These volcanic edifices were successively covered by three lapilli pumice fall deposits associated with ash and pumice beds possibly related to the Torre di Franco Tuffs of

Campi Flegrei (outcropping at the base of Camaldoli hill, Fig. 1b). Rolandi et al. (2004) do not recognize the paleosols between the different lapilli pumice fall deposits and attribute all this thick sequence to a single plinian event, vented in this area, that predate the Campanian Ignimbrite eruption of almost 1 ka. Our interpretation, based on the presence of paleosols and the good sorting of the lapilli pumice fall deposits, is that these deposits were the products of different eruptions and that their source is possibly within the Campi Flegrei. We suggest that only the uppermost and coarser fall deposit is related to the onset of the Campanian Ignimbrite eruption. The grading features and the thickness of this deposit are not easily comparable with that defined for distal locations (>30 km from the presumed source, see details in Rosi et al., 1999; Perrotta and Scarpati, 2003), but this is possibly due to the combined effect of deep erosion and the emplacement in a proximal environment. During the Campanian Ignimbrite eruption a thick sequence of welded tuff, spatter deposit and lithic breccia was emplaced in this area. The large average size of the clasts, their lithic nature and the welding feature suggest the proximal character of these deposits. A caldera collapse cut through the Campanian Ignimbrite and Ancient Tuffs forming the steep scarps that border the south and east sides of Vomero-S.Martino hill and south side of Capodimonte hill. This collapse possibly produced a scarp also west of the Vomero-S.Martino hill, linking this structural high with the well-known Piperno-Breccia Museo outcrop of Camaldoli, that is supposed to be completely buried by recent volcanic products (e.g. Neapolitan yellow Tuff). It is noteworthy that few tens of metres from the previously described proximal deposits of the Campanian Ignimbrite, we have found a grey welded tuff, 5 m thick, with reverse graded, black scoriae, embedded in an ashy matrix. We speculate that this deposit could represent the lateral transition between the proximal coarse and welded products and the typical facies of the Campanian Ignimbrite. The volcanic activity post-Campanian Ignimbrite is represented by the Chiatamone volcano which, with the Trentaremi tuff ring located on the west side of the bay of Naples (Cole and Scarpati, 1993) testify of an explosive activity inside the city of Naples after the Campanian Ignimbrite caldera collapse. The thicker pyroclastic sequence present, at the same stratigraphic height, in the intra-caldera boreholes should be related to remobilized deposits during the prolonged (24 ka) erosion of these scarps. Around 15 ka, the Neapolitan Yellow Tuff erupted, producing about 50 km^3 DRE (Scarpati et al., 1993) of material and forming a second major caldera collapse in the Campi Flegrei. The eruption produced up to 150 m thick deposit in proximal areas, which draped the erosive remnants of the Campanian Ignimbrite rim, in the Campi Flegrei and Naples. The seaward side of the structural heights were deeply eroded again to the local exhumation of the Campanian Ignimbrite caldera wall. The primary (i.e. volcanic) post-Neapolitan Yellow Tuff activity produced several thin ash and pumice lapilli layers that do not contribute significantly to the structural and morphological features of the study area with the exception of Mt. Echia volcano (Cole and Scarpati, 1993). On the contrary, volcanoclastic hydrologic remobilization and resedimentation processes were capable of transporting a voluminous sediment load to the level part of the city.

5. Conclusions

(1) The recovery in the city of Naples of coarse-lithic breccia (Breccia Museo) and welded deposits (Piperno) associated with the grey facies of the Campanian Ignimbrite testifies the co-genetic nature of these deposits.

(2) The Breccia-Piperno succession found at Naples is exactly similar to that outcropping along the Campi Flegrei caldera rim suggesting that these structures crosses the city of Naples.
(3) The occurrence of the proximal Campanian Ignimbrite deposits 2 km south of the previous limit proposed by Orsi et al. (1996) allows a better localization of the caldera boundary.

Acknowledgements

We are very grateful to many people for help with access to exposures in Naples. Thanks are due to Valerio Acocella for his comments on an earlier version of this manuscript. The constructive comments of Roberto Scandone and Angus Duncan are appreciated.

References

Aramaki, S., 1984. Formation of the Aira caldera, southern Kyushu, 22000 years ago. J. Geophys. Res. 89B10, 8485–8501.

Barberi, F., Cassano, E., La Torre, P., Sbrana, A., 1991. Structural evolution of Campi Flegrei in light of volcanological and geophysical data. J. Volcanol. Geotherm. Res. 48, 33–50.

Branney, M.J., Kokelaar, P., 1994. Rheomorphism and soft-state deformation of tuffs induced by volcanotectonic faulting at a piecemeal caldera, English Lake District. Bull. Soc. Geol. Am. 106, 507–530.

Cole, P.D., Perrotta, A., Scarpati, C., 1994. The volcanic history of the southwestern part of the city of Naples. Geol. Mag. 131, 785–799.

Cole, P.D., Scarpati, C., 1993. A facies interpretation of the eruption and emplacement mechanisms of the upper part of the Neapolitan Yellow Tuff, Campi Flegrei, southern Italy. Bull. Volcanol. 55, 311–326.

de'Gennaro, M., Cappelletti, P., Langella, A., Perrotta, A., Scarpati, C., 2000. Genesis of zeolites in the Neapolitan Yellow Tuff: geological, volcanological and mineralogical evidences. Contrib. Mineral. Petrol. 139, 17–35.

Deino, A.L., Orsi, G., de Vita, S., Piochi, M., 2004. The age of the Neapolitan Yellow Tuff caldera-forming eruption (Campi Flegrei caldera, Italy) assessed by $^{40}Ar/^{39}Ar$ dating method. J. Volcanol. Geotherm. Res. 133, 157–170.

De Lorenzo, G., 1904. L'attività vulcanica nei Campi Flegrei. Rend. Acc. Sc. Fis. Mat., Napoli, serie 3(10), 203–211.

D'Erasmo, G., 1931. Studio geologico dei pozzi profondi della Campania. Boll. Soc. Nat. 43, 15–130.

De Vivo, B., Rolandi, G., Gans, P.B., Calvert, A., Bohrson, W.A., Spera, F.J., Belkin, H.E., 2001. New constraints on the pyroclastic eruptive history of the Campanian volcanic plain (Italy). Mineral. Petrol. 73, 47–65.

Druitt, T.H., Sparks, R.S.J., 1982. A proximal ignimbrite breccia facies on Santorini, Greece J. Volcanol. Geotherm. Res. 13, 147–171.

Fridrich, C.J., Smith, R.P., DeWitte, E., McKee, E.H., 1991. Structural, eruptive, and intrusive evolution of the Grizzly Peak caldera, Sawatch range, Colorado Geol. Soc. Am. Bull. 103, 1160–1177.

Insinga, D., Calvert, A., D'Argenio, B., Fedele, L., Lanphere, M., Morra, V., Perrotta, A., Sacchi, M., Scarpati, C., 2004. $^{40}Ar/^{39}Ar$ Dating of the Neapolitan Yellow Tuff eruption (Campi Flegrei, southern Italy): Volcanological and Chronostratigraphic Implications. EGU Assembly, Nice.

Johnston-Lavis, H.J., 1888. Report of the committee appointed for the investigation of the volcanic phenomena of Vesuvius and its neighbourhood, London, pp. 1–7.

Johnston-Lavis, H.J., 1889. On a remarkable sodalite trachyte lately discovered in Naples, Italy. Geol. Mag. 6, 74–77.

Lirer, L., Luongo, G., Scandone, R., 1987. On the volcanological evolution of Campi Flegrei. EOS 68(16), 226–233.

Ono, K., Watanabe, K., 1983. Aso caldera. Earth Monthly, 46, 73–82.

Orsi, G., De Vita, S., Di Vito, M., 1996. The restless, resurgent Campi Flegrei nested caldera (Italy): constraints on its evolution and configuration. J. Volcanol. Geotherm. Res. 74, 179–214.

Perrotta, A., Scarpati, C., 1994. The dynamics of the Breccia Museo eruption (Campi Flegrei, Italy) and the significance of spatter clasts associated with lithic breccias. J. Volcanol. Geotherm. Res. 59(4), 335–355.

Perrotta, A., Scarpati, C., 2003. Volume partition between the plinian and co-ignimbrite air-fall deposits of the Campanian Ignimbrite eruption. Mineral. Petrol. 79, 67–78.

Rittmann, A., 1950. Rilevamento geologico della collina dei camaldoli nei Campi Flegrei. Boll. Soc. Geol. It. 69, 129–177.

Rolandi, G., Bellucci, F., Heizler, M.T., Belkin, H.E., De Vivo, B., 2003. Tectonic controls on the genesis of ignimbrites from the Campanian Volcanic Zone, southern Italy. Mineral. Petrol. 79, 3–31.

Rosi, M., Sbrana, A., 1987. The Phlegrean Fields. Quad. Ric. Sci. 9, 1–175.

Rosi, M., Vezzosi, L., Castelmenzano, A., Grieco, G., 1999. Plinian pumice fall deposit of the Campanian Ignimbrite eruption (Phlegrean Fields, Italy). J. Volcanol. Geotherm. Res. 91, 179–198.

Scandone, R., Bellucci, F., Lirer, L., Rolandi, G., 1991. The structure of the Campanian Plain and the activity of the Neapolitan volcanoes (Italy). J. Volcanol. Geotherm. Res. 48, 1–32.

Scarpati, C., Cole, P.D., Perrotta, A., 1993. The Neapolitan Yellow Tuff – A large volume multiphase eruption from Campi Flegrei, southern Italy. Bull. Volcanol. 55, 343–356.

Scherillo, A., 1957. I "tufi antichi" tra S. Maria Apparente e via Parco Grifeo in Napoli. Boll. Soc. Nat. 66, 69–89.

Volcanism in the Campania Plain:
Vesuvius, Campi Flegrei and Ignimbrites
edited by B. De Vivo

Chapter 6

The Late-Holocene evolution of the Miseno area (south-western Campi Flegrei) as inferred by stratigraphy, petrochemistry and $^{40}Ar/^{39}Ar$ geochronology

Donatella Insinga[a,], Andrew T. Calvert[b], Marvin A. Lanphere[b], Vincenzo Morra[c], Annamaria Perrotta[d], Marco Sacchi[a], Claudio Scarpati[d], James Saburomaru[b] and Lorenzo Fedele[c]*

[a]*Istituto per l'Ambiente Marino Costiero (IAMC) CNR, Napoli, Calata P.ta di Massa, Porto di Napoli, 80133-Napoli, Italy*
[b]*USGS, 345 Middlefield Road, Menlo Park, MS-937, 94025 CA, USA*
[c]*Dipartimento di Scienze della Terra, Università degli Studi Federico II, via Mezzocannone 8, 80138-Napoli, Italy*
[d]*Dipartimento di Geofisica e Vulcanologia, Università degli Studi Federico II, Largo San Marcellino, 80134-Napoli, Italy*

Abstract

This study on terrestrial and marine successions increases the understanding of the Late-Holocene volcanological and stratigraphical evolution of the south-western part of Campi Flegrei caldera.

Stratigraphic data derived from field studies of two major tuff vents located along the coastal zone, namely Porto Miseno and Capo Miseno, clearly indicate that the Porto Miseno tuff ring slightly predates the Capo Miseno tuff cone. $^{40}Ar/^{39}Ar$ step-heating experiments, carried out on fresh sanidine separates from pumice samples, yielded a plateau age of 5090 ± 140 yr BP for Capo Miseno and 6490 ± 510 yr BP for Porto Miseno vent, thus confirming field observations.

The volcanoclastic input derived from this recent and intense eruptive activity played a major role in the inner-shelf stratigraphic evolution of the Porto Miseno Bay deposits that have been drilled up to 40 m depth off the crater rim. The cored succession is characterised by transgressive marine deposits (mostly volcanic sand) with two intercalated peat layers (t_1 and t_2), dated at 3560 ± 40 yr BP and 7815 ± 55 yr BP (^{14}C), respectively, interbedded with a 1–5 m thick pumice layer (tephra C). Peat layers have been chronostratigraphically correlated with two widespread paleosols onland while petrochemical analyses allowed us to correlate tephra C with the Capo Miseno tuff cone deposits.

The results presented in this study imply a Late-Holocene volcanic activity that is also well preserved in the marine record in this sector of the caldera where a new chronostratigraphic reconstruction of the eruptive events is required in order to better evaluate the hazard assessment of the area.

1. Introduction

The Campi Flegrei (CF) coastline is characterised by the occurrence of several explosive vents active during the Holocene (Di Girolamo et al., 1984; Rosi and Sbrana, 1987; Di Vito

*Corresponding author. *E-mail address*: donatella.insinga@iamc.cnr.it (D. Insinga).

et al., 1999). They produced large volumes of volcaniclastic materials, which contributed to the stratigraphic record of Pozzuoli Bay. These pyroclastic deposits, which can be found interbedded within marine sediments, are widespread in the inner continental shelf and they also occur in the middle to outer continental shelf (Carbone et al., 1984; Milia, 1996, 1998; Insinga, 2003; D'Argenio et al., 2004) and onland due to recent tectonic uplift (La Starza marine terrace, Cinque et al., 1985; Barra, 1991).

Particularly, hydrovolcanic centres, such as tuff rings and tuff cones have significantly contributed to sediment accumulation in the coastal areas of Pozzuoli Bay (Bagnoli-Fuorigrotta depression, Calderoni and Russo, 1998; Porto Miseno, Insinga et al., 2002). These vents acted as local but significant sediment sources to marine depositional systems in response to relative sea-level changes and landward migration of coastline during the last 10,000 years (Cinque et al., 1985, 1997; Milia, 1998).

The study of these coastal successions in active volcanic settings requires an integrated stratigraphic approach, which takes into account continental and transitional deposits interbedded with marine sediments (Lajoie and Stix, 1992). The recognition and dating with appropriate techniques of distinct event horizons in these types of successions, allow to establish a chronostratigraphic correlation both with other coastal deposits and with the inland sectors of the volcanic district.

In this paper, we apply a multidisciplinary study to the Capo Miseno and Porto Miseno volcanic centres and the marine epiclastic succession of Porto Miseno in the south-western sector of CF caldera (Fig. 1) including $^{40}Ar/^{39}Ar$ age determinations, stratigraphic analysis and petrochemical characterisation. The results presented in this work aim to complement previous research on the effects that volcanism had on Holocene coastal sedimentation of CF, and to provide new insights into the recent eruptive history of the south-western sector of this volcanic district.

2. Geological setting

2.1. Campi Flegrei caldera

The CF is a major volcanic district of southern Italy (Fig. 1) that formed during the latest Pleistocene as a consequence of lithospheric extension across the eastern Tyrrhenian margin (Beccaluva et al., 1991). The CF caldera, located on the coastal margin of the Campanian Plain, is an active volcanic area as indicated by its last eruption in AD 1538 (Di Vito et al., 1987) and by the evidence of hydrothermal activity, seismic activity and bradyseismic episodes over the last 30 years (Corrado et al., 1977; Barberi et al., 1984; Rosi and Sbrana, 1987; Allard et al., 1991). Since its onset, (around 60,000 yr BP, Pappalardo et al., 1999) volcanism in the CF has been essentially explosive with subordinate effusive episodes (Di Girolamo et al., 1984; Rosi and Sbrana, 1987; Di Vito et al., 1999). The Holocene volcanic activity took place inside a nested caldera structure (Lirer et al., 1987; Orsi et al., 1996) related to the Campanian Ignimbrite (Rosi and Sbrana, 1987) and the Neapolitan Yellow Tuff (NYT) eruptions (Scarpati et al., 1993) which have been $^{40}Ar/^{39}Ar$ dated, respectively, at 39,000 yr BP (Ricci et al., 2000; De Vivo et al., 2001) and 15,000 yr BP (Insinga, 2003; Deino et al., 2004). The recent volcanism has been subdivided into three epochs in a chronostratigraphic reconstruction of CF events essentially based on stratigraphic relations and ^{14}C

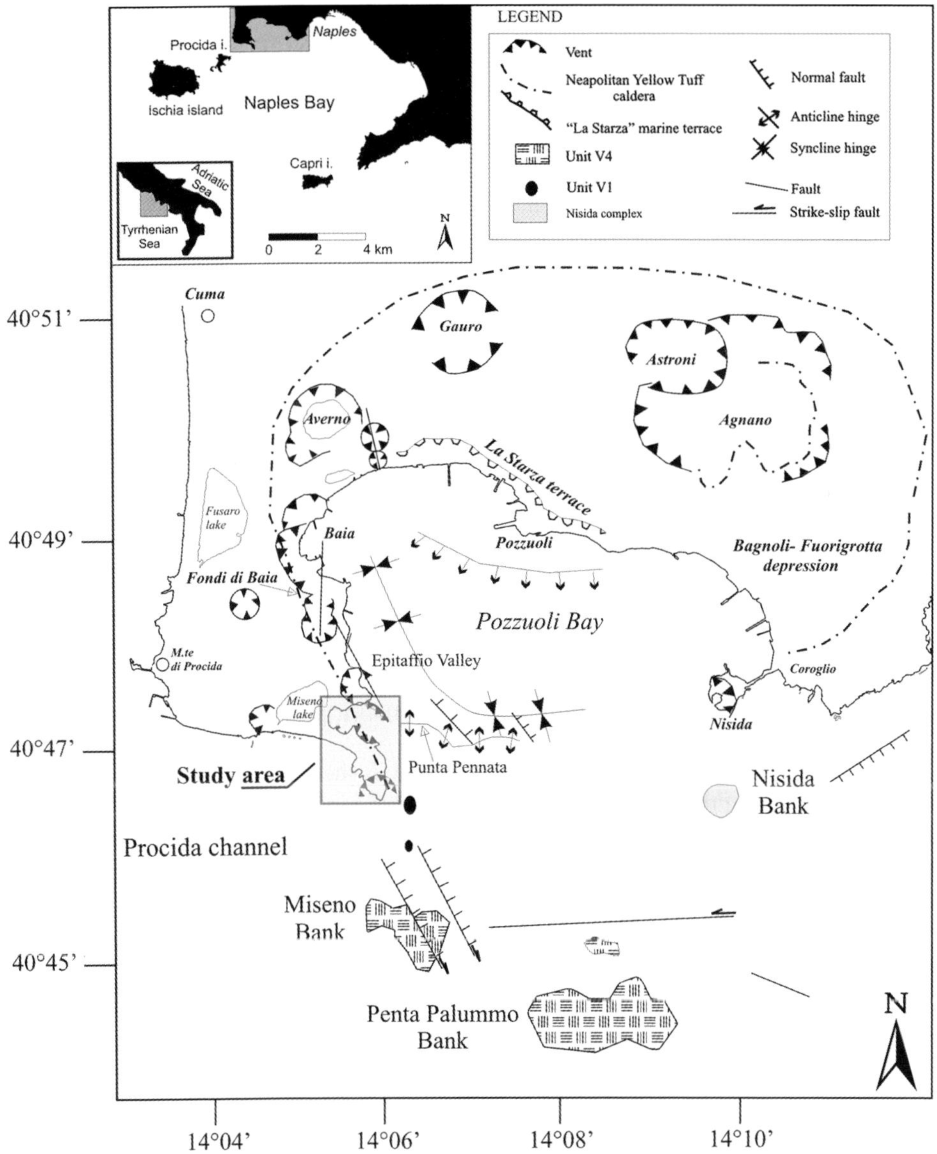

Figure 1. Volcano-tectonic sketch map of Campi Flegrei caldera and Pozzuoli Bay (after Scarpati et al., 1993; Milia and Torrente, 2000). The study area is remarked.

ages (Di Vito et al., 1999; Orsi et al., 2004). According to the above authors, the volcanic activity was concentrated in three main periods, ranging from 15,000 to 9500 yr BP (I Epoch), from 8600 to 8200 yr BP (II Epoch) and finally from 4800 yr BP to 3800 yr BP (III Epoch). The periods of quiescence separating these phases of intense volcanism lasted 1000 and 3500 years, respectively, and are recorded onland by two major paleosoils horizons (A and B).

Tens of monogenetic phreato-magmatic volcanoes including tuff rings, tuff cones, cinder and spatter cones (Di Girolamo et al., 1984; Di Vito et al., 1987; Rosi and Sbrana, 1987; de Vita et al., 1999) formed after the NYT eruption and a migration of the eruptive vents from the structural boundary of the caldera (I Epoch) towards the north-eastern sector of caldera floor (III Epoch) has been suggested (Di Vito et al., 1999).

From a petrographic point of view, the CF products belong to the Roman Magmatic Province; they range in composition from shoshonitic basalts to rare phonolites (Di Girolamo et al., 1984; Rosi and Sbrana, 1987; Melluso et al., 1995). Most of the outcropping rocks (90%) are represented by differentiated products (trachytes and trachyphonolites).

2.2. The Pozzuoli Bay

The Pozzuoli Bay (eastern Tyrrhenian Sea) represents the offshore counterpart of the CF volcanic field (Fig. 1). It is bounded to the south by several submerged volcanic banks and is characterised by Holocene epiclastic deposits (Colantoni et al., 1972; Pescatore et al., 1984; Insinga, 2003; D'Argenio et al., 2004), which overlie a widespread volcanic unit that has been interpreted as NYT by Milia and Torrente (2000). Holocene deposits have been grouped by seismostratigraphic analysis into two volcanic units, Nisida Complex (NC) and unit V1, interbedded within three depositional units: G1, G2 and G3 (Milia, 1998; Milia and Torrente, 2000) which have been correlated with the marine succession outcropping at La Starza terrace near Pozzuoli (Cinque et al., 1985; Barra, 1991). The unit V1 is interpreted as dikes and sills associated with deformation of the host marine deposits of the most recent unit G1.

The present-day morphology of the Pozzuoli Bay is the result of tectonic deformation that occurred during the deposition of units G1 and G2 in the last 8000 years (Milia and Torrente, 2000). Particularly, morphologic highs and areas of recent uplift, such as Punta Pennata and La Starza terrace, have been interpreted as the surficial expression of deeper anticlinal folds, while the subsiding depocentre of the central Pozzuoli Bay would correspond to a syncline structure (Epitaffio Valley; Fig. 1).

3. Hydrovolcanic centres of Miseno area: previous studies

The western sector of CF is characterised by the occurrence, along the Pozzuoli Bay coastline, of several tuff cones and tuff rings (De Lorenzo, 1905) aligned on a N-S structure, which has been repeatedly reactivated through time (Fig. 2) (Di Vito et al., 1999). Porto Miseno and Capo Miseno volcanic centres, in particular, originated from phreatomagmatic activity and were subject to intense zeolitisation as shown by the transition from yellow lithified to grey unlithified pyroclastic facies along their flanks (Rosi and Sbrana, 1987; de'Gennaro et al., 1999). The eruptive products have a trachytic and trachyphonolitic composition (Di Girolamo et al., 1984; D'Antonio et al., 1999). Stratigraphic, geochronologic and petrochemical studies on volcanic vents, which form and shape the surface landform in the Miseno area, have been generally rare (Di Girolamo et al., 1984; Rosi and Sbrana, 1987; Di Vito et al., 1999), hampering proper understanding of the eruptive history and geological evolution of the south-western sector of CF caldera.

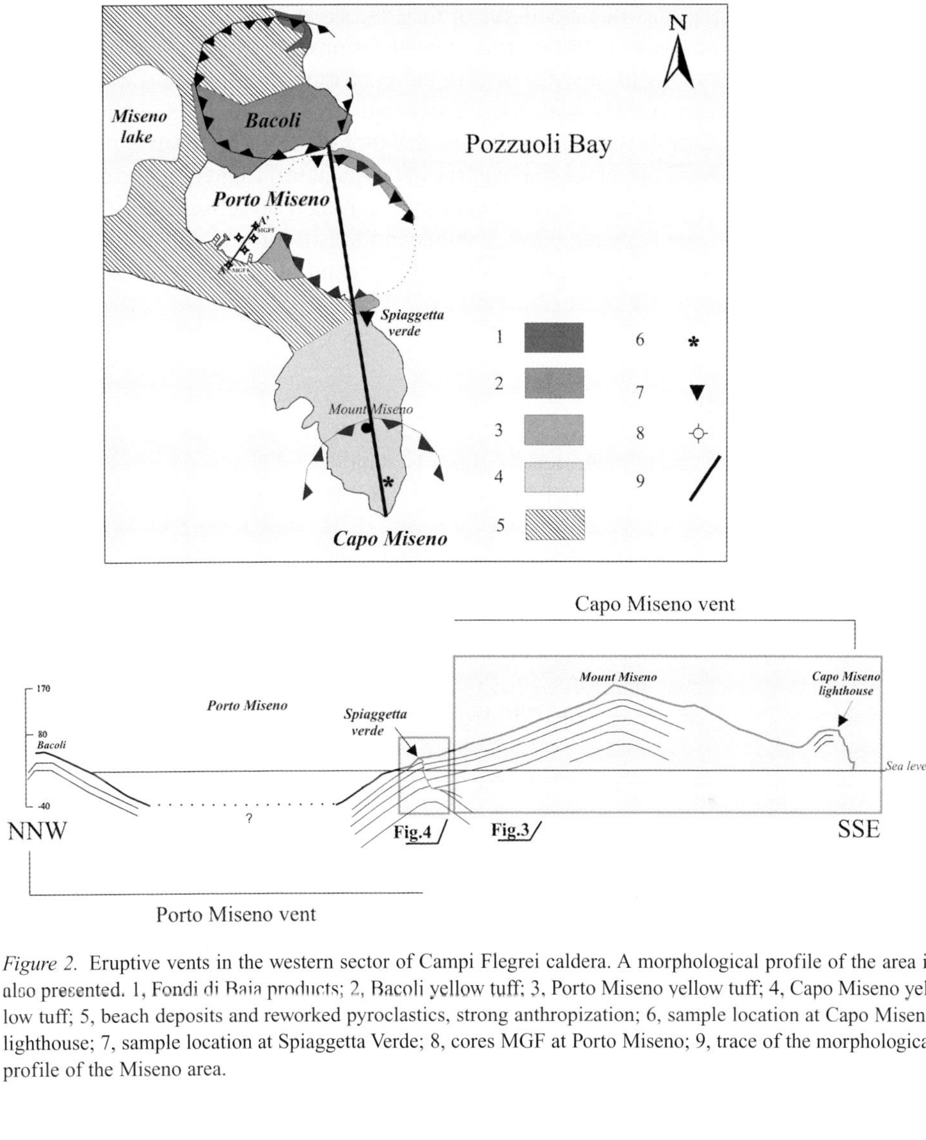

Figure 2. Eruptive vents in the western sector of Campi Flegrei caldera. A morphological profile of the area is also presented. 1, Fondi di Baia products; 2, Bacoli yellow tuff; 3, Porto Miseno yellow tuff; 4, Capo Miseno yellow tuff; 5, beach deposits and reworked pyroclastics, strong anthropization; 6, sample location at Capo Miseno lighthouse; 7, sample location at Spiaggetta Verde; 8, cores MGF at Porto Miseno; 9, trace of the morphological profile of the Miseno area.

A K/Ar age of 4 ka proposed by Di Girolamo et al. (1984) for Capo Miseno and generally adopted by archeologists (Albore Livadie, 1986), was considered unreliable by Rosi and Sbrana (1987), based on the presumed incongruence arising from the observation that a significant coastal erosion had substantially modified the original volcanic morphology, thus suggesting an older age for this vent. In line with the above interpretation, the activity of Capo Miseno and Porto Miseno vents is regarded by the same authors as slightly predating the NYT eruption (last stages of phase A). On the basis of outcrop evidence, Scandone et al. (1991) and Scarpati et al. (1993) suggested that several eruptive centres, among which Capo Miseno and Porto Miseno, postdate the NYT and erupted on the NYT caldera rim (Fig. 2). More recently, Di Vito et al. (1999) proposed that Capo Miseno and

Porto Miseno are among the oldest edifices of the I Epoch at 10,500 yr BP and 10,550 yr BP, respectively, on the basis of stratigraphic considerations.

The original morphology of the Miseno area has been dramatically modified as a result of the interplay between volcano-tectonic activity and sea-level changes, which affected the coastline of the Pozzuoli Bay during the last few thousand years. However, the embayment of Porto Miseno still preserves the features of a typical tuff ring vent. The occurrence here of a submerged Roman harbour (I–IV century C.e.) suggests subsidence of about 9 m during the last 2000 years (Caputo, 1989; Dvorak and Mastrolorenzo, 1991) with a maximum subsidence of 11–12 m between the VII and XI century C.e. as a result of historical bradyseismic movements (Cinque et al., 1991).

4. Materials and methods

This research is based on an integrated stratigraphic analysis of five deep cores drilled off the submerged yellow tuff volcano of Porto Miseno and field studies of selected outcrops of both the Capo Miseno and Porto Miseno vents. Boreholes were located close to the relic rim of Porto Miseno tuff ring and penetrated a Holocene shallow marine succession (Fig. 2). Coring stations have been denoted MGF1 through MGF5. Core MGF1 was drilled on the docks of Porto Miseno while MGF2–MGF5 were drilled in a few meters of water depth off the docks. Drilling operations have been carried out using a simple rotary well-drilling rig and corer with external diameter of 101 mm. Cores reached a maximum depth of 40 m for a total core length of 170 m and a 76% recovery.

4.1. $^{40}Ar/^{39}Ar$ dating

$^{40}Ar/^{39}Ar$ incremental-heating experiments were performed on five aliquots of feldspar phenocrysts, which were separated from pumice of Capo Miseno and Porto Miseno vents sampled at Capo Miseno lighthouse (samples DI7A and DI7B) and at Spiaggetta Verde (sample SV8), respectively (Fig. 2). Capo Miseno samples were analysed in duplicate. Analytical techniques for incremental-heating experiments and data-handling procedures were described by Lanphere (2000). In this study, all stated analytical errors for Ar analyses are 1σ errors. Fresh sanidine was packaged in copper foil and placed in cylindrical quartz vials together with TCR-2 sanidine from the Taylor Creek Rhyolite of New Mexico (Duffield and Dalrymple, 1990) whose age is 27.87 Myr. The samples were irradiated for 1 h in the central thimble of the U.S. Geological Survey TRIGA reactor in Denver, Colorado (Dalrymple et al., 1981). Monitors bracketed unknowns during irradiation and were analysed by laser fusion using a system described by Dalrymple (1989) that includes a continuous 5 W Ar ion laser, a getter clean-up system, and an ultrasensitive, ultralow-background, 15 –cm radius, 90°-sector mass spectrometer (MAP 216). Gas liberated from both monitors and unknowns was cleaned with SAES AP-10 getters before spectrometry. *J* values for the TCR-2 sanidine standard were collected from the weighted mean of six analyses of 3–4 grains each of TCR-2 sanidine. TCR-2 sanidine has been calibrated against a primary standard mineral, SB-3 biotite, whose age of 162.9±0.9 Myr was determined using first-principle calibrations (Lanphere and Dalrymple, 2000). Negligible interference of ^{40}Ar was produced by thermal-neutron reaction with K because the samples were shielded with Cd foil; experiments on artificial K glass shielded with Cd yielded $^{40}Ar/^{40}K = 0$. $^{40}Ar/^{39}Ar$ incremental-heating experiments were performed with a resistance furnace attached to the same mass spectrometer.

4.2. Petrochemical analysis

Pumice samples have been analysed for major and trace elements by XRF (Philips PW 1400). Petrochemical composition has been obtained on pressed powder pellets and the data were corrected according to the reduction methods of Franzini et al. (1972) and Leoni and Saitta (1976). Calibration curves were obtained using 35 international standards. Precision is better than 5% for major elements (excluding P_2O_5) and for Rb, Sr, Y, Zr, Nb, Zn and V, and better than 10% for the remaining trace elements excluding Sc, for which precision is closer to the XRF detection limits. MgO and Na_2O have been analysed with atomic absorption spectrophotometry. LOI (weight loss on ignition) was determined with standard gravimetric techniques.

5. Results

5.1. Porto Miseno tuff ring and Capo Miseno tuff cone

5.1.1. Morphology and volcanological features

Both Capo Miseno and Porto Miseno volcanoes are composed of mostly lithified pyroclastic deposits. These pyroclastic sequences consist of ash, lapilli and block clasts in various proportions and some accidental components. The deposits are dominantly composed of grey rounded juvenile pumice clasts that are poorly vesicular to vesicular. Lithic fragments are mainly composed of trachytic lava clasts (Fig. 3).

The main morphometric data related to these two edifices are reported in Table 1. The shape of Porto Miseno volcano appears nearly circular in plan-view, although the SE and NW sectors are missing due to marine erosion. The morphology of this volcanic centre is that of a tuff ring (Cas and Wright, 1987) of moderate size (volume approximately 10×10^6 m^3) formed by a lower lithified succession overlain by non-lithified thinner deposits. The boundary between these two facies is quite sharp. Both facies are bedded tuffs with planar and wavy structures. This deposit is interpreted to reflect deposition from pyroclastic density currents derived from phreatomagmatic explosions. Large ballistic blocks, typically up to 40 cm in diameter, form numerous bomb sags at different stratigraphic heights.

Capo Miseno volcano has a smaller crater and a larger height to width ratio than the Porto Miseno tuff ring. According to morphometric data (Table 1), Capo Miseno edifice is a tuff cone (Heiken, 1971; Cas and Wright, 1987) with an original volume of about 100×10^6 m^3. Marine erosion has removed most of the volcanic edifice exposing a remnant of the plumbing system (vertical dykes cutting subhorizontal pyroclastic layers) along its southern cliff. The Capo Miseno tuff cone comprises predominantly of coarse grained deposits with planar to wavy stratification. Individual layers range from poorly stratified to massive

Figure 3. Capo Miseno tuff cone seen from the Procida channel and (A) particular of its deposits, (B) size of sanidine crystals in pumiceous scoria blocks.

Table 1. Morphometric data for Capo Miseno tuff cone and Porto Miseno tuff ring.

	Inner slope	Outer slope	Height max asl (m)	Crater diameter (m)	Cone diameter (m)	Volume (m^3)
Capo Miseno	50°	30°	163	300	1500	98×10^6
Porto Miseno	40°	10°	33	550	850	9.4×10^6

layers. Most of the pyroclastic succession is lithified; the uppermost few metres, formed mainly by fallout deposits, are generally non-lithified. This feature possibly reflects fluctuating emplacement conditions and thermal dispersion as suggested for the NYT by de'Gennaro et al. (2000).

No paleosols or reworked materials are interbedded within Capo Miseno and Porto Miseno pyroclastic sequences suggesting that each volcano was built up by a single eruptive phase.

The contact between Porto Miseno and Capo Miseno deposits is well exposed at the Spiaggetta Verde site (Figs. 2–4) along the southern rim of Porto Miseno volcanic edifice buried under the Capo Miseno deposits that rest on a deeply eroded surface in the Porto Miseno pyroclastic sequence (Fig. 4).

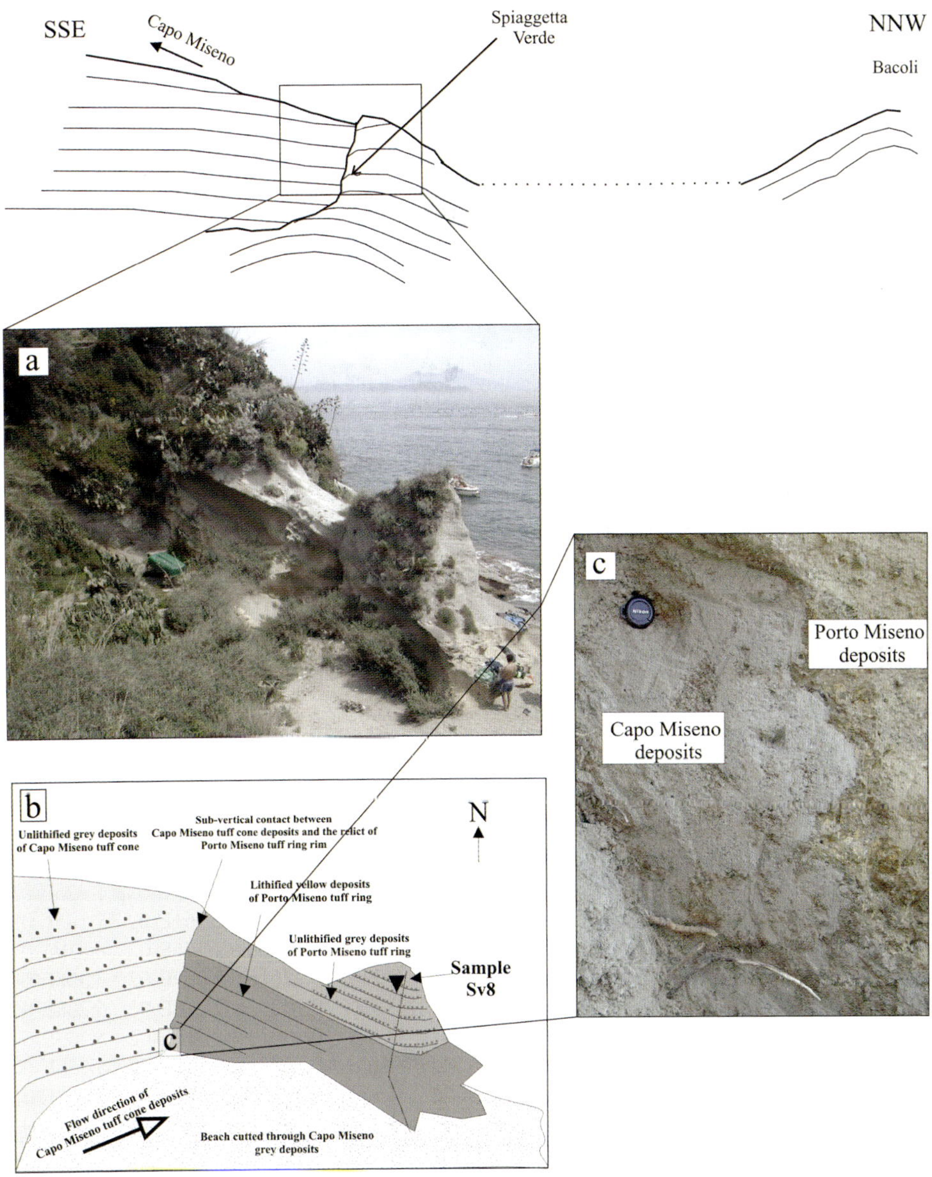

Figure 4. Stratigraphic relations between Capo Miseno tuff cone and Porto Miseno tuff ring deposits as they have been observed at Spiaggetta Verde locality. The section is from Figure 2, reversed. (a) part of the Spiaggetta Verde; (b) sketch map of geometrical relations between the two formations; (c) particular of the erosive surface along which the most recent deposits of Capo Miseno overlay Porto Miseno products.

5.1.2. $^{40}Ar/^{39}Ar$ geochronology

Step-heating results are listed in full in Table 2 and summarised in Table 3. Sample SV8, collected from the Porto Miseno tuff ring, yielded a weighted mean plateau age of 6490 ± 510 yr BP using 100% of the ^{39}Ar released. The SV8 isochron age of 6710 ± 800 yr BP is concordant with the plateau age and yields a $(^{40}Ar/^{36}Ar)_{trapped}$ composition within error of atmosphere ($^{40}Ar/^{36}Ar = 295.5$) (Fig. 5). The K/Ca ratio is roughly constant throughout the experiment indicating no xenocrystic contamination. We analysed four splits of samples from the Capo Miseno tuff cone, two each from samples DI7A and DI7B. All of the DI7 samples have consistent K/Ca ratios over the step-heating experiment, so there is no obvious contamination. Both splits from sample DI7B yielded simpler age spectra and concordant isochron ages (Figs. 6 and 7). A weighted mean of the two DI7B plateau ages yields 5090 ± 140 yr BP Both splits from sample DI7A yielded total gas ages similar to DI7B but with inconsistent apparent ages and isotopic ratios. We conclude that the weighted mean of the two DI7B experiments yields the best age for the Capo Miseno tuff cone and that DI7A contains heterogeneously distributed non-radiogenic ^{40}Ar that has complicated the step-heating experiment.

Table 2. Analytical data for incremental-heating experiments on Capo Miseno (DI7A-1, DI7A-2, DI7B-1, DI7B-2) and Porto Miseno (SV8) sanidines.

DI7A-1 Sanidine, Capo Miseno
J = 0.000199621

Temperature (°C)	Age(ka)	K/Ca	%rad	40∗(mol)	Σ39Ar	40/39	37/39	36/39
550	4924.62 ± 757.21	3.7	3.5	1.94E-15	0.000	393.523	1.336E-01	1.285E+00
575	−87.10 ± 35.68	7.4	−3.8	−3.20E-16	0.001	6.28249	6.592E-02	2.210E-02
625	34.51 ± 11.73	8.3	2.7	4.08E-16	0.005	3.5358	5.877E-02	1.166E-02
675	3.73 ± 6.42	9.9	0.4	9.05E-17	0.012	2.46486	4.966E-02	8.318E-03
725	−3.94 ± 3.73	13.0	−0.8	−1.48E-16	0.023	1.30137	3.777E-02	4.451E-03
775	5.22 ± 2.37	15.8	1.9	3.18E-16	0.041	0.79883	3.096E-02	2.662E-03
825	6.41 ± 1.88	17.7	2.9	4.89E-16	0.063	0.63064	2.776E-02	2.081E-03
875	6.56 ± 1.19	19.1	4.2	8.08E-16	0.100	0.43771	2.569E-02	1.426E-03
925	4.10 ± 0.90	20.8	3.4	6.82E-16	0.150	0.348	2.361E-02	1.145E-03
955	2.16 ± 0.79	22.5	2.4	4.02E-16	0.200	0.26506	2.176E-02	8.819E-04
985	4.10 ± 0.79	23.7	5.3	7.46E-16	0.250	0.22616	2.069E-02	7.308E-04
1015	3.73 ± 0.73	24.3	5.6	7.13E-16	0.310	0.1925	2.016E-02	6.208E-04
1065	7.53 ± 0.56	24.3	11.4	1.96E-15	0.380	0.19023	2.014E-02	5.763E-04
1090	7.08 ± 0.58	25.1	11	1.74E-15	0.460	0.18273	1.954E-02	5.558E-04
1120	7.01 ± 0.51	24.9	10.6	2.01E-15	0.540	0.18667	1.971E-02	5.702E-04
1160	5.59 ± 0.76	24.9	8.3	9.63E-16	0.590	0.18281	1.969E-02	5.726E-04
1190	8.20 ± 0.61	25.0	14.8	1.84E-15	0.660	0.15556	1.962E-02	4.540E-04
1200	4.62 ± 0.51	25.1	10.3	1.23E-15	0.740	0.12625	1.953E-02	3.887E-04
1210	6.71 ± 0.48	25.0	18.6	1.88E-15	0.820	0.10146	1.958E-02	2.849E-04
1220	3.73 ± 0.48	25.1	10.8	1.08E-15	0.900	0.10144	1.955E-02	3.117E-04

Table 2. *(Continued)*

Temperature (°C)	Age(ka)	K/Ca	%rad	40*(mol)	Σ39Ar	40/39	37/39	36/39
1250	2.98 ± 0.49	24.8	6.7	8.55E-16	0.980	0.12997	1.979E-02	4.160E-04
1350	9.99 ± 2.82	23.3	2.4	5.30E-16	1.000	1.14364	2.105E-02	3.782E-03
DI7A-2 Sanidine, Capo Miseno								
J = 0.000191989								
550	1605.79 ± 471.82	8.7	8.5	1.21E-15	0.000	54.33812	5.658E-02	1.682E-01
600	−225.26 ± 84.61	6.2	−7.6	−9.24E-16	0.001	8.60629	7.902E-02	3.135E-02
650	101.07 ± 24.39	8.7	5.6	1.48E-15	0.004	5.26159	5.617E-02	1.683E-02
700	−3.41 ± 10.40	10.5	−0.4	-1.17E-16	0.011	2.33619	4.677E-02	7.952E-03
750	−19.71 ± 13.54	13.2	−3.5	-5.05E-16	0.017	1.60535	3.726E-02	5.635E-03
800	1.64 ± 3.52	16.5	0.4	1.81E-16	0.039	1.13884	2.970E-02	3.846E-03
850	7.08 ± 2.78	18.7	3.4	9.50E-16	0.066	0.62775	2.625E-02	2.060E-03
900	6.19 ± 2.01	20.3	3.9	1.13E-15	0.100	0.46315	2.418E-02	1.512E-03
950	10.51 ± 1.56	22.0	8.9	2.47E-15	0.150	0.35224	2.225E-02	1.092E-03
1000	5.59 ± 1.28	23.0	5.7	1.55E-15	0.210	0.28177	2.127E-02	9.053E-04
1050	3.28 ± 1.06	23.9	4.4	1.16E-15	0.280	0.2268	2.047E-02	7.397E-04
1100	13.71 ± 0.92	24.8	15.5	5.41E-15	0.360	0.2557	1.976E-02	7.365E-04
1150	0.67 ± 0.91	24.8	0.7	2.68E-16	0.450	0.26145	1.973E-02	8.839E-04
1200	2.09 ± 0.62	24.8	2.9	1.25E-15	0.570	0.20496	1.972E-02	6.789E-04
1225	5.07 ± 0.53	25.2	11.8	3.49E-15	0.710	0.12583	1.947E-02	3.812E-04
1250	7.38 ± 0.59	25.4	20.8	4.54E-15	0.840	0.10487	1.928E-02	2.863E-04
1275	5.07 ± 0.80	25.1	14.2	2.30E-15	0.930	0.1068	1.949E-02	3.155E-04
1300	9.47 ± 1.44	24.9	18.5	2.30E-15	0.980	0.14861	1.971E-02	4.153E-04
1350	−1.45 ± 3.96	23.3	−1	−1.22E-16	1.000	0.38308	2.107E-02	1.316E-03
DI7B-1 Sanidine, Capo Miseno								
J = 0.000184224								
550	5446.64 ± 2404.83	3.9	2.7	9.00E-16	0.000	597.5426	1.249E-01	1.967E+00
600	−227.01 ± 29.91	5.6	−10.6	−1.00E-15	0.001	6.44723	8.791E-02	2.415E-02
650	−255.56 ± 43.53	6.8	−11.3	−7.63E-16	0.002	6.78599	7.238E-02	2.559E 02
700	−6.36 ± 4.17	9.1	−0.9	−2.44E-16	0.013	2.14122	5.370E-02	7.326E-03
750	1.64 ± 2.55	14.2	0.5	1.04E-16	0.029	1.08897	3.461E-02	3.677E-03
800	2.91 ± 1.95	17.2	0.9	2.37E-16	0.051	0.95773	2.846E-02	3.219E-03
850	4.92 ± 1.38	18.9	2.8	5.13E-16	0.080	0.53788	2.590E-02	1.777E-03
900	7.53 ± 1.04	20.7	5.6	1.06E-15	0.120	0.41138	2.373E-02	1.320E-03
950	2.46 ± 1.07	22.0	1.5	3.60E-16	0.160	0.50734	2.232E-02	1.698E-03
1000	5.22 ± 0.67	23.2	6	1.19E-15	0.220	0.27572	2.116E-02	8.833E-04
1050	4.92 ± 0.58	23.4	6.4	1.27E-15	0.290	0.23674	2.098E-02	7.556E-04
1075	6.41 ± 0.62	23.9	8.4	1.45E-15	0.350	0.23043	2.052E-02	7.204E-04
1105	3.80 ± 0.73	24.0	3.7	8.35E-16	0.410	0.32867	2.042E-02	1.077E-03
1135	4.92 ± 0.61	23.2	4.8	1.33E-15	0.480	0.31674	2.115E-02	1.026E-03
1160	5.89 ± 0.64	23.1	7.1	1.38E-15	0.540	0.25817	2.121E-02	8.176E-04
1190	5.74 ± 0.62	23.9	5.5	1.47E-15	0.610	0.31395	2.052E-02	1.010E-03
1220	5.74 ± 0.40	24.3	7.9	2.48E-15	0.730	0.22208	2.014E-02	6.976E-04
1225	5.59 ± 0.42	24.8	8.1	2.04E-15	0.830	0.20576	1.976E-02	6.452E-04

Table 2. (Continued)

Temp-erature (°C)	Age(ka)	K/Ca	%rad	40*(mol)	Σ39Ar	40/39	37/39	36/39
1230	3.80 ± 0.55	24.7	5.6	1.10E-15	0.910	0.22539	1.987E-02	7.258E-04
1250	4.10 ± 0.77	24.7	4	7.96E-16	0.960	0.32003	1.984E-02	1.046E-03
1300	4.40 ± 1.26	24.0	2.6	5.31E-16	0.990	0.52829	2.038E-02	1.747E-03
1350	18.19 ± 5.72	21.2	1.5	6.37E-16	1.000	3.73432	2.308E-02	1.246E-02
DI7B-2 Sanidine, Capo Miseno								
J = 0.00017613								
550	2673.68 ± 8805.45	13.8	0.6	7.96E-16	0.000	1529.355	3.552E-02	5.147E+00
600	−139.69 ± 63.29	7.1	−7	−7.64E-16	0.002	6.28185	6.875E-02	2.276E-02
650	−152.61 ± 49.26	7.8	−12.5	−1.07E-15	0.004	3.84486	6.273E-02	1.465E-02
700	7.38 ± 9.95	10.4	0.7	2.81E-16	0.017	3.23317	4.716E-02	1.087E-02
750	36.97 ± 10.08	16.0	8.7	1.28E-15	0.028	1.33054	3.059E-02	4.118E-03
800	28.47 ± 5.13	18.7	9.5	1.96E-15	0.051	0.94182	2.625E-02	2.891E-03
850	4.40 ± 4.69	20.4	2	3.32E-16	0.075	0.71482	2.405E-02	2.378E-03
900	11.11 ± 2.75	22.1	7	1.42E-15	0.120	0.50106	2.222E-02	1.583E-03
950	5.22 ± 2.43	23.2	4.3	7.85E-16	0.160	0.39996	2.114E-02	1.301E-03
1000	5.07 ± 2.04	24.0	4.6	8.78E-16	0.220	0.34709	2.044E-02	1.126E-03
1050	2.98 ± 1.61	24.1	3.5	7.29E-16	0.290	0.29548	2.034E-02	9.702E-04
1100	6.56 ± 1.24	24.1	5.6	1.93E-15	0.390	0.37061	2.032E-02	1.190E-03
1150	6.19 ± 1.28	24.6	7.7	1.73E-15	0.480	0.25637	1.995E-02	8.064E-04
1200	4.62 ± 1.09	24.7	6	1.55E-15	0.590	0.24846	1.987E-02	7.958E-04
1250	4.40 ± 0.62	24.9	9.2	2.55E-15	0.780	0.15135	1.966E-02	4.706E-04
1300	4.40 ± 0.68	24.9	11.5	2.29E-15	0.950	0.12176	1.965E-02	3.701E-04
1350	1.64 ± 2.30	24.5	3.5	2.74E-16	1.000	0.16305	2.001E-02	5.380E-04
SV8 Sanidine, Porto Miseno								
J = 0.0001259								
600	-8.66 ± 28.57	21.5	−0.9	-6.90E-17	0.004	4.33114	2.283E-02	1.479E-02
675	0.97 ± 15.15	23.7	0.1	1.48E-17	0.013	3.24836	2.068E-02	1.098E-02
750	4.40 ± 7.31	25.6	1.2	1.40E-16	0.030	1.56269	1.910E-02	5.228E-03
825	3.43 ± 4.17	25.7	2.8	2.07E-16	0.060	0.58366	1.909E-02	1.925E-03
900	9.17 ± 2.52	26.0	8.8	8.88E-16	0.110	0.45924	1.884E-02	1.423E-03
950	2.76 ± 2.44	25.3	1.2	3.01E-16	0.170	1.00991	1.939E-02	3.381E-03
1000	6.56 ± 1.90	25.7	13.2	9.02E-16	0.240	0.22007	1.910E-02	6.515E-04
1050	7.08 ± 1.46	26.2	19.4	1.47E-15	0.350	0.16752	1.872E-02	4.621E-04
1100	5.89 ± 1.37	25.9	10.4	1.35E-15	0.470	0.25525	1.892E-02	7.792E-04
1150	5.74 ± 1.39	25.9	11.3	1.26E-15	0.590	0.22551	1.889E-02	6.819E-04
1200	5.59 ± 1.60	25.7	12.2	9.59E-16	0.690	0.19928	1.906E-02	5.973E-04
1250	6.56 ± 1.15	25.4	11.5	2.20E-15	0.870	0.25601	1.931E-02	7.725E-04
1300	9.17 ± 1.55	24.7	7.4	1.77E-15	0.970	0.54917	1.983E-02	1.726E-03
1350	8.35 ± 5.35	25.7	5.4	3.72E-16	0.990	0.69945	1.905E-02	2.244E-03
1400	6.56 ± 18.26	28.4	2.4	8.38E-17	1.000	1.26217	1.725E-02	4.176E-03

J= neutron fluence monitor. Errors quoted are given at one standard error (1σ). K/Ca = $0.49*{}^{39}Ar/{}^{37}Ar$, % rad = 40radiogenic/40total*100, and measured ratios are corrected for mass discrimination, blank and background. Plateaus are defined as 50% or more of the ^{39}Ar released within 2σ error.

Table 3. Summary of $^{40}Ar/^{39}Ar$ incremental-heating experiments.

Miseno $^{40}Ar/^{39}Ar$ ages										
Sample	Description	Separate	Total gas age (ka)	Plateau age (ka)	Steps used) (%39 released)	Isochron age (ka)	Steps used (%39 released)	40/36 Intercept	MSWD	Comment
SV8	Porto Miseno tuff ring	Sanidine	6.41 ± 0.58	**6.49 ± 0.51**	600–1400 (100%)	6.71 ± 0.80	600–1400 (100%)	293.4 ± 2.8	0.63	Excellent plateau
DI7A-1	Capo Miseno tuff cone	Sanidine	5.89 ± 0.21	—	—	5.07 ± 0.30	775–1350 (97.7%)	297.1 ± 1.3	8.44	Disturbed spectrum, discordant isochron
DI7A-2	Capo Miseno tuff cone	Sanidine	5.74 ± 0.33	—	—	5.59 ± 0.47	800–1300 (96.5%)	293.4 ± 2.0	14.2	Disturbed spectrum, discordant isochron
DI7B-1	Capo Miseno tuff cone	Sanidine	4.92 ± 0.22	**5.16 ± 0.16**	700–1300 (98.8%)	5.59 ± 0.27	550–1350 (100%)	294.5 ± 0.7	9.49	Short plateau, recoil spectrum
DI7B-2	Capo Miseno tuff cone	Sanidine	5.59 ± 0.99	**4.75 ± 0.35**	850–1350 (94.9%)	4.40 ± 0.42	550–1350 (100%)	297.6 ± 1.0	3.45	Excellent plateau
	Weighted mean of Capo Miseno DI7B samples			**5.09 ± 0.14**						Best age

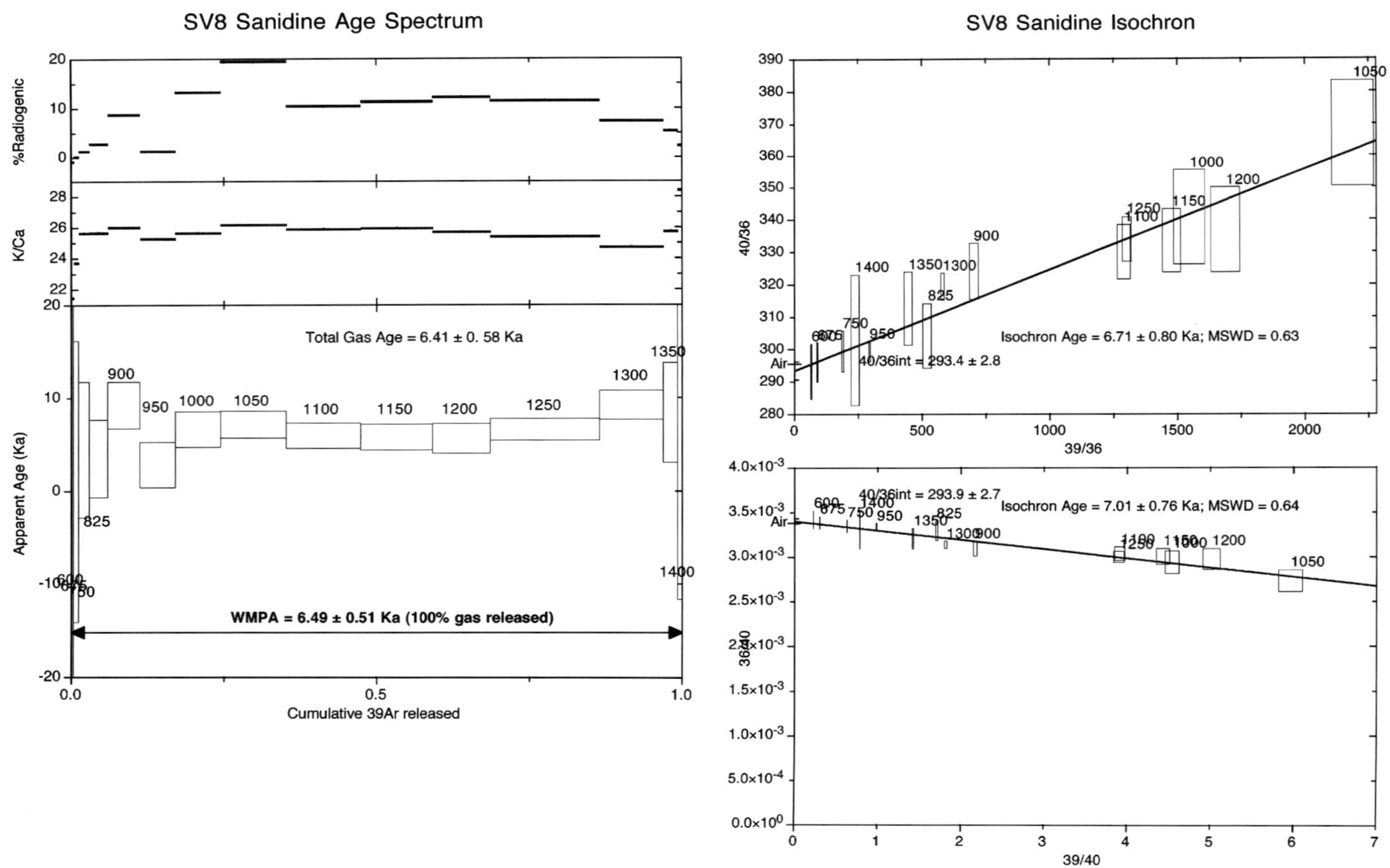

Figure 5. $^{40}Ar/^{39}Ar$ age spectrum and isochron diagram for sample DI7B-1 from Capo Miseno tuff cone. Half of the vertical dimension of the increment boxes is the estimated standard deviation of precision of the increment age. The error in the weighted mean plateau age is the weighted standard deviation.

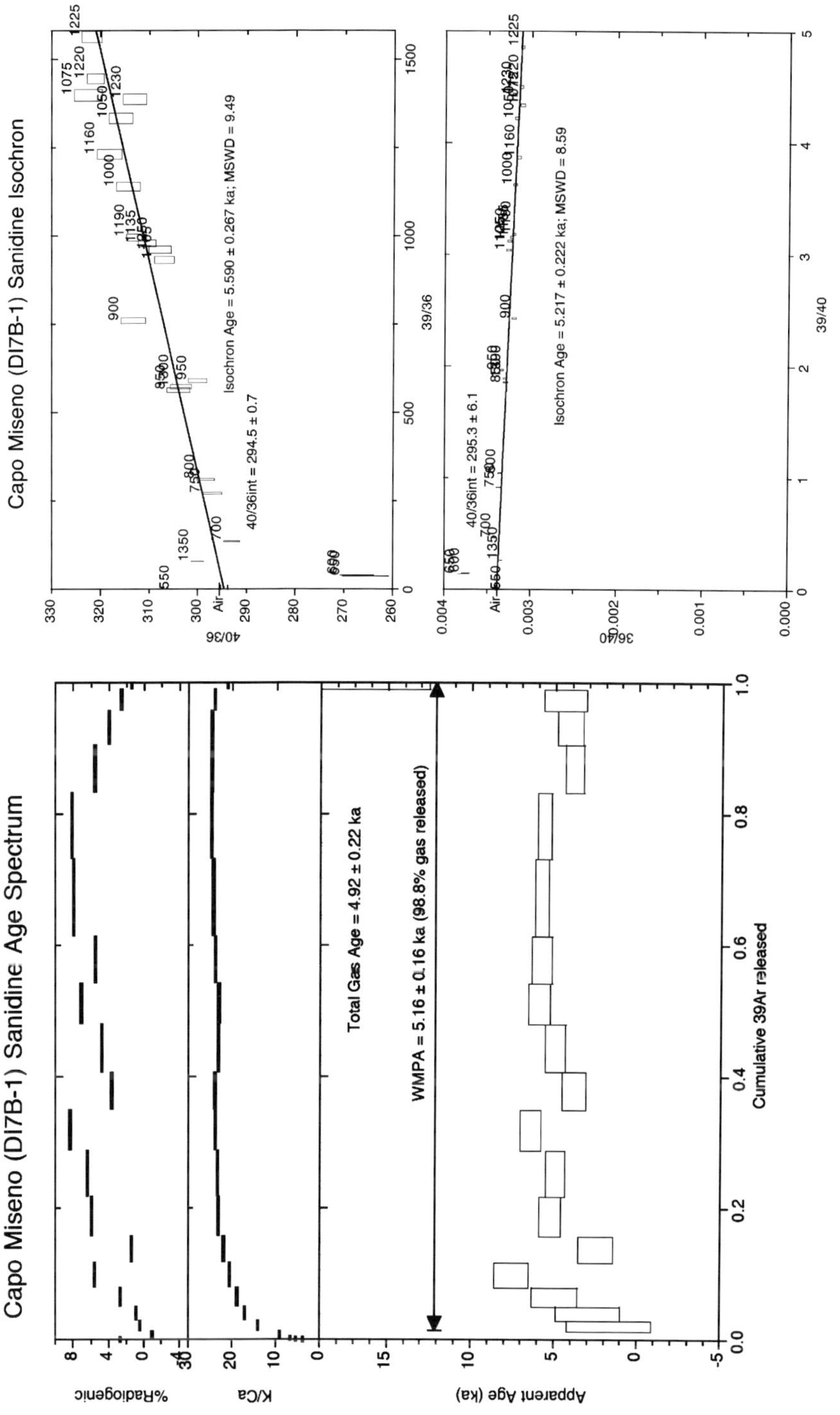

Figure 6. $^{40}Ar/^{39}Ar$ age spectrum and isochron diagram for sample DI7B-2 from Capo Miseno tuff cone.

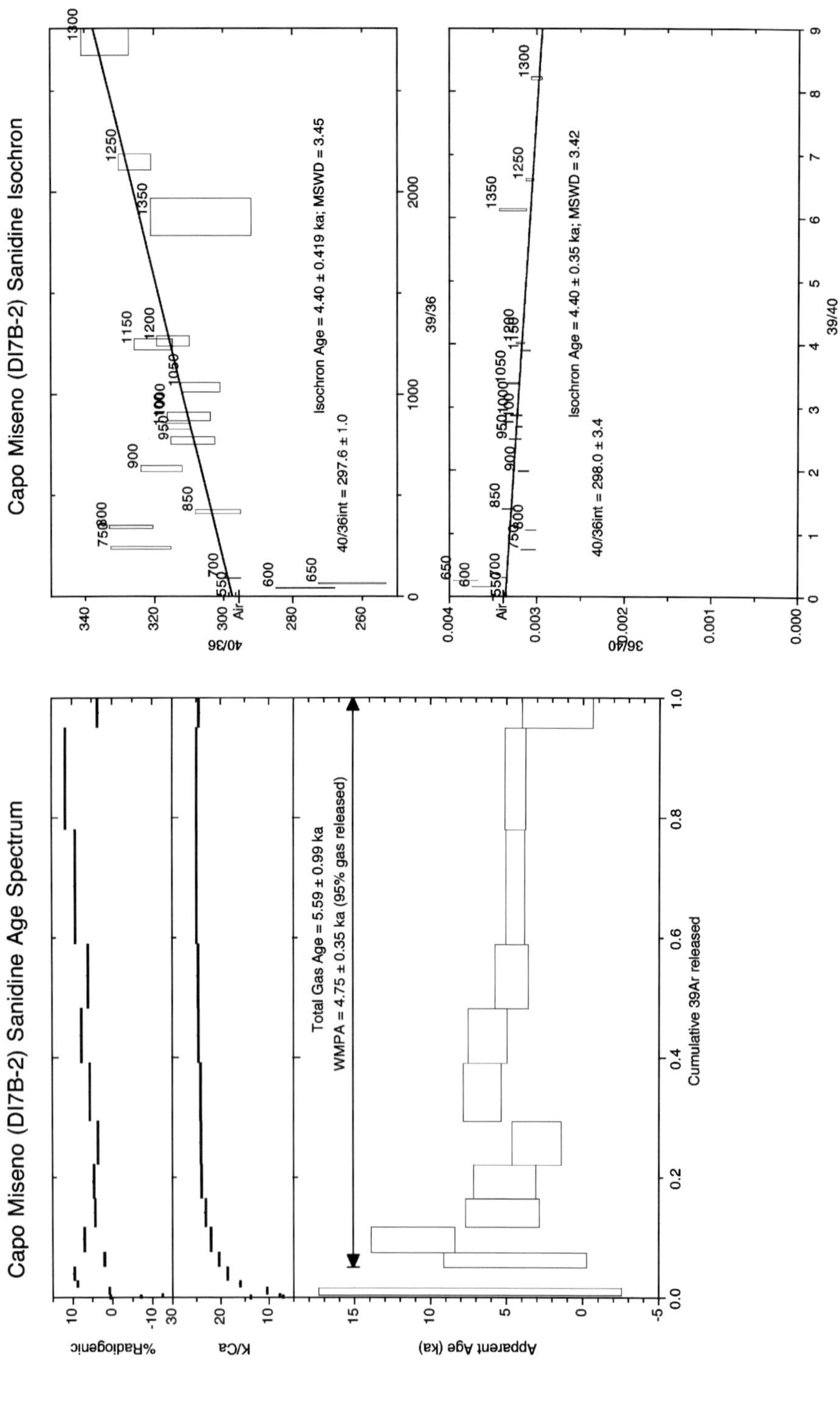

Figure 7. $^{40}Ar/^{39}Ar$ age spectrum and isochron diagram for sample SV8 from Porto Miseno tuff ring.

In sum, the best data for Porto Miseno tuff ring is 6490 ± 510 yr BP and for Capo Miseno tuff cone is at 5090 ± 140 yr BP.

5.2. The marine succession of Porto Miseno

5.2.1. Stratigraphy

The marine succession of Porto Miseno has been sampled by MGF cores down to a maximum depth of 40 m bsl (Fig. 8). Detailed logging allows the recognition of five stratigraphic units including marine deposits and a 1–5 m thick pyroclastic layer (tephra C). Marine sediments are represented, from the bottom to the top of the succession, by units E and D (gravel and fine sand, respectively) and units B and A (coarse and very fine sand, respectively) (Insinga et al., 2002); two peat layers have also been recovered.

Unit E: Very coarse-gravelly sand. The lowermost part of the succession is characterised by at least 15 m (core MGF5) of very coarse sand with subordinate gravel, including pumice and angular clasts. The sediment texture and grain size suggest a shoreface depositional setting for this unit.

Unit D: Fine volcanic sand. This unit reaches up to 13 m of thickness and is represented by fine sand with a significant volcaniclastic component. The upper part of this deposit is rich in mollusc shells (gastropods and bivalve fragments) while the lower part is characterised by low-matrix content. The depositional setting is typical of a low-energy foreshore area.

Tephra C: Pumice in sandy matrix. Tephra C was cored at depths between 14 m and 19 m and reaches a maximum thickness of 5 m in core MGF1. The deposit is characterised by coarse grey pumice (5 cm of maximum diameter) with sanidine phenocrysts in a sandy volcanic matrix. The occurrence of well-rounded pebbles in this pyroclastic deposit and red laminae on pumice suggests a shoreface depositional environment.

Unit B: Medium-coarse size sand. This unit forms a normally graded succession characterised by medium to coarse sand passing to fine sand towards the top. These deposits are considered to be the bottom of a short transgressive–regressive marine cycle.

Unit A: Fine sand. The uppermost unit is characterised by volcanic fine sand with abundant plant remains and bioclasts. In particular, at the base of this unit (6 m beneath the sea floor), bivalve shell (*Cardium* sp) beds locally occur (cores MGF2, MGF3 and MGF5). Facies analysis suggests a low-energy foreshore area for unit A.

5.2.2. Peat layers

Two peat layers have been recovered in the succession within unit B (layer t_1), at a depth ranging from 8 to 12 m, and at the transition between units D and E (layer t_2) at a depth between 26 and 31.5 m (Fig. 8). The thickness of layer t_1 ranges from 50 cm in cores MGF1 and MGF5 to 20 cm in core MGF3; layer t_2, recovered in cores MGF1, MGF4 and MGF5, is a lens about 80 cm thick which may be locally correlated to pumice (core MGF4).

Radiocarbon measurements yielded ages of 3560 ± 40 yr BP for the layer t_1, sampled in core MGF1 at a depth of 31.5 m, and 7815 ± 55 yr BP for the layer t_2 sampled in core MGF3 at a depth of 9.50 m (Insinga et al., 2002).

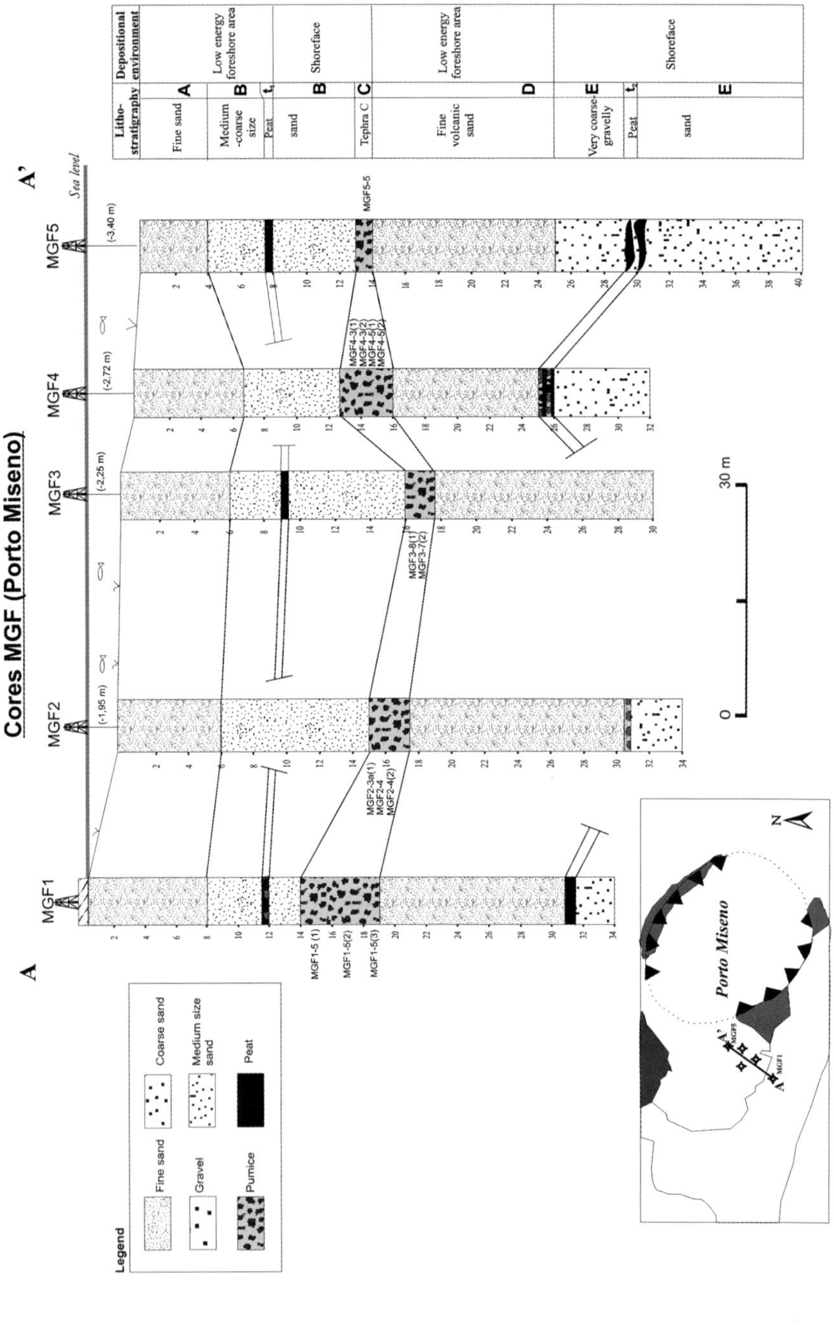

Figure 8. Stratigraphic correlation among deep five cores at Porto Miseno. Pumice from tephra C, sampled for petrochemical analysis, are marked.

5.2.3. Petrochemistry and source vent of tephra C

Pumice from tephra C range in composition from trachyphonolites to phonolites with a maximum occurrence in the trachyphonolite field (Fig. 9 and Table 4). No evidence of significant chemical variation among samples has been found. The trachyphonolitic composition of tephra C is in agreement with the average composition of CF products which are characterised by relatively homogenenous major element contents for the Late-Holocene products. Discrimination among the possible sources can be made on the basis of trace elements analysis. Particularly, the strongly incompatible nature in these volcanic series of some trace elements like Zr and Nb (Melluso et al., 1995) and their scarce mobilisation during secondary processes, such as diagenesis in marine environments, make them among the most helpful tools in land–sea tephrostratigraphic correlations. The sedimentary texture, particularly the grain size and the poor sorting, as well as the occurrence in the marine succession of Porto Miseno, suggests, moreover, that the eruptive centre of tephra C is local. This petrochemical and stratigraphic approach allow us to correlate the studied tephra with Capo Miseno deposits (Table 4) (5090 yr BP, this paper) and to exclude other flegrean events occurred during the III Epoch of activity.

5.2.4. Depositional environment and relative sea level

The Holocene stratigraphy of Porto Miseno is characterised by abundant volcanic sand including a 1–5 m thick pyroclastic layer (tephra C).

Facies analysis indicates that the fine-grained deposits accumulated in a shallow marine-paralic environment (shoreface, coastal lagoon) due to low-energy conditions and the lack

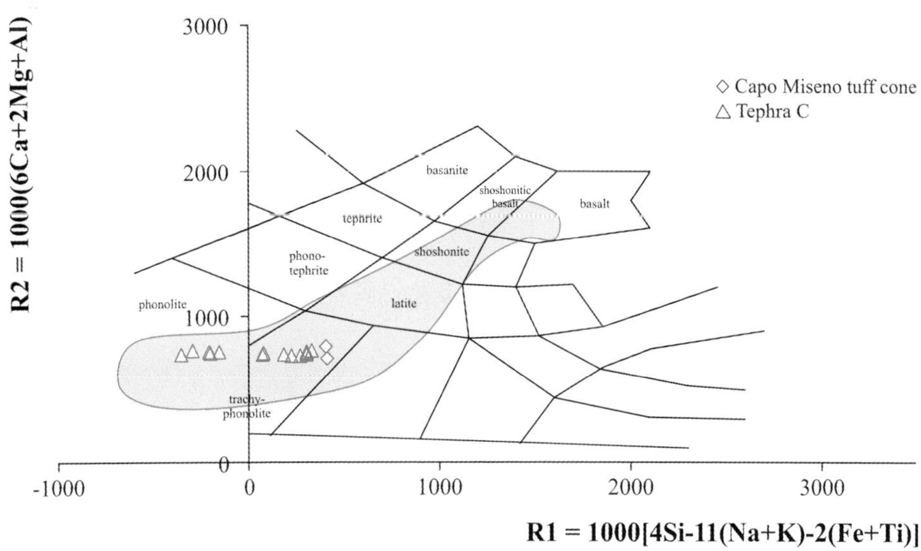

Figure 9. Classification of pumice samples from tephra C and Capo Miseno tuff cone in the R1–R2 diagram (De La Roche et al., 1980), see Table 4 for analysis. Average composition of the Campi Flegrei products have been plotted for comparison. (Database from Di Girolamo et al., 1984; Rosi and Sbrana, 1987; D'Antonio et al., 1999; Pappalardo et al., 2002 and references therein.)

Table 4. Composition of tephra C recovered in the Porto Miseno marine succession.

	Tephra C														Capo Miseno	
Sample Composition	MGF1-5(2) TP	MGF1-5(2) TP	MGF1-5(3) TP	MGF2-3a(1) TP	MGF 2-4 TP	MGF2-4(2) TP	MGF2-4(2) TP	MGF3-7(2) TP	MGF3-8(1) TP	MGF4-3(1) TP	MGF4-3(2) TP	MGF4-5(1) TP	MGF4-5(2) TP	MGF5-5 TP	DI7 TP	97122 MNa TP
SiO_2	59.02	58.56	58.85	59.12	58.19	58.70	58.87	58.15	58.51	58.98	58.53	57.73	57.86	58.11	59.76	58.87
TiO_2	0.52	0.53	0.52	0.52	0.51	0.51	0.52	0.51	0.50	0.51	0.51	0.51	0.52	0.52	0.49	0.49
Al_2O_3	18.68	19.08	18.94	18.72	17.86	18.64	18.72	18.37	18.71	18.76	18.69	18.23	18.35	18.26	18.52	19.16
Fe_2O_3	4.47	4.74	4.45	4.45	4.38	4.69	4.71	4.45	4.45	4.44	4.48	4.41	4.51	4.51	4.45	4.34
MnO	0.16	0.15	0.16	0.16	0.16	0.15	0.16	0.16	0.15	0.16	0.16	0.16	0.17	0.16	0.13	0.11
MgO	1.25	1.37	1.30	1.14	1.42	1.31	1.33	1.36	1.38	1.09	1.30	1.56	1.48	1.38	0.79	0.91
CaO	2.85	3.01	2.99	2.98	2.95	2.93	2.93	2.98	2.82	2.93	2.95	3.12	3.06	3.13	2.93	3.52
Na_2O	4.31	4.43	4.42	4.36	6.20	4.78	4.44	5.89	4.86	4.49	5.03	5.91	5.68	5.63	3.80	4.09
K_2O	8.58	7.95	8.20	8.39	8.15	8.11	8.16	7.96	8.45	8.47	8.18	8.19	8.21	8.12	8.96	8.31
P_2O_5	0.16	0.17	0.16	0.16	0.17	0.17	0.17	0.17	0.17	0.17	0.17	0.19	0.17	0.19	0.15	0.19
sum	100.00	100.00	100.00	100.00	100.00	100.00	100.00	100.00	100.00	100.00	100.00	100.00	100.00	100.00	100.00	100.00
LOI	3.00	3.40	2.56	2.41	4.41	3.50	3.00	3.30	2.76	2.12	3.00	3.22	3.75	4.46	2.18	3.19
AI	0.88	0.83	0.85	0.87	1.06	0.89	0.86	1.00	0.92	0.88	0.92	1.02	0.99	0.99	0.86	0.82
Alk	12.89	12.38	12.62	12.75	14.35	12.88	12.59	13.85	13.31	12.96	13.20	14.10	13.89	13.75	12.77	12.40
R1	270	337	309	305	−353	188	307	−203	73	229	78	−291	−208	−153	411	405
R2	733	764	757	742	737	745	747	747	737	735	747	768	760	761	716	798
Sc	6	4	2	4	3	4	3	2	4	1	4	2	3	1	4	4.5
V	76	79	75	72	74	71	78	75	75	75	76	83	75	80	70	80
Zn	101	82	82	84	78	79	79	78	74	81	82	74	77	84	75	
Rb	324	258	306	309	299	287	288	285	317	322	296	299	308	304	306	324
Sr	664	671	641	637	674	640	652	663	667	632	651	731	680	680	658	702
Y	37	35	39	35	37	34	37	35	38	34	39	36	38	35	26	29
Zr	348	346	362	339	307	338	352	338	340	343	349	314	330	318	349	342
Nb	47	52	54	50	47	47	51	46	48	51	51	47	50	49	50	49.0
Ba	973	861	846	840	995	863	881	869	877	884	833	1092	931	915	899	901
La	76	82	72	61	69	72	76	73	68	71	67	65	78	72	—	73
Ce	140	150	162	140	127	156	129	151	159	138	139	132	131	149	—	139
Nd	50	56	58	59	48	59	59	58	60	57	51	51	51	49	—	50
Pb	46	49	45	46	50	49	46	42	49	68	48	34	60	40	—	51
Th	32	34	34	37	30	34	37	30	35	37	37	28	35	29	—	30.5

Major (%) and trace (ppm) elements analysed with XRF. Two samples from Capo Miseno tuff cone are reported for comparison: DI7 from this work and 97122MNa (analysed with ICP-MS) from D'Antonio et al. (1999).

LOI, loss on ignition; TP, trachyphonolite; AI, molar $(Na_2O+K_2O)/Al_2O_3$; R1-R2 parameters from De La Roche et al. (1980).

of long-shore currents. The general transgressive trend of the lower part of this succession is locally interrupted by short-regressive aggradational episodes, possibly induced by clastic input from the surrounding hydrovolcanic vents. Local increase of pyroclastic input ostensibly resulted in a forced progradation of the coastline, where the degree of syneruptive reworking is very high (Thorarinsson et al., 1964). Pyroclastic deposits, in fact, practically fill up the available accomodation space and create favourable conditions for the occurrence of peat layers. In particular, the rapid accumulation up to 5 m (core MGF1) of tephra C may have induced the transition from unit D (shoreface) to unit B (beach) through a phase of low-rate deposition and the formation of layer t_1. Angular clasts and red laminae at the top of tephra C can be interpreted as the result of oxidation under vadose and/or subaerial conditions which characterised this evolutionary stage of the coastline transition.

6. The Miseno area

6.1. Late-Holocene volcanic activity

The $^{40}Ar/^{39}Ar$ ages of 6490 ± 510 yr BP and of 5090 ± 140 yr BP obtained in this work from collected samples of Porto Miseno tuff ring and Capo Miseno tuff cone, respectively, suggest an intense volcanic activity in this area during the Late Holocene i.e. much earlier than reported in the previous literature.

The $^{40}Ar/^{39}Ar$ ages are not in agreement with those reported in the previous chronostratigraphic reconstruction of CF volcanism where the Miseno events are located in the I Epoch (15,000–9500 yr BP; Di Vito et al., 1999; Orsi et al., 2004). Following our results, the Porto Miseno tuff ring and Capo Miseno tuff cone erupted in the time interval spanning from 6500 to 5000 yr BP which corresponds, in that reconstruction, to a period of quiescence of the entire CF caldera between the II and the III Epoch of volcanism (8000–4800 yr BP). Furthermore, the occurrence in the Pozzuoli Bay of a tephra layer petrochemically correlated with Capo Miseno tuff cone deposits (Insinga, 2003) evidences a dispersion of products not limited to proximal areas such as Porto Miseno where tephra C has been recovered.

The new $^{40}Ar/^{39}Ar$ age of Capo Miseno tuff cone is close to the K/Ar age of 4000 yr BP proposed by Di Girolamo et al. (1984). Their radiometric result was regarded as incompatible with the apparently high-erosion rates that have modified the original shape of the cone (Rosi and Sbrana, 1987). The volcano-tectonic activity, documented in the Pozzuoli Bay (Milia and Torrente, 2000) and along its coastal zone with the uplift of La Starza terrace (Cinque et al., 1985), as well as the relative sea-level changes, may have affected the relatively soft deposits of these hydrovolcanic centres controlling their morphologic evolution. Moreover, as evident at the Spiaggetta Verde site, strong marine erosion affected the southern flank of Porto Miseno vent in the brief period (~1400 yr) before the emplacement of the Capo Miseno products. This suggests for the Miseno area that major erosional processes might have occurred in very short periods and this has been frequently observed where tuff cones and tuff rings occur (e.g. Capelinhos; Cole et al., 2001).

We can speculate at this point that the Holocene transgression, that inundated the present coastal areas, might have triggered or enhanced this hydromagmatic activity as reported for other volcanic district (Sohn et al., 2002). It is now becoming apparent on a global scale that large changes in sea level can influence the internal-stress regimes of coastal and island volcanoes acting through a range of mechanisms including water-table changes and variations in confining pressures (McGuire et al., 1997).

6.2. Volcaniclastic sedimentation

The marine succession of Porto Miseno covers at least the last 8000 yr and the overall sedimentary characteristics suggests that volcanoclastic input from the surrounding vents of the Capo Miseno tuff cone and Porto Miseno tuff ring strongly influenced the types of deposits, their depositional environments and sedimentation rates. No evidences have been found of volcanic inputs coming from other eruptive centres of CF caldera active during this time span. Our geochronologic data on tephra and peat deposits in the Porto Miseno succession, permit an estimate of average non-decompacted depositional rates of about 6 m/ka in the time span ranging from 7800 (peat layer t_2) to 5100 yr BP (tephra C), of 3.3 m/ka from 5100 to 3600 yr BP (peat layer t_1) and finally of 2.2 m/ka for the last 3600 yr at Porto Miseno drilling site (Fig. 10).

The value of 6 m/1000 yr, much higher than the average calculated for the last 5000 yr, is explained here as the result of erosional processes which affected Porto Miseno vent after its formation at 6490 yr BP. The tuff ring acted as a local but significant volcaniclastic source to the marine depositional system until the Capo Miseno eruption occurred causing the syneruptive arrival of tephra C at 5090 yr . Unit D, resulting in more than 10 m of fine volcanic sand, represents the marine stratigraphic signature left by the Porto Miseno event. The deposition of pumice layer C appears to have caused rapid aggradation and/or progradation of the coastline, resulting in the formation of peat layer t_1. The 20–50 cm thickness of this layer implies that the transitional setting likely covered a period not longer than 1 ka: according to Cameron et al. (1989), in fact, Holocene peats in temperate regions typically accumulate at rates of about 20–200 cm/1000 yr.

Based on radiocarbon ages, we can correlate peat layer t_1 (3560 $\pm$ 40 years BP) and t_2 (7815±55 yr BP) with two paleosol horizons developed during periods of volcanic quiescence within the CF caldera. In particular, layer t_1 is correlated with the quiescent phase between Senga (3.7 ka BP) and Monte Nuovo eruption (AD 1538) while layer t_2 may be correlated with paleosoil B (8200–4800 yr BP, Di Vito et al., 1999) that is interbedded with volcanic deposits erupted during the II and the III Epoch of activity. It is evident, however, that in the Miseno area layer t_1 formed as a response of the sedimentary environment to local volcanic activity.

Finally, the absolute age determinations of event horizons in the Porto Miseno succession made possible a chronostratigraphic correlation with other Holocene epiclastic successions recovered in the Pozzuoli Bay and along its coastal area (Fig. 11). Accordingly, unit D and units C–B–A, described here, can be correlated with the distinct seismostratigraphic units G2 and G1, respectively, mapped in the central and eastern sectors of Pozzuoli Bay (Milia, 1996, 1998; Milia and Torrente, 2000). They can be also correlated with the section cropping out onshore, along the erosional slope of "La Starza" terrace (Cinque et al., 1985; Barra, 1991), and with the stratigraphic units filling the "Bagnoli–Fuorigrotta depression" (Calderoni and Russo, 1998).

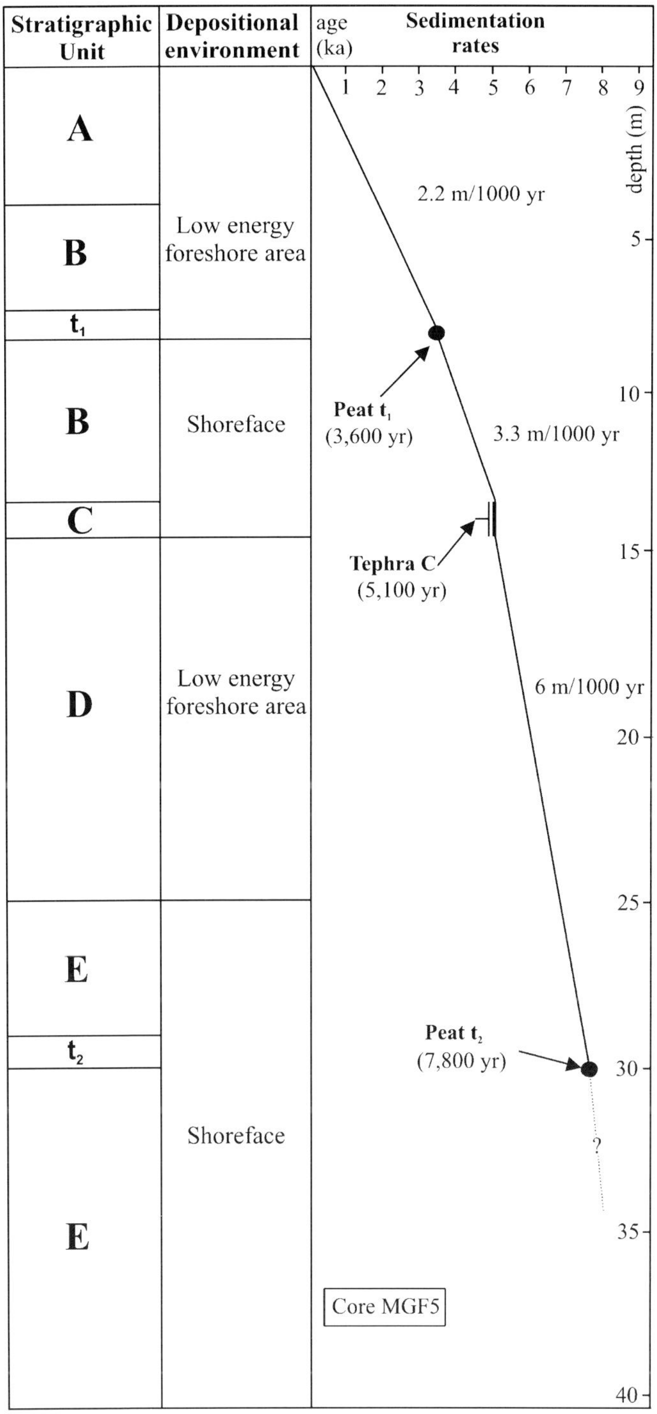

Figure 10. Stratigraphic units, depositional environments and non-decompacted sedimentation rates for the Porto Misero marine succession.

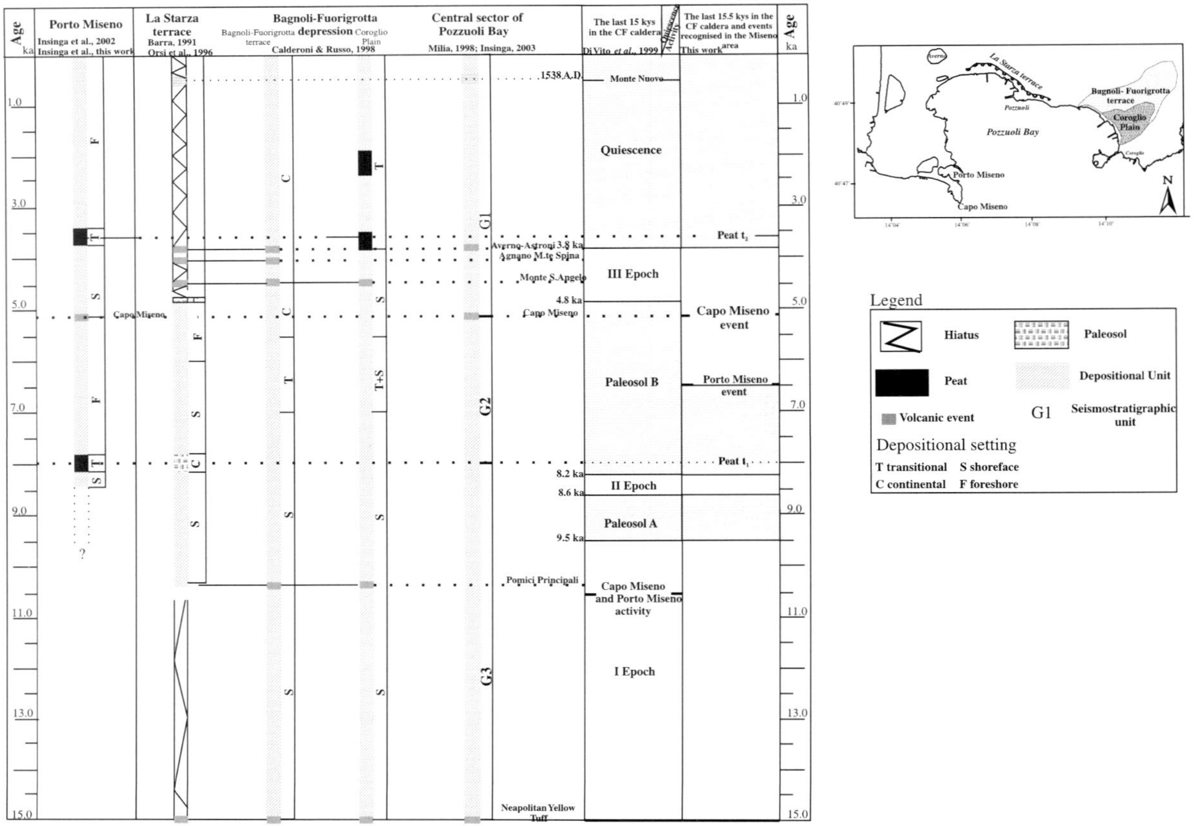

Figure 11. Event stratigraphy for the Porto Miseno succession and others epiclastic deposits of Pozzuoli Bay coastline and offshore area. The chronostratigraphic reconstruction of flegrean activity by Di Vito et al. (1999) has been modified on the basis of Ar/Ar ages obtained in this work for Capo Miseno tuff cone and Porto Miseno tuff ring in the western sector of CF caldera.

7. Concluding remarks

Stratigraphic, petrochemical and geochronologic data presented in this work permit the reconstruction of the evolution of the Miseno area in the south-western sector of CF caldera during the last 8000 yr. The most important remarks and implications on the CF chronostratigraphy can be summarised as follows:

- The chronostratigraphic reconstruction of the events in the Miseno area, as inferred by $^{40}Ar/^{39}Ar$ dating of Capo Miseno tuff cone and Porto Miseno tuff ring deposits, indicates that volcanic activity affected the CF caldera rim until the Late Holocene and the area is ostensibly younger than previously reported. Pyroclastic deposits of Capo Miseno tuff cone are well preserved both within nearshore successions (Porto Miseno) and offshore marine deposits (unit G1) where they represent a distinctive marker horizon.
- Volcaniclastic input from Miseno vents played a major role in the inner shelf stratigraphic evolution off the south-western CF which is characterised by transgressive sandy deposits. The marine sedimentation is locally interrupted by two short regressive episodes bracketed by peat layers (t_1 and t_2) dated at 7800 yr BP and 3560 yr BP These levels have been correlated with two major paleosols of CF caldera, although at least layer t_1 formed as the response of sedimentary environment to the Capo Miseno tuff cone event and the arrival of its proximal deposits (tephra C).

Acknowledgements

We wish to thank Bruno D'Argenio and Leone Melluso for suggestions and critical review of an early version of the manuscript. Roberto Scandone and Paul Cole greatly improved the paper with their comments and suggestions.

This work was supported by D. Insinga's PhD, PRIN (2003 to Leone Melluso) and Regione Campania L.5-2005 (to Vincenzo Morra) grants and by a national project aimed to the geological mapping of the coastal zone of Campania at 1:50000 scale (CARG project).

References

Albore Livadie, C., 1986. Tremblements de terre, eruptions volcaniques et vie des hommes dans la Campanie antique. Institut Français de Naples, Naples.

Alessio, M., Bella, F., Improta, S., Belluomini, G., Cortesi, C., Turi, B., 1971. University of Rome Carbon-14 Dates IX. Radiocarbon 13, 395–411.

Allard, P., Maiorani, A., Tedesco, D., Cortecci, G., Turi, B., 1991. Isotopic study of the origin of sulfur and carbon in Solfatara fumaroles, Campi Flegrei caldera. J. Volcanol. Geotherm. Res. 48, 139–159.

Barberi, F., Corrado, G., Innocenti, F., Luongo, G., 1984. Phlegraean Fields 1982–1984: brief chronicle of a volcano emergency in a densely populated area. Bull. Volcanol. 47, 175–185.

Barra, D., 1991. Studio del Pleistocene superiore-Olocene delle aree vulcaniche campane. PhD thesis, University of Naples Federico II, Naples, 298pp.

Beccaluva, L., Di Girolamo, P., Serri, G., 1991. Petrogenesis and tectonic setting of Roman Volcanic Province, Italy. Lithos 26, 191–221.

Calderoni, G., Russo, F., 1998. The geomorphological evolution of the outskirts of Naples durino the Holocene: a case study of the Bagnoli–Fuorigrotta depression. Holocene 8, 581–588.

Cameron, C.C., Esterle, J.S., Palmer, C.A., 1989. The geology, botany and chemistry of selected peat forming environments from temperate and tropical latitudes. In: Lyons, P.C., Alpern, B. (Eds), Peat and Coal: Origin, Facies, and Depositional Models. Elsevier, Amsterdam, pp. 105–156.

Caputo, P., 1989. Attività di tutela della Soprintendenza Archeologica di Napoli e Caserta, Atti II Conv. Naz. Archeologia subacquea. Ministero Beni AA.CC., Roma.

Carbone, A., Lirer, L., Munno, R., 1984. Caratteri petrografici dei livelli piroclastici rinvenuti in alcuni gravity cores nel Golfo di Pozzuoli e di Napoli. Mem. Soc. Geol. It. 27, 195–204.

Cas, R.A.F., Wright, J.V., 1987. Volcanic Successions, Ancient and Modern. Unwin Hyman, London, 528pp.

Cinque, A., Augelli, P.P.C., Brancaccio, L., Mele, R., Milia, A., Robustelli, G., Romano, P., Russo, F., Russo, M., Santangelo, N., Sgambati, D., 1997. Volcanism, tectonic and recent geomorphological change in the Bay of Napoli. Suppl. Geogr. Fis. Dinam. Quat. 3, 123–141.

Cinque, A., Rolandi, G., Zamparelli, V., 1985. L'estensione dei depositi marini olocenici nei Campi Flegrei in relazione alla vulcano-tettonica. Boll. Soc. Geol. It. 104, 327–348.

Cinque, A., Russo, F., Pagano, M., 1991. La successione dei terreni di età post-romana delle terme di Miseno (Napoli): nuovi dati per la storia e la stratigrafia del bradisisma puteolano. Boll. Soc. Geol. It. 110, 231–244.

Colantoni, P., Del Monte, M., Fabbri, A., Gallignani, P., Selli, R., Tomadin, L., 1972. Ricerche geologiche nel Golfo di Pozzuoli. In: Versino, L. (Ed.), Relazione sui rilievi effettuati nell'area flegrea nel 1970—71. CNR, Quaderni de "La Ricerca Scientifica" 83, pp. 23–76.

Cole, P.D., Guest, J.E., Duncan, A.M., Pacheco, J.-M., 2001. Capelinhos 1957–1958, Faial, Azores: deposits formed by an emergent surtseyan eruption. Bull. Volcanol. 63, 201–220.

Corrado, G., Guerra, I., Lo Bascio, A., Luongo, G., Rampoldi, R., 1977. Inflation and microearthquake activity of Phlegraean Fields, Italy. Bull. Volcanol. 40, 169–188.

Dalrymple, G.B., 1989. The GLM continuous laser system for $^{40}Ar/^{39}Ar$ dating: description and performance characteristics. U.S. Geol. Surv. Bull. 1890, 89–96.

Dalrymple, G.B., Alexander, E.C., Jr., Lanphere, M.A., Kraker, G.P., 1981. Irradiation of samples for $^{40}Ar/^{39}Ar$ dating using the Geological Survey TRIGA reactor. U.S. Geol. Surv. Professional Paper 1176.

D'Antonio, M., Civetta, L., Orsi, G., Pappalardo, L., Piochi, M., Carandente, A., de Vita, S., Di Vito, M.A., Isaia, R., 1999. The present state of the magmatic system of the Campi Flegrei caldera based on a reconstruction of its behavior in the past 12 ka. J. Volcanol. Geotherm. Res. 91, 247–268

D'Argenio, A., Pescatore, T.S., Senatore, M.R., 2004. Sea-level change and volcano tectonic interplay. The Gulf of Pozzuoli (Campi Flegrei, Eastern Tyrrhenian Sea) during the last 39 ka. J. Volcanol. Geotherm. Res. 133, 105–121.

de Vita, S., Orsi, G., Civetta, L. Carandente, A., D'Antonio, M., Deino, A., di Cesare, T., Di Vito, M.A., Fisher, R.V., Isaia, R., Marotta, E., Necco, A., Ort, M., Pappalardo, L., Piochi, M., Saitton, J., 1999. The Agnano-Monte spina eruption (4100 years B.P.) in the restless Campi Flegrei caldera (Italy). J. Volcanol. Geotherm. Res. 91, 269–301.

de'Gennaro, M., Cappelletti, P., Langella, A., Perrotta, A., Scarpati, C., 2000. Genesis of zeolites in the Neapolitan Yellow Tuff; geological, volcanological and mineralogical evidence. Contrib. Mineral. Petrol. 139, 17–38.

de'Gennaro, M., Incoronato, A., Mastrolorenzo, G., Adabbo, M., Spina, G., 1999. Depositional mechanisms and alteration processes in different types of pyroclastic deposits from Campi Flegrei volcanic field (Southern Italy). J. Volcanol. Geotherm. Res. 91, 303–320.

Deino, A.L., Orsi, G., de Vita, S., Piochi, M., 2004. The age of the Neapolitan Yellow Tuff caldera-forming eruption (Campi Flegrei caldera, Italy) assessed by $^{40}Ar/^{39}Ar$ dating method. J. Volcanol. Geotherm. Res. 133, 157–170.

De La Roche, H., Leterrier, P., Grandclaude, P., Marchal, E., 1980. A classification of volcanic and plutonic rocks using R1-R2 diagram and major element analyses. Its relationships with current nomenclature. Chem. Geol. 29, 183–210.

De Lorenzo, G., 1905. I crateri di Miseno nei Campi Flegrei. Atti Rend. Acc. Sc. Fis. Mat. 13, 1–25.

De Vivo, B., Rolandi, G., Gans, P.B., Calvert, A., Bohrson, W.A., Spera, F.J., Belkin, H.E., 2001. New constraints on the pyroclastic eruptive history of the Campanian volcanic Plain (Italy). Mineral. Petrol. 73, 47–65.

Di Girolamo, P., Ghiara, M.R., Lirer, L., Munno, R., Rolandi, G., Stanzione, D. 1984. Vulcanologia e petrologia dei Campi Flegrei. Boll. Soc. Geol. It. 103, 349–413.

Di Vito, M., Isaia, R., Orsi, G., Southon, J., de Vita, S., D'Antonio, M., Pappalardo, L., Piochi, M., 1999. Volcanism and deformation since 12,000 years at the Campi Flegrei caldera (Italy). J. Volcanol. Geotherm. Res. 91, 221–246.

Di Vito, M., Lirer, L., Mastrolorenzo, G., Rolandi, G., 1987. The Monte Nuovo eruption (Campi Flegrei, Italy). Bull. Volcanol. 49, 608–615.

Duffield, W.A., Dalrymple, G.B., 1990. The Taylor Creek Rhyolite of New Mexico: a rapidly emplaced field of lava domes and flows. Bull. Volcanol. 52, 475–487.

Dvorak, J.J., Mastrolorenzo, G., 1991. The mechanisms of recent vertical crustal movements in Campi Flegrei caldera, southern Italy. Geol. Soc. Am. Bull., Special Paper, p. 47.

Franzini, M., Leoni, L., Saitta, M., 1972. Revisione di una metodologia analitica per fluorescenza-X, basata sulla correzione completa degli effetti di matrice. Rendiconti della Società Italiana di Mineralogia e Petrologia 31, 365–378.

Heiken, G.H., 1971. Tuff rings: examples from the Fort Rock-Christmas Lake valley basin, South Central Oregon. J. Geophys. Res. 75, 5615–5626.

Insinga, D., 2003. Tefrostratigrafia dei depositi tardo-quaternari della fascia costiera campana. Ph.D. thesis, University of Naples Federico II, Naples, 202pp.

Insinga, D., Di Meglio, A., Molisso, F., Sacchi, M., 2002. Stratigrafia e caratteristiche fisiche dei depositi olocenici del porto di Miseno, Golfo di Pozzuoli (Tirreno centro-orientale). Il Quaternario 15, 9–19.

Lajoie, J., Stix, J., 1992. Volcaniclastic rocks. In: Walzer, R.G., James, N.P. (Eds), Facies Models: Response to Sea Level Change. Geol. Assoc. Can., Toronto pp. 101–118.

Lanphere, M.A., 2000. Comparison of conventional K-Ar and $^{40}Ar/^{39}Ar$ dating of young mafic volcanic rocks. Quaternary Res. 53, 294–301.

Lanphere, M., Dalrymple, G.B., 2000. First principles calibration of ^{38}Ar tracers: implications for the ages of $^{40}Ar/^{38}Ar$ fluence monitors. US Geol. Surv. Professional Paper 1621.

Leoni, L., Saitta, M., 1976. X-ray fluorescence analysis of 29 trace elements in rock and mineral standards. Rend. Soc. It. Mineral. Petrol. 32, 497–510.

Lirer, L., Luongo, G., Scandone, R., 1987. On the volcanological evolution of Campi Flegrei. EOS 68 (16), 226–233.

McGuire, W.J., Howarth, R.J., Firth, C.R., Solow, A.R., Pullen, A.D., Saunders, S.J., Stewart, I.S., Vita-Finzi, C., 1997. Correlation between rate of sea-level change and frequency of esplosive volcanism in the Mediterranean. Nature 389, 473–476.

Milia, A., 1996. Evoluzione tettono-stratigrafica di un bacino peritirrenico: Il Golfo di Napoli, Ph.D. thesis, University of Naples Federico II.

Milia, A., 1998. Stratigrafia, strutture deformative e considerazioni sull'origine delle unità deposizionali oloceniche del Golfo di Pozzuoli (Napoli). Boll. Soc. Geol. It. 117, 777–787.

Milia, A., Torrente, M.M., 2000. Fold uplift and synkinematic stratal architectures in a region of active transtensional tectonics and volcanism, eastern Tyrrhenian Sea. Geol. Soc. Am. Bull. 112, 1531–1542.

Milia, A., Torrente, M.M., Giordano, F., 2000. Active deformation and volcanism offshore Campi Flegrei, Italy: new data from high-resolution seismic reflection profiles. Mar. Geol. 171, 61–73.

Melluso, L., Morra, V., Perrotta, A., Scarpati, C., Adabbo, M., 1995. The eruption of the Breccia Museo (Campi Flegrei, Italy): fractional crystallization processes in a shallow, zoned magma chamber and implications for the eruptive dynamics. J. Volcanol. Geotherm. Res. 68, 325–339.

Orsi, G., de Vita, S., Di Vito, M., 1996. The restless resurgent Campi Flegrei nested caldera (Italy): constraints on its evolution and configuration. J. Volcanol. Geotherm. Res. 74, 179–214.

Orsi, G., Di Vito, M., Isaia, R., 2004. Volcanic hazard assessment at the restless Campi Flegrei caldera. Bull. Volcanol. 66, 514–530.

Pappalardo, L., Civetta, L., D'Antonio, M., Deino, A., Di Vito, M., Orsi, G., Carandente, A., de Vita, S., Isaia, R., Piochi, M., 1999. Chemical and Sr-isotopical evolution of the Phlegrean magmatic system before the Campanian Ignimbrite and the Neapolitan Yellow Tuff eruptions. J. Volcanol. Geotherm. Res. 91, 141–166.

Pappalardo, L., Piochi, M., D'Antonio, M., Civetta, L., Petrini, R., 2002. Evidence for multi-stage magmatic evolution during the past 60 kyr at Campi Flegrei (Italy) deduced from Sr, Nd and Pb isotope data. J. Petrol. 43, 1415–1434.

Pescatore, T.S., Diplomatico, G., Senatore, M.R., Tramutoli, M., Mirabile, L., 1984. Contributi allo studio del Golfo di Pozzuoli: aspetti stratigrafici e strutturali. Mem. Soc. Geol. It. 27, 133–149.

Ricci, G., Lanphere, M., Morra, V., Perrotta, A., Scarpati, C, Melluso, L., 2000. Volcanological, geochemical and geochronological data from ancient pyroclastic successions of Campi Flegrei (Italy). AGU 2000 Fall Meeting, Eos, Transactions, American Geophysical Union, 81 (48).

Rosi, M., Sbrana, A., 1987. Phlegrean Fields. CNR, Quaderni de "La Ricerca Scientifica", Vol. 114–119, 176pp.

Scandone, R., Bellucci, F., Lirer, L., Rolandi, G., 1991. The structure of the Campanian Plain and the activity of the Neapolitan volcanoes (Italy). J. Volcanol. Geotherm. Res. 48 (1/2), 1–31.

Scarpati, C., Cole, P., Perrotta, A., 1993. The Neapolitan Yellow Tuff – A large volume multiphase eruption from Campi Flegrei, Southern Italy. Bull. Volcanol. 55, 343–356.

Sohn, Y.K., Park, J.B., Khim, B.K., Park, K.H., Koh, G.W., 2002. Stratigraphy, petrochemistry and Quaternary depositional record of the Songaksan tuff ring, Jeju Island, Korea. J. Volcanol. Geotherm. Res. 119, 1–20.

Thorarinsson, S., Einarsson, T., Sigvaldason, G., Elisson, G., 1964. The submarine eruption off the Vestmann Islands 1963–1964. Bull. Volcanol. 27, 435–445.

Volcanism in the Campania Plain:
Vesuvius, Campi Flegrei and Ignimbrites
edited by B. De Vivo

Chapter 7

Magmatic–hydrothermal fluid interaction and mineralization in alkali-syenite nodules from the Breccia Museo pyroclastic deposit, Naples, Italy

Luca Fedele[a,*], *Maurizio Tarzia*[a], *Harvey E. Belkin*[b], *Benedetto De Vivo*[a], *Annamaria Lima*[a] *and Jacob B. Lowenstern*[c]

[a]*Dipartimento di Scienze della Terra, Università degli Studi di Napoli, Federico II, Via Mezzocannone 8, Napoli 80134, Italy*
[b]*U.S. Geological Survey, 956 National Center, Reston, VA 20192, USA*
[c]*U.S. Geological Survey, Bldg 15, McKelvey Building, Menlo Park, CA 94025, USA*

Abstract

The Breccia Museo, a pyroclastic flow that crops out in the Campi Flegrei volcanic complex (Naples, Italy), contains alkali-syenite (trachyte) nodules with enrichment in Cl and incompatible elements (e.g., U, Zr, Th, and rare-earth elements). Zircon was dated at ~52 ka, by U–Th isotope systematics using a SHRIMP. Scanning electron microscope and electron microprobe analysis of the constituent phases have documented the mineralogical and textural evolution of the nodules of feldspar and mafic accumulations on the magma chamber margins. Detailed electron microprobe data are given for alkali and plagioclase feldspar, salite to ferrosalite clinopyroxene, pargasite, ferropargasite, magnesio-hastingsite hornblende amphibole, biotite mica, Cl-rich scapolite, and a member (probable davyne-type) of the cancrinite group. Detailed whole rock, major and minor element data are also presented for selected nodules. A wide variety of common and uncommon accessory minerals were identified such as zircon, baddeleyite, zirconolite, pollucite, sodalite, titanite, monazite, cheralite, apatite, titanomagnetite and its alteration products, scheelite, ferberite, uraninite/thorianite, uranpyrochlore, thorite, pyrite, chalcopyrite, and galena. Scanning electron microscope analysis of opened fluid inclusions identified halite, sylvite, anhydrite, tungstates, carbonates, silicates, sulfides, and phosphates; most are probably daughter minerals. Microthermometric determinations on secondary fluid inclusions hosted by alkali feldspar define a temperature regime dominated by hypersaline aqueous fluids. Fluid-inclusion temperature data and mineral-pair geothermometers for coexisting feldspars and hornblende and plagioclase were used to construct a pressure–temperature scenario for the development and evolution of the nodules. We have compared the environment of porphyry copper formation and the petrogenetic environment constructed for the studied nodules. The suite of ore minerals observed in the nodules supports a potential for mineralization, which is similar to that observed in the alkaline volcanic systems of southern Italy (Pantelleria, Pontine Archipelago, Mt. Somma-Vesuvius).

1. Introduction

Ore-forming fluids are either directly derived from magma or strongly influenced by the hydrothermal system attending the cooling magma body, or variable mixtures of the two. Recent studies have addressed the metallogenetic potential related to the extensive

*Corresponding author. *E-mail address*: lufedele@unina.it (L. Fedele).

alkaline volcanism of the Campanian Province, southern Italy. Paone (1999) and Paone et al. (2001) pointed out the similarity between the Mt. Somma-Vesuvius volcanic system and mineralized alkaline volcanic systems in general (e.g., Kelley et al., 1998). Detailed studies carried out by De Vivo et al. (1995) and Belkin et al. (1996) on cognate syenite nodules from the nearby islands of Ventotene and Ponza (Pontine Archipelago) have documented the existence of immiscibility between hydrosaline fluids and silicate melt, proving the presence of hydrothermal fluids of magmatic origin that are very similar to those reported from porphyry copper systems (Roedder, 1984; Cline and Bodnar, 1994). Occurrences of this type are also reported from other subvolcanic systems of southern Italy such as Pantelleria (De Vivo et al., 1992, 1993; Lowenstern, 1994), Mt. Somma-Vesuvius (Fulignati et al., 1997; Gilg et al., 2001), and Campi Flegrei (CF) (Tarzia et al., 1999, 2000).

The CF volcanic system is the largest alkaline volcanic complex in the Campanian Province where the last eruption occurred in 1538 AD and that created the small cone, Monte Nuovo. However, the CF system, according to Rosi et al. (1983, 1991), Fisher et al. (1993), and Orsi et al. (1996), is also the source of the eruption of the Campanian Ignimbrite, considered as a unique event at 37 ka. In contrast, according to De Vivo et al. (2001) and Rolandi et al. (2003), the Campanian Ignimbrite (CI) has an age of 39 ka with an estimated volume of about 180 km^3 (dry rock equivalent) and is only one of the different ignimbrite events that originated from fissures (see also Bellucci et al., this volume), active at various times and places in the Campania Plain in the period from 300 to 18 ka. Large volumes of magma have also been emplaced prior to eruption(s) and some magma fraction is currently cooling at depth, driving the abundant geothermal manifestations observed in the CF volcanic system.

The Breccia Museo (BM), in the CF system, is a volcanic breccia of complex origin which contains abundant fragments of juvenile lava, country rock, hydrothermally altered rock, and feldspar-dominated cumulate nodules (syenites) that we interpret to represent portions of a magma chamber margin. These alkali-syenite nodules illustrate the processes, mineralogy, and chemistry of the magma–host rock–hydrothermal system interaction and show the metallogenetic potential of the CF volcanic system. Here we emphasize our study of the trapped fluids and present a general discussion of the observed mineralogy. More detailed presentations of the feldspathoid and zirconium minerals are in preparation.

1.1. The Breccia Museo eruption (geologic setting)

The BM, named by Johnston-Lavis (1889), is a volcanic breccia which crops out in the southwest sector of the CF, a large and active volcanic field located west of Naples, Italy (Fig. 1). Lirer et al. (1991) dated charcoal from the BM by ^{14}C at ~21 ka and considered it unrelated to the 39 ka CI. Many authors (Di Girolamo et al., 1984; Perrotta, 1992; Perrotta and Scarpati, 1994) propose an origin by explosive activity from distinct monogenetic vents, while several other studies (Rosi et al., 1983, 1991; Rosi and Sbrana, 1987; Rosi and Vezzoli, 1989) suggest that the BM is indeed associated with the CI and it is an unique event dated ~37 ka, originating from the CF caldera. Two recent comprehensive studies by De Vivo et al. (2001) and Rolandi et al. (2003) dated the CI at ~39 ka by $^{40}Ar/^{39}Ar$ from sanidine and suggested that the ignimbrite events originated from fissure emissions related to regional fault systems in the Campanian Plain.

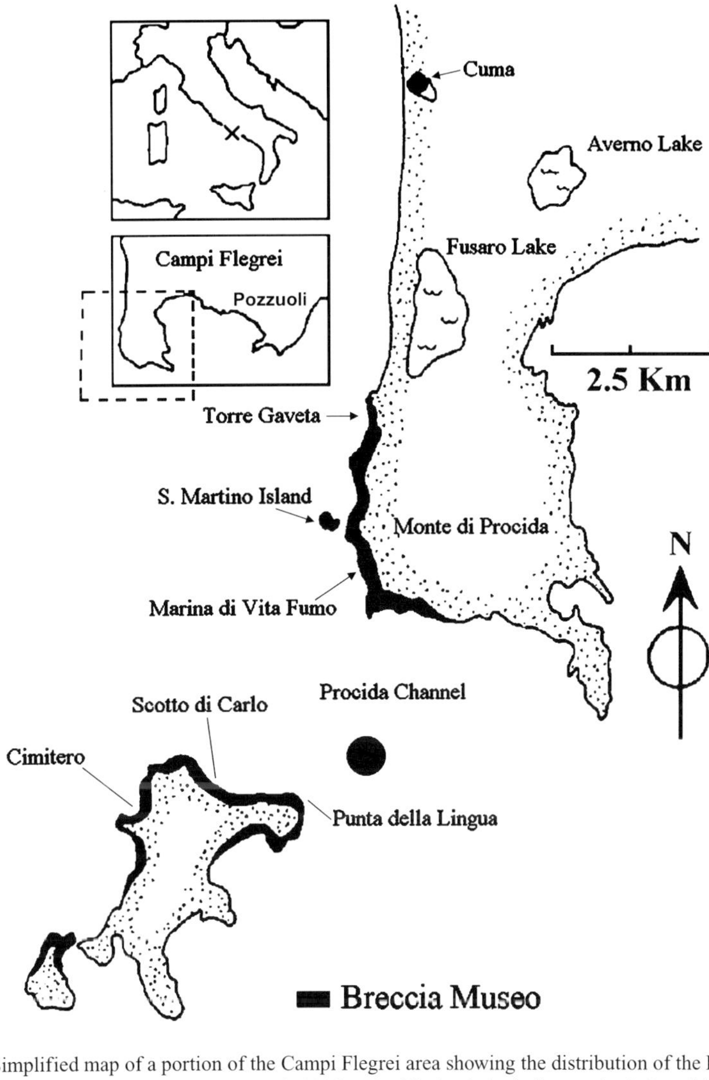

Figure 1. Simplified map of a portion of the Campi Flegrei area showing the distribution of the Breccia Museo deposit (modified from Perrotta and Scarpati, 1994). The black circle indicates the inferred location of the eruption vent.

Perrotta and Scarpati (1994) divide the BM deposit into four stratigraphic units grouped into two overlapping depositional units: (1) Lower Depositional Unit (LDU), a poorly sorted, matrix-supported pyroclastic flow deposit; (2) Upper Depositional Unit, consisting of layers of lithic breccias, coarse welded spatter clasts, and pumice flow(s) [Breccia Unit (BU), Spatter Unit (SU), Upper Pumice Flow Unit (UPFU); Fig. 2]. Perrotta and Scarpati (1994) propose the formation of the BM by an unsteady density-stratified flow (e.g., Branney and Kokelaar, 1992) controlled by morphology.

Melluso et al. (1995) classified the products of the BM, which range from trachyte to trachyphonolite and are characterized by an assemblage dominated by sanidine (Or_{88-63}) ± Na-plagioclase (An_{33-27}) ± biotite ± titanomagnetite ± apatite. Chemical variations

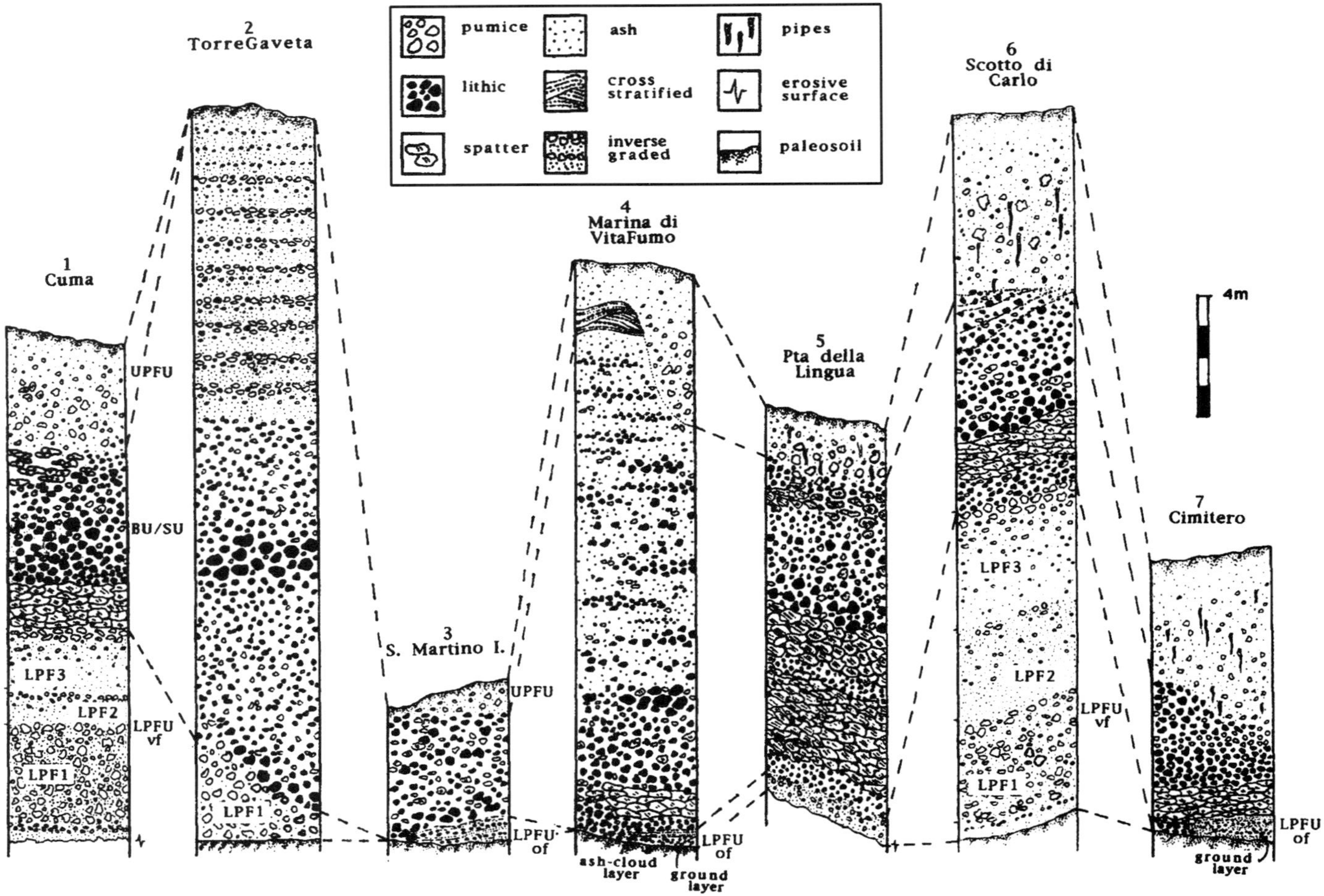

Figure 2. Stratigraphic sections of the BM member showing in detail the single units and their facies variations (after Perrotta and Scarpati, 1994). Locations are noted in Figure 1. LPFU, Lower Pumice Flow Unit; of, overbank facies; vf, valley facies. LPFU-associated layers are also indicated. BU (Breccia Unit) and SU (Spatter Unit) are grouped owing to the interfingering of the two units. For every unit, the maximum thickness is shown.

(major and trace elements) suggest an evolution from a less-differentiated trachyte by fractional crystallization of a sanidine-rich assemblage. Stratigraphic features and chemistry suggest the formation of the BM by progressive tapping of a thermally and chemically zoned shallow magma chamber.

1.2. Campi Flegrei hydrothermal system

De Vivo et al. (1989), carried out a detailed fluid inclusion (FIs) study on hydrothermal minerals from the CF geothermal system and identified an extensive alteration of the basement volcanic rocks and the existence of a shallow low-salinity fluid (~4% NaCl eq.) and deeper hypersaline fluids (>26% NaCl eq.) that were generated either by a continuous boiling process at depth, near a magmatic body, or by addition of magmatic fluids. Caprarelli et al. (1997) suggest the existence of two distinct reservoirs and origins for these geothermal fluids: (1) seawater infiltrated at relatively shallow depth (< 2000 m) and mixed with steam-heated groundwater, and (2) a deeper (>2000 m) hypersaline fluid of probable magmatic origin mixed with meteoric water. The two reservoirs show little mixing (if any) probably due to fluid density contrast.

All the data strongly suggest a similarity between the aqueous fluids associated with the CF–BM magma chamber(s) and the mineralized brines related to porphyry-type systems (Roedder, 1984). In these systems, there is increasing evidence that metal transport occurs in high-salinity brines or in hydrosaline melts which exsolved from silicate magmas (Roedder, 1971; Kilinc and Burnham, 1972; Cline and Bodnar, 1991, 1994; Kamenetsky et al., 1999).

The hydrothermal system of the BM (CF) magma chamber margin resembles the characteristics of other hydrothermal systems associated with the Italian alkaline volcanism, such as Pantelleria, Pontine islands, and Mt. Somma-Vesuvius (De Vivo et al., 1989, 1992, 1993, 1995; Lowenstern, 1994; Belkin et al., 1996; De Vivo, 1999; Paone, 1999; Gilg et al., 2001; Paone et al., 2001).

2. Studied samples

The nodules, generally light-colored, medium-grained, and ranging in size from 1 to 15 cm, were collected from the lithic horizon (BU) of the BM formation (Figs. 1 and 2). Sample MT14 was collected from Torre Franco-Verdolino (Soccavo); samples BL2, BL3, CFNA, VFB, BL8, MT17, MT19, MT20, MT21, MT22, MT24, and MT26 from Punta della Lingua (Procida), and sample MT27 from Scoglio cannone (Procida). Polished thin sections were prepared for microbeam analysis, crystals were separated for FI study, and a representative split of five selected samples was crushed and powdered for chemical analysis.

2.1. Radiometric age

In order to evaluate the timing and genetic relationships between the studied nodules, the BM eruption, and the CF volcanic system, we dated one sample using U–Th isotopic systematics of single zircon crystals extracted by crushing and hand picking from nodule

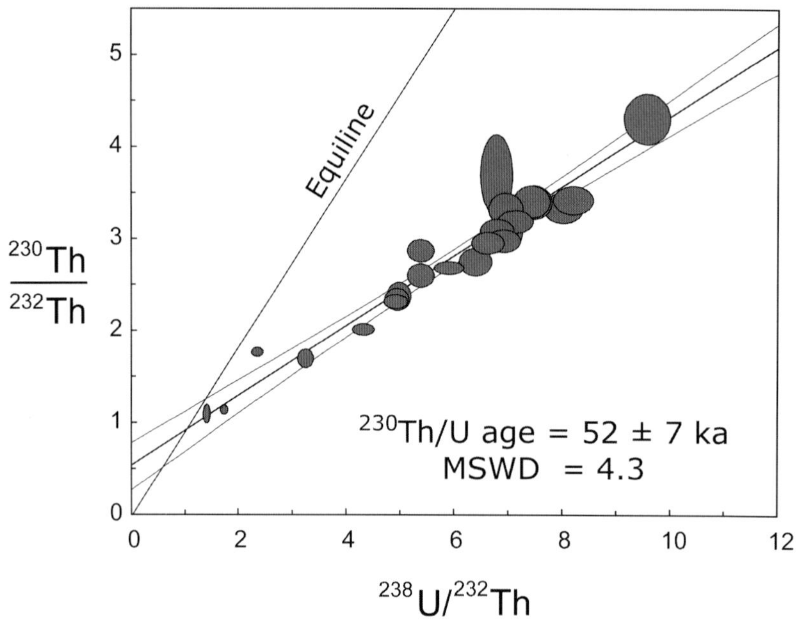

Figure 3. Plots of (^{238}U)/(^{232}Th) versus (^{230}Th)/(^{232}Th) for zircon crystals from a nodule (MT24) of the BM. The line through the data represents the least-squared regression (with MSWD) as calculated by ISOPLOT (Lowenstern et al., 2000). The age is calculated from the slope of the regression line. Data-point error ellipses represent 68.3% confidence.

MT24. Uranium and Th data were collected by one of us (Lowenstern) on the Stanford-USGS SHRIMP-RG SIMS, using analytical techniques and data reduction similar to those described in Lowenstern et al. (2000). Eighteen different zircons were analyzed (one of the 18 was analyzed seven times to check for data consistency). Analyses give model ages between 39 and ~100 ka, but the oldest and youngest two have much larger errors (Fig. 3). The oldest zircon has low U/Th and thus is more subject to errors as it is closer to the intercept. Owing to the fact that all the data fall very close to a straight line, we conclude that all the zircons were crystallized within a very narrow window of a few thousand years. However, some zircons could be older. The calculated model age is 52 ± 7 ka (Fig. 3), which is compatible with one of the proposed ages for the BM (≈37 ka, Rosi et al., 1983, 1991; Rosi and Sbrana, 1987; Rosi and Vezzoli, 1989).

2.2. Analytical methods

FI measurements were carried out on a Linkam THS600 heating–freezing stage at the Dipartimento di Scienze della Terra, Università degli Studi di Napoli, Federico II. The stage was calibrated for the temperature range of interest using synthetic FIs. Precision can be estimated at ±1°C in the range 200–350°C.

The nodules were studied at the U.S. Geological Survey, Reston, VA, USA using a JEOL-840 scanning electron microscope (SEM) equipped with a Princeton Gamma-Tech energy-dispersive X-ray fluorescence analyzer (SEM-EDS) and a JEOL JXA-8900 fully automated, five-spectrometer electron microprobe for wavelength-dispersive analysis (EMPA). Electron microprobe operating conditions were 20 kV voltage and 30 nA beam current for oxides and sulfides and 15 kV voltage and 20 nA beam current for other phases. Reported beam currents were measured with a Faraday cup. Minor elements were counted for 60 s and major elements for 20 s. Standardization was done before each analytical session on synthetic and natural silicates, oxides, glasses, and phosphates. Count data were reduced online by a ZAF or phi-rho-z correction algorithm as supplied by JEOL USA, Inc. Analytical chemistry was done by Activation Laboratories, Ancaster, Ontario, Canada. Inductively coupled plasma mass spectrometry (ICP-MS) and inductively coupled plasma-atomic emission spectrometry (ICP-AES) were done on solutions using a lithium metaborate/tetraborate fusion procedure before sample dissolution with multiple acids. Other analytical techniques were: loss on ignition (LOI) by gravimetric (GRAV) techniques, infrared spectrometry (INFR), ion selective electrode (ISE), prompt gamma neutron activation analysis (PGNAA), and instrumental neutron activation analysis (INAA).

2.3. Chemistry of the samples

Table 1 gives the detailed chemical data for major, trace, and volatile elements in five representative samples. The nodules, medium-grained alkali-feldspar accumulations, compositionally are classified as syenites (trachyte equivalents) (Le Maitre, 2002). Samples containing scapolite and cancrinite-group minerals have a noticeable enrichment in Cl. Table 2 compares the mean of our five analyses of the BM nodules with the CI. The data show a strong chemical similarity supporting our assumption that the studied nodules are cognate and represent a medium-grained equivalent of the common products erupted from the CF volcanic system. Figure 4A shows the range of compositions on the total alkali-silica diagram (Le Maitre, 2002). The compositional range reflects the observed mineralogy. Figure 4B shows the co-variation of K_2O with CaO, Na_2O, and FeO (total). The good correlation (CaO, $r^2 = 0.82$; Na_2O, $r^2 = 0.43$; FeO, $r^2 = 0.96$) illustrates that the compositions result from a two-component mixture of A – magmatic alkali feldspar with oxides and mafics and B – late-stage Na-rich plagioclase, scapolite, and cancrinite-group minerals. The details of the mineralogy (see below) are different from typical CI or other CF rock compositions, which reflects their different formation histories; the nodules have been affected by pervasive late-stage magmatic fluids. Some nodules show high values of certain elements, for example, Zr in sample MT26 or Zn in sample BL2 (Table 1). This reflects the fact that each nodule represents a unique sample of the magma chamber margin where different volumes have been affected by somewhat different conditions with regard to incompatible elements and volatiles. The similarity in SiO_2, Al_2O_3, and total alkalis plus CaO confirm that the nodules are essentially accumulations of alkali feldspar with minor plagioclase and mafic minerals.

2.4. Petrography and mineralogy of the samples

The syenite nodules are primarily composed of potassium feldspars (up to ~80%) with subordinate plagioclase, scapolite, a S- and Cl-rich member of the cancrinite group, amphibole, pyroxene, biotite, magnetite, titanite, apatite, and uncommon sodalite. SEM-EDS and

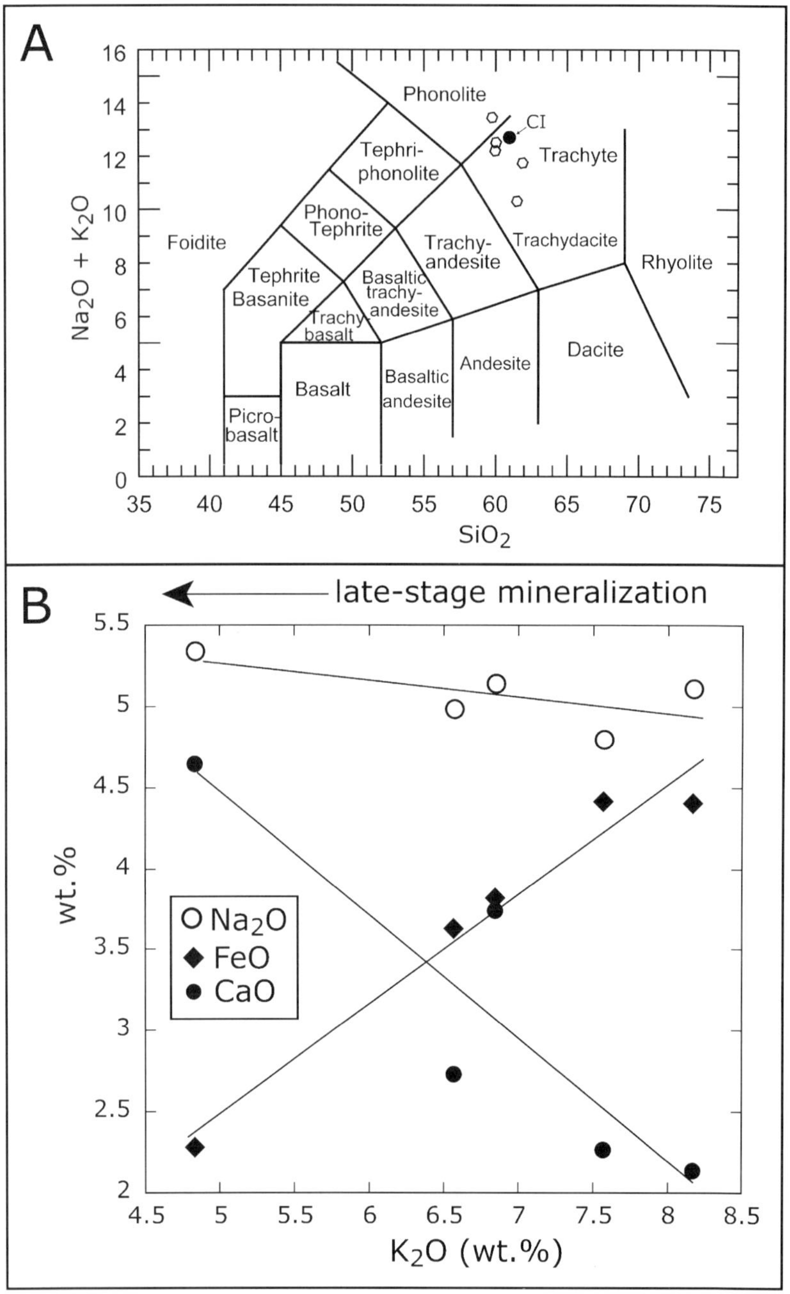

Figure 4. (A) Plot of nodule whole-rock chemistry on the total alkali versus silica (TAS) diagram of Le Maitre (2002). The five nodules are shown as open symbols and the average of 40 CI pumices (H.E. Belkin, unpublished data) also is shown as the black filled circle (CI). (B) Co-variation of selected chemistry of the five analyzed nodules. The linear regression curves for ◆K_2O–FeO, $r^2=0.96$; ○K_2O–Na_2O, $r^2=0.43$; and ●K_2O–CaO, $r^2=0.82$ are shown. The nodule compositions can be modeled as a two-component mixture of magmatic alkali feldspar, oxides, and mafics affected by late-stage mineralization (arrow), which contains less FeO, more CaO, and slightly more Na_2O.

Table 1. Chemical analyses of BM nodules. The following elements were analyzed but were below the detection limit; element (detection limit): CO_2 (0.05%), Ag (0.5 ppm), In (0.1 ppm). The particular method of analysis is given in the right column (definitions in text).

Sample	BL2 (%)	BL8 (%)	MT14 (%)	MT19 (%)	MT26 (%)	Method
SiO_2	59.19	60.98	61.10	59.16	59.54	ICP-AES
TiO_2	0.44	0.62	0.42	0.44	0.44	ICP-AES
Al_2O_3	18.71	19.47	18.04	18.59	19.30	ICP-AES
FeO*	4.40	2.28	3.63	3.82	4.41	ICP-AES
MnO	0.29	0.10	0.19	0.06	0.18	ICP-AES
MgO	0.45	0.82	0.86	0.51	0.52	ICP-AES
CaO	2.14	4.64	2.73	3.75	2.27	ICP-AES
Na_2O	5.11	5.34	4.99	5.15	4.80	ICP-AES
K_2O	8.17	4.83	6.57	6.85	7.57	ICP-AES
P_2O_5	0.10	0.02	0.11	0.21	0.09	ICP-AES
F	0.19	0.08	0.20	0.07	0.30	ISE
Cl	1.10	0.05	0.08	1.64	0.25	INAA
SO_3**	0.10	0.02	<0.01	0.05	0.02	INFR
LOI	1.17	<0.01	0.61	1.11	0.58	GRAV
−O=F,Cl	0.33	0.04	0.10	0.40	0.18	
Total	101.23	99.21	99.42	101.01	100.09	
ppm						
As	11	<5	9	10	27	ICP-MS
B	30	16	39	13	29	PGNAA
Ba	16	143	339	235	30	ICP-AES
Be	13	17	11	19	14	ICP-AES
Bi	0.31	<0.06	0.16	0.14	0.23	ICP-MS
Ce	198	176	182	217	230	ICP-MS
Co	2	<1	4	5	3	ICP-MS
Cr	<20	<20	<20	30	36	ICP-MS
Cs	25.5	1.7	37.3	10.4	103	ICP-MS
Cu	<10	12	<10	68	20	ICP-MS
Dy	8.73	8.75	7.86	10.4	10.1	ICP-MS
Er	5.38	4.86	4.43	5.80	6.16	ICP-MS
Eu	1.92	2.46	1.76	2.63	2.01	ICP-MS
Ga	22	23	21	22	24	ICP-MS
Gd	9.36	10.5	9.20	13.1	10.8	ICP-MS
Ge	1.5	1.3	1.4	1.3	1.4	ICP-MS
Hf	13.0	7.0	10.4	11.3	14.3	ICP-MS
Ho	1.74	1.67	1.47	1.95	1.94	ICP-MS
La	107	72.2	98.1	95.5	122	ICP-MS
Lu	0.863	0.641	0.660	0.780	1.04	ICP-MS
Mo	11	2	4	4	6	ICP-MS
Nb	90.3	94.4	72.8	100	109	ICP-MS
Nd	67.7	74.6	63.7	89.3	79.2	ICP-MS
Ni	<15	16	<15	44	46	ICP-MS
Pb	59	8	26	24	32	ICP-MS
Pr	20.3	21.0	18.9	25.3	23.8	ICP-MS

Table 1. Continued

Sample	BL2 (%)	BL8 (%)	MT14 (%)	MT19 (%)	MT26 (%)	Method
Rb	499	183	255	301	379	ICP-MS
Sb	0.9	<0.2	1.4	0.3	1.6	ICP-MS
Sc	2	3	4	2	2	ICP-AES
Sm	11.5	12.6	10.8	15.2	13.2	ICP-MS
Sn	7	11	4	11	8	ICP-MS
Sr	40	329	233	310	102	ICP-AES
Ta	5.2	5.0	4.4	5.5	6.3	ICP-MS
Tb	1.59	1.60	1.42	2.01	1.81	ICP-MS
Th	40.2	20.0	37.0	34.7	48.0	ICP-MS
Tl	2.25	0.44	0.52	0.70	3.83	ICP-MS
Tm	0.829	0.689	0.670	0.833	0.970	ICP-MS
U	17.5	2.89	12.8	6.98	16.2	ICP-MS
V	29	22	41	32	34	ICP-AES
W	21	1.9	5.5	3.2	2.8	ICP-MS
Y	56	52	46	61	63	ICP-AES
Yb	5.36	4.27	4.18	5.04	6.49	ICP-MS
Zn	112	<30	69	<30	53	ICP-MS
Zr	658	285	487	539	735	ICP-AES

FeO*, total iron; SO_3**, total sulfur.

Note: Less than values equal not detected at the lower limit.

electron microprobe analysis of these nodules has identified U- and Zr-bearing phases such as uranpyrochlore, U-bearing thorite, U-and REE-bearing phosphate (monazite group), U-bearing zircon, baddeleyite and probable zirconolite and, less abundant, sulfides, carbonates, and tungstates.

The interlocking nature of the feldspar crystals gave rise to small angular cavities/vugs. No glass was identified in these cavities (except in one sample, VFB) suggesting that the phases found therein precipitated from a dense, supercritical aqueous fluid. Thus, the nodules are a mix of major phases precipitated from a silicate liquid and minor phases precipitated from a late-stage aqueous fluid permeating and moving through a volume of earlier-formed crystals.

3. Nodules: mineral chemistry and petrography

Feldspars. Sanidine is by far the most abundant phase (Fig. 7A,B) with an approximate composition of $Or_{73}Ab_{26}An_1$, whereas early plagioclase has a typical composition of $Or_5Ab_{55}An_{35}$ (Table 3). The sanidine composition is fairly homogeneous from core to rim, suggesting equilibrium with the host liquid; neither Sr nor Ba were detected by EMPA. Early-formed plagioclase, characterized by euhedral to subhedral crystal shapes, is more calcic than late-stage plagioclase (Fig. 5A,B). Late-stage plagioclase is found in and around crystals of scapolite and cancrinite group minerals (Fig. 7E). The feldspars contain aqueous FIs and melt inclusions (see below). Crystallization temperatures were estimated with the two-feldspar geothermometer, using the software SOLVCALC (Wen and

Table 2. Comparison between the mean chemical analysis of five selected BM nodules and the CI. Also shown is the maximum value.

	Nodules (%) (n = 5)	Campanian Ignimbrite[a] (%) (n = 40)
SiO_2	59.99	59.70
TiO_2	0.47	0.44
Al_2O_3	18.82	18.53
FeOT	3.71	3.32
MnO	0.16	0.18
MgO	0.63	0.70
CaO	3.11	2.43
Na_2O	5.08	4.75
K_2O	6.80	7.66
P_2O_5	0.11	0.14
F	0.17	0.16
Cl	0.62	0.40
S	0.02	0.01
LOI	0.69	1.09
Selected trace elements Maximun values = (x)		
ppm		
Ba	170.1 (339)	339.9
Ce	196.8 (230)	185.4
Cu	21.6 (68)	20.0
Eu	2.1 (2.63)	1.6
Hf	11.0 (14.3)	10.0
Lu	0.8 (1.04)	0.6
Nb	90.1 (109)	77.3
Pb	45.0 (59)	47.0
Rb	312.3 (499)	347.0
Sm	12.3 (15.2)	11.9
Sr	210.3 (329)	234.3
Th	36.1 (48)	38.7
U	11.4 (17.5)	11.4
W	6.6 (21)	4.6
Y	54.3 (63)	52.0
Yb	4.9 (6.49)	4.3
Zn	58.9 (112)	100.5
Zr	537 (735)	484

[a]H.E. Belkin (unpublished data).

Nekvasil, 1994), and ranged from 650°C to 725°C for late-stage phases and from 810°C to 890°C for early-formed phases (see discussion below).

Pyroxene. The clinopyroxene, typically salite to ferrosalite (Table 4) is slightly pleochroic, subhedral to anhedral, and sometimes zoned (Fig. 7B). It also occurs as euhedral crystals in

Table 3. Representative EMPA analyses of alkali and plagioclase feldspars in the BM nodules.

Sample	MT17.A	MT17.F	MT17.I	MT21.B	MT21.G	MT19.E	MT19.G	MT24.G	MT24.C	MT24.K	VFB.A	VFB.H	CFNA.A	CFNA.D	BL3.E	BL3.K
n	3	3	3	3	3	2	2	3	3	3	3	3	3	3	3	3
Comment	1×2 mm subhedral	200 μm euhedral	200 μm intersertal	2 μm euhedral	300 μm euhedral	1×2 mm euhedral	50 μm in S/C	600 μm euhedral	100 μm in S/C	1 μm euhedral	1 μm euhedral	500 μm euhedral	1×3 mm euhedral	2×4 mm euhedral	600 μm euhedral	300 μm subhedral
SiO_2	64.83	57.60	58.80	64.70	58.44	64.93	60.61	65.54	60.23	65.48	64.83	59.62	65.22	58.99	65.19	59.77
TiO_2	0.02	0.04	0.02	0.04	0.04	0.08	0.01	0.02	0.00	0.04	0.05	0.02	0.05	0.03	0.03	0.03
Al_2O_3	18.92	25.90	25.36	18.82	25.42	19.01	23.84	18.95	25.20	18.90	19.15	24.84	19.32	25.36	18.92	25.08
FeO	0.09	0.25	0.35	0.10	0.35	0.10	0.27	0.12	0.30	0.09	0.15	0.26	0.18	0.35	0.08	0.28
MgO	0.00	0.01	0.01	0.01	0.01	0.00	0.01	0.00	0.00	0.00	0.01	0.01	0.00	0.00	0.01	0.00
CaO	0.25	7.78	7.17	0.22	7.21	0.17	5.15	0.27	6.48	0.21	0.45	6.26	0.55	6.87	0.21	6.78
BaO	0.00	0.00	0.00	0.00	0.00	0.00	0.00	0.02	0.00	0.00	0.00	0.00	0.01	0.00	0.00	0.00
SrO	0.01	0.03	0.02	0.03	0.03	0.03	0.05	0.03	0.09	0.02	0.02	0.02	0.02	0.02	0.01	0.01
Na_2O	2.68	6.74	7.04	2.60	6.94	2.87	7.48	3.05	7.01	2.92	3.76	6.97	3.86	6.69	2.66	7.23
K_2O	12.40	0.44	0.49	12.85	0.59	12.29	1.09	12.06	0.60	12.30	10.78	1.18	10.63	1.13	12.86	0.60
P_2O_5	0.01	0.00	0.01	0.00	0.01	0.01	0.01	0.00	0.01	0.01	0.01	0.01	0.01	0.01	0.00	0.01
Sum	99.21	98.78	99.28	99.37	99.04	99.49	98.51	100.06	99.92	99.96	99.20	99.19	99.86	99.46	99.97	99.79

Number of ions on the basis of 32 (O)																
Si	11.920	10.441	10.584	11.910	10.554	11.904	10.944	11.935	10.733	11.942	11.867	10.732	11.855	10.611	11.920	10.688
Ti	0.003	0.005	0.003	0.006	0.006	0.011	0.001	0.002	0.000	0.005	0.006	0.003	0.007	0.004	0.004	0.004
Al	4.099	5.533	5.381	4.083	5.410	4.107	5.073	4.067	5.292	4.061	4.130	5.269	4.139	5.376	4.078	5.285
Fe^{3+}	0.014	0.037	0.053	0.015	0.053	0.016	0.041	0.018	0.045	0.014	0.023	0.040	0.028	0.052	0.013	0.042
Mg	0.001	0.002	0.002	0.001	0.003	0.000	0.002	0.001	0.001	0.000	0.002	0.002	0.000	0.001	0.002	0.000
Ca	0.050	1.512	1.383	0.043	1.396	0.033	0.995	0.053	1.236	0.041	0.089	1.207	0.108	1.323	0.042	1.299
Ba	0.000	0.000	0.000	0.000	0.000	0.000	0.000	0.001	0.000	0.000	0.000	0.000	0.000	0.000	0.000	0.000
Sr	0.001	0.003	0.002	0.003	0.003	0.003	0.006	0.003	0.009	0.002	0.002	0.002	0.002	0.003	0.002	0.001
Na	0.954	2.369	2.457	0.928	2.430	1.018	2.617	1.076	2.422	1.032	1.334	2.432	1.359	2.334	0.942	2.508
K	2.908	0.102	0.112	3.018	0.136	2.874	0.252	2.802	0.136	2.861	2.516	0.272	2.465	0.260	2.999	0.137
P	0.001	0.001	0.001	0.001	0.001	0.001	0.001	0.001	0.001	0.001	0.001	0.001	0.002	0.001	0.001	0.001
Sum	19.951	20.003	19.979	20.008	19.990	19.967	19.931	19.958	19.876	19.959	19.972	19.961	19.964	19.965	20.000	19.965
An	1.3	38.0	35.0	1.1	35.2	0.8	25.8	1.3	32.6	1.0	2.3	30.9	2.7	33.8	1.1	32.9
Ab	24.4	59.5	62.2	23.3	61.3	25.9	67.7	27.4	63.8	26.2	33.9	62.2	34.6	59.6	23.6	63.6
Or	74.3	2.5	2.8	75.7	3.4	73.2	6.5	71.3	3.6	72.7	63.9	6.9	62.7	6.6	75.3	3.5

FeO*, total iron; in S/C, an inclusion in scapolite or a cancrinite group mineral; 0.00, below detection limit.

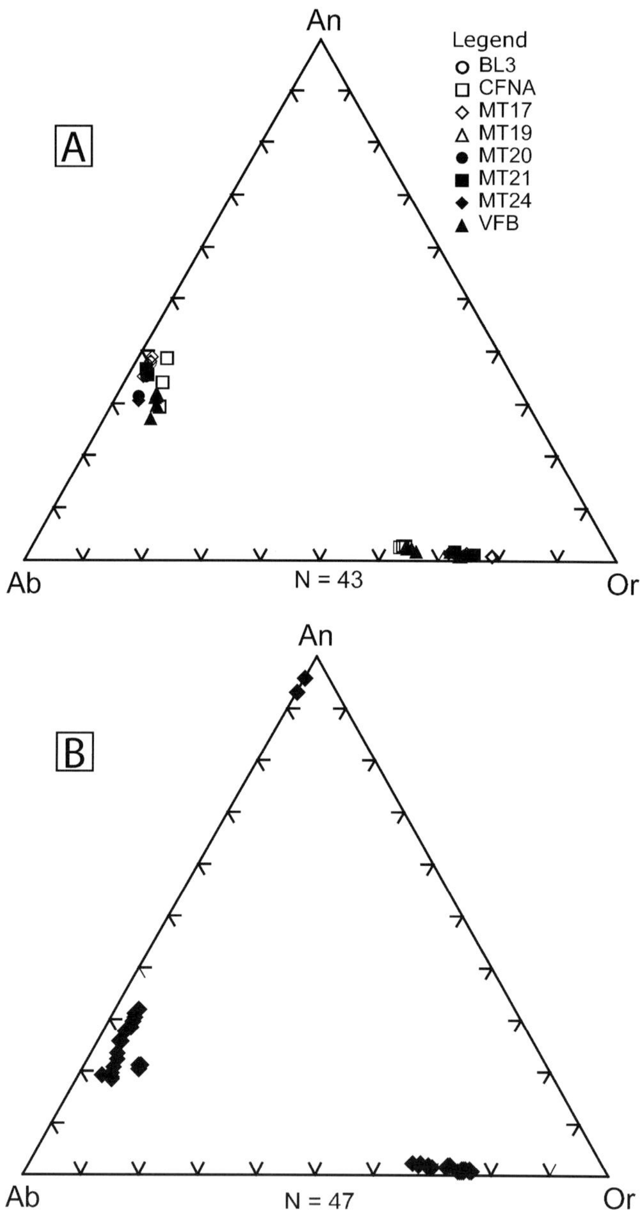

Figure 5. (A) Composition of early-stage alkali feldspar and plagioclase plotted on the mole % An–Ab–Or ternary diagram. (B) Composition of late-stage alkali feldspar and plagioclase plotted on the mole % An–Ab–Or ternary diagram. Legend applies to both diagrams. The two data points closest to the An apex in Figure 5B from sample MT24 are plagioclase crystals found in a mixture of scapolite and a cancrinite-group mineral.

Table 4. Representative EMPA analyses of clinopyroxenes in the BM nodules.

Sample	MT24.D	CFNA.F	MT17.D	MT20.E-rim	MT20.E-core	MT14.15	MT19.D
n	3	2	3	1	1	1	3
Comment	1 μm anhedral, zoned	100 μm subhedral	200 μm subhedral	1 μm subhedral, zoned	1 μm subhedral, zoned	200 μm subhedral	300 μm subhedral in vein
SiO_2	50.49	50.03	50.86	48.34	50.48	51.41	51.28
TiO_2	0.29	0.77	0.61	0.70	0.77	0.60	0.25
Al_2O_3	2.25	3.29	2.58	3.48	3.14	2.04	1.69
Cr_2O_3	0.01	0.00	0.01	0.01	0.00	0.00	0.01
FeO*	12.76	10.39	9.60	14.38	9.82	9.81	11.44
MgO	9.55	11.11	11.74	7.78	11.39	11.97	10.16
CaO	21.66	21.95	21.80	20.44	22.20	20.86	22.24
MnO	0.96	0.91	0.76	1.77	0.75	0.97	1.41
K_2O	0.00	0.00	0.00	0.00	0.02	0.05	0.00
Na_2O	0.69	0.71	0.54	1.17	0.63	0.69	0.76
Total	98.66	99.14	98.50	98.07	99.18	98.40	99.25
Formulae on the basis of 6 (O)							
Si	1.954	1.909	1.941	1.908	1.918	1.962	1.967
Al^{4+}	0.046	0.091	0.059	0.092	0.082	0.038	0.033
Al^{6+}	0.057	0.057	0.057	0.069	0.059	0.053	0.044
Fe^{3+}	0.035	0.064	0.013	0.106	0.039	0.006	0.046
Cr	0.000	0.000	0.000	0.000	0.000	0.000	0.000
Ti	0.008	0.022	0.017	0.021	0.022	0.017	0.007
Fe^{2+}	0.377	0.266	0.292	0.367	0.272	0.305	0.320
Mn	0.032	0.029	0.025	0.059	0.024	0.031	0.046
Mg	0.551	0.632	0.668	0.458	0.645	0.681	0.581
Ca	0.898	0.897	0.891	0.864	0.904	0.853	0.914
Na	0.052	0.053	0.040	0.090	0.046	0.051	0.057
K	0.000	0.000	0.000	0.000	0.001	0.002	0.000
Total	4.010	4.020	4.003	4.034	4.012	4.001	4.014
Mg	28.3	32.6	34.6	23.6	33.4	35.3	29.6
ΣFe**	22.8	18.5	17.1	27.4	17.3	17.8	20.9
Ca	46.2	46.2	46.2	44.5	46.8	44.2	46.6

FeO*, total iron; ΣFe** = (Fe^{2+} + Fe^{3+} + Mn).

Note: Fe^{2+}, Fe^{3+} calculated using the method of Droop (1987).

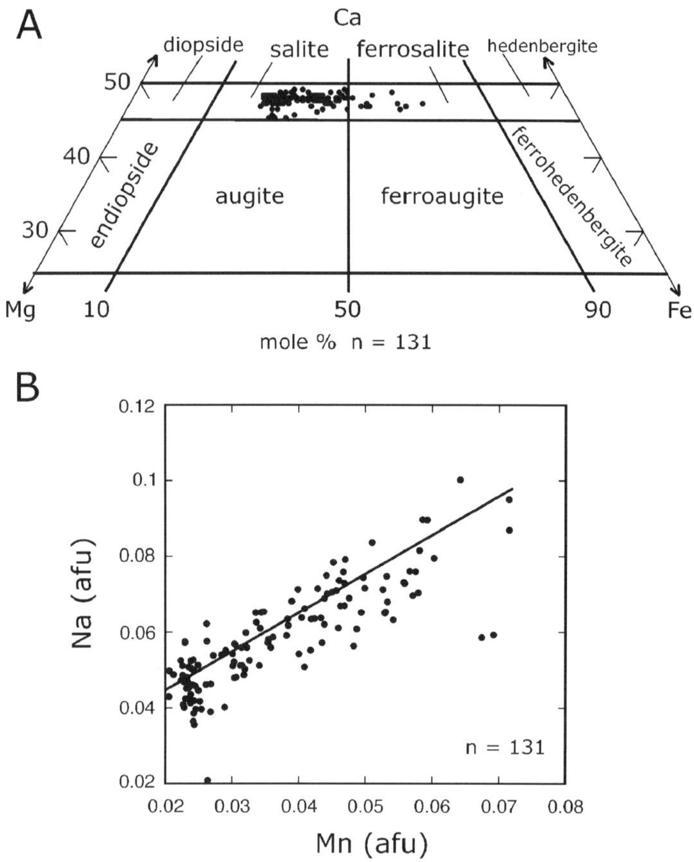

Figure 6. (A) A portion of the pyroxene quadrilateral showing the compositional range of the clinopyroxenes measured by EMPA in all the studied nodules. (B) Co-variation of Na and Mn (afu – atom formula unit) in the clinopyroxenes shown in the quadrilateral. The linear regression of the Na and Mn data is positive with $r^2 = 0.70$.

crosscutting veins or cavities/vugs. Pyroxene is often associated with amphibole, biotite, titanite, and Fe–Ti oxides. Figure 6A shows that these pyroxenes have essentially constant Ca but vary in Mg and Fe. Rims, visible in transmitted light and SEM analysis, typically are enriched in Fe, Na, and Mn (Table 4, sample MT20.E). Figure 6B shows the positive co-variation between Na and Mn. The acmite ($NaFe^{3+}Si_2O_6$) component in the studied clinopyroxenes varies from 1.1 to 5.5. Clinopyroxenes in the BM pyroclastics typically are somewhat more Mg-rich and range from diopside to ferrosalite (Melluso et al., 1995). The studied clinopyroxenes in the nodules, enriched in Mn and Na, probably represent later-stage magmatic crystallization instead of early-magmatic crystallization.

Amphibole. The amphibole is a hornblende that varies in composition from ferropargasite, pargasite, to magnesio-hastingsite/pargasite (Table 5). This phase occurs in large euhedral to subhedral crystals, is slightly to moderately pleochroic, and is commonly zoned (Fig. 7A). The amphibole has moderate F, variable TiO_2, and a low Cl content (Table 5).

Table 5. Representative EMPA analyses of amphiboles in the BM nodules.

Sample	MT24.F (fp)	VFB.C (p)	BL3.B (p)	MT17.B (p)	MT21.C (p)	MT21.F (p)	MT19.E (mh-p)	MT20.D (mh-p)
Comment	2 mm anhedral, zoned	600 μm suhedral	1 mm euhedral	1 mm euhedral	1 mm subhedral, zoned	4 mm euhedral	300 μm euhedral in vein	600 μm euhedral
n	3	3	3	3	3	3	3	3
SiO_2	39.43	40.19	42.85	40.19	41.72	40.57	41.38	41.39
TiO_2	1.81	2.04	0.86	2.60	0.77	2.72	1.50	0.78
Al_2O_3	11.62	11.27	10.02	11.50	10.55	11.40	10.97	10.04
Cr_2O_3	0.00	0.00	0.00	0.00	0.00	0.00	0.00	0.00
FeO*	18.22	16.44	15.10	17.03	15.90	15.85	15.74	16.00
MnO	1.56	1.70	1.22	0.96	1.37	0.86	1.28	1.17
MgO	9.12	9.92	11.74	9.72	11.62	10.51	11.31	11.81
CaO	11.02	10.58	11.65	11.34	11.40	11.58	11.26	11.08
Na_2O	2.32	2.43	2.51	2.37	2.50	2.39	2.43	2.27
K_2O	1.86	1.91	1.92	1.99	1.86	1.91	1.88	2.18
F	2.40	1.87	2.61	2.28	2.61	1.83	2.54	3.31
Cl	0.10	0.10	0.24	0.14	0.09	0.13	0.11	0.14
−O = F,Cl	1.03	0.81	1.15	0.99	1.12	0.80	1.09	1.43
Total	98.42	97.65	99.57	99.12	99.28	98.95	99.29	98.75
Formulae calculated assuming 13 cations excluding Ca, Na, and K								
Si	5.990	6.110	6.304	6.018	6.189	6.060	6.131	6.161
Ti	0.207	0.234	0.095	0.293	0.086	0.305	0.167	0.087
Al	2.081	2.020	1.737	2.029	1.845	2.007	1.916	1.762
Cr	0.000	0.000	0.000	0.000	0.000	0.000	0.000	0.000
Fe^{3+}	0.000	0.000	0.000	0.000	0.000	0.000	0.000	0.000
Fe^{2+}	2.315	2.090	1.858	2.133	1.972	1.979	1.950	1.991
Mn	0.201	0.219	0.152	0.121	0.173	0.109	0.160	0.147
Mg	2.065	2.247	2.576	2.170	2.569	2.340	2.499	2.622
Ca	1.793	1.723	1.836	1.819	1.812	1.853	1.787	1.767
Na	0.682	0.715	0.716	0.687	0.720	0.693	0.697	0.656
K	0.361	0.370	0.361	0.380	0.352	0.363	0.355	0.415
F	0.126	0.098	0.137	0.120	0.137	0.096	0.134	0.174
Cl	0.003	0.003	0.007	0.004	0.003	0.004	0.003	0.004

fp, ferro-pargasite; p, pargasite; mh-p, magnesio-hastingsite/pargasite.

FeO* = total iron, Cr_2O_3=below detection limit.

Note: Fe^{3+} calculated using the method of Droop (1987).

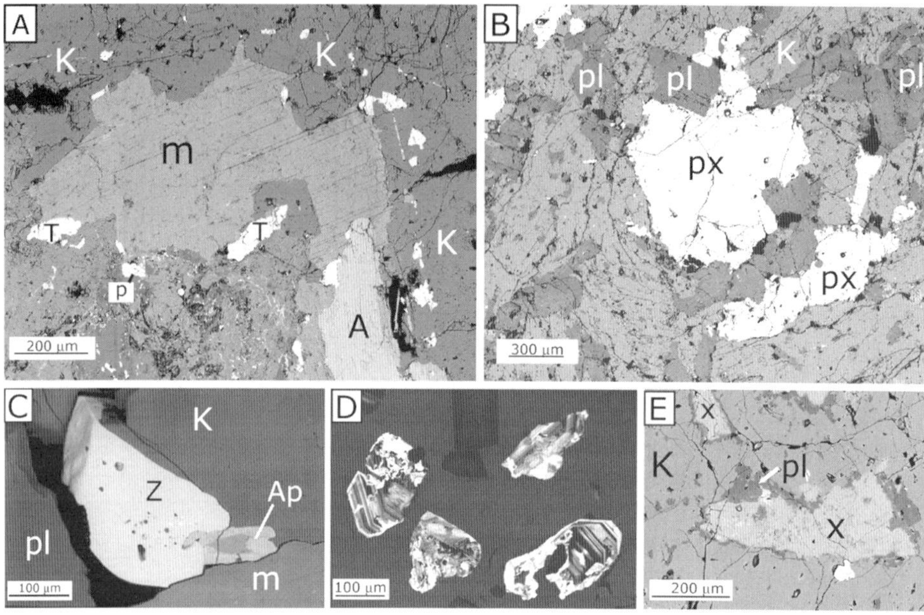

Figure 7. (A) SEM back-scattered image of sample MT21 showing late-stage biotite mica (m) and earlier amphibole (A) filling a vug formed by earlier crystallized alkali feldspars (K). Later phases, monazite (p) and titanite (T), enriched in incompatible elements, formed after the mafics. (B) SEM back-scattered image of sample MT20 showing clinopyroxene (px) and plagioclase (pl) forming with alkali feldspar (K). (C) SEM back-scattered image of sample CFNA illustrating the texture of late-stage zircon (Z) forming in a vug of alkali feldspar (K), plagioclase (pl), and biotite mica (m). The zircon has formed subsequent to the zoned apatite (Ap). (D) Cathodoluminescence image of hand-picked zircons used for U–Th age dating illustrating their complex zoning. (E) SEM back-scattered image of sample BL3 showing late-stage formation of scapolite plus a cancrinite-group mineral (X) in a vug formed by earlier alkali feldspar (K). Late-stage plagioclase (pl) appears to have formed just prior to the scapolite and cancrinite-group mineral.

Biotite. This phase was observed in only about half of the studied nodules. Typically, it fills open spaces formed by feldspar accumulations (Fig. 7A). Some are very F-rich (about 5–6 wt.%) but they are all low in Cl (less than 0.1 wt.%) and Ba (Table 6). We observed two compositional groups, one with higher TiO_2, F, and MgO and the other with lower TiO_2, F, and higher FeO.

Scapolite. Scapolite occurs in large crystals or inter-grown with cancrinite-group minerals (Fig. 7E) and contains FIs as well as vugs. Scapolite crystals are often associated with cancrinite-group minerals, titanite, apatite and Zr-bearing minerals, or subordinately, with amphibole, pyroxene, mica, and Fe–Ti oxides. The scapolite is marialitic (Table 7), the percentage of Me varies from 17 to 22, and is relatively enriched in K_2O. It is one of the two observed phases rich in Cl and from the textural relations, it is a later-stage phase in the nodules paragenesis.

Cancrinite-group minerals. The presence of these phases accounts for the enrichment of sulfur in the bulk composition of some of the analyzed samples (Table 1). This phase is also enriched in Cl and Na (Table 7). On the basis of the microprobe data (Table 7), these minerals are interpreted as members of the davyne-type cancrinite; most close to microsommite

Table 6. Representative EMPA analyses of biotite micas in the BM nodules.

Sample	CFNA.B (wt%)	CFNA.D (wt%)	CFNA.E (wt%)	MT21.A (wt%)	MT21.D (wt%)	MT21.E (wt%)
n	3	3	3	3	3	3
Comment	1 mm	200 μm	600 μm	600 μm	900 μm	1 mm
SiO_2	37.61	37.20	37.91	40.35	41.85	40.66
TiO_2	3.44	3.67	3.44	1.66	0.74	1.82
Al_2O_3	13.65	14.18	13.78	12.15	11.35	11.82
Cr_2O_3	0.00	0.00	0.00	0.00	0.00	0.00
FeO	13.41	13.90	13.49	10.80	9.47	10.39
MnO	0.93	0.95	0.95	0.65	0.65	0.63
MgO	16.32	15.69	16.34	19.27	20.69	19.60
CaO	0.02	0.01	0.01	0.01	0.00	0.01
BaO	0.03	0.02	0.00	0.00	0.00	0.03
Na_2O	0.58	0.63	0.64	0.32	0.32	0.42
K_2O	9.35	9.38	9.52	10.05	10.25	10.08
F	3.55	3.39	3.71	5.70	6.76	6.19
Cl	0.04	0.04	0.05	0.05	0.05	0.05
−O = F,Cl	1.51	1.44	1.57	2.41	2.85	2.62
H_2O calc	2.32	2.40	2.28	1.37	0.90	1.15
Sum	99.76	100.01	100.54	99.96	100.17	100.25
Formulae on the basis of 11 (O)						
Si	2.806	2.776	2.808	2.964	3.048	2.976
Al^{4+}	1.201	1.247	1.203	1.052	0.974	1.020
SumT	4.000	4.000	4.000	4.000	4.000	3.996
Al^{6+}	0.007	0.023	0.011	0.017	0.022	0.000
Ti	0.193	0.206	0.192	0.092	0.041	0.100
Cr	0.000	0.000	0.000	0.000	0.000	0.000
Fe	0.837	0.867	0.835	0.664	0.577	0.636
Mn	0.059	0.060	0.059	0.041	0.040	0.039
Mg	1.815	1.745	1.805	2.110	2.246	2.139
Sum*R*	2.911	2.901	2.902	2.923	2.926	2.915
Ca	0.002	0.001	0.001	0.001	0.000	0.001
Ba	0.001	0.001	0.000	0.000	0.000	0.001
Na	0.084	0.091	0.092	0.045	0.046	0.060
K	0.890	0.893	0.899	0.942	0.952	0.941
Sum*A*	0.977	0.986	0.992	0.988	0.998	1.003
F	0.839	0.801	0.868	1.323	1.556	1.433
Cl	0.005	0.005	0.006	0.006	0.006	0.006
OH calc	1.156	1.194	1.126	0.671	0.438	0.561
Mg/(Mg + Fe* + Mn)	0.67	0.65	0.67	0.75	0.78	0.76

Table 7. Representative EMPA analyses of cancrinite-group minerals and scapolite in the BM nodules.

Sample	MT17.B (C)	MT21.D (C)	BL3.G (C)	MT24.H (C)
n	3	2	3	3
SiO_2	31.92	31.87	32.37	31.89
Al_2O_3	27.99	27.56	28.53	27.42
FeO*	0.04	0.08	0.08	0.09
MgO	0.00	0.01	0.01	0.02
MnO	0.01	0.00	0.00	0.01
CaO	10.97	9.84	9.85	9.90
SrO	0.03	0.02	0.01	0.02
K_2O	4.02	3.64	3.59	2.82
Na_2O	11.35	13.38	13.22	13.99
P_2O_5	0.02	0.01	0.02	0.01
SO_3	1.22	3.56	3.32	4.81
Cl	10.00	8.34	9.36	8.29
−O = Cl	2.26	1.88	2.12	1.87
Sum	95.30	96.41	98.25	97.40
Number of ions on the basis of 12 (Si, Al, Fe)				
Si	5.899	5.939	5.883	5.957
Al	6.098	6.054	6.111	6.036
Fe	0.003	0.006	0.006	0.007
Mg	0.000	0.002	0.002	0.005
Mn	0.002	0.000	0.000	0.001
Ca	2.172	1.964	1.919	1.982
Sr	0.003	0.002	0.001	0.002
K	0.948	0.864	0.833	0.672
Na	4.069	4.833	4.659	5.066
P	0.003	0.002	0.003	0.001
S	0.169	0.498	0.453	0.675
Cl	3.132	2.634	2.884	2.625

Sample	MT24.D (S)	MT19.I (S)	MT20.G (S)
n	3	1	3
SiO_2	57.37	56.28	56.62
Al_2O_3	22.02	22.80	22.12
FeO*	0.15	0.14	0.13
MgO	0.00	0.01	0.00
MnO	0.00	0.01	0.01
CaO	4.57	5.38	4.76
SrO	0.05	0.03	0.02
K_2O	1.99	2.00	2.10
Na_2O	9.97	9.61	9.72
P_2O_5	0.00	0.01	0.00
SO_3	0.01	0.03	0.01
Cl	3.95	4.06	3.99
CO_2**	0.17	0.02	0.09
−O = Cl	0.89	0.92	0.90
Sum	99.37	99.46	98.66
Number of ions on the basis of 12 (Si, Al)			
Si	8.262	8.122	8.217
Al	3.738	3.878	3.783
Fe	0.018	0.016	0.015
Mg	0.000	0.002	0.000
Mn	0.000	0.002	0.001
Ca	0.705	0.832	0.740
Sr	0.005	0.002	0.001
K	0.366	0.368	0.388
Na	2.784	2.689	2.735
P	0.000	0.001	0.000
S	0.001	0.003	0.001
Cl	0.965	0.992	0.981
Me%	18.7	21.8	19.5

C, cancrinite group; S, scapolite. FeO*, total iron; CO_2**, calculated by stoichiometry.

in composition (Ballirano et al., 1996; Deer et al., 2004). These cancrinite-group phases occur in late-stage masses, filling open spaces or vugs (Fig. 7E). The crystals are associated with Na-rich plagioclase and appear to be one of the last major silicate phases to be formed.

Titanite. This phase is common throughout the nodules (Fig. 7A). It occurs in a wide range of shapes, with subhedral to irregular habit and often shows moderate zonation. Some crystals are slightly enriched in REE and Zr. In growth around other crystals or filling cavities, titanite often is associated with Fe/Ti oxides and/or Zr-bearing minerals such as zircon and baddeleyite.

Apatite. Apatite occurs as small inclusions in feldspars, or in vugs as a late-stage mineral. It ranges in habit from anhedral masses to euhedral crystals of up to 100 μm in length.

Apatite sometimes is zoned and is frequently associated with zircon, monazite, and cheralite (Fig. 7C).

3.1. Uranium, REE, and zirconium-bearing minerals

The nodules have an exceptional assemblage of U (Table 8), REE, and zirconium-bearing minerals. The mineral habit and mode of occurrence strongly suggests deposition from dense, supercritical fluids, that separated from the crystallizing magma, and were enriched in incompatible elements. We have identified these phases by quantitative electron microprobe and SEM-EDS analysis. A separate paper in preparation will detail these phases.

Zircon typically is found associated with titanite, Fe–Ti oxides and, to a lesser extent, pyroxene, amphibole, and mica. Large crystals are zoned (Fig. 7D) and frequently host small uraninite/thorianite crystals. On the basis of petrography, we distinguish, two types of zircon. One type occurs as euhedral to subhedral grains and appears to have crystallized under magmatic conditions (Fig. 7D), whereas the other type fills cavities, is irregularly shaped, and appears to be hydrothermal in origin (Fig. 7C). The first type was separated from sample MT24 and was dated using the U–Th isotope method (see above).

Baddeleyite is found either in groups or isolated and shows a common bladed habit. Gianfagna (1985) reports baddeleyite from an ejected block from Colle Cimino, in the Alban Hills volcanic complex located just south of Rome, Italy. There, baddeleyite has a composition similar to that found in the BM nodules except for the lack of UO_2, and the host rock was also thought to represent a piece of magma chamber. Zircon and baddeleyite are enriched in UO_2 (Table 8), and baddeleyite is enriched in Nb_2O_5, up to 2.5 wt.%. The occurrence of both late-stage zircon and baddeleyite poses an interesting problem in mineral paragenesis, which we will not address in this presentation. However, we suggest that this is a worthy subject for further study, especially concerning the influence of F- and Cl-rich fluids on the relative stability of these two phases.

Zirconolite is a relatively rare accessory mineral, ideally $CaZrTi_2O_7$ in composition, but generally accommodating a large number of elements in its structure: in natural samples of zirconolite, 30 or more elements may be present at the 0.1–1.0 wt.% concentration level (Williams and Gieré, 1996). The main substitutions are: the REE and actinide elements for Ca; Hf for Zr; and Nb, Ta, Fe, Mn, Mg, and W for Ti. Zirconolite occurs in a wide range of rock types and paragenesis. The majority are from carbonatites (Gieré et al., 1998). Only four syenite occurrences have been reported out of a total of more than 50 terrestrial localities (Williams and Gieré, 1996): (1) at Glen Dessarry, Scotland (Fowler and Williams, 1986), (2) from the alkali intrusion of the Arbarastakh Massif, Russia (Borodin et al., 1960), (3) in syenite pegmatites in the Oslo region of Norway at Fredicksvärn (Brøgger, 1890) and Langesundfjord (Larsen, 1996), and (4) from a syenitic ejecta enclosed in a rock known as "sanidinite" found at Monte di Procida at Campi Flegrei, Italy (Mazzi and Munno, 1983). The analyses of the BM nodules show that some zirconolite crystals are enriched with thorium and uranium (Table 8) and have low to moderate REE content.

Pyrochlore, ideally $(Na,Ca)_2Nb_2O_6(OH,F)$, is a relatively common accessory phase in the nodules. This mineral shows a large compositional variation and is the main Nb-bearing mineral in the nodules, and some crystals that contain significant UO_2 are termed uranpyrochlore (Table 8).

Table 8. Uranium content of minerals in the studied nodules measured by EMPA.

Uranium content of minerals (given as wt.% UO_2 from EMPA data)	
Uraninite/thorianite	~29
Uranpyrochlore	~15
Thorite	~5
Zircon	0.1–2.0
Baddeleyite	0.06–1.2
Zirconolite	0.7–5.5
Monazite/cheralite	0.13–1.0
Apatite	dl–0.1
Titanite	dl–0.1

Detection limit (dl) = 0.05 wt.%.

Other phosphates. Monazite and cheralite also are common accessory minerals (Fig. 7A); often they show zoning and are typically enriched in ThO_2, light REE, and UO_2 (Table 8).

Other minerals. We observed a plethora of minor phases either identified by electron microprobe or SEM-EDS analyses; sulfides (pyrite, pyrrhotite; Pb-Cu-Fe- and Zn-sulfides), carbonates (calcite and cerussite), tungstates (scheelite and ferberite), and silicates (rare pollucite, sodalite, and celsian). Typically, these minerals occur in a variety of habits as single crystals commonly embedded in feldspar or associated with other phases growing in cavities. There are also common oxide phases ranging from magmatic titanomagnetite to various secondary Fe oxides resulting from subsequent alteration of earlier phases.

3.2. Fluid and melt inclusions

We used two approaches to study fluid and melt inclusions: (1) host crystals were fractured and a detailed SEM-EDS examination of these surfaces identified the wide variety of daughter crystals (or accidentally trapped crystals) observed with transmitted visible light and (2) microthermometric data were collected to constrain the conditions and composition of the trapped fluids. Fluid and melt inclusions in K-feldspar host crystals were examined. Thick sections (~300 μm thickness) were prepared for optical observation, whereas microthermometry was performed on K-feldspar crystals hand picked after crushing a selected subset of the available nodules. Inclusions that were found in healed fractures of the host minerals vary in shape and size but seldom exceed 30 mm in length.

Observations with transmitted light at room temperature revealed the presence of FIs containing numerous solids. The common occurrence of similar crystal habits, relative index of refraction, and phase proportions suggests that most are daughter minerals (i.e., precipitated from the trapped fluid during cooling). In order to identify these crystals, some small pieces of nodules were carefully fractured and the fresh surface was examined by SEM techniques.

Spectra collection and EDS analysis of the daughter minerals presented many difficulties owing to their small size and the random orientation of the analyzed samples. We have assumed that (1) we have adequately corrected for any spectral additions or interferences from the host phase, (2) the presence or degree of hydration is not known, and (3) the mineral assemblage of any individual inclusion may be incomplete owing to accidental loss during opening. Nevertheless, SEM-EDS examination has provided valuable information regarding the composition and phase assemblages of the daughter minerals.

3.2.1. Daughter mineral analyses

The SEM-EDS analytical studies performed on the daughter crystals in FIs revealed the presence of the following minerals.

Chlorides. Na (halite) and K (sylvite) chlorides were identified in many opened inclusions (Fig. 8A,B). They were often associated and together occupied approximately 25% of the volume of the exposed portion of the inclusion. Both phases were cubic and some crystals had rounded edges. Fe- and Mn-chlorides commonly were identified and these often occurred as irregular masses or coated the wall of the inclusions.

Sulfides. Minerals showing Fe, Cu, and S EDS spectral peaks were identified as pyrite, (some were perhaps pyrrhotite), galena, and chalcopyrite, based on morphology and standardless EDS spectral analysis. These minerals often were found in association (Fig. 8D,E). We observed the uncommon occurrence of sphalerite and Ag_2S (argentite/acanthite). Monoclinic acanthite, the stable Ag_2S phase below 179°C, transforms to isometric argentite above 179°C. The habit of this phase was botryoidal and/or arborescent.

Sulfates. We identified the oxygen-bearing Ca and S phase as anhydrite ($CaSO_4$) based on apparent orthorhombic morphology (Fig. 8B,F).

Carbonates. Uncommon calcite was observed with no Mg noted above the SEM-EDS detection limit (~1000 ppm). Other carbonate phases are $ZnCO_3$ (smithsonite) (Fig. 8F) and a Pb-bearing phase (probable cerussite).

Tungsten and molybdenum-bearing phases. Uncommon scheelite (with a minor powellite component), identified on the basis of morphology and composition, occurs separately or in association with carbonate or sulfides (Fig. 8C).

Other minerals. Collected spectra frequently identified mixed Fe- and Mn-oxygen-bearing minerals which we identify as oxides or hydroxides. Apatite occurs as single, euhedral crystals, or in multiphase inclusions. A small Ca, Th, and REE phosphate was identified in one inclusion hosted by zircon. Two inclusions contain a very Mn-rich pyroxene (tentatively identified as johannsenite). We observed no major F-bearing phases in opened FIs.

3.2.2. Microthermometry

Since microthermometric data (Table 9) were collected on a limited number of samples and FI assemblages, results presented here should be considered preliminary but are nevertheless important to constrain the temperature and pressure of the nodule-forming environment.

FIs in potassium feldspar were found in healed fractures and are considered secondary; their dimensions seldom exceed 30 mm in length. The following types can be distinguished:

(1) one-phase inclusions ["Vapor only" (V), Type 1];
(2) three-phase inclusions [liquid + vapor + solid (Halite), Type 2].

In addition to FIs, we also identified several melt inclusions (MIs), partially or totally crystallized, which occur isolated or in small clusters, and are never related to any apparent

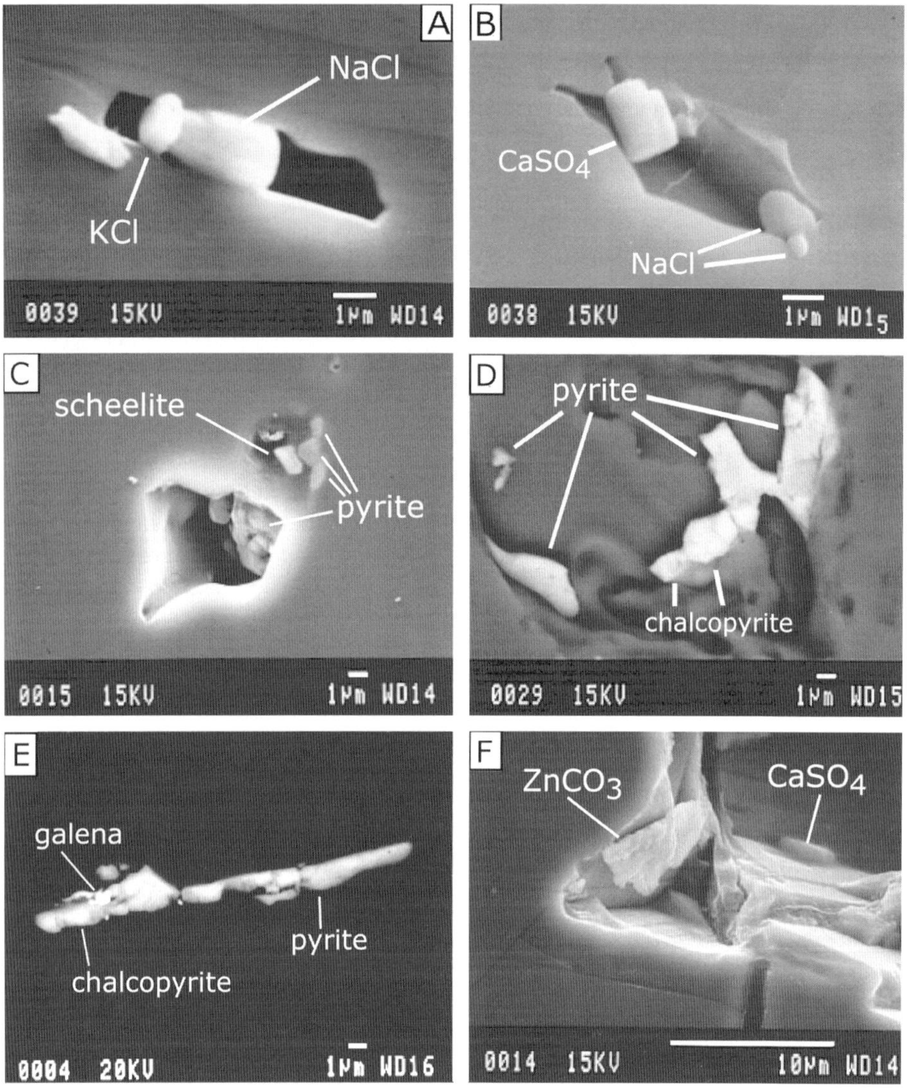

Figure 8. SEM images of opened fluid inclusions in host alkali feldspar. (A) Opened inclusion containing chlorides. NaCl (halite) and KCl (sylvite) chlorides are commonly found in opened inclusions in sample MT19 (secondary-electron SEM image). (B) $CaSO_4$ (anhydrite) and halite (NaCl) occurring in an opened fluid inclusion from sample MT19 (secondary-electron SEM image).(C) Opened fluid inclusion containing pyrite and scheelite from sample BL8 (secondary-electron SEM image). (D) SEM back-scattered image of a complex opened fluid inclusion which contains pyrite and chalcopyrite crystals. (E) Back-scattered SEM image of base-metal sulfides, galena, pyrite, and chalcopyrite along the edge of an opened inclusion from sample MT19. (F) SEM secondary-electron image of a complex opened inclusion containing a Zn carbonate and $CaSO_4$ (anhydrite) from sample BL8.

Table 9. Microthermometric data, Th, Tf, and NaCl (wt.% equivalent).

Th	Tf	NaCl (wt.%)
330	324	40.1
296	308	38.8
294	302	38.3
263	272	36.1
293	293	37.6
293	293	37.6
293	278	36.6
301	329	40.5
299	306	38.6
300	300	38.2
300	300	38.2
290	295	37.8
295	304	38.5
300	299	38.1
287	287	37.2
300	300	38.2
289	294	37.7
288	294	37.7
293	303	38.4
299	311	39.0
295	295	37.8
292	287	37.2
287	282	36.8
292	306	38.7
292	308	38.8
329	273	36.2
291	288	37.3
283	284	37.0
298	313	39.2
302	307	38.7
307	313	39.2
288	293	37.7
298	≥ Th[a]	N/A
298	≥ Th[a]	N/A
299	≥ Th[a]	N/A
294	≥ Th[a]	N/A
340	≥ Th[a]	N/A

[a]In this case Tf is estimated to be slightly higher or equal to Th since poor visibility made detection of final halite melting impossible.

fracture or cleavage plane. They are thought to be primary. Tentative thermometric experiments (on a Linkam THS1500 heating stage calibrated in the temperature range of interest using the melting point of gold) found that the MIs start melting at ~950°C and are completely homogenized (i.e., total melting and complete bubble disappearance) at ~1150°C. This high temperature range may reflect previous loss of volatiles during cooling.

Type 1 FIs are closely associated with Type 2 and they appear either completely dark or "gray" with a thick rim (Fig. 9) in transmitted light. When heated or frozen they do not show any detectable phase change. We consider Type 1 FIs to be vapor-rich inclusions having a homogenization temperature (Th) ≤ 350°C. According to known phase relations, these inclusions would appear as single phase, vapor only, at room temperature (Bodnar et al., 1985).

Type 2 FIs contain a single daughter mineral of halite (Fig. 9A,B). Microthermometry gives homogenization temperatures [Th (L + V → L)] in the range 263–310°C, and temperatures of final disappearance (Tf) of the daughter mineral in the range 271–328°C (Table 9). Based on different homogenization behavior, Type 2 FIs can be further subdivided into:

Type 2a: Tf > Th (in 49% of FIs halite disappears after the bubble);
Type 2b: Tf < Th (in 20% of FIs halite disappears before the bubble);
Type 2c: Tf ≈ Th (in 31% of FIs halite and bubble disappear at the same time).

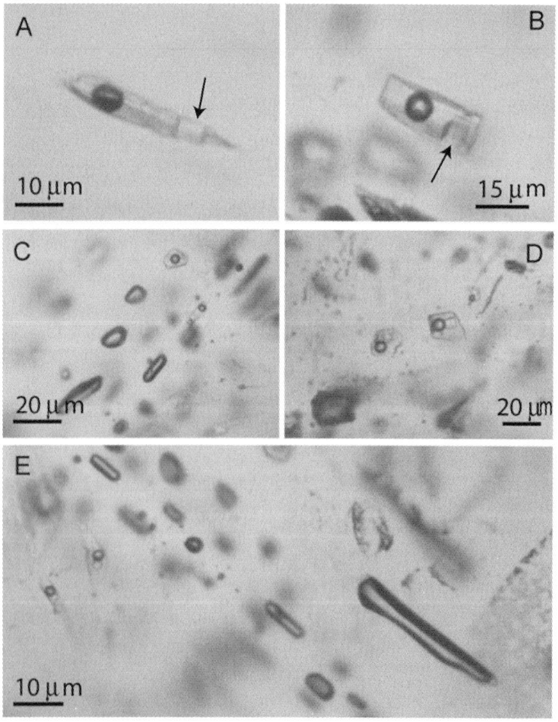

Figure 9. Transmitted light photomicrographs of fluid inclusions in BM nodules. (A, B) three-phase inclusions, the arrow indicates halide daughter crystals; (C, E) details of a plane of coexisting three-phase and vapor inclusions; (D) three-phase inclusions.

The coexistence of different homogenization behaviors with a relatively consistent salinity, independent from the homogenization mode, for halite-bearing FIs, is a well-known phenomenon reported in the porphyry copper literature (e.g., Bodnar and Beane, 1980).

Tf can be used to estimate salinity using the equation by Sterner et al. (1988; $\Psi = T°C/100$):

$$\text{Salinity (wt.\%)} = 26.242 + 0.4928\,\Psi + 1.42\,\Psi^2 - 0.223\,\Psi^3 + 0.04129\,\Psi^4 + 6.295 \times 10^{-3}\,\Psi^5 - 1.967 \times 10^{-3}\,\Psi^6 + 1.1112 \times 10^{-4}\,\Psi^7$$

This equation theoretically is valid only for FIs where $Tf \approx Th$; however it can be used to approximate salinity where $Tf > Th$, keeping in mind that the actual NaCl content will be over-or underestimated. The magnitude of the error will depend on both pressure in the inclusion at Tf and salinity. For our Tf range, and assuming that FIs internal pressure does not exceed 2 kbar (a reasonable assumption, considering that otherwise the inclusion would likely decrepitate), the error can be estimated at ± 1 wt.%. In this manner, the corresponding calculated salinity range for our FIs is 36–40 wt.% NaCl equivalent.

The coexistence of Type 1 and Type 2 FIs suggests that, at a certain point in the system evolution, boiling has occurred. However, it must be recognized that FIs that trapped a hypersaline fluid in the NaCl–H_2O system and that show homogenization behavior like Type 2a and 2c, cannot coexist with Type 1 owing to phase equilibrium constraints (Bodnar, 1994).

It is possible to explain the formation of these inclusions by calling upon two quite different scenarios:

Boiling did not occur (according to the phase equilibrium constraints in the NaCl-H_2O system; Bodnar, 1994). In this case, it is possible that high-salinity fluids were generated by direct exsolution from the magma during the late stages of crystallization. Type 2b inclusions would have been trapped at high P–T in field A (Fig. 10). Later, with the system cooling, P–T conditions would cross the isochore B where inclusions Type 2c could be trapped. Finally, further cooling would bring the system in the field C where Type 2a inclusions can be trapped. It is also possible to produce inclusions Type 2a, 2b, and 2c by significant pressure fluctuations at nearly constant T, though in this case the measured Th probably would have been spread over a wider range (Bodnar, 1992) than the range of our data (Fig. 11). Trapping temperatures for Type 2 inclusions can be tentatively inferred by relating the available data on pressures of formation of xenoliths at Mt. Somma-Vesuvius and Ventotene (1–3.5 kbar for Vesuvius, Belkin et al., 1985; Belkin and De Vivo, 1993; and 0.2–0.4 kbar for Ventotene, De Vivo et al., 1995) with the isochores obtained by the experimental data on the H_2O-NaCl system (Bodnar and Vityk, 1994). Calculations suggest that trapping temperatures are in the range 325–525°C.

Boiling did occur. In our case, the studied system is too complex to be adequately modeled by the phase relationships in the NaCl-H_2O system. This is compatible with the presence of varied and unusual hydrothermal mineral assemblages and also indicates that the fluids circulating in the system at the moment of the inclusion trapping were compositionally complex. Furthermore, the petrographic evidence of the coexistence of Type 1 and 2 inclusions is quite compelling. If we admit boiling, homogenization temperatures (Th) are to be considered as real trapping temperatures (Tt = 263–310°C), and trapping must have occurred under hydrostatic pressure at shallow depth.

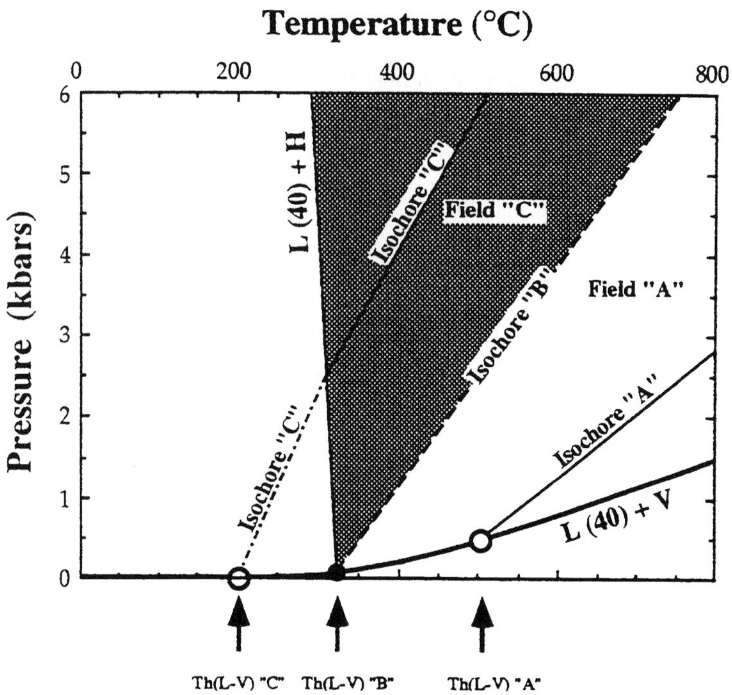

Figure 10. P–T diagram showing the three different modes of homogenization for a 40 wt.% NaCl FI (Bodnar and Vityk, 1994).

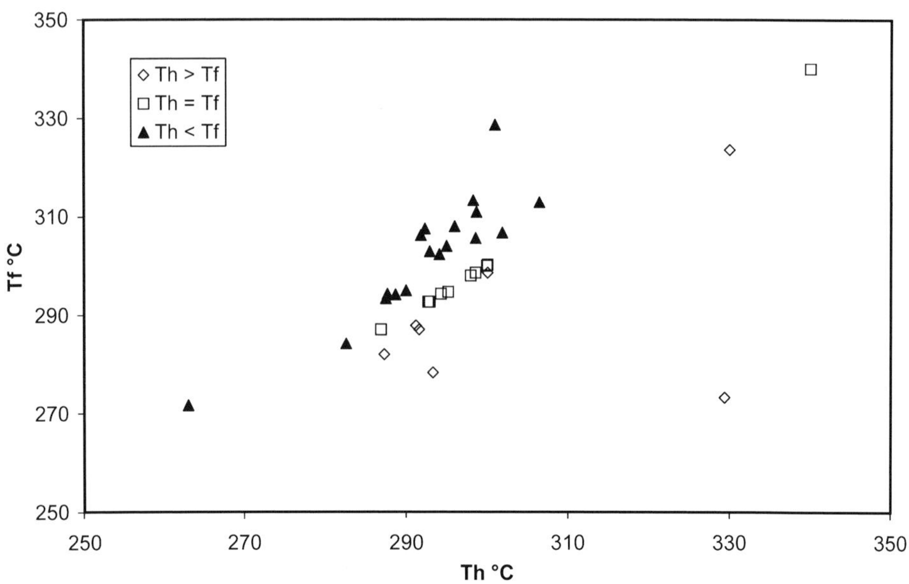

Figure 11. Th versus Tf for Fluid inclusions.

4. Discussion

4.1. Alkali-syenite nodules as samples of a magma chamber margin

Fundamental to our interpretation of the various textures, mineral phases, and microthermometry from the studied samples is that they are cognate with respect to the BM–CF magmatic system. All our evidence from chemistry, mineralogy, and petrography point to the fact that the studied samples are portions of an alkalic magma chamber margin that was broken up, entrained, and erupted with the BM deposits. Thus, features observed in these nodules allow us to tentatively describe the peripheral part of the magma chamber where the transition from a silicate liquid melt to a hypersaline aqueous fluid occurs.

The bulk of each nodule is potassium feldspar, precipitated from a trachytic magma and accumulated around the wall or bottom of the chamber. During pre-eruption crystallization, aqueous immiscible fluids may exsolve and become enriched in various incompatible elements. This may be especially important in the marginal region of the magma chamber where cooling and magma–wall rock interaction may be significant.

4.2. Textural development of the nodules

The mineral paragenesis and their petrographic relationships have recorded a transition from silicate melt-magmatic to hydrothermal aqueous fluid conditions. It is reasonable to assume that this transition was essentially isobaric with the major differences occurring in temperature and fluid composition. The particular unique volume that any nodule represents, would be progressively further away from the interface of magmatic melt/crystal accumulation as more and more crystals settled or migrated to the chamber margins. Figure 12 (A–C) illustrates this development whereby a loosely packed accumulation of early magmatic feldspar and mafic minerals becomes progressively mineralized, first with precipitation from a volatile-enriched silicate melt, followed by mineralization from exsolved, dense brines. Figure 13 shows a simplified paragenesis of the major mineral phases forming from the evolution of a magmatic (silicate melt) to a hydrothermal (aqueous brine) fluid. All high-temperature crystallization and/or alteration processes operating in the chamber margin would be quickly and effectively quenched during eruption. However, cooling of the erupted products may result in some further alteration.

This temperature transition from higher-temperature magmatic processes to lower-temperature aqueous processes has been documented during our study. Figure 14 shows the general pressure–temperature regime estimated for the studied nodules using measured temperature data and assumptions regarding pressure. In Figure 14, we have plotted mineral-pair geothermometers and the liquidus curve derived from the program MELTS (Ghiorso and Sack, 1993). For the two-feldspar geothermometer we used Nekvasil and Burnham (1987) from the program SOLVCALC, and for the hornblende–plagioclase geothermometer, we used Holland and Blundy (1994). Our assumption is that the mineral pairs were in equilibrium, which is supported by textural evidence. The recognition of equilibrium pairs is not a trivial matter, as pointed out by Mottana (1998). For any mineral pair (especially alkali feldspar and plagioclase) in these nodules, we observed overlapping petrogenetic stages in a single thin section.

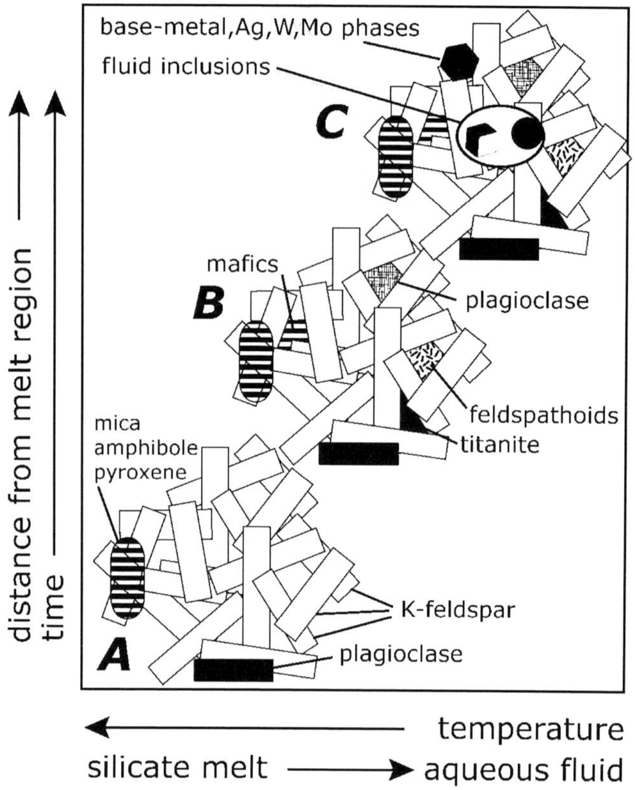

Figure 12. Schematic cartoon illustrating the mineralogical and textural evolution of the nodules, simplified into three main stages. The arrow-marked trends along the base and left side indicate, in general, the evolution of the average studied nodule. Each regime, A, B, and C, corresponds to the same small volume of magma chamber margin which was dislodged, entrained, and erupted with the BM products. (A) early magmatic accumulation of alkali feldspar, plagioclase, and mafics (mostly mica, amphibole, and clinopyroxene). This accumulation volumetrically was mostly K-feldspar and contained many vugs and "open" spaces. (B) The vugs and "open" spaces of the K-feldspar accumulation were filled with more mafics, Na-rich plagioclase, feldspathoids (scapolite and a cancrinite-group mineral), and phases rich in incompatible elements such as titanite, apatite, monazite, and Zr- and U-bearing minerals. (C) The last features we observe are the precipitation of chlorides, sulfates, base-metal, Ag, W, and Mo phases in fluid inclusions or in fractures.

Each nodule represents a small volume of chamber margin, which is different from any other nodule. The differences result from the particular location of the nodule volume and the differences that would be expected to occur around the margins of a crystallizing magma. Nevertheless, all the nodules display the same petrogenetic theme – early-formed magmatic-melt phases evolving to late-stage hypersaline brine precipitated phases.

4.3. Porphyry copper system analogue

The comparison between modern geothermal/hydrothermal systems and porphyry copper systems was studied to develop and enhance an exploration model (e.g., Burnham, 1979;

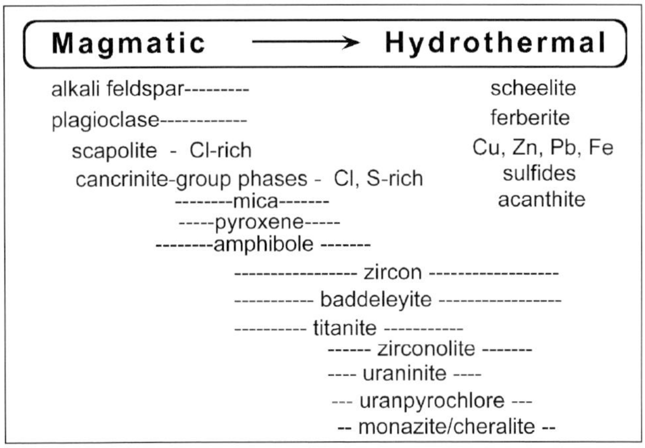

Figure 13. A simplified paragenetic scheme showing the main mineral phases as formed from a silicate-melt magmatic environment evolving to formation from a relatively high-temperature aqueous brine. We have used the general texture and observed mineralogical and fluid inclusion characteristics to classify the mineral phases. The general sequence of deposition is shown with more common phases at the top descending to less common ones toward the bottom. Phases such as ferberite, sulfides, and acanthite are late-stage but the details of their sequence of deposition are difficult to decipher as these phases are uncommon.

Beane, 1983). The research on active geothermal/hydrothermal systems was directed to understanding the processes, although the ultimate criterion for a "porphyry copper ore deposit" will be the mass of the metal deposited. Sasada (2000) discusses the geothermal porphyry copper analogue and concludes that the major alteration difference, at least in the context of Japanese geothermal systems, is that porphyry copper systems have a pervasive potassium-rich alteration. The BM–CF system is developed in an high-potassium magmatic environment. Further evolution of the hydrothermal system may lead to potassic alteration. The Burnham (1979) model, whereby crystallizing magmas become progressively enriched in volatiles and exsolve an aqueous fluid rich in incompatible elements, such as Cl, REE, metals, etc., is compatible with our mineralogical, textural, and chemical observations of the nodules.

De Vivo and Lima (this volume) have addressed the problem of frequent ground movements (bradyseisms) in the CF volcanic region. They suggest that hydrothermal processes driven by heat evolved from the cooling magma at depth are responsible for the ground movements.

4.4. *Magmatic–hydrothermal transition*

In the alkali-syenite nodules of the BM we found textural, mineralogical, and chemical evidence that suggests they recorded a transition from magmatic to hydrothermal conditions. The nodules primarily are composed of alkali feldspars (up to 80%) with subordinate plagioclase, scapolite, a S- and Cl-rich member of the cancrinite group, amphibole, clinopyroxene, biotite, titanomagnetite, titanite, and apatite.

SEM-EDS and electron microprobe analyses of the nodules provide evidence for the presence of U, Zr, Nb, Th, and REE elements in late-stage minerals such as apatite, Zr-bearing minerals (i.e., zircon and baddeleyite), pyrochlore-group minerals, thorite, and

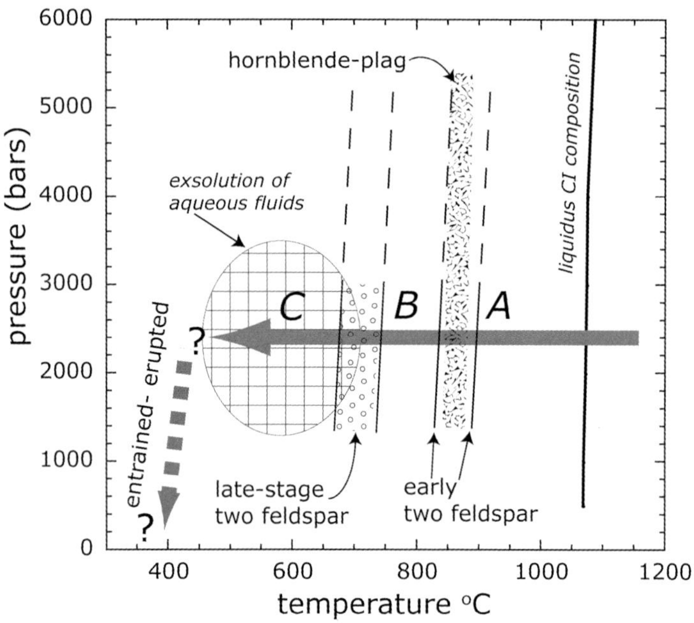

Figure 14. A schematic cartoon in pressure/temperature space that illustrates the evolution of the studied nodules. Here we have used various mineral-pair geothermometers, fluid inclusion data, and liquidus temperatures to construct a reasonable PT evolution of the nodules. The liquidus curve is derived from MELTS (Ghiorso and Sack, 1993) using the average CI whole-rock composition given in Table 2. The general range of the temperatures from early-alkali feldspar–plagioclase pairs is shown by two curves and the field (small open circles) is shown for texturally late-stage alkali feldspar–plagioclase pairs (both using Nekvasil and Burnham, 1987). The temperature is extrapolated to higher pressures by the dashed lines. The field for the hornblende–plagioclase geothermometer data (Holland and Blundy, 1994) is shown by a pattern. The general, estimated PT field for the exsolution of aqueous fluid is shown. The horizontal arrow traces, in general, the PT evolution of the average nodule from the melt environment to the PT time just before entrainment and eruption. We have chosen a constant pressure of about 2200 bars, corresponding to a depth of 8 km (assuming an overburden density of 2.7 g/cm^3). With regard to the liquidus and mineral-pair geothermometer data, their curves are very steep so the exact pressure regime assumption is not critical. The letters, A, B, and C mark the general PT areas as outlined by Figure 12.

phosphate (monazite group). Many of these accessory minerals occur with typical hydrothermal textures (i.e., as replacement, in vugs or as infillings). Microprobe data show the partition of fluorine, chlorine, and sulfur in the syenite nodules. Fluorine is incorporated in biotite and subordinately in amphibole, while chlorine and sulfur are incorporated in scapolite, cancrinite-group minerals and trapped in as FIs. Complex assemblages of daughter minerals found in multiphase FIs in the nodules of the BM are evidence of the trapping of high-solute fluids. Abundance of chlorides, sulfides, and to a lesser extent sulfates and carbonates, suggest that FIs trapped a hypersaline/sulfur-rich fluid (possibly with minor CO_2), which might have been exsolving from a crystallizing magma.

Preliminary microthermometry and observations on secondary hypersaline FIs in K-feldspars suggest two possible scenarios for fluid trapping: (1) circulation of non-boiling, high-temperature (up to 525°C), high-salinity fluids which possibly were trapped under decreasing P–T conditions; and (2) circulation of boiling, hypersaline fluids, trapped at low pressure and temperatures up to 300°C.

The strong petrographic evidence for the coexistence of Type 1 (vapor) and Type 2 (hypersaline) FIs, seems to suggest that, at least, these particular inclusions are trapped boiling fluids and formed in a later hydrothermal stage.

In the last decade, many studies of magma chamber margins in alkali-enriched magmatic systems (Turbeville, 1992; Federico et al., 1994; Belkin et al., 1994, 1996; Renzulli et al., 1995; Fulignati et al., 1997; Tarzia et al., 1999, 2000; Gilg et al., 2001) demonstrated that these peripheral areas can show selective enrichment of incompatible elements (i.e., U, Th, Zr, REE) with features similar to those found in the BM nodules.

4.5. Nature of the exsolved aqueous fluid

Many studies on magmatic systems (Roedder, 1984 and references therein) report the presence of high-salinity FIs (i.e., the presence of a Cl-rich, high-density aqueous phase). Similar occurrences are also reported from Italian subvolcanic systems (De Vivo et al., 1992, 1993, 1995; Lowenstern, 1994; Belkin et al., 1996).

Experimental studies have pointed out that many metals tend to partition into a hypersaline chloride-bearing brine exsolved from a silicate melt (Candela, 1986, 1989; Cline and Bodnar, 1991; Shinohara, 1994; Candela and Piccoli, 1995; Kamenetsky et al., 2003). Similarly, trace and REE elements can be efficiently extracted from magma by Cl-rich fluids (Kravchuk and Keppler, 1994; Haas et al., 1995). Sulfate-rich fluids are reported in peralkaline silicic magmatic intrusions and in orthomagmatic fluids related to porphyry copper deposits (Roedder, 1984). Furthermore, under oxidizing conditions, S fugacity constraints yield a sulfur-rich vapor exsolving from a crystallizing magma (Carroll and Rutherford, 1985).

The occurrence of sulfides together with chlorides, sulfates, and Fe-, Mn-oxides in the nodules and as daughter minerals suggest that the redox state of the nodule-forming environment shifted progressively toward more oxidized conditions. Concerning the partition of REE into S-rich fluid, Wood (1990) states that sulfate–REE complexes can predominate over aqueous species in the absence of other ligands. Furthermore, temperature strongly affects the REE-complex stability constant: increasing temperatures produce an increase in the stability constant values for fluoride, sulfate, and chloride (in that order). Therefore, chloride and sulfates, which were present in the peripheral magma chamber system of the BM magma, could have played an important role in the selective enrichment of REE in the accessory minerals of the nodules.

4.6. W (Mo) enrichment

The detailed nodule petrography and mineralogy revealed many occurrences of W (Mo) mineral species (scheelite–powellite, ferberite). Although these nodules are not exceptionally mineralized in comparison to typical W-skarn deposit samples, they do contain an interesting assemblage of W (Mo) mineral species. Manning and Henderson (1984) demonstrated that chloride complexes can efficiently extract tungsten from magmatic melts into an aqueous phase. This suggests that an exsolving aqueous phase most likely extracted W and Mo from the magma and transported these elements to the nodule environment where cooler conditions or some particular reaction caused precipitation.

5. Concluding remarks

The alkali-syenite nodules found in the BM deposit display convincing evidence of a transition from a magma-dominated system to a fluid-dominated hydrothermal system. This transition took place along the margins of a magma chamber where a magma of trachytic composition was sufficiently evolved to exsolve an aqueous fluid carrying a complex solute, containing high amounts of incompatible elements. Textures, mineralogy, and mineral chemistry of the nodules and nature of FI of daughter minerals all point toward this interpretation. The various sulfides and tungstates observed in the nodules suggest the potential for an ore deposit somewhere (or sometime) in this volcanic system.

Acknowledgements

The authors wish to thank Carter Hearn (Smithsonian, USGS retired) and Paul C. Hackley (USGS) for their constructive reviews. The work benefited from partial funds from GNV-INGV to B. De Vivo (2003).

References

Ballirano, P., Maras, A., Buseck, P.R., 1996. Crystal chemistry and IR spectroscopy of Cl- and SO_4-bearing cancrinite-like minerals. Am. Mineral. 81, 1003–1012.

Beane, R.E., 1983. The magmatic–meteoric transition. Geotherm. Res. Counc. Spec. Rep. 13, 245–253.

Belkin, H.E., Cavarretta, G., Tecce, F., 1994. Petrogenesis of leucite-bearing nodules from Colli Albani Volcano ejecta, Latium, Italy. Abstracts International Mineralogical Association, 16th General Meeting, 4–9 September, Pisa, Italy, p. 35.

Belkin, H.E., De Vivo, B., 1993. Fluid inclusion studies of ejected nodules from plinian eruptions of Mt. Somma-Vesuvius. J. Volcanol. Geotherm. Res. 58, 89–100.

Belkin, H.E., De Vivo, B., Lima, A., Török, K., 1996. Magmatic silicate/saline/sulfur rich/CO_2 immiscibility and zirconium and REE enrichment from alkaline magma chamber margins: evidence from Ponza Island and Pontine archipelago, Italy. Eur. J. Mineral. 8, 1401–1420.

Belkin, H.E., De Vivo, B., Roedder, E., Cortini, M., 1985. Fluid inclusion geobarometry from ejected Mt. Somma-Vesuvius nodules. Am. Mineral. 70, 288–303.

Bellucci, F., Milia, A., Rolandi, G., Torrente, M.M., this volume. Structural control on the Upper Pleistocene ignimbrite eruptions in the Neapolitan area (Italy): volcano tectonic faults versus caldera faults.

Bodnar, R.J., 1992. Can we recognize magmatic fluid inclusions in fossil systems based on room temperature phase relations and microthermometric behavior? Geol. Surv. Jpn. Rep. 279, 26–30.

Bodnar, R.J., 1994. Synthetic fluid inclusions: XII. The system H_2O-NaCl. Experimental determination of the halite liquidus and isochores for a 40 wt.% NaCl solution. Geochim. Cosmochim. Acta 58, 1053–1063.

Bodnar, R.J., Beane, R.E., 1980. Temporal and spatial variations in hydrothermal fluid characteristics during vein filling in preore cover overlying deeply buried porphyry copper-type mineralization at Red Mountain, Arizona. Econ. Geol. 75, 876–893.

Bodnar, R.J., Burnham, C.W., Sterner, S.M., 1985. Synthetic fluid inclusions in natural quartz. III. Determination of phase equilibrium properties in the system H_2O-NaCl to 1000°C and 1500 bars. Geochim. Cosmochim. Acta 49, 1861–1873.

Bodnar, R.J., Vityk, M.O., 1994. Interpretation of microthermometric data for H_2O-NaCl fluid inclusions. 117-130. In: De Vivo, B., Frezzotti, M.L. (Eds), Fluid Inclusions in Minerals: Methods and Applications. Short Course of the Working Group (IMA) "Inclusions in minerals", Pontignano, Siena, 1–4 September, 376pp.

Borodin, L.S., Bykova, A.B., Kapitanova, T.A., Pyatenko, Y.U.A., 1960. New data on zirconolite and its niobium variety. Dokl. Acad. Sci, Earth Sci. Sec. 134, 1022–1024.

Branney, M.J., Kokelaar, B.P., 1992. A reappraisal of ignimbrite emplacement: progressive aggradation and particulate to non-particulate flow transition during emplacement of high-grade ignimbrite. Bull. Volcanol. 54, 504–520.

Brøgger, W.C., 1890. Die Mineralien der Syenitpegmatitgänge der Südnorwegischen Augit- und Nephelinsyenite. 46. Polymignyt. Z. Krystall. Mineral. 16, 387–396.

Burnham, C.W., 1979. Magmas and hydrothermal fluids. In: Barnes, H.L. (Ed.), Geochemistry of Hydrothermal Ore Deposits. Wiley, New York, pp. 71–136.

Candela, P.A., 1986. Towards a thermodynamic model for the halogens in magmatic systems; an application to melt-vapor-apatite equilibria. Chem. Geol. 57, 289–301.

Candela, P.A., 1989. Calculation of magmatic fluid contributions to porphyry-type ore systems: predicting fluid inclusion chemistries. Geochem. J. 23, 295–305.

Candela, P.A., Piccoli, P.M., 1995. Model oremetal partitioning from melts into vapor and vapor/brine mixtures. MAC Short Course Ser. 23, 101–127.

Caprarelli, G., Tsutsumi, M., Turi, B., 1997. Chemical and signatures of the basement rocks from the Campi Flegrei geothermal field (Naples, southern Italy): inferences about the origin and evolution of its hydrothermal fluids. J. Volcanol. Geotherm. Res. 76, 63–82.

Carroll, M.R., Rutherford, M.J., 1985. Sulfide and sulfate saturation in hydrous silicate melts. J. Geophys. Res. 90, C601–C612.

Cline, J.S., Bodnar, R.J., 1991. Can economic porphyry copper mineralization be generated by a typical calcalkaline melt? J. Geophys. Res. 96, 8113–8126.

Cline, J.S., Bodnar, R.J., 1994. Direct evolution of brine from crystallizing silicic melt at Questa, New Mexico, molybdenum deposit. Econ. Geol. 89, 1780–1802.

Deer, W.A., Howie, R.A., Wise, W.S., Zussman, J., 2004. Rock-Forming Minerals, Volume 4B: Framework Silicates, Silica minerals, Feldspathoids, and the Zeolites. The Geological Society, London, 982pp.

De Vivo, B., 1999. Fluids in volcanic and plutonic environments: evidence from fluid inclusions. In: Marshall, C.P., Fairbridge, E.W. (Eds), Encyclopedia of Geochemistry. Kluwer Academic Publishers, Dordrecht, NL, pp. 250–252.

De Vivo, B., Belkin, H.E., Barbieri, M., Chelini, W., Lattanzi, P., Lima, A., Tolomeo, L., 1989. The Campi Flegrei (Italy) geothermal system: a fluid inclusion study of the Mofete and san Vito fields. J. Volcanol. Geotherm. Res. 36, 303–326.

De Vivo, B., Frezzotti, M.L., Lima, A., 1993. Immiscibility in magmatic differentiation and fluid evolution in granitoid xenoliths at Pantelleria: fluid inclusion evidence. Acta Vulcanol. 3, 195–202.

De Vivo, B., Frezzotti, M.L., Mahood, G., 1992. Fluid inclusions in xenoliths yield evidence for fluid evolution in peralkaline granitic bodies at Pantelleria (Italy). J. Volcanol. Geotherm. Res. 52, 295–301.

De Vivo, B., Lima, A., this volume. An hydrothermal model to explain the ground movements (bradyseism) at Campi Flegrei.

De Vivo, B., Rolandi, G., Gans, P.B., Calvert, A., Bohrson, W.A., Spera, F.J., Belkin, H.E., 2001. New constraints on the eruptive history of the Campanian volcanic plain (Italy). Mineral. Petrol. 73, 47–65.

De Vivo, B., Török, K., Ayuso, R.A., Lima, A., Lirer, L., 1995. Fluid inclusion evidence for magmatic silicate/saline/CO_2 immiscibility and geochemistry of alkaline xenoliths from Ventotene island (Italy). Geochim. Cosmochim. Acta 59, 2941–2953.

Di Girolamo, P., Ghiara, M.R., Lirer, L., Munno, R., Rolandi, G., Stanzione, D., 1984. Vulcanologia and petrologia dei Campi Flegrei. Boll. Soc. Geol. It. 103, 349–413.

Droop, G.T.R., 1987. A general equation for estimating Fe^{3+} concentrations in ferromagnesian silicates and oxides from microprobe analyses, using stoichiometric criteria. Mineral. Mag. 51, 431–435.

Federico, M., Peccerillo, A., Barbieri, A., Wu, T.W., 1994. Mineralogical and geochemical study of granular xenoliths from the Alban Hill, Central Italy: bearing on evolutionary processes in potassic magma chambers. Contrib. Mineral. Petrol. 115, 384–401.

Fisher, R.V., Orsi, G., Ort, M., Heiken, G., 1993. Mobility of large volume pyroclastic flow – emplacement of the Campanian Ignimbrite, Italy. J. Volcanol. Geotherm. Res. 56, 205–220.

Fowler, M.B., Williams, C.T., 1986. Zirconolite from the Glen Dessarry syenite: a comparison with other Scottish zirconolites. Mineral. Mag. 50, 326–328.

Fulignati, P., Gioncada, A., Maineri, C., Sbrana, A., 1997. A first insight into the magmatic–hydrothermal system related to 79 AD eruption magmatic chamber of Vesuvius (Italy). XIV ECROFI, Nancy, Abstract, 115–116.

Ghiorso, M.S., Sack, R.O., 1993. MELTS; software for the thermodynamic analysis of phase equilibria in magmatic systems. Geol. Soc. Am. Abstr. Pr. 25, 96.

Gianfagna, A., 1985. Occurrence of baddeleyite – ZrO_2 – in an ejected block from Colle Cimino, Marino (Alban Hills, Italy). Periodico di Mineralogia 54, 129–133.

Gieré, R., Williams, C.T., Lumpkin, G.R., 1998. Chemical characteristics of natural zirconolite. Schweiz. Mineral. Petrogr. Mitt. 78, 433–459.

Gilg, H.A., Lima, A., Somma, R., Belkin, H.E., De Vivo, B., Ayuso, R.A., 2001. Isotope geochemistry and fluid inclusion study of skarns from Vesuvius. Mineral. Petrol. 73, 145–176.

Haas, J.R., Shock, E.L., Sassani, D.A., 1995. Rare earth elements in hydrothermal systems: estimates of standard partial molal thermodynamic properties of aqueous complexes of the rare earth elements at high pressures and temperatures. Geochim. Cosmochim. Acta 59(21), 4329–4350.

Holland, T., and Blundy, J., 1994. Non-ideal interactions in calcic amphiboles and their bearing on amphibole-plagioclase thermometry. Contrib. Mineral. Petrol. 116(4), 433–447.

Johnston-Lavis, H.J., 1889. Report of the committee appointed for the investigation of the volcanic phenomena of Vesuvius and its neighbor. The Royal Society, London.

Kamenetsky, V.S., Wolfe, R.C., Eggins, S.M., Mernagh, T.P., Bastrakov, E., 1999. Volatile exsolution at the Dinkidi Cu–Au porphyry deposit, Philippines: a melt-inclusion record of the initial ore-forming process. Geology 27, 691–694.

Kamenetsky, V.S., De Vivo, B., Naumov, V.B., Kamenetsky, M.B., Mernagh, T.P., Van Achterbegh, E., Ryan, C.G., Davidson, P., 2003. Magmatic inclusions in the search for natural silicate-salt melt immiscibility: methodology and examples. In: De Vivo, B., Bodnar, R.J. (Eds), Melt Inclusions in Volcanic Systems. Methods, Applications and Problems. Series: Developments in Volcanology, Vol. 5, Elsevier, Amsterdam, pp. 65–82.

Kelley, K.D., Romberger, S.D., Beaty, D.W., Pontius, J.A., Snee, L.W., Stein, H.J., Thompson, T.B., 1998. Geochemical and geochronological constraints on the genesis of Au-Te deposits at Cripple Creek, Colorado. Econ. Geol. 93, 981–1012.

Kilinc, I.A., Burnham, C.W., 1972. Partitioning of chloride between a silicate melt and coexisting aqueous phase from 2 to 8 kilobars. Econ. Geol. 67, 231–235.

Kravchuk, I.F., Keppler, H., 1994. Distribution of chloride between aqueous fluid and felsic melts at 2 kbars and 800°C. Eur. J. Mineral. 6, 913–923.

Larsen, A.O., 1996. Rare earth minerals from the syenite pegmatites in the Oslo region, Norway. In: Jones, A.P., Wall, F., Williams, C.T. (Eds), Rare Earth Minerals: Chemistry, Origin and Ore Deposits. Mineralogical Society Series 7. Chapman and Hall, London, UK, pp. 151–166.

Le Maitre, R.W. (Ed.), 2002. Igneous Rocks, A Classification and Glossary of Terms, 2nd ed. Cambridge University Press, Cambridge, p. 236.

Lirer, L., Perrotta, A., Rolandi, G., Rubin, M., Donahue, D., 1991. The Campi Flegrei "Museum Breccia" deposits: eruptive events younger than Campanian Ignimbrite Formation. International Conference on Active Volcanic Risk Mitigation, Abstracts.

Lowenstern, J.B., 1994. Chlorine, fluid immiscibility, and degassing in peralkaline magmas from Pantelleria, Italy. Am. Mineral. 79, 353–369.

Lowenstern, J.B., Persing, H.M., Wooden, J.L., Lanphere, M., Donnelly-Nolan, J., Grove, T.L., 2000. U–Th dating of single zircons from young granitoid xenoliths: new tools for understanding volcanic processes. EPSL 183, 291–302.

Manning, D.A.C., Henderson, P., 1984. The behavior of tungsten in granitic melt vapor system. Contrib. Mineral. Petrol. 86, 286–293.

Mazzi, F., Munno, R., 1983. Calciobetafite (new mineral from the pyrochlore group) and related minerals from Campi Flegrei, Italy: crystal structures of polymignyte and zirkelite: comparison with pyrochlore and zirconolite. Am. Mineral. 68, 262–276.

Melluso, L., Morra, V., Perrotta, A., Scarpati, C., Adabbo, M., 1995. The eruption of the Breccia Museo (Campi Flegrei, Italy): fractional crystallization processes in a shallow, zoned magma chamber and implications for the eruptive dynamics. J. Volcanol. Geotherm. Res. 68, 325–339.

Mottana, A., 1998. Sanidine holocrystalline ejecta from central Sabatini Volcanic District, Latium (Italy), II Intergranular ejecta and minerogenetic deductions. Rend. Fis. Acc. Lincei. 9, 131–143.

Nekvasil, H., Burnham, C.W., 1987. The calculated individual effects of pressure and water content on phase equilibria in the granite system. In: Mysen, B.D. (Ed), Magmatic Processes, Physicochemical Principles; A Volume in Honor of Hatten S. Yoder, Jr., Special Publication – The Geochemical Society, Vol. 1, pp. 433–445.

Orsi, G., de Vita, S., Di Vito, M., 1996. The restless, resurgent Campi Flegrei nested caldera (Italy): constraints on its evolution and configuration. J. Volcanol. Geotherm. Res. 74, 179–214.

Paone, A., 1999. Modellizzazione dei processi metallogenetici associati a vulcanismo di tipo alcalino. Potenziale della provincia vulcanica Napoletana. Ph.D. thesis, Università degli studi di Napoli "Federico II".

Paone, A., Ayuso, R.A., De Vivo, B., 2001. A metallogenic survey of alkalic rocks of Mt. Somma-Vesuvius volcano. Mineral. Petrol. 73, 201–233.

Perrotta, A., 1992. Evoluzione vulcanologica dei Campi Flegrei tra 20,000 and 12,000 anni and dinamica dell'eruzione della Breccia Museo. Ph.D. thesis, Università di Napoli, 103pp.

Perrotta, A., Scarpati, C., 1994. The dynamics of the Breccia Museo eruption (Campi Flegrei. Italy) and the significance of spatter clasts associated with lithic breccias. J. Volcanol. Geotherm. Res. 59, 335–355.

Renzulli, A., Upton, B.G.J., Nappi, G., 1995. Magma chamber processes preceding the Pitigliano Formation eruption (Latera volcanic complex, central Italy): evidence from cognate plutonic clasts. Acta Vulcanol. 7, 55–74.

Roedder, E., 1971. Fluid inclusion studies on the porphyry-type ore deposits at Bingham, Utah, Bute, Montana and Climax, Colorado. Econ. Geol. 66, 98–120.

Roedder, E., 1984. Fluid inclusions. Reviews in Mineralogy, Vol. 12. Min. Soc. Am., 644pp.

Rolandi, G., Bellucci, F., Heisler, M.T., Belkin, H.E., De Vivo, B., 2003. Tectonic controls on the genesis of ignimbrites from the Campanian Volcanic Zone, Southern Italy. Mineral. Petrol. 79, 3–31.

Rosi, M., Sbrana, A., 1987. Phlegrean Fields. CNR, Quaderni della Ricerca Scientifica, 114(9), 175pp.

Rosi, M., Sbrana, A., Principe, C., 1983. The Phlegrean Fields: structural evolution, volcanic history and eruptive mechanisms. J. Volcanol. Geotherm. Res. 17, 273–288.

Rosi, M., Vezzoli, L., 1989. I depositi prossimali di breccia associati all'eruzione dell'Ignimbrite Campana (Campi Flegrei). Boll. GNV 979–998.

Rosi, M., Vezzoli, L., Aleoti, P., De Censi, M., 1991. Campanian Ignimbrite eruption: proximal deposits. CEV Field Guide, pp. 41–74.

Sasada, M., 2000. Igneous-related active geothermal system versus porphyry copper hydrothermal system. Proceedings World Geothermal Congress 2000, Kyushu-Tohoku, Japan, 1691–1693.

Shinohara, H., 1994. Exsolution of immiscible vapor and liquid phases from a crystallizing silicate melt: implications for chlorine and metal transport. Geochim. Cosmochim. Acta 58, 5215–5221.

Sterner, S.M., Hall, D.L., Bodnar, R.J., 1988. Synthetic fluid inclusions. V. Solubility relations in the system NaCl-KCl-H_2O under vapor-saturated conditions. Geochim. Cosmochim. Acta 52, 989–1005.

Tarzia, M., De Vivo, B., Belkin, H.E., 2000. Occurrence of uranium, zirconium and REE bearing minerals in alkali syenite nodules from the Breccia Museo pyroclastic deposit, Naples, Italy. EOS 81(48), 1272.

Tarzia, M., Lima, A., De Vivo, B., Belkin, H.E., 1999. Uranium, zirconium and rare earth element enrichment in alkali syenite nodules from the Breccia Museo Deposit, Naples, Italy. Geol. Soc. Am. Abst. Pr. 31(7), A-69.

Turbeville, B.N., 1992. Relationship between chamber margin accumulates and pore liquids: evidence from arrested in situ processes in ejecta, Latera caldera, Italy. Contrib. Mineral. Petrol. 100, 429–441.

Wen, S., Nekvasil, H., 1994. SOLVCALC; an interactive graphics program package for calculating the ternary feldspar solvus and for two-feldspar geothermometry. Comput. Geosci. 20, 1025–1040.

Williams, C.T., Gieré, R., 1996. Zirconolite: a review of localities worldwide, and a compilation of its chemical composition. Nat. Hist. Mus. London Bull. 52, 1–24.

Wood, S.A., 1990. The aqueous geochemistry of the rare earth elements and yttrium 2. Theoretical predictions of speciation in hydrothermal solutions to 350°C at saturation water vapor pressure. Chem. Geol. 88, 99–125.

Chapter 8

Structural control on the Upper Pleistocene ignimbrite eruptions in the Neapolitan area (Italy): volcano tectonic faults versus caldera faults

F. Bellucci[a], A. Milia[b], G. Rolandi[a,] and M.M. Torrente[c]*

[a]*DST, University Federico II, via Mezzocannone 8, I-80138, Naples, Italy*
[b]*IAMC, CNR, Calata Porta di Massa, Porto di Napoli, I-80100, Naples, Italy*
[c]*DSGA, University of Sannio, via Portarsa 11, I-82100, Benevento, Italy*

Abstract

In this paper, we present an interdisciplinary study of the Upper Pleistocene ignimbrites of the Campanian margin in the Neapolitan area performed using outcrops, cores and seismic reflection data. We established a physical correlation between onshore and offshore stratigraphic units and reconstructed three regional geological sections. The stratigraphic succession in Naples is formed, from older to younger, by: Middle Pleistocene marine sediments; ancient ignimbrites reaching a maximum thickness of 200 m; 100–150 m thick tuffs of the 39 ka-old Campanian Ignimbrite; and products of 15 ka-old Neapolitan Yellow Tuff. The whole ignimbrite succession is thicker in Naples city and thins progressively towards the Bay of Naples, thus suggesting that some of the vents of these ignimbrites were possibly located near the city of Naples. NW-SE and NE-SW normal faults were recognized in the Neapolitan area. In particular, NW-SE trending normal faults displace both the pre-CI tuffs and the Campanian Ignimbrite downthrowing the blocks towards the Bay of Naples in the order of hundreds of metres and feature ignimbrite thicknesses that are higher in the footwall than in the hangingwall blocks, whereas NE-SW normal faults formed after the Neapolitan Yellow Tuff eruption. In conclusion, the key result of our work is that NW-SE volcano tectonic normal faults were active during the Upper Pleistocene when ignimbrites were emplaced in the Neapolitan area in conforming with a model of fissure emission related to regional fault systems.

1. Introduction

The Campanian Plain is an approximately 2000-km^2-wide region, bounded on the west by the Tyrrhenian Sea and on the east by the Apennines (Fig. 1). In the last 600 ka, it has been affected by intense volcanism alternating with periods of marine sedimentation (Ballini et al., 1989; Scandone et al., 1991). Recent studies indicate that at least five ignimbrites were emplaced over the entire Campanian Plain in the last 300 ka (De Vivo et al., 2001; Rolandi et al., 2003), but the relationships between these Upper Pleistocene ignimbrite eruptions and the structure of the Campanian margin (formed by the Campanian Plain and Naples Bay) is still a matter of debate. Some authors (Rosi and Sbrana, 1987; Orsi et al., 1996) argue that the a major ignimbrite eruption (39 ka-old Campanian Ignimbrite) is associated with a caldera located in the Campi Flegrei and northern Bay of Naples (Fig. 2),

*Corresponding author. *E-mail address:* rolandi@unina.it (G. Rolandi).

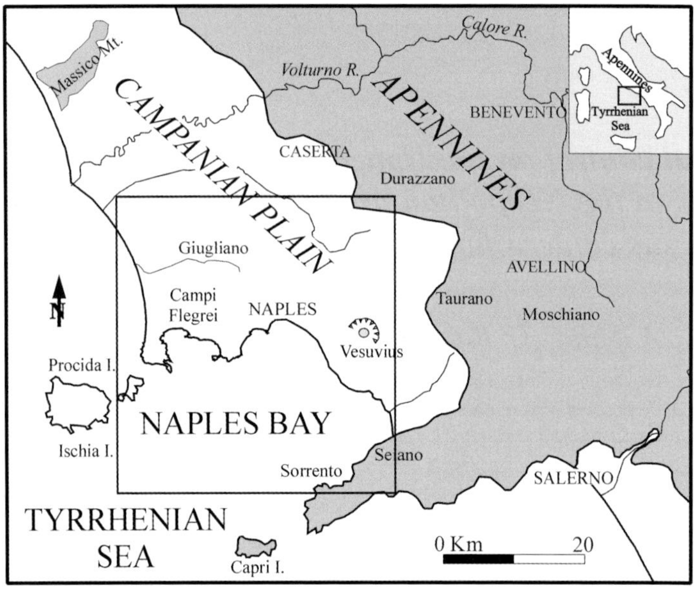

Figure 1. Location map of the Campanian margin.

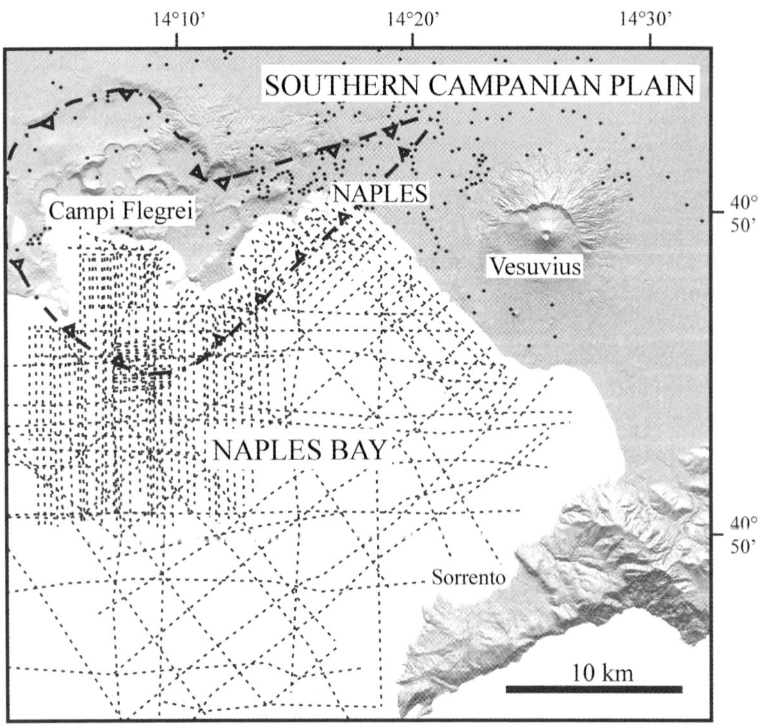

Figure 2. Index map of seismic profiles and bore holes. The presumed Campanian Ignimbrite caldera is also shown (after Orsi et al., 1996).

whereas others propose that regional faults controlled its emission in the Bay of Naples (Milia, 2000; Milia and Torrente, 2003) and the Campanian Plain (Rolandi et al., 2003).

High-volume ignimbrite eruptions are frequently associated with the collapse of a coherent crustal block, corresponding to the top of a magma chamber, along a ring fault that produces a superficial circular depression, or caldera (Smith and Bailey, 1968). However, recent studies indicate that ignimbrite emission can also be controlled by regional fault systems clearly documenting a fissural origin (Cole, 1990; Burkart and Self, 1995; De Rita and Giordano, 1996; Aguirre-Diaz and Labarthe-Hernandez, 2003; Rolandi et al., 2003). In some cases, the regional fault system develops incrementally producing a late stage piecemeal caldera by several differential subsidences of independent fault blocks (Branney and Kokelaar, 1994; Moore and Kokelaar, 1998).

This work is an interdisciplinary study of the Upper Pleistocene ignimbrites of the Campanian margin surrounding Naples (Fig. 2). In order to locate the source area of these eruptions, we acquired seismic reflection profiles in the Bay of Naples, undertook a geological survey in the Neapolitan urban area and analysed core data collected from literature, drilling companies and local authorities. Our goal was to reconstruct the three-dimensional geometry of the Upper Pleistocene ignimbrite units and their associated volcano tectonic faults. A key result of our study supports the view that many of the Campanian ignimbrites (including the Campanian Ignimbrite) may have been fed from volcano tectonic faults.

2. Geological setting

The Campanian continental margin (southern Italy) displays the typical features of a continental crust and lithosphere extensional domain: numerous normal faults, a very shallow Moho (Ferrucci et al., 1989), high heat flow values (Della Vedova et al., 2001) and large volume ignimbrite eruptions.

The structure of the Campanian margin is characterized by upper Miocene thrusts of the southern Apennines displaced by numerous Quaternary fault systems linked to the last stages of the Tyrrhenian Sea opening. Structural analyses performed on the structural highs and downthrown zones of the Campanian margin reveal that NW-SE normal faults, Lower Pleistocene in age, pre-date NE-SW faults post-700 ka in age (Gars and Lippman, 1984; Milia and Torrente, 1997; Milia, 1999). In detail, the geometry of the NE-trending fault system, based on the interpretation of high-resolution seismic reflection profiles in Naples Bay, is seen to be characterized by tilted block, half graben and multiple fault segments linked by relay zones (Milia ,1999; Milia and Torrente, 1999). In addition, the basin infill architecture of the Bay of Naples (a Lower Pleistocene basal marine unconformity-bounded unit, covered by seven fourth-order depositional sequences that form a Middle Pleistocene Transgressive-Regressive sedimentary cycle) reflects slip rate changes of the NE-trending faults. A NW-SE trending regional geologic section across the Campanian margin displays Quaternary half graben bounded by a linked asymmetrical normal fault system and a main decollement surface located at depths of 10–12 km (Milia et al., 2003a). This balanced and restored section indicates a total NW-SE elongation of 0.25.

The Late Quaternary fault pattern of the Bay of Naples is characterized by E-W trending left-lateral faults, NE-trending normal faults and NW-trending transtensional faults (Milia, 2000; Milia and Torrente, 2000). This fault pattern was interpreted as being the result of block rotation associated with a transtensional regime along an E-W left-lateral fault zone (Milia and Torrente, 2003).

Numerous trachitic ignimbrites dated at 290 ka and 240 ka (Seiano Ignimbrites), 157 ka (Taurano Ignimbrite), 116 ka (Durazzano Ignimbrite), 39 ka (Campanian Ignimbrite) and 23–18 ka (Giugliano Ignimbrite) are inferred to have been emplaced across the whole Campanian Plain (Rolandi et al., 2003). The products of the older ignimbrites correspond to distal ash flows and are exposed in the Appennine valleys.

The Campanian Ignimbrite (hereinafter CI) eruption at 39 ka (De Vivo et al., 2001) produced the most widespread (about 6000 km^2) and largest (200 km^3 Dense Rock Equivalent) volcanic deposits in the Campanian margin (Rolandi et al., 2003). The products of the Campanian Ignimbrite crop out on the Campanian Plain, in the city of Naples, the Sorrento coastal slope and Capri Island. In addition, a thick seismically chaotic unit, interpreted as the CI, has been recognized on the Bay of Naples continental shelf (Fusi et al., 1991; Milia, 1998; Milia et al., 1998). It is important to remark that both the CI and Giugliano Ignimbrite have been recognized in the Mediterranean Sea as Y-5 and Y-3 ash layers, respectively (Keller et al., 1978; Tunnel et al., 1978).

Finally, the Neapolitan Yellow Tuff (hereinafter NYT), dated at 15 ka (Deino et al., 2004) with an estimated total dense-rock-equivalent volume of 49.3 km^3 (Scarpati et al., 1993), is the only pyroclastic flow deposit that was fed from the Campi Flegrei area. The NYT crops out as both lithified and unlithified facies (Scherillo, 1955); basically the lithified facies occurs in Campi Flegrei and Naples city, whereas the unlithified facies is present in the distal areas (De Gennaro et al., 2000). The NYT forms a thick and widespread pyroclastic unit in the Naples area and reaches a thickness of approximately 150 m at Posillipo Hill (Guadagno, 1928). It overlies older pyroclastic deposits and marine sediments. The NYT has also been documented offshore Naples where it forms a wedge thickening towards Posillipo Hill (Milia, 1998; Milia et al., 1998; Milia and Torrente, 2003).

The products of the CI are overlain in the southern Campanian Plain by the Somma-Vesuvius volcanic complex (Brocchini et al., 2001). The ancient Monte Somma volcano was built up between 39 and 25 ka through effusive and moderately explosive eruptions. From 25 ka to 472 AD, the Somma volcano gave rise to seven plinian eruptions (Santacroce, 1987; Rolandi et al., 1993a, b). Since 3550 (Avellino eruption) each plinian eruption has been followed by small-scale inter-plinian eruptions and a repose period (Rolandi et al., 1998).

3. Onshore stratigraphy

The geology of southern Campanian Plain, corresponding to the urban area of Naples, has been investigated in the past through the description of outcropping sections (Johnston-Lavis, 1889; De Lorenzo, 1904; Scherillo and Franco, 1967) and deep drill holes (Guadagno, 1924; D'Erasmo, 1931). The stratigraphic reconstructions performed in hilly and coastal area of Naples reported, from older to younger, marine sediments devoid of volcanic clasts, tufaceous deposits referred to as "primo periodo flegreo" ("first Phlegrean period", De Lorenzo, 1904) and NYT. The uppermost part of the "primo periodo flegreo" deposits (breccia deposit) corresponds to the CI unit (De Vivo et al., 2001). Orsi et al. (1996), by contrast, interpreted the tufaceous deposits underlying the NYT in the coastal area of Naples as pyroclastics younger than the CI. On the basis of this interpretation the authors located the CI caldera depression partly in the Neapolitan area.

We investigated the city of Naples and southern Campanian Plain by means of detailed geological surveys, an analysis of outcropping stratigraphic sections and a stratigraphic study of deep boreholes collected from literature (Johnston-Lavis, 1889; Guadagno, 1924; D'Erasmo, 1931), drilling companies and local authorities (Fig. 1). Our stratigraphic study focused on the oldest tufaceous deposits and the overlying CI unit.

The oldest volcanic units crop out in Naples at St. Martino hill. A good stratigraphic correlation exists between St. Martino hill and the adjacent Montesanto funicular tunnel (Johnston-Lavis, 1889) located on the hill's eastern flank. The overall stratigraphic sequence is composed, from older to younger, of (Figs. 3–5):

(*A*) *Basal fine tuff unit.* This is a yellow lithified, very fine, homogeneous tuff that passes to a grey and less-compacted tuff at the top. Stratified structures with pumice lenses and pisolites are present at the base indicating a phreatomagmatic character (Figs. 3 and 4A). This unit outcrops along the south cliff of St. Martino hill and corresponds to pre-CI tuffs. The top of these deposits are brown in colour due to humification.

(*B*) *Pumice fall unit.* This consists of a sequence of alternating pumice fall layers and ash layers 10–30 m thick, locally intercalated by pumice flow lenses (Fig. 4B). At the top of this sequence a layer of very coarse pumice (15–20 cm of diameter) is present; pumices are sharp, vesicular and characterized by a greenish core. This unit outcrops only on the south flank of St. Martino hill (Fig. 3) and has been interpreted as the pumice fall that preceded the CI eruption.

(*C*) *Piperno–breccia–reddish tuff unit.* The Piperno, consisting of a dark grey welded tuff with eutaxitic structure (Fig. 4C), corresponds to the CI unit 1 (Rolandi et al., 2003). It reaches a maximum thickness of 2–4 m at St. Martino hill and Montesanto. The Piperno is overlain by an approximately 4–12 m thick breccia deposit (Figs. 3 and 4D) that gradually passes upward to a reddish tuff. The breccia deposit consists of etherometric fragments mainly composed of xenolites (lavas and obsidian) and subordinate green vesicular pumice fragments. The reddish tuff consists of a massive pumice flow with large black scoria fragments embedded in a reddish sandy matrix (Fig. 4E). Both breccia and reddish tuff correspond to the CI unit 2 (Rolandi et al., 2003). Typically CI unit 1 and unit 2 crop out towards the eastern Neapolitan area (Ponti Rossi, Poggioreale zones) (Fig. 4F).

(*D*) *NYT.* This consists of thick banks (20–50m) of zeolitized pyroclastic deposits, characterized at the base by a succession of rhythmic fall pumice layers.

In places between CI and NYT a succession of alternating pumice and ash layers occurs (Fig. 5).

Deep drill holes located near the St. Martino, Montesanto and Poggioreale outcrops display the same stratigraphic sequence. In fact they encounter, from older to younger, Middle Pleistocene marine deposits, pre-CI tuffs, the CI unit and the NYT; the former are made up of coarse sands passing upwards to a fossils-rich clay layer, followed by a fine to coarse coastal sandy deposit with a sandstone layer at the top.

The correlation between outcropping sections and borehole successions allow us to better interpret the stratigraphic data of boreholes located along the coast (Fig. 6). Here we can recognize the same stratigraphic section covered by alluvial and coastal deposits.

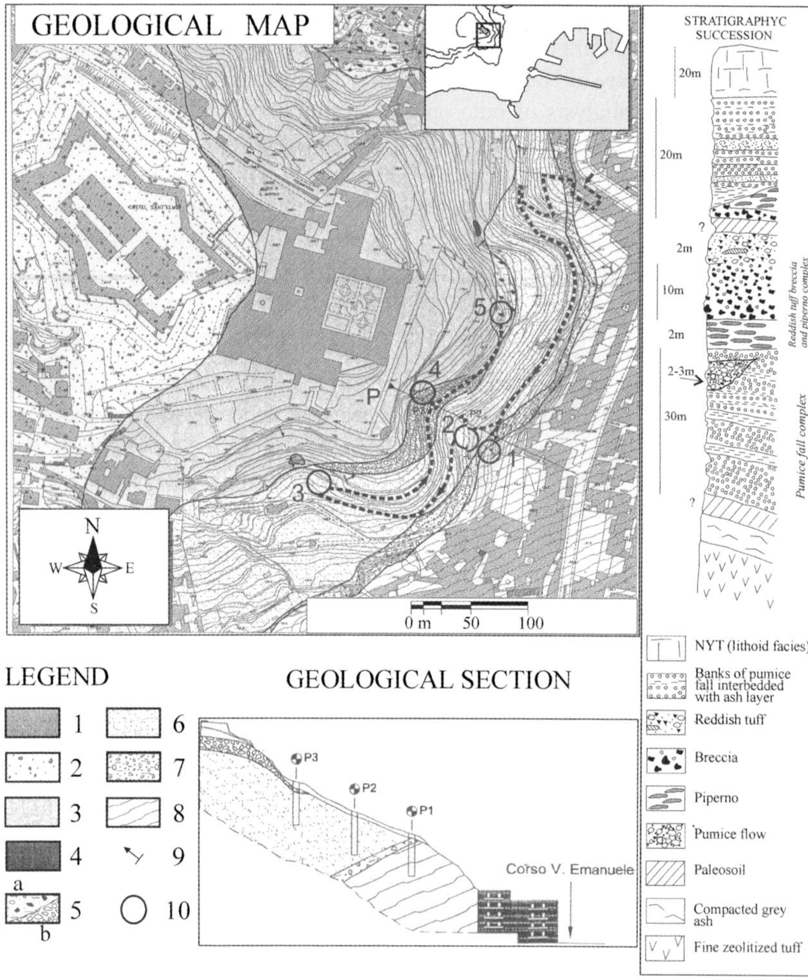

Figure 3. Geological map of St. Martino hill. (1) Pumice fall and surge deposits of the Agnano eruption (10–8 ka). (2) Neapolitan Yellow Tuff (15 ka). (3) Stratified pumice fall deposits interbedded with grey ash layer. (4) Pipernoid layer. (5) Reddish tuff, breccia and piperno complex (39 ka). (6) Pumice fall complex constituted by pumice layers passing upwards to ash layers. (7) Coarse pyroclastic deposits interbedded with fine grey pyroclastic layers. (8) Fine yellow lithoid tuff. (9) Strata attitude. (10) Outcrop sites.

Moreover the pre-CI units are interbedded with marine sediments containing fossils. The entire succession also lies on the Middle Pleistocene marine deposits.

The stratigraphic succession of the southern Campanian Plain was investigated by means of drill holes collected for water exploitation in the area around Vesuvius (Bellucci, 1998) together with the 1835-m-deep Trecase geothermal drill hole that reached the Mesozoic Cenozoic carbonate substrate (Brocchini et al., 2001). This stratigraphic sequence is made up, from older to younger, of *pleistocene marine deposits*. *Pre-CI units* consisting of very coarse to fine ash layers alternating with black tuffs. *The basal pumice from the CI unit* (in the northern area). *CI unit* consisting of grey tuffs. *Somma unit*

Figure 4. Stratigraphic units outcropping at St. Martino hill. (A) fine yellow tuff; (B) basal pumice complex; (C) piperno; (D) breccia; (E) reddish tuff with black scoriae; (F) Poggioreale units: basal CI unit 1, CI unit 2 breccia passing upwards to reddish tuff.

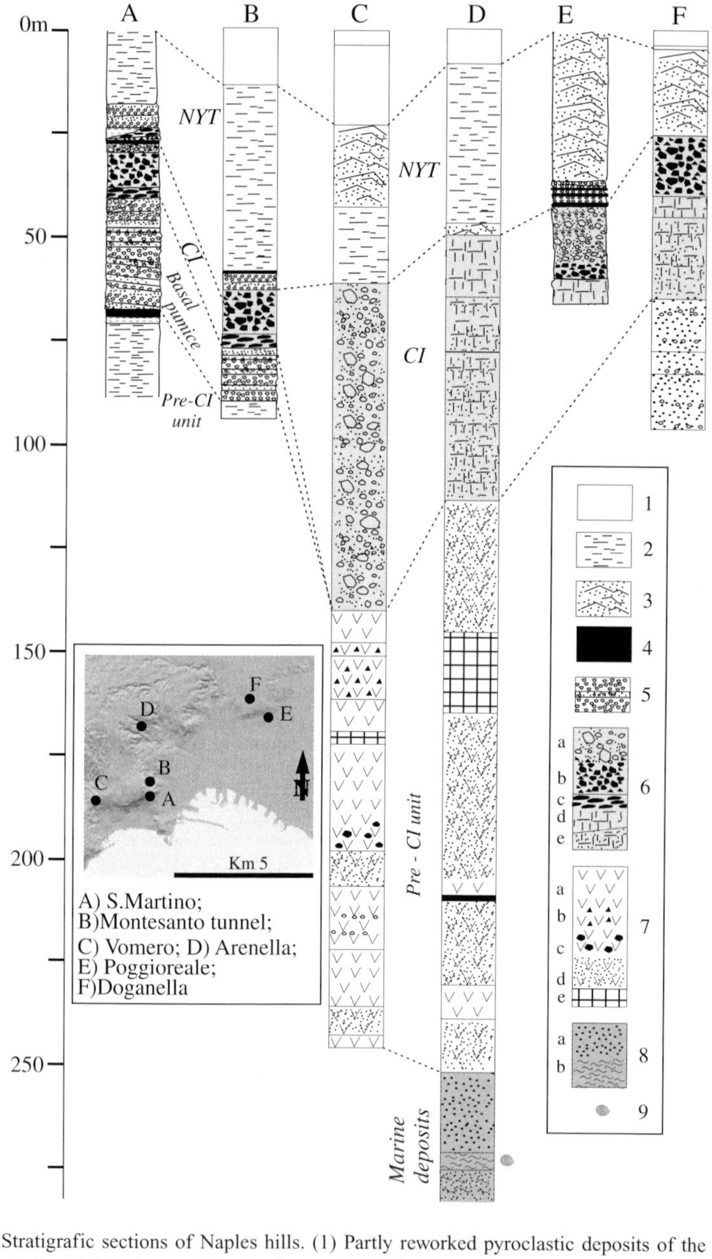

Figure 5. Stratigrafic sections of Naples hills. (1) Partly reworked pyroclastic deposits of the Campi Flegrei post-NYT activity. (2) Massive zeolitized yellow tuff. (3) Sand wave to massive ash (pozzolana). (4) Paleosoil. (5) Pumice deposit alternated with ash beds. (6) Reddish tuff with big pumice (a); breccia (b); piperno (c); welded to partly welded grey tuff (d); incoherent grey tuff (e). (7) Compacted tuff varying from yellow to reddish, to grey colour (a); lava fragments (b); black pumice (c); incoherent tuff (d); Foam lava (e); (8) Sand (a) and clay (b) marine deposits; (9) Marine shells.

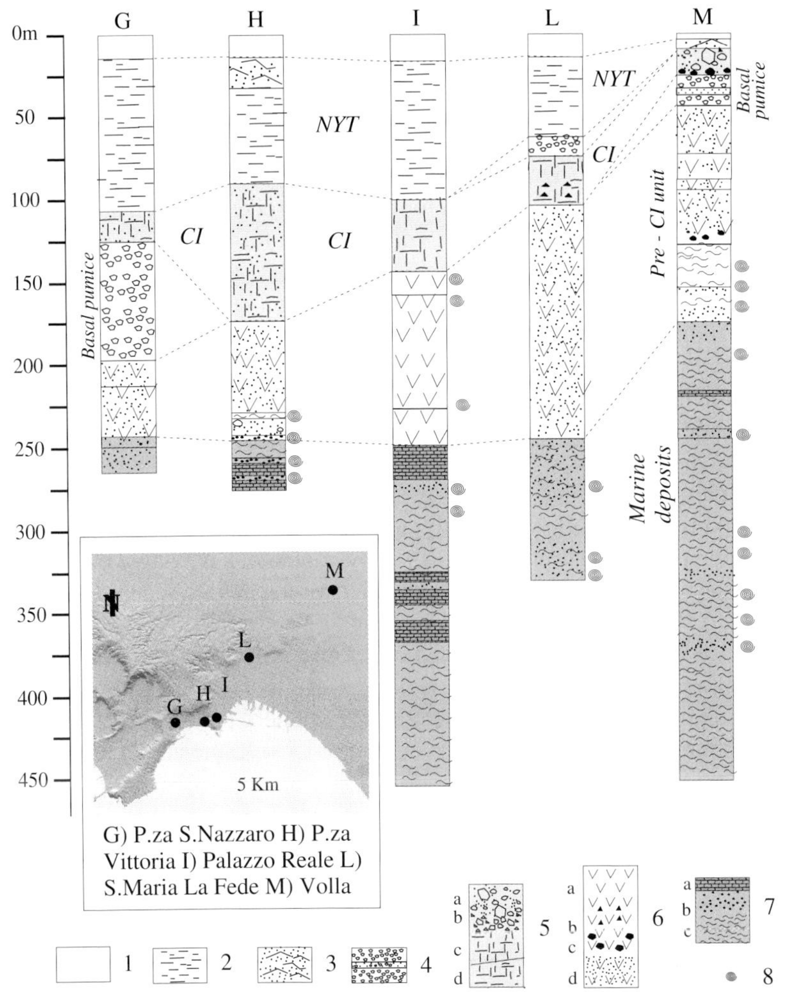

Figure 6. Stratigraphic sections of the Naples coast and plain. (1) Partly reworked pyroclastic deposits of the Campi Flegrei post-NYT activity. (2) Massive zeolitized yellow tuff. (3) Sand wave to massive ash (pozzolana). (4) Pumice deposit alternated with ash beds. (5) Reddish tuff with big pumice (a); lava fragments (b); welded to partly welded grey tuff (c); incoherent grey tuff (d). (7) Sandstone (a), sand (b) and clay (c) marine deposits; (8) Marine shells.

(39–17 ka) made up of compacted and fractured lava flows intercalated by scoriaceous layers. *Somma-Vesuvius pyroclastic unit* made up of pyroclastic deposits produced from the explosive activity of Somma and Vesuvius post-17 ka. *Vesuvius unit* corresponding to the historic lava flows located only in the southern sector of the volcano.

4. Offshore stratigraphy

The Bay of Naples was investigated by means of closely spaced single-channel seismic reflection profiles (Fig. 2) that were acquired using a 16 kJ Multispot Extended Array

Sparker (MEAS) system, a 1 kJ Surfboom system and a 0.2 kJ multi-electrode Sparker system. All seismic sections were recorded graphically on continuous paper sheets with vertical recording scales of 0.25 s for multi-electrode Sparker, 0.25 and 0.5 s for Surfboom and 1 and 2 s for MEAS.

Ship positioning was determined using LORAN C for MEAS, Micro-Fix Racall for Surfboom (with a position accuracy of 1 m) and a differential Global Poisoning System (GPS) (with a position accuracy of 1 m) for multi-electrode Sparker. The best vertical resolution was approximately 6 m for MEAS data and 1 m for Surfboom and multi-electrode Sparker data.

The stratigraphic framework was reconstructed using a sequence stratigraphic approach (e.g. Thorne and Swift, 1991; Posamentier and Vail, 1988). Volcanic and sedimentary units were delineated on the basis of strata termination and internal and external seismic configuration (Mitchum et al., 1977).

Seven seismic units have been recognized south of Naples (Fig. 7). They consist of three southward-thinning chaotic wedges, interlayered with marine sediments and a chaotic lens-shaped unit. Correlations with the onshore stratigraphic sequence from the Piazza Vittoria well indicate the following stratigraphic succession, from older to younger:

Seismic unit A. This basal unit features poorly continuous reflectors of variable amplitude and frequency. It corresponds to Middle Pleistocene marine sediments.

Seismic unit B. Characterized by chaotic seismic facies, this unit is wedge-shaped and thins southward where it terminates. It can be correlated with the pre-CI tuffs.

Seismic unit C. Made up of parallel reflectors of low-to-high amplitude, high-to-low frequency and good continuity. It is present in the southern part of the seismic line and is associated with marine sediments that cover the pre-CI volcanic unit.

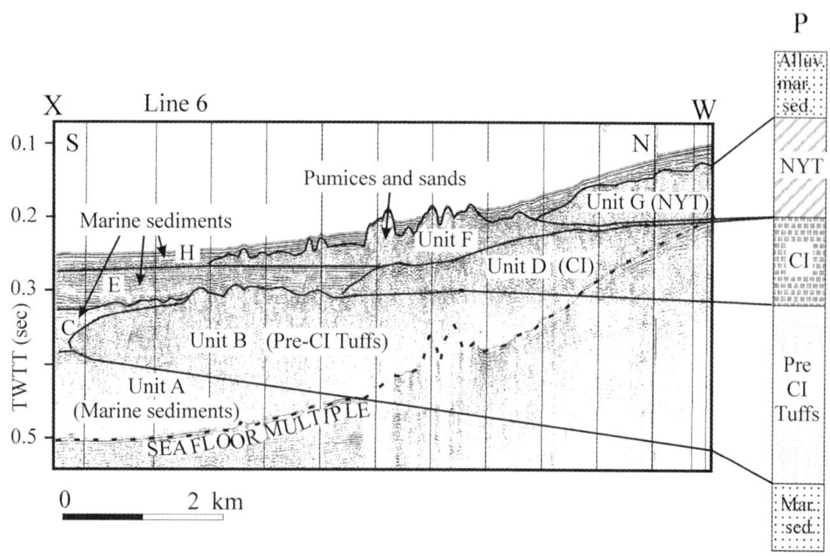

Figure 7. Surfboom profile XW in Naples Bay and interpretation. P, Piazza Vittoria well. For profile location see Figure 8.

Seismic unit D. Wedge-shaped chaotic unit. It has a maximum thickness of 75 m, thins southward and corresponds to the CI unit.

Seismic unit E. This unit is bounded by a basal erosional unconformity and is characterized by a variable amplitude and frequency with moderate-to-good continuity parallel reflectors. This unit is interpreted as marine sediments.

Seismic unit F. It is characterized by a chaotic seismic facies and a lenticular external form. Its southern margin is bounded by a basal flat surface and an upper mounded surface. A gravity core across one of these mounds reveals that this unit is made up of upper Pleistocene marine sands and pumices that are characterized by a chaotic sedimentary structure suggesting diapirism (D'Argenio et al., in press). This unit witnesses volcanic activity in the time span between the CI and the NYT.

Seismic unit G. Featuring a chaotic seismic facies and wedge-like external form this unit thins southward and corresponds to the NYT (Milia et al., 1998).

Seismic unit H. This unit features parallel seismic reflectors and extends across the whole profile. It corresponds to the marine sediments deposited over the last 15 ka.

5. Geological sections in the Neapolitan area

Previous work dedicated to the structure of the southern Campanian Plain suggests the occurrence of: NE-trending normal faults southeast of Posillipo Hill (Milia and Torrente, 2000, 2003); a major NW-SE normal fault controlling the Vesuvian coast recognized on the basis of the wedge geometry of the Campanian Ignimbrite and a tilting of 1° of the platform block (Milia, 2000; Milia and Torrente, 2003); a NW-SE trending fault across the Somma-Vesuvius volcano corresponding to a gravimetric anomaly (Santacroce, 1987); an E-W trending fault mapped north of the Monte Somma scarp in correspondence to a self potential lineament (Di Maio et al., 1998).

Using three marker horizons, the Middle Pleistocene marine sediments, the pre-CI tuffs and the CI unit, the onshore and offshore data were combined into three geologic sections (for location see Fig. 8) that permitted the authors to evaluate the distribution of the Upper Pleistocene ignimbrites and to locate volcano tectonic structures associated with ignimbrite emplacement.

5.1. N-S geologic section from Naples to the Bay of Naples

An 18-km-long regional section (Figs. 8 and 9), running N-S offshore from Naples, displays a substrate made up of coastal marine sediment that dips slightly seaward. These older marine sediments are covered by the pre-CI tuffs, which form a volcanic relief (up to 200 m thick) in Naples and gradually thin towards the bay (where they show a maximum thickness of 75 m) before terminating approximately 11 km from the coast. These tuffs are in turn covered by younger marine sediments in the southern part of the section and by the CI unit. The latter unit also features a relief with a maximum thickness of approximately 150 m in Naples, becoming thinner seaward before disappearing approximately 7 km from the coast. The CI is overlain in the southern part of the section by Upper Pleistocene pumices and marine sands that were interpreted as pyroclastic diapirs by D'Argenio et al. (in press). The pumices can be correlated with the products of the eruptions occurred between the CI and

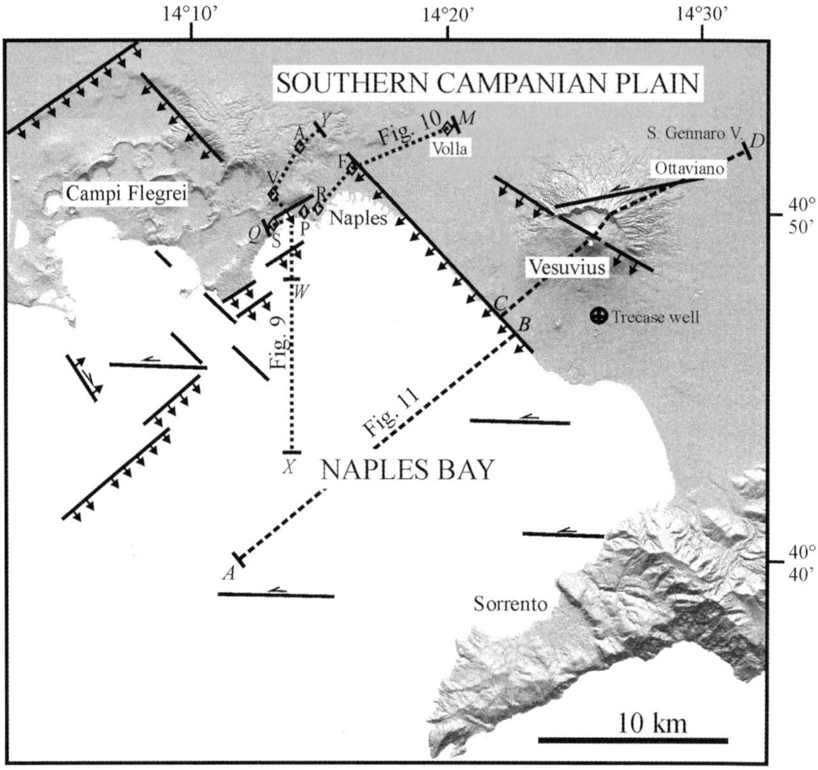

Figure 8. Upper Pleistocene and Holocene faults in the Neapolitan area. Location of seismic profile XW and geological sections AD, QM and XY are also shown. P, Piazza Vittoria well; V, Vomero well; A, Arenella well; S, Sannazzaro well; R, Palazzo Reale well; F, S. Maria La Fede well.

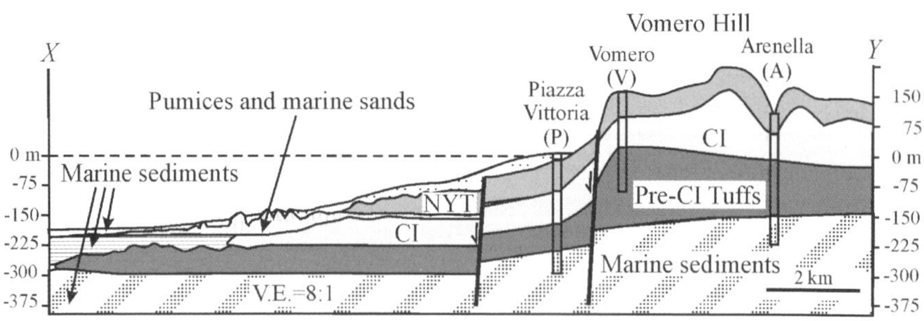

Figure 9. Geological section XY going from Naples Bay to Naples city. For section location see Figure 8.

the NYT. Finally, the NYT overlies, with an almost constant thickness, the CI onshore and the pumices and marine sands offshore, terminating seaward 4 km from the coast.

This geologic section is characterized by two NE-SW trending normal faults downthrown to the southeast (Figs. 8 and 9). The cumulative throw is approximately 50 m.

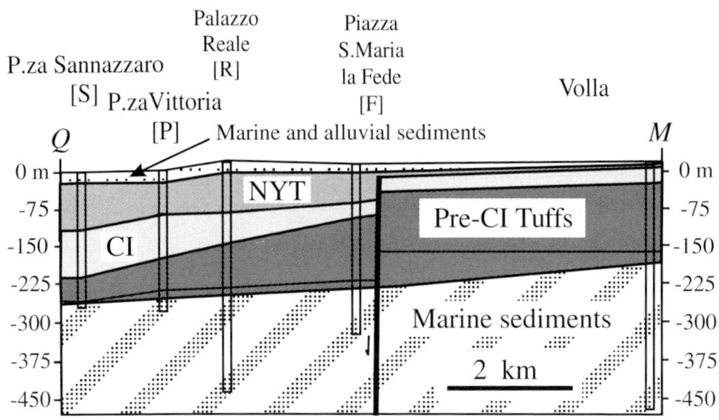

Figure 10. Geological section QM going from Naples coast to Volla. For section location see Figure 8.

The first fault is located onshore and bounds the Vomero Hill, whereas the second represents the offshore continuation of the Posillipo Hill fault. The faulting is post-15 ka because it offsets the NYT. The faults are also part of a NE-trending fault swarm previously detected southeast of Posillipo Hill off the coast of Campi Flegrei (Fig. 8; Milia and Torrente, 2000, 2003).

5.2. ENE-WSW geologic section from Naples to Volla

A 10-km-long ENE-WNW section in the southern Campanian Plain, from the coast at Naples to Volla (Figs. 8 and 10), shows Middle Pleistocene marine sediments overlain by Upper Pleistocene ignimbrites (pre-CI tuffs and CI unit), NYT products and Holocene marine and alluvial deposits. It should be noted that the pre-CI tuffs thins westward, whereas the CI thins eastward. Tracing of the Middle Pleistocene marine sediments and Upper Pleistocene ignimbrites reveals an important normal fault located in the middle part of the geologic section. This structure downthrows to the west. The thickness of the pre-CI tuffs is greater in the footwall than in the hangingwall block and fault throws differ according to the different stratigraphic levels (55 m for the base of pre-CI, 50 m for the CI base, 45 m for the CI top). The occurrence of a higher stratigraphic thickness of the footwall block together with the higher fault throws of the older marker levels suggest that fault activity occurred contemporaneously with the emplacement of these Upper Pleistocene ignimbrites. This fault is buried by the NYT.

5.3. NE-SW geologic section from southern Campanian Plain to the Bay of Naples

A 34-km-long NE-SW trending regional section (Figs. 8 and 11) from the slope and shelf of the Bay of Naples to the southern Campanian Plain was reconstructed by fitting offshore seismic stratigraphy (Milia et al., 1998; Milia, 2000; Milia et al., 2003b), drill hole stratigraphy (Bellucci, 1998; Brocchini et al., 2001) and potential field data (Santacroce, 1987; Di Maio et al., 1998) in the southern Campanian Plain. The sequence consists of

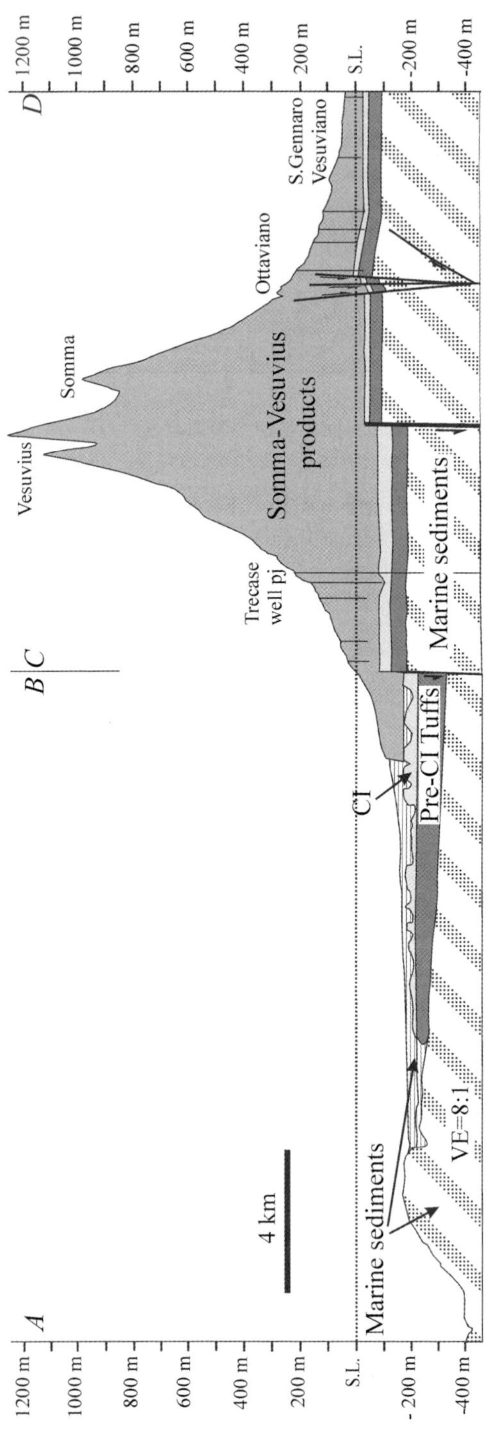

Figure 11. Geological section AD going from Naples Bay to the southern Campanian Plain. For section location see Figure 8.

Middle Pleistocene marine sediments overlain by Upper Pleistocene ignimbrites (pre-CI tuffs and CI) and post-25 ka Somma-Vesuvius products.

Tracing of the Upper Pleistocene ignimbrite marker horizons from the shelf to the southern Campanian Plain (Fig. 11) permitted the imaging of two normal faults (one located on the coast and another below the Vesuvius cone) and a strike-slip structure (located on the northern flank of the volcano). The coastal fault forms the southeastward extension (Fig. 8) of the normal fault previously imaged on the geological section from Naples to Volla (Fig. 10). Both coastal and Vesuvius cone normal faults displace the pre-CI tuffs and Campanian Ignimbrite and are buried by the post-25 ka Somma-Vesuvius products. They downthrow to the southwest. The thicknesses of the pre-CI tuffs and CI unit is greater in the footwall than in the hangingwall block and fault throws are greater for older marker levels. For the coastal and Vesuvius cone fault throws are respectively of 135-97 m for the base of pre-CI, 88-80 m for the CI base and 82-49 m for the CI top were calculated. On the basis of these features we maintain that normal fault activity occurred during the emplacement of these Upper Pleistocene ignimbrites.

The recognition of the strike slip structure located on the northern flank of the volcano was based on a peculiar structural style. Indeed, the northern part of the section displays four faults merging in depth at a single fault surface; in addition three of these faults feature normal separation and the fourth one reverse separation (Fig. 11). These features are typical of a negative flower structure (Harding, 1990). We argue that this negative flower structure corresponds to a left-lateral fault zone because of the occurrence of many Late Quaternary E-W trending sinistral strike-slip faults in the Bay of Naples off Vesuvius and Campi Flegrei (Milia and Torrente, 2003).

6. Discussion and conclusions

This is the first time that a geological study of the Neapolitan area, based on the integration of onshore and offshore stratigraphic data pertaining to the Upper Pleistocene ignimbrites, has been presented. The latter are stratigraphic units peculiar to the Campanian Plain basin fill.

Great importance was placed on the existence of ancient ignimbrites in Naples city. These pyroclastic units are locally inter-layered with marine strata and reach a maximum total thickness of 200 m at Vomero Hill and near the coast at Palazzo Reale (Fig. 9). Indeed, detailed stratigraphic analyses reveal that these volcanic sequences are mainly formed by pre-CI tuffs, covered by 100–150 m thick ignimbrite products of the CI (39 ka) and NYT (15 ka). The whole ignimbrite succession rests on a Middle Pleistocene substrate formed by marine sediments the top of which is a marker level for the interpretation of subsequent deformations.

Another relevant finding is that the whole ignimbrite succession is thicker in Naples city and progressively thins towards the Bay of Naples, thus suggesting that some of the vents of these ignimbrites were possibly located near the city of Naples. Our study also documents that NE-SW trending normal faults bound the Posillipo and Vomero hills downthrowing the southeastern blocks towards the Bay of Naples. These faults feature a constant throw at different stratigraphic levels, thus implying a unique post-NYT tectonic activity.

It is well established that an ash-flow caldera (Lipman, 1997) is a piston-like or piecemeal subsiding block filled with ignimbrite sheets with thicknesses in the order of thousands of metres (Smith and Bailey, 1968). Assuming the existence of a caldera related to the CI eruption, a volume of 200 km^3 and an area of approximately 180 km^2 (Orsi et al., 1996) would imply a mean thickness of 1100 m within the subsided block. In the Neapolitan area, however, several elements exclude the occurrence of such a caldera: (i) the thickness of the CI in this study area reaches moderate values (up to 100 m); (ii) the volcanic tuffs underlying the CI deposit occur at maximum depths of 200 m; (iii) in the St. Martino-Vomero area the CI products cover a pre-existing volcanic relief and not a caldera depression; (iv) the Middle Pleistocene marine substrate has been detected at maximum depths of approximately 300 m confirming the lack of large collapses in the overall study area; (v) linear normal faults, and not the presumed caldera ring fault, were mapped south of Naples city that downthrow the blocks towards the Bay of Naples (Fig. 8).

It is here proposed that Upper Pleistocene ignimbrite eruptions in the Neapolitan area can be explained using a model of fissure emission related to extensional fault systems (e.g. Aguirre-Diaz and Labarthe-Hernandez, 2003). In fact two NW-SE trending normal faults (one located on the coast and another below the Vesuvius cone; Fig. 11) have been recognized near the Vesuvius coast in the southern Campanian Plain (Fig. 8). These volcano tectonic faults were active during the emplacement of both the pre-CI tuffs and the Campanian Ignimbrite and possibly controlled the vent location. These Upper Pleistocene ignimbrites feature thicknesses that are higher in the footwall than in the hangingwall blocks and fault throws in the order of hundreds of metres. Such faults close to the Vesuvian coast possibly correspond northwestard to another NW-SE normal fault located north of Campi Flegrei. The Posillipo and Vomero hills can thus be explained as being a transfer zone during the volcano tectonic deformations coeval with ignimbrite emplacement (Fig. 8). A common triggering mechanism of magma injection along NW-SE extensional faults was also suggested for large Pleistocene ignimbrite eruptions of the Latium margin by Marra (2001).

As was mentioned earlier, an important finding of this research is the documentation within the Neapolitan area of many trachytic ignimbrites (including the CI unit). These Campanian ignimbrites were also reported throughout the whole Campanian Plain (e.g. Romano et al., 1994; Rolandi et al., 2003). Based on the large map distribution and time span (post-300 ka) of the Campania ignimbrites we argue that this activity cannot be explained solely in terms of a unique magmatic system underlying the Campi Flegrei region as suggested by Rosi and Sbrana (1987) and Civetta et al. (1997). Our data support a model featuring a large trachytic magma body that underlies the Campanian margin owing to the partial melting of the crystalline basement as a consequence of the crustal thinning and the rising magma along regional normal faults (Milia et al., 2003a; Rolandi et al., 2003).

Acknowledgements

The present manuscript benefited significantly from the constructive reviews of an earlier version provided by C. Kilburn and F. Stoppa.

References

Aguirre-Diaz, G., Labarthe-Hernandez, G., 2003. Fissure ignimbrites: fissure-source origin for voluminous ignimbrites of the Sierra Madre Occidental and its relationship with basin and range faulting. Geology 31, 773–776.

Ballini, A., Barberi, F., Laurenzi, M.A., Mezzetti, F., Villa, I.M., 1989. Nuovi dati sulla stratigrafia del vulcano di Roccamonfina. Boll. Gruppo Nazionale Vulcanologia 2, 533–556.

Bellucci, F., 1998. Nuove conoscenze stratigrafiche sui depositi effusivi ed espolosivi nel sottosuolo dell'area del Somma-Vesuvio. Boll. Soc. Geol. It. 117, 385–405.

Branney, M.J., Kokelaar, P., 1994. Volcanotectonic faulting, soft-state deformation, and reomorphism of tuffs during development of a piecemeal caldera, English Lake District. Geol. Soc. Am. Bull. 106, 507–530.

Brocchini, D., Principe, C., Castradori, D., Laurenzi, M.A., Gorla, L., 2001. Quaternary evolution of the southern sector of the Campanian Plain and early Somma-Vesuvius activity: insigths from the Trecase 1 well. Mineral. Petrol. 73, 67–91.

Burkart, B., Self, S., 1995. Extension and rotation of crustal blocks in northern Central America and effects on the volcanic arc. Geology 13, 22–26.

Civetta, L., Orsi, G., Pappalardo, L., Fisher, R.V., Heiken, G.H., Hort, M., 1997. Geochemical zoning, mixing, eruptive dynamics and depositional processes – the Campanian Ignimbrite, Campi Flegrei, Italy. J. Volcanol. Geotherm. Res. 75, 183–219.

Cole, J.W., 1990. Structural control and origin of volcanism in the Taupo volcanic zone, New Zealand. Bull. Volcanol. 52, 445–459.

D'Argenio, B., Aiello, G., de Alteriis, G., Milia, A., Sacchi, M., Tonielli, R., Angelino, A., Budillon, F., Chiocci, F., Conforti, A., De Lauro, M., Di Martino, G., d'Isanto, C., Esposito, E., Ferraro, L., Innangi, S., Insinga, D., Iorio, M., Marsella, E., Molisso, F., Morra, V., Passaro, S., Pelosi, N., Porfido, S., Raspini, A., Ruggirei, S., Sarnacchiaro, G., Terranova, C., Vilardo, G., Violante, C., in press. Digital elevation model of the Naples Bay and adjacent areas, eastern Tyrrhenian Sea. In: Pasquarè, G., Venturini, C. (Eds), Mapping Geology in Italy. APAT Dipartimento Difesa del Suolo-Servizio Geologico d'Italia, pp. 21–28.

D'Erasmo, G., 1931. Studio geologico dei pozzi profondi della Campania. Boll. Soc. Nat. 43, 15–130.

De Gennaro, M., Cappelletti, P., Langella, A., Perrotta, A., Scarpati, C., 2000. Genesis of zeolites in the Neapolitan Yellow Tuff: geological, volcanological and mineralogical evidence. Contrib. Mineral. Petrol. 139, 17–35.

Deino, A.L., Orsi, G., de Vita, S., Piochi, M., 2004. The age of the Neapolitan Yellow Tuff caldera-forming eruption (Campi Flegrei caldera-Italy) assessed by $^{39}Ar/^{40}Ar$ dating method. J. Volcanol. Geotherm. Res. 133, 157–170.

Della Vedova, B., Bellini, S., Pellis, G., Squarci, P., 2001. Deep temperatures and surface heat flow distribution. In: Vai, G.B., Martini, I.P. (Eds), Anatomy of an Orogen: The Apennines and Adjacent Mediterranean Basin. Kluwer, Dordrecht, the Netherlands, pp. 65–67.

De Lorenzo, G., 1904. L'attività vulcanica dei Campi Flegrei. Rend. Acc. Sci. Fis. Mat. Napoli 10, 203–221.

De Rita, D., Giordano, G., 1996. Volcanological and structural evolution of Roccamonfina volcano (Italy): origin of the summit caldera. In: McGuire, W.J., Jones, A.P., Neuberg, J. (Eds), Volcano Instability on the Earth and Other Planets. Geol. Soc. Spec. Publ., Vol. 110, pp. 209–224.

De Vivo, B., Rolandi, G., Gans, P.B., Calvert, A., Bohrson, W.A., Spera, F.J., Belkin H.E., 2001. New constraints on the pyroclastic eruptive history of the Campanian volcanic Plain (Italy). Mineral. Petrol. 73, 47–65.

Di Maio, R., Mauriello, P., Patella, D., Petrillo, Z., Piscitelli, S., Siniscalchi, A., 1998. Electric and electromagnetic outline of the Mount Somma-Vesuvius structural setting. J. Volcanol. Geotherm. Res. 82, 219–238.

Ferrucci, F., Gaudiosi, G., Pino, N.A., Luongo, G., Hirn, A., Mirabile, L., 1989. Seismic detection of a major Moho upheaval beneath the Campania volcanic area (Naples, Southern Italy). Geophys. Res. Lett. 16, 1317–1320.

Fusi, N., Mirabile, L., Camerlenghi, A., Ranieri, G., 1991. Marine geophysical survey of the Gulf of Naples (Italy): relationship between submarine volcanic activity and sedimentation. Mem. Soc. Geol. It. 47, 95–114.

Gars, G., Lippman, M., 1984. Nouvelle donneés néotectonique dans l'Apennin campanien (Italie du Sud). CR Acad. Sci. Paris. 298(II-11), 495–500.

Guadagno, M., 1924. Notizie sul pozzo artesiano recentemente trivellato nella piazza S. Maria la Fede in Napoli. Contributo alla conoscenza del sottosuolo cittadino e delle sue acque sotterranee. Boll. Soc. Nat. 36, 120–128.

Guadagno, M., 1928. Il tufo giallo trachitico nel sottosuolo della città di Napoli. Atti Reale Istituto d'Incoraggiamento. Napoli 3–36.

Harding, T.P., 1990. Identification of wrench faults using subsurface structural data: criteria and pitfalls. Am. Assoc. Petrol. Geol. Bull. 74, 1590–1604.

Johnston-Lavis, H.F., 1889. Report for the investigation of the volcanic phenomena of Vesuvius and neighbourhood. Tyne Meeting of the British Association, pp. 1–12.

Keller, J., Ryan, W.B.F., Ninkonvich, D., Altherr, R., 1978. Explosive volcanic activity in the Mediterranean over the past 200,000 yr as recorded in deep-sea sediments. Geol. Soc. Am. Bull. 89, 591–604.

Lipman, P.W., 1997. Subsidence of ash flow calderas: relation to Caldera size and magma-chamber geometry. Bull. Volcanol. 59, 198–218.

Marra, F., 2001. Strike-slip faulting and block rotation: a possible triggering mechanism for lava flows in the Alban Hills. J. Struct. Geol. 23, 217–141.

Milia, A., 1998. Le unità piroclastiche tardo-quaternarie nel Golfo di Napoli. Geogr. Fis. Dinam. Quat. 21, 147–153.

Milia, A., 1999. Aggrading and prograding infill of a pery-tyrrhenian basin (Naples Bay, Italy). Geo-Mar. Lett. 19, 237–244.

Milia, A., 2000. The Dohrn Canyon formation: a response to the eustatic fall and tectonic uplift of the outer shelf (Eastern Tyrrhenian Sea margin, Italy). Geo-Mar. Lett. 20, 101–108.

Milia, A., Torrente, M.M., 1997. Evoluzione tettonica della Penisola Sorrentina (margine peritirrenico campano). Boll. Soc. Geol. It. 116, 487–502.

Milia, A., Torrente, M.M., 1999. Tectonics and stratigraphic architecture of a pery-Tyrrhenian half-graben (Bay of Naples, Italy). Tectonophysics 315, 297–314.

Milia, A., Torrente, M.M., 2000. Fold uplift and syn-kinematic stratal architectures in a region of active transtensional tectonics and volcanism, Eastern Tyrrhenian Sea. Geol. Soc. Am. Bull. 112, 1531–1542.

Milia, A., Torrente, M.M., 2003. Late Quaternary volcanism and transtensional tectonics in the Bay of Naples, Campanian continental margin, Italy. Mineral. Petrol. 79, 49–65.

Milia, A., Mirabile, L., Torrente, M.M., Dvorak, J.J., 1998. Volcanism offshore of Vesuvius volcano in Naples Bay. Bull. Volcanol. 59, 404–413.

Milia, A., Torrente, M.M., Russo, M., Zuppetta, A., 2003a. Tectonics and crustal structure of the Campania continental margin: relationships with volcanism. Mineral. Petrol. 79, 33–47.

Milia, A., Torrente, M.M., Zuppetta, A., 2003b. Offshore debris avalanches at Somma-Vesuvius volcano (Italy): implications for hazard evaluation. J. Geol. Soc. London 160, 309–317.

Mitchum, R.M., Vail, P., Sangree, J.B., 1977. Seismic stratigraphy and global changes of sea level, part 6: stratigraphic interpretation of seismic reflection patterns in depositional sequences. In: Payton, C.E. (Ed.), Seismic Stratigraphy – Application to Hydrocarbon Exploration. Am. Assoc. Petrol. Geol. Memoir. 26, 205–212.

Moore, I., Kokelaar, P., 1998. Tectonically controlled piecemeal caldera collapse: a case study of Glencoe volcano, Scotland. Geol. Soc. Am. Bull. 110, 1448–1466.

Orsi, G., de Vita, S., Di Vito, M., 1996. The restless resurgent Campi Flegrei nested Caldera (Italy): constraints on its evolution and configuration. J. Volcanol. Geotherm. Res. 17, 273–288.

Posamentier, H.W., Vail, P.R., 1988. Eustatic control on clastic deposition. II. Sequence and system tract models. In: Wilgus, C.K., Hastings, B.S., Posamentier, H.W., Van Wagoner, J., Ross, C.A., Kendall, C.G.C. (Eds), Sea Level Changes – An Integrated Approach. Soc. Econ. Paleontol. Mineral. Spec. Publ., Vol. 42, pp. 125–154.

Rolandi, G., Bellucci, F., Heizler, M.T., Belkin, H.E., De Vivo, B., 2003. Tectonic controls on the genesis of ignimbrites from the Campanian Volcanic Zone, southern Italy. Mineral. Petrol. 79, 3–31.

Rolandi, G., Maraffi, S., Petrosino, P., Lirer, L., 1993a. The Ottaviano eruption of Somma-Vesuvius (8000 y B.P.): a magmatic alternatine fall and flow-forming eruption. J. Volcanol. Geotherm. Res. 58, 43–65.

Rolandi, G., Mastrolorenzo, G., Barrella, A.M., Borrelli, A., 1993b. The Avellino plinian eruption of Somma-Vesuvius (3679 y B.P.): the progressive evolution from magmatic to hydromagmatic style. J. Volcanol. Geotherm. Res. 58, 67–81.

Rolandi, G., Petrosino, P., McGeehin, I., 1998. The interplinian activity at Somma-Vesuvius in the last 3500 years. J. Volcanol. Geotherm. Res. 82, 19–52.

Romano, P., Santo, A., Voltaggio, M., 1994. L'evoluzione geomorfologia della Pianura del Fiume Volturno (Campania) durante il tardo Quaternario (Pleistocene medio-superiore-Olocene). Il Quaternario 7(1), 41–56.

Rosi, M., Sbrana, A., (Eds), 1987. Phlegrean Fields. CNR Quad. Ric. Sci., 114, 1–175.

Santacroce, R. (Ed.), 1987. Somma-Vesuvius. CNR Quad. Ric. Sci., 114, 1–251.

Scandone, R., Bellucci, F., Lirer, L., Rolandi, G., 1991. The structure of the Campanian Plain and the activity of the Neapolitan volcanoes (Italy). J. Volcanol. Geotherm. Res. 48, 1–31.

Scarpati, C., Cole, P., Perrotta, A., 1993. The Neapolitan Yellow Tuff – a large volume multiphase eruption from Campi Flegrei, Southern Italy. Bull. Volcanol. 55, 343–356.

Scherillo, A., 1933. Petrografia chimica dei tufi flegrei 2 tufo giallo, mappamonte, pozzolana. Rend. Acc. Sc. Fis Mat. 22, 317–330.

Scherillo, A., Franco, E., 1967. Introduzione alla carta stratigrafica del suolo di Napoli. Atti Acc. Pont. 16, 5–15.

Smith, R.L., Bailey, R.A., 1968. Resurgent claudrons. Mem. Geol. Soc. Am. 116, 613–662.

Thorne, J.A., Swift, D.J.P., 1991. Sedimentation on continental margins, VI: a regime model for depositional sequences, their components systems tracts, and bounding surfaces. In: Swift, D.J.P., Oertel, G.F., Tillman, R.W., Thorne, J.A. (Eds), Shelf Sand and Sandstone Bodies: Geometry, Facies and Sequence Stratigraphy. Int. Assoc. Sedim. Spec. Publ., Vol. 14, pp. 189–255.

Tunnel, R., Federman, A., Sparks, S., Williams, D., 1978. The age, origin and volcanological significance of the Y-5 ash layer in the Mediterranean. Quat. Res. 12, 241–253.

Volcanism in the Campania Plain:
Vesuvius, Campi Flegrei and Ignimbrites
edited by B. De Vivo

Chapter 9

The magma feeding system of Somma-Vesuvius (Italy) strato-volcano: new inferences from a review of geochemical and Sr, Nd, Pb and O isotope data

M. Piochi[a,*], *B. De Vivo*[b] *and R.A. Ayuso*[c]

[a]*Istituto Nazionale di Geofisica e Vulcanologia, Osservatorio Vesuviano, Napoli, Italy*
[b]*Dipartimento di Scienze della Terra, Università Federico II, Napoli, Italy*
[c]*U.S. Geological Survey, MS 954 National Center, Reston, VA, USA*

Abstract

A large database of major, trace and isotope (Sr, Nd, Pb, O) data exists for rocks produced by the volcanic activity of Somma-Vesuvius volcano. Variation diagrams strongly suggest a major role for evolutionary processes such as fractional crystallization, contamination, crystal trapping and magma mixing, occurring after magma genesis in the mantle. Most mafic magmas are enriched in LILE (Light Ion Lithophile Elements; K, Rb, Ba), REE (Ce, Sm) and Y, show small Nb–Ta negative anomalies, and have values of Nb/Zr at about 0.15. Enrichments in LILE, REE, Nb and Ta do not correlate with Sr isotope values or degree of both K enrichment and silica undersaturation. The results indicate mantle source heterogeneity produced by slab-derived components beneath the volcano. However, the Sr isotope values of Somma-Vesuvius increase from 0.7071 up to 0.7081 with transport through the uppermost 11–12 km of the crust. The Sr isotope variation suggests that the crustal component affected the magmas during ascent through the lithosphere to the surface. Our new geochemical assessment based on chemical, isotopic and fluid inclusion data points to the existence of three main levels of magma storage. Two of the levels are deep and may represent long-lived reservoirs; the uppermost crustal level probably coincides with the volcanic conduit. The deeper level of magma storage is deeper than 12 km and fed the 1944 AD eruption. The intermediate level coincides with the seismic discontinuity detected by Zollo et al. (1996) at about 8 km. This intermediate level supplies magmas with $^{87}Sr/^{86}Sr$ values between 0.7071 and 0.7074, and $\delta O^{18} < 8‰$ that typically erupted both during inter-plinian (i.e. 1906 AD) and sub-plinian (472 AD, 1631 AD) events. The shallowest level of magma storage at about 5 km was the site of magma chambers for the Pompei and Avellino plinian eruptions. New investigations are necessary to verify the proposed magma feeding system.

1. Introduction

Somma-Vesuvius (Fig. 1a) has long attracted intense scrutiny because of its recent activity, enormous hazard potential to the Campanian region and immediate proximity to the city of Naples. Plinian eruptions from the Somma-Vesuvius volcano were first described during the eruption of 79 AD. The erupted silica-undersaturated potassium-rich rocks have been the object of petrological studies (Rittmann, 1933; Savelli, 1967; Cortini and Hermes, 1981; Joron et al., 1987; Civetta and Santacroce, 1992; Belkin et al., 1993; Cioni et al., 1995, 1998; Ayuso et al., 1998; Cioni, 2000; Peccerillo, 2001; Paone, 2005; Piochi et al., 2006;

*Corresponding author. Fax: 39-81-6100811. *E-mail address*: piochi@ov.ingv.it (M. Piochi).

and references therein) aimed at evaluating how the erupted magmas reflect the contributions of mantle sources, how their compositions have been affected during transport, and to what extent they can be used to deduce their geodynamic setting. Recently, a large major, trace and isotope (Sr, Nd, Pb, O) database has been published (De Vivo et al., 2003) and can be downloaded at the Internet site http://www.dgv.unina.it/ricerca/de_vivo.htm. The summary of results shows that rocks produced during major plinian and sub-plinian eruptions, and during the last interplinian period of activity which started in 1631 AD, are relatively well characterized on the basis of mineralogy, chemistry and isotopes. Adequate data also exist for some rocks from interplinian periods of volcanism occurring before the last sub-plinian eruption in 1631 AD.

In this paper, we briefly present a description of the chemical and isotopic database and a synthesis of previous petrological studies in order to summarize the main evidence for mantle source heterogeneity associated with the Somma-Vesuvius magmas, and highlight the results supporting the importance of shallow-level evolution. Particularly, our brief review of existing data points to a magma feeding system formed by multi-depth storage levels; the magma storage level at 8 km imaged by seismic tomography (Zollo et al., 1996) fed both low- and large-magnitude eruptions. Significant progress has been made in the last 20 years of research focused on Somma-Vesuvius volcano (Civetta and Santacroce, 1992; Belkin et al., 1993; Villemant et al., 1993; Cioni et al., 1995; Ayuso et al., 1998; Del Moro et al., 2001; Peccerillo, 2001; Fulignati et al., 2004, 2005; Pappalardo et al., 2004; Piochi et al., 2006), and it is now possible to combine the results of previous studies to produce a framework for more detailed investigations of the behaviour of magma and of the definition of magma feeding system in Somma-Vesuvius volcano.

2. Volcanological and magmatological background

Somma-Vesuvius is a strato-volcano (Fig. 1a) that consists of an older collapsed edifice (Somma), and a younger cone (Vesuvius). The volcano has been active at least since 300 ky bp (Brocchini et al., 2001 and references therein) up to the major eruption of 1944 AD. Presently, the volcano is the site of fumaroles, diffuse degassing (Chiodini et al., 2001; Federico et al., 2002; Frondini et al., 2004) and low-magnitude seismicity (Bianco et al., 1999; Vilardo et al., 1999). Volcanism has been characterized by high explosive sub-plinian and plinian eruptions that followed long periods of quiescence, and by intermediate and small-scale explosive and explosive/effusive eruptions that occurred during continuous periods of activity (interplinian period) (Fig. 1b) (Arnò et al., 1987; Civetta and Santacroce, 1992; Rolandi et al., 1998; Principe et al., 2004). Sub-plinian and plinian eruptions have always produced larger volumes of rocks (one to a few cubic kilometres DRE, i.e. Dense Rock Equivalent) (Rosi and Santacroce, 1983; Arnò et al., 1987; Civetta and Santacroce, 1992; Rolandi et al., 1993; Cioni et al., 1995; Landi et al., 1999) than the intermediate and small-scale events (0.01–0.1 km^3 DRE) (Scandone et al., 1986; Mastrolorenzo et al., 1993; Rolandi et al., 1998; Arrighi et al., 2001).

The volcano rests on a sequence of Mesozoic and Cenozoic carbonates overlain by Miocene sediments outcropping in the surrounding Apennine chain (D'Argenio et al., 1973; Ippolito et al., 1975) and encountered at a depth of around 2 km (Brocchini et al., 2001). The Moho discontinuity has been detected at about 30 km of depth (Corrado and

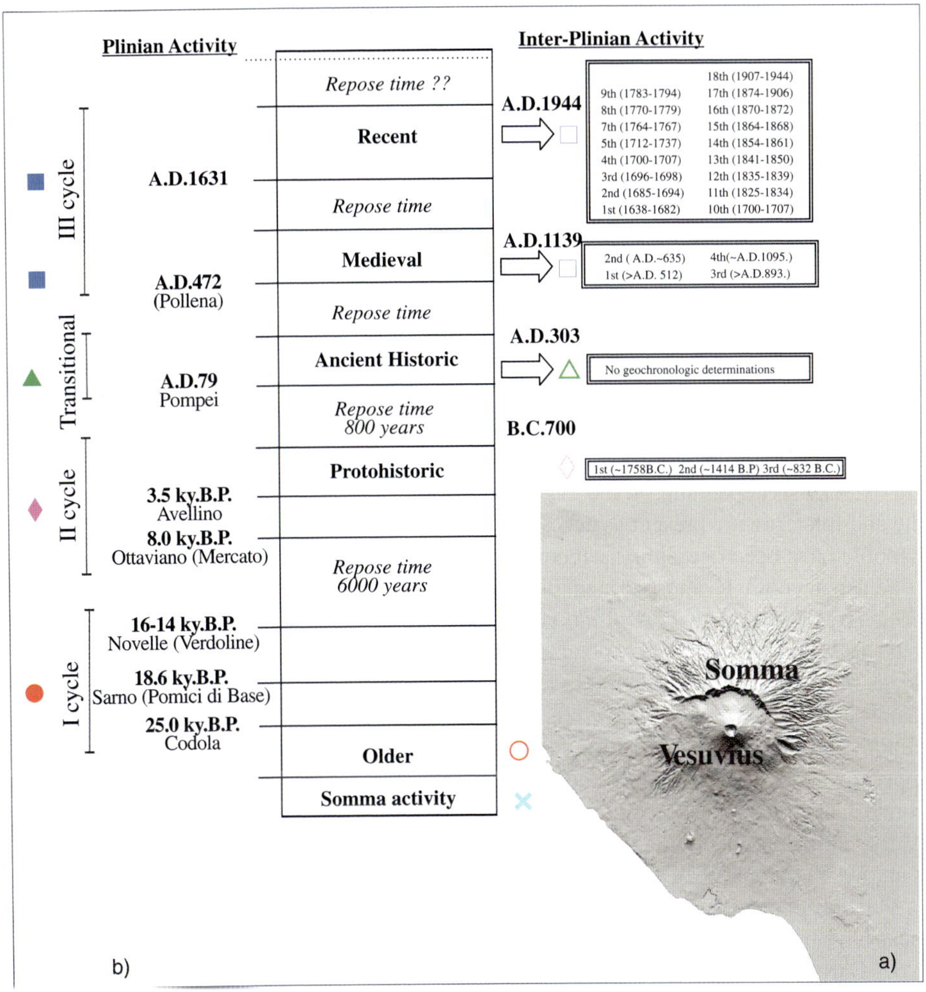

Figure 1. (a) DTM of the Somma-Vesuvius strato-volcano; (b) Reconstructed stratigraphy of volcanic activity during the last 25 ka. *Source*: Arnò et al. (1987); Arrighi et al. (2001); Ayuso et al. (1998); Landi et al. (1999); Rolandi et al. (1993, 1998); Rosi and Santacroce (1983). Symbols as used in the following figures. Names of eruptions in parenthesis are from Arnò et al. (1987).

Rapolla, 1981; Ferrucci et al., 1989; Chiarabba et al., 2005). A high-velocity body dipping westward from 65 km down to 285 km was interpreted as a plate within the mantle (De Natale et al., 2001). Furthermore, an active, large magma chamber is located at about 8–10 km (Zollo et al., 1996; Di Maio et al., 1998) and has been proposed to extend up to 30 km (De Natale et al., 2001). However, based on fluid and melt inclusion evidence, magma storage is indicated at 3.5–5, 8–10 and > 12 km (Belkin et al., 1985; Belkin and De Vivo, 1993; Cioni et al., 1998; Marianelli et al., 1999; Cioni, 2000; Lima et al., 2003). At present, no geophysical evidence for magma chambers of significant lateral extension has been found at < 8 km (Zollo et al., 1996; Di Maio et al., 1998). This may be due to the fact that resolution for the method used in tomography investigations is "blind" for magma chambers with lateral extension < 1 km.

3. Mineral, chemical and isotopic data: description and previous interpretations

3.1. Mineralogy and classification

Somma-Vesuvius volcanic rocks are poorly (lava) to highly (scoria to pumice) vesiculated, and nearly aphyric (mostly in the plinian eruptions) to strongly porphyritic (up to 50%; in 472 AD eruption and in the products younger than 1631 AD) (Joron et al., 1987; Villemant et al., 1993). Two rock types are generally distinguishable on the basis of occurrence of leucite minerals. In leucite-free rocks, olivine and Mg-rich diopside, plagioclase, Fe-rich diopside, K-feldspar, magnetite and biotite can also occur, depending on the degree of evolution. Leucite-bearing rocks contain olivine, Fe-rich and Mg-poor diopside, plagioclase and oxide, also depending on the degree of evolution. Apatite, amphibole, garnet, phlogopite and forsterite are present as accessory phases. Nepheline, as the only feldspathoid, and scapolite have been occasionally recovered (e.g. 472 AD and Avellino rocks).

Feldspar (both K-feldspar and plagioclase) is the most abundant mineral phase in leucite-free rocks, such as Avellino and Sarno (Pomici di Base) (Joron et al., 1987; Landi et al., 1999), as well as in 79 AD leucite-bearing pumices (Cioni et al., 1998). Instead, diopside is the most common mineral in the products younger than 1631 AD, whose abundance changes as function of the degree of vesicularity of rocks (Villemant et al., 1993). Clinopyroxenes have compositions indicative of multiple stages of crystallization in the upper (< 10 km) crust (Trigila and De Benedetti, 1993; Marianelli et al., 1995). Olivines from 1944 and 1906 AD eruptions show compositions similar to olivine from peridotite (Marianelli et al., 1995) and high pressure (> 400 MPa) of volatile entrapment (Marianelli et al., 1999) indicative of very early stage of magma crystallization.

Metamorphosed carbonates, skarns, lavas, cumulates, hornfels, sub-volcanic igneous rocks have been generally recovered as xenolith ejecta within pyroclastic deposits (Savelli, 1967; Barberi and Leoni, 1980; Hermes and Cornel, 1981; Belkin et al., 1985; Del Moro et al., 2001; Gilg et al., 2001; Fulignati et al., 2004, 2005). Metamorphosed carbonate ejecta are considered to be representative of the carbonate basement modified during contact metamorphism under the pressure of 1500–2000 bars. Skarn xenoliths consist of calc-silicate and carbonatic components and contain fassaitic pyroxene, forsterite ($Fo_{>90}$), spinel, calcite, phlogopite, nepheline, garnet, periclase, brucite, calcite, and dolomite. They are considered as representative of the crystallizing margins of the magma chamber (Del Moro et al., 2001; Gilg et al., 2001; Fulignati et al., 2004, 2005). However, these xenoliths were also interpreted to represent highly metasomatized blocks of stopped carbonates incorporated into the magma (Hermes and Cornell, 1981). Silicate melt inclusions from skarns show homogenization temperatures (*Th*) of 1000 ± 50°C and trapping pressures between 925 and 3550 bars (Belkin et al., 1985; Fulignati et al., 2004). Hornfels are characterized by rhyolitic vesiculated glass and minerals of wollastonite, anorthite, calcite, pyroxene and quartz, and have been considered the products of high-grade thermometamorphism from marly siltite rocks (Del Moro et al., 2001; Fulignati et al., 2005). Cumulates are dunites, wherlites and biotite-bearing pyroxenites (Joron et al., 1987; Belkin and De Vivo, 1993). The cumulus phases are clinopyroxene, phlogopite, biotite, apatite, plagioclase and olivine with Fo_{80-90}; glass also occurs between individual crystal grains or within cavities. Leucite is rare in cumulate nodules. Spinel and chromite can occur as accessory phases. *Th* and trapping pressure of silicate melt inclusions in cumulates are 1200 ± 50°C and in the range 1200–3050 bars, respectively (Belkin et al., 1985).

3.2. Major and trace elements

It is well known that rocks from Somma-Vesuvius are characterized by large compositional variations. These rocks show variable alkali contents (Fig. 2a), and, in particular, show variable degree of K_2O enrichment. These rocks are slightly, mildly and highly silica undersaturated, following Peccerillo (2003). Slightly silica-undersaturated volcanic rocks are leucite-free and range in composition from shoshonites to trachy-phonolites; mildly to highly undersaturated, nepheline- or, more commonly, leucite-bearing rocks, range from alkali-basalt to phonolite.

Plinian and sub-plinian deposits are generally characterized by the most evolved compositions and chemical gradients through the stratigraphic sequence. The basal part of deposits (white pumices) always shows the more sialic compositions, and the evolution degree decreases upwards (grey pumices) (Arnò et al., 1987; Civetta et al., 1991; Civetta and Santacroce, 1992; Rolandi et al., 1993; Cioni et al., 1995; Landi et al., 1999). These features possibly reflect the progressive withdrawal of a chemically (and density) stratified magma chamber located at shallow depth beneath the volcano. The variable layers can be linked through simple chemical differentiation of unique parental magma (Landi et al., 1999) or can be generated due to the arrival of diverse magma batches from deeper reservoirs (e.g. Civetta et al., 1991; Cioni, 2000). Sometimes, the occurrence of products with compositions intermediate between that of the different layers indicates syn-eruptive mingling of magmas or the existence of a double-diffusive interface between the two magmatic layers within the magma chamber (Landi et al., 1999).

Because of the occurrence of carbonate and metamorphic ejecta (see previous section), it has been suggested that plinian and sub-plinian chambers formed within the carbonate basement, between 5 and 8 km depth (Barberi and Leoni, 1980; Belkin and De Vivo, 1993; Landi et al., 1999; Cioni, 2000) during the long time of quiescence that precedes the eruption (Fig. 1b) and that allows reaching the high evolution degree of these rocks.

Magmas erupted during interplinian periods are characterized by low degree of evolution (Fig. 2) and depths of storage at < 5 km, 8–10 km and > 12 km (Belkin et al., 1985; Belkin and De Vivo, 1993; Cioni et al., 1998; Marianelli et al., 1999; Cioni, 2000; Lima et al., 2003). Owing to the occurrence of deeply crystallized olivines (see previous section), the existence of CO_2-bearing melt inclusions and the brief repose time between two eruptions (not more than 7 years) (Arnò et al., 1987), the various authors indicate that during interplinian periods magmas can rapidly rise to the surface in open-conduit conditions. The last 1944 AD eruption was fed by a magma directly rising from a depth of > 12 km. After 61 years of volcanic quiet, this latter eruption probably closes the third, last mega-cycle of volcanism (Ayuso et al., 1998) and marks the transition to the closed-conduit condition (Rosi et al., 1987). This situation of repose might last for centuries, heading towards the starting of new, fourth, mega-cycle of volcanism, with a new plinian–sub-plinian eruption

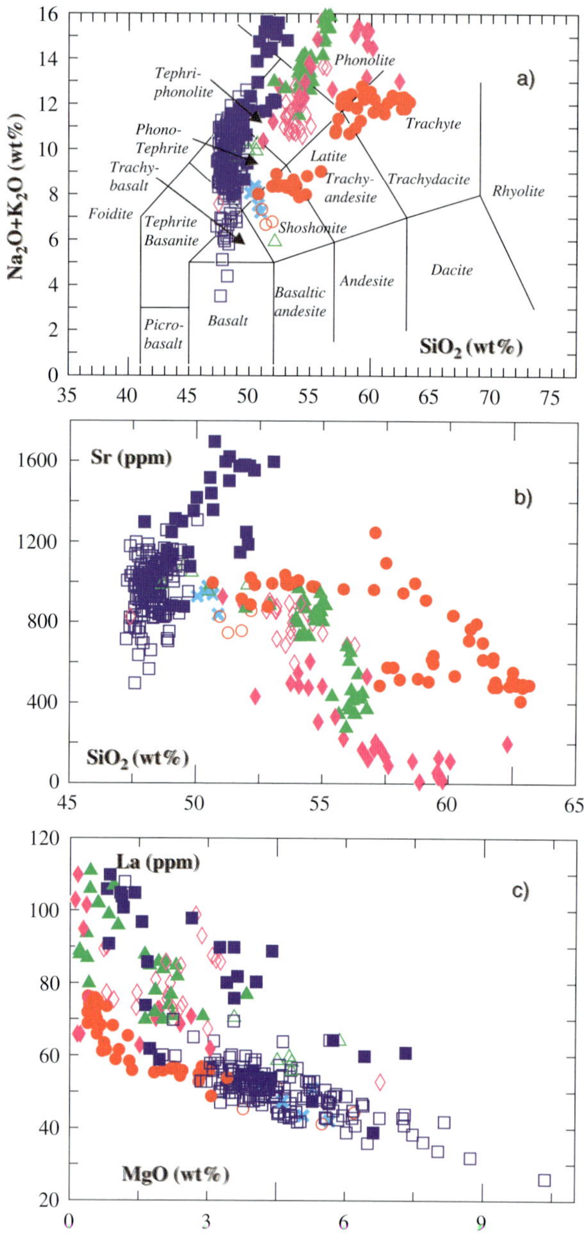

Figure 2. (a) T.A.S. (Le Bas et al., 1986); (b) Sr versus SiO_2 contents; and (c) La versus MgO for SV rocks. Symbols as in Figure 1: bold crosses are dykes from Somma activity; closed symbols are rocks from plinian and sub-plinian events; and open symbols rocks from interplinian periods. Circles, first magmatic cycle; rhombus, second magmatic cycle; triangles, transitional magmatic cycle; squares, third magmatic cycle. *Source*: Cioni et al. (1995); Civetta et al. (1991); Civetta and Santacroce (1992); De Vivo et al. (2003); Marianelli et al. (1999); Santacroce et al. (1993).

(Lima et al., 2003). In this context, to predict the behaviour of the volcano, acquiring a better understanding of magma evolution processes, of the magma feeding system, and of the precise role of the volatiles are crucial (Raia et al., 2000; Webster et al., 2001, 2003, 2006).

In many studies, SiO_2 or MgO have been utilized as differentiation indices. Variations in SiO_2 seem to adequately illustrate the evolution of intermediate-to-most fractionated rocks (Fig. 2b), but not the least-fractionated rocks. In contrast, MgO, appears more adequate for the least-fractionated rocks (Fig. 2c). In any case, based on major and trace elements variations (Fig. 2a–c), diverse evolutionary trends characterized by variable K, P, Ti, some trace elements (i.e. Th, U, Sr) and LREE (Light Rare Earth Elements; La, Ce) enrichment have been found (Joron et al., 1987; Ayuso et al., 1998; Piochi et al., 2006). Within each trend, the role of crystal fractionation processes in magma evolution has been widely accepted (Joron et al., 1987; Civetta et al., 1991; Ayuso et al., 1998; Piochi et al., 2006 and references therein).

Feldspar and clinopyroxene are the main crystallizing minerals, in agreement with petrographic data reported in previous section. Chemical trends in Sr and CaO/Al_2O_3 versus K_2O diagrams (Fig. 3a,b) suggest clinopyroxene associated with feldspar (mostly plagioclase and subordinately K-feldspar) crystallization during evolution of magmas older than 472 AD eruption (Piochi et al., 2006). The Sr versus Th diagram (Fig. 3c) highlights plagioclase fractionation. In contrast, clinopyroxene crystallization dominated during evolution of highly undersaturated magmas of the post-1631 AD interplinian period, as also suggested by Belkin et al. (1993), Villemant et al. (1993) and Trigila and De Benedetti (1993). In these younger rocks, the variable abundance of clinopyroxene affects major- and REE-elements variation (Belkin et al., 1993; Villemant et al., 1993). REE showing fairly homogeneous patterns and variable LREE enrichment support the above data. In particular, the Eu anomaly is not a typical feature of primary magmas from Somma-Vesuvius. It seems to be correlated with the degree of evolution; it is mostly present in highly evolved rocks, such as 79 AD, Avellino, and probably reflects feldspar fractionation (Joron et al., 1987).

In the MORB (Mid Oceanic Ridge Basalt)- and OIB (Oceanic Island Basalt)-multi-elements normalized diagrams (Fig. 4a,b) rocks from Somma-Vesuvius show similar trace elements distribution, regardless of the degree of silica undersaturation and K enrichment. The least evolved rocks ($MgO > 3$ wt%) are characterized by high LILE (Rb, Ba, Th, K) and slight HFSE (High Field Strenght Elements; Zr, Nb) enrichment, and slight Nb and Ta trough with respect to MORB (Fig. 4a), similarly to other potassic magmas (Peccerillo and Manetti, 1985; Peccerillo, 2001, 2003). Furthermore, these rocks have higher Cs, K, Pb, Rb, Th, Ba and lower Nb and Ti contents compared to OIB (Fig. 4b). A heterogeneous mantle source(s) has been therefore proposed to explain the variable undersaturation degree of the rocks and, in particular, the occurrence of different parental magmas and different evolutionary trends as shown in Figure 2 (Civetta et al., 1991; Civetta and Santacroce, 1992; Ayuso et al., 1998; Piochi et al., 2006). Other authors (Rittmann, 1933; Pappalardo et al., 2004; Piochi et al., 2006) have also speculated that crustal contamination processes contributed to the enrichment in K and in various other trace elements.

3.3. Sr, Nd, Pb, Hf, O and He isotope ratios

The variable silica-undersaturated Somma-Vesuvius volcanic rocks show similar range of Sr, Nd, Pb and O isotopic compositions, with large variability within each cycle. $^{87}Sr/^{86}Sr$

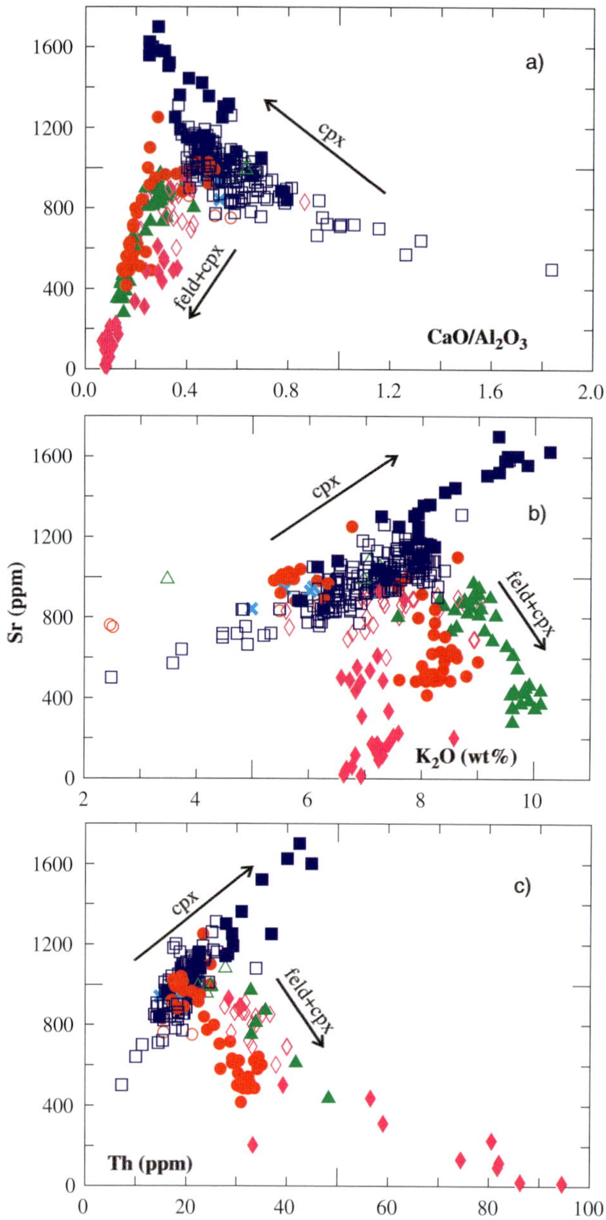

Figure 3. (a) Sr versus Al_2O_3; (b) Sr versus K_2O; and (c) Sr versus Th contents for Somma-Vesuvius rocks. Symbols and source of data as in Figure 2.

isotopic values span from 0.706283 to 0.708070 (Cortini and Hermes, 1981; Civetta and Santacroce, 1992; Caprarelli et al., 1993; Cioni et al., 1995; Ayuso et al., 1998; De Vivo et al., 2003; Piochi et al., 2006). The $^{143}Nd/^{144}Nd$ values range from 0.51225 to 0.51226 (Fig. 5a). Pb isotopic compositions have a moderate variation (Fig. 5b): $^{206}Pb/^{204}Pb$ values

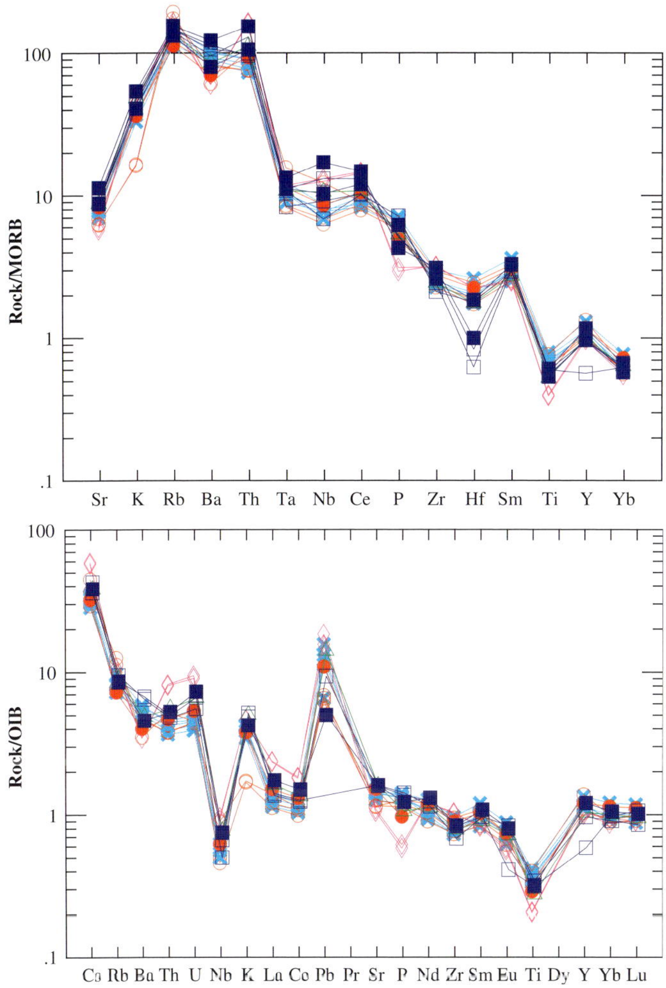

Figure 4. Spider diagrams for selected Somma-Vesuvius rocks with MgO > 3 wt%. Source of data and symbols as in Figure 2.

vary from 18.94 to 19.09, $^{208}Pb/^{204}Pb$ from 38.7 to 39.3 and $^{207}Pb/^{204}Pb$ from 15.61 to 15.71 (Somma et al., 2001; De Vivo et al., 2003; Cortini et al., 2004). Pb isotope variations are not correlated to Sr and Nd isotope variations. δO^{18} values obtained on whole-rocks range from 7.5% to 10‰, showing no correlation with Nd and Pb isotopic compositions, and defines no typical correlation with the $^{87}Sr/^{86}Sr$ ratio (Fig. 5c) (Wilson, 1989). Among the isotopes, only δO^{18} correlates (positively) with degree of chemical evolution (Fig. 6a,b). He isotope composition is about 2.4 Ra (where Ra is the $^{3}He/^{4}He$ of the atmosphere equal to 1.40×10^{-6}) (Graham et al., 1993) for 1944 AD olivines and pyroxenes, indicating a source within the lithosphere or in a slab-enriched mantle source. Similar He-isotopic values have been measured in fumarole gases suggesting a magmatic contribution to the degassing observed at the surface (Graham et al., 1993). $^{176}Hf/^{177}Hf$ ratios determined on two Somma-Vesuvius rocks

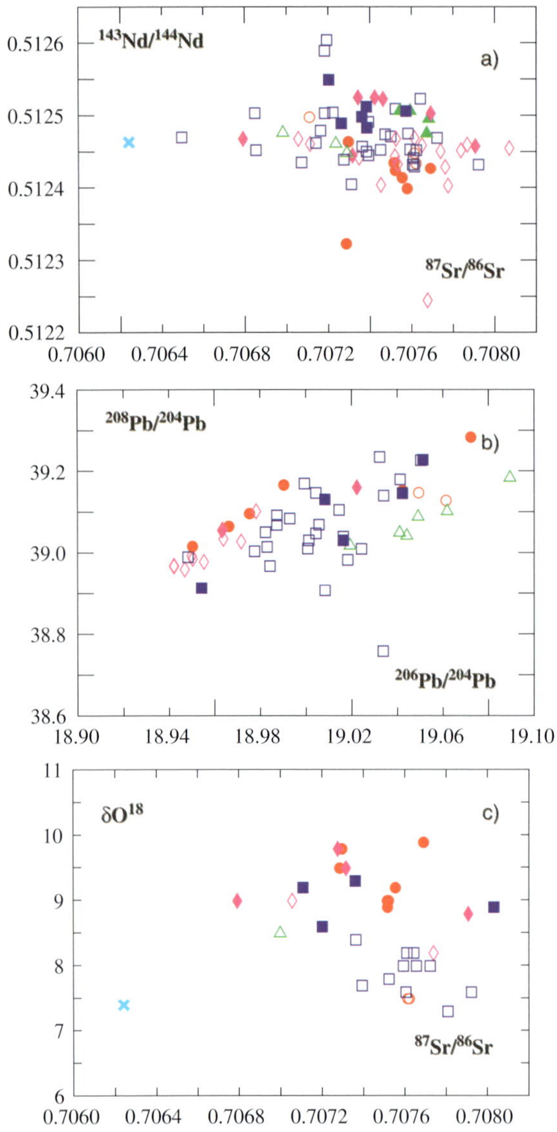

Figure 5. Isotopic diagrams for Somma-Vesuvius rocks: (a) $^{87}Sr/^{86}Sr$ versus $^{143}Nd/^{144}Nd$ ratios; (b) $^{208}Pb/^{204}Pb$ versus $^{206}Pb/^{204}Pb$; and (c) δO^{18} versus $^{87}Sr/^{86}Sr$ ratio. Symbols and source of data as in Figure 2.

characterized by Sr isotopic values lower than 0.7072 are 0.282784 and 0.282786, suggesting a pelagic component added to HIMU (HIgh μ) and DM (Depleted Mantle) mantle sources (Gasperini et al., 2002).

The Sr isotope compositions of products from plinian and sub-plinian eruptions follow a systematic trend through the stratigraphic sequence, consistent with the previously recognized chemostratigraphy (see previous section) though to represent magmas residing in a shallow and chemically stratified chamber (Civetta et al., 1991). For example, the Avellino and the 79 AD pyroclastic sequences consist of white pumices, at the base, overlain by grey pumice deposits. White and grey pumices have different chemical and Sr isotope compositions. However, both pumice types contain feldspars with a constant Sr isotopic composition, similar to that of white pumices, suggesting Sr isotopic disequilibrium in rocks upwards in the sequence and mingling of magmas during eruption. Moreover, the lowermost part of the 79 AD eruption and the uppermost part of Avellino have similar $^{87}Sr/^{86}Sr$ values, suggesting that magma remnants can be left behind within the chamber after large magnitude events (Civetta et al., 1991; Civetta and Santacroce, 1992). Such a type of incomplete magma removal has also been suggested by evidence showing that events following plinian or sub-plinian eruptions produced magmas that have isotopic characteristics comparable to those of previous eruptions (Civetta and Santacroce, 1992; Piochi et al., 2006) (Fig. 7).

The $^{87}Sr/^{86}Sr$ isotopic variations have been attributed to the arrival of isotopically diverse magma batches generated in a variable mantle source(s) (Cortini and Hermes, 1981; Civetta and Santacroce, 1992; Caprarelli et al., 1993; Cioni et al., 1995; Ayuso et al.,

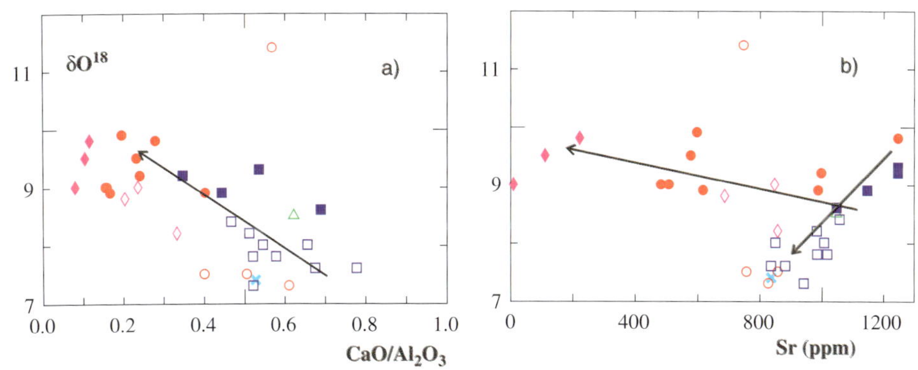

Figure 6. δO^{18} versus CaO/Al_2O_3 ratio (a) and Sr (b) for Somma-Vesuvius rocks. Lines indicate trend of magma evolution. Symbols and source of data as in Figure 2.

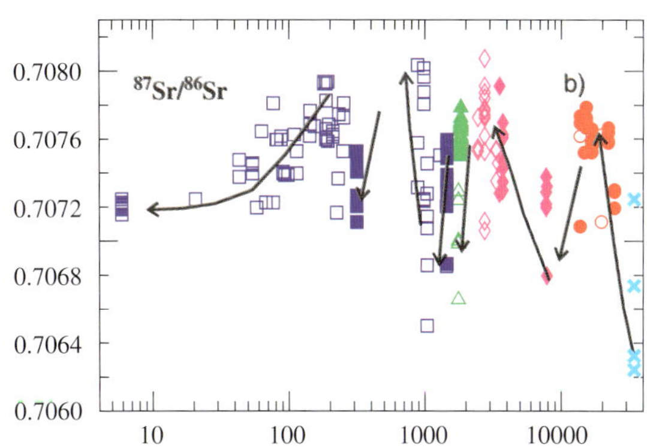

Figure 7. $^{87}Sr/^{86}Sr$ versus age of rocks from Somma-Vesuvius. Symbols and source of data as in Figure 2.

1998; Piochi et al., 2006). Recently, as first recognized by Rittmann (1933), various authors (Civetta et al., 2004; Pappalardo et al., 2004; Paone, 2005; Piochi et al., 2006) suggested the fundamental role of crustal contamination in modifying the isotopic composition of erupted magmas at Somma-Vesuvius. Civetta et al. (2004) and Paone (2005) proposed that contamination occurred within a Hercynian-like basement, similarly to what happens at the Campi Flegrei (Pappalardo et al., 2002). Pappalardo et al. (2004) and Piochi et al. (2006) suggested that carbonate was the main contaminant. In particular, based on Sr isotope variations through time, Pappalardo et al. (2004) suggested that between 1631 and 1944 AD the degree of magma contamination decreased owing to rapid magma rising from a deep reservoir in open-conduit conditions.

4. Discussion

The relationship between magma compositions and tectonic setting depends on reliably distinguishing among geochemical features that image the source region and those that resulted from magma evolution during transport. Processes affecting magmas after their genesis are important in characterizing the behaviour of the magmatic supply system. Such processes, for example, *fractional crystallization*, can produce highly evolved magmas, which when associated with long-lived magma storage in the crust can generate high-magnitude explosive events. *Recharge* of distinct magma batches from deeper levels within the feeding reservoir may be required to trigger volcanic eruptions. *Crustal contamination* requires chemical exchange between magma and wall rocks that can lead to fluid enrichment, increasing the possibility of highly explosive eruptions, or that can induce quick cooling and/or crystallization of magma limiting its further mobility. Properly identifying the exact mechanism of magma evolution, i.e. magma mixing or crustal contamination, can be a useful tool for hazard assessment studies. For the Somma-Vesuvius volcano, it would be important to determine to what extent the evolution of the magmas depend on involvement of the crust during magma genesis (with heterogeneously slab-enriched mantle sources) or during magma evolution (Rittman, 1933; Savelli, 1967, 1968; Turi and Taylor, 1976; Vollmer, 1976; Civetta and Santacroce, 1992; Santacroce et al., 1993; Cioni et al., 1995; Ayuso et al., 1998; Peccerillo, 2001; Pappalardo et al., 2004; Piochi et al., 2006), and how the geochemical evolution exactly triggers sub-plinian and plinian eruptions.

4.1. The role of crustal component on magma composition

The role of the crust on magma composition at the Somma-Vesuvius volcano is suggested from both mineralogical and compositional data. For example, phlogopite occurs among mineral phases. Th/Yb is always higher than 2 (Peccerillo and Manetti, 1985; Peccerillo, 2001). Ce/Pb ratios, being significantly lower than those of mantle sources free of subduction influences (≈ 25; Hofmann et al., 1986), tend towards the upper crustal value (≈ 3.5; Taylor and Mc Lennan, 1985). Similarly, Nb/U value mostly falls within the continental crustal range (< 12; Rudnick and Fountain, 1995) (Fig. 8).

In addition, the role of the crust is also suggested from Sr, Pb and O (as well as Hf) isotope ratios. In fact, these isotope ratios, although highly scattered, show rough correlations with the above chemical ratios: Ce/Pb negatively correlates with $^{87}Sr/^{86}Sr$ and δO^{18},

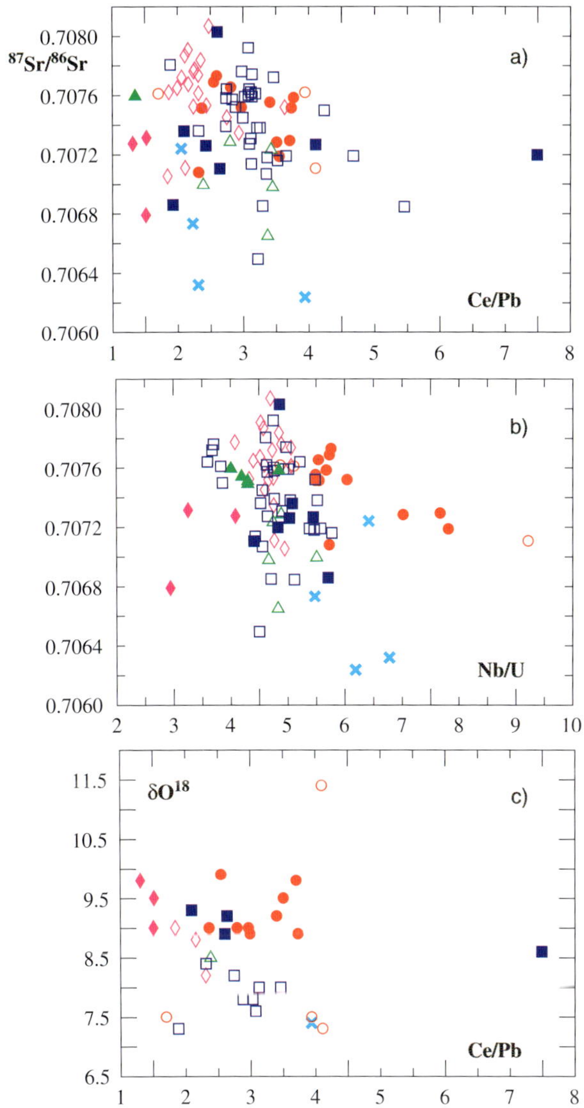

Figure 8. (a) $^{87}Sr/^{86}Sr$ versus Ce/Pb ratios; (b) $^{87}Sr/^{86}Sr$ versus Nb/U ratios; and (c) δO^{18} versus Ce/Pb ratios for Somma-Vesuvius rocks. Symbols and source of data as in Figure 2.

Nb/U positively correlates with Sr isotope composition (Fig. 8a–c). These ratios do not depend on the stage of evolution of the rocks because Ce and Pb, as well as Nb and U, show almost comparable behaviour with respect to SiO_2 or MgO, suggesting a similar partition coefficient in the melt. The observed correlations can be attributed to the variable contributions of the crustal component to the magma.

One important problem is to establish if the crustal component was involved at the time of melting of the source or subsequently during ascent. Generally, radiogenic and stable isotopes can be used to define the site at which contamination occurs. Nevertheless, available O and Sr isotopic data do not conclusively provide information about input of crustal materials/fluids to the magma (either in the mantle source or during shallow-levels differentiation processes), although we know that higher O isotope compositions are found in plinian-type rocks. Below we report some evidence that can be helpful to deal with this fundamental question.

The generally low Mg, Ni and Cr (most values are < 40 and 100 ppm, respectively) contents, and high crystallinity suggest the importance of processes occurring in magmas during crustal storage and ascent. Chemical exchange processes between magmas and carbonate wall rocks are indicated by garnet and phlogopite (Belkin et al., 1985; Joron et al., 1987) and by Ca–Mg-silicate-rich ejecta (skarns) (Savelli, 1968; Fulignati et al., 1995, 1998, 2005; Gilg et al., 1999, 2001; Del Moro et al., 2001). Oxygen isotope studies (Turi and Taylor, 1976; Ayuso et al., 1998), U-disequilibria (Black et al., 1998) and Pb isotope data (Cortini et al., 2004) document shallow-level evolution of Somma-Vesuvius magmas as open systems. Nevertheless, the strongest evidence for the dominating role of shallow-level (crust) processes subsequent to high-pressure (mantle) processes derives from a synthesis of Sr isotope and fluid inclusion data that suggests a positive correlation between $^{87}Sr/^{86}Sr$ values and the estimated depths of mineral crystallization (Fig. 9). The suggestion is that products enriched in radiogenic Sr formed during later stages of magma evolution (Pappalardo et al., 2004; Piochi et al., 2006).

The lower $^{87}Sr/^{86}Sr$ ratios (mostly around 0.7071–0.7072 with few spikes at 0.7062–0.7068) are associated with the highly silica-undersaturated rocks from the 1944 AD eruption containing primitive olivine compositions (Marianelli et al., 1995). These ratios partially overlap the Campi Flegrei Sr-isotope range (0.7068–0.7086) (Pappalardo et al., 2002), differ from values recovered at the nearby Procida (0.70523–0.70678) (De Astis et al., 2004) and are higher than the Tyrrhenian Sea basalts (0.70733–0.7056) (Beccaluva et al., 1990). In addition, they are associated with $^{176}Hf/^{177}Hf$ ratios of 0.282785 (two 1944 AD

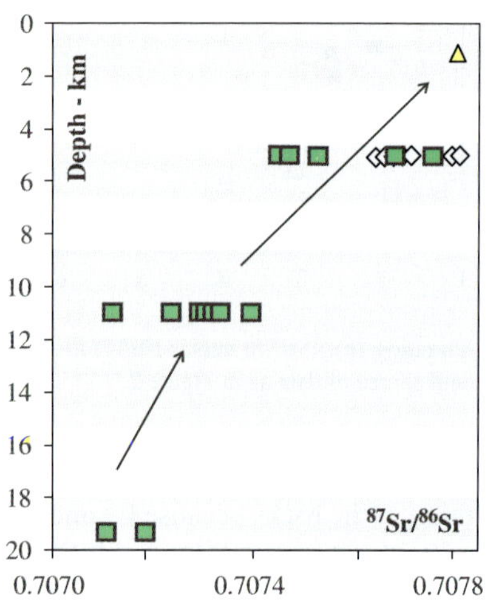

Figure 9. $^{87}Sr/^{86}Sr$ versus depth of crystallizing phases from SV rocks. Squares, clinopyroxene; rhombus, feldspar; and triangles, leucite. Source of data as in Figure 2 (modified from Pappalardo et al. (2004). Grey areas indicate probably levels of magma storage, based on fluid inclusion, volcanological and seismic data (see text).

samples reported in Gasperini et al., 2002) and He isotope ratio lower than MORB-like magmas (Graham et al., 1993). Moreover, the 1944 AD eruption, and other rocks that are generally poorly evolved (MgO > 3 wt%), are enriched in LILE, LREE and other incompatible trace elements (e.g. Th, Nb, Ta), as well as in more compatible elements such as HREE (High Rare Earth Elements) and Y (Fig. 4a). These geochemical features are usually related to magmas erupted along subduction zones, implying the involvement of a crustal component in the mantle source beneath Somma-Vesuvius.

4.2. The mantle source

The least-evolved Somma-Vesuvius rocks (MgO > 3 wt%) belong to the within-plate type in term of Zr (> 100 ppm) and Zr/Y (> 4) (Pearce and Norry, 1979) (Fig. 10), in agreement with evidence from the multi-element normalized diagram (Fig. 4b) showing a certain similarity to the OIB basalts. The positive correlation in Figure 10 points to a decrease in degree of partial melting or (fluid-controlled) source heterogeneity. Based on the Cs–Pb enrichment in Figure 4b, the LILE enrichment and the slight Nb–Ta negative anomalies in Figure 4a, and Nb/Zr at about 0.15, as well as on the isotope features discussed in the previous section, we suggest that the mantle source of Somma-Vesuvius contains a slab-derived component. This conclusion is consistent with the general idea that enriched potassium-rich magmas are generated by partial melting of phlogopite-rich garnet peridotite (Gupta and Fyfe, 2003).

Poorly evolved rocks (MgO > 3 wt%) with a high degree of silica undersaturation show significant constancy of Th/Zr (0.05–0.08), Ta/Yb (0.7) and Cs/Rb (< 0.06), as well as Th/Yb, Th/Ta and other ratios, that are independent of fractional crystallization and/or partial melting. These relatively unevolved rocks, as well as the slightly and mildly silica-undersaturated rocks, have comparable trace elements distributions, showing similar enrichment in LILE, Ce and other incompatible trace elements (e.g. Th, Nb, Ta), as well as in more compatible elements such as Sm and Y (Fig. 4) independent of their Sr isotope values and K-enrichment degree. Therefore, in a general sense, these data suggest the

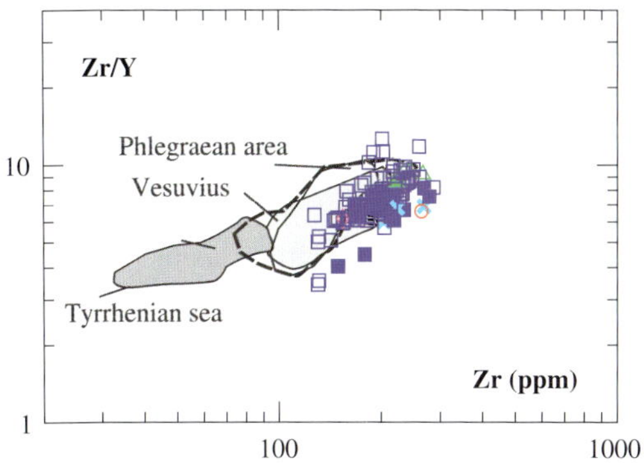

Figure 10. Zr and Zr/Y for SV rocks with MgO >3 wt%. Source of data and symbols as in Figure 2. Data from Phlegraean Fields (D'Antonio et al., 1999; Pappalardo et al., 1999; Piochi et al., 1999) and Tyrrhyenian Sea (Beccaluva et al., 1990) are also reported for comparison.

existence of an invariable mantle source during the life of the Somma-Vesuvius volcano. In agreement with Peccerillo and Manetti (1985), we suggest that diverse degrees of silica undersaturation in potassic "mafic" rocks was linked to small degrees of partial melting at different pressures in a phlogopite-bearing potassium-rich peridotitic mantle source containing CO_2 and small amounts of water. Sr, Nd, Pb, O, He and Hf isotopes were likely affected by processes in the mantle source. However, with our hypothesis, the absence of relationships between Sr–Nd isotope compositions and degree of both alkali enrichment and silica undersaturation of "mafic" rocks suggests that mantle source processes mostly influence the chemical composition of parental magmas, but it cannot be the main cause of the large isotopic variability of Somma-Vesuvius rocks with $^{87}Sr/^{86}Sr$ ratios higher than 0.7071.

4.3. The behaviour of the magmatic feeding system

Based on the variation of the $^{87}Sr/^{86}Sr$ values, contamination of Somma-Vesuvius magmas was attributed to a Hercynian-like basement (Civetta et al., 2004; Paone, 2005) or to rocks in the overlying sedimentary series (Rittmann, 1933; Pappalardo et al., 2004; Piochi et al., 2006). However, on the basis of data in Figure 9 we suggest that the increase in Sr isotope values from 0.7071-3 to 0.7081 mostly occurs within the uppermost 11–12 km of the crust and points to these sedimentary rocks as the main crustal contaminant. However, we cannot exclude that magma contamination could have also occurred in crustal rocks underlying the carbonate basement. We stress the fact that no xenolith of possible Hercynian origin has been found at Somma-Vesuvius, contrary to what happened at the nearby Campi Flegrei (Pappalardo et al., 2002; Paone, 2005).

Contamination of magma ($^{87}Sr/^{86}Sr \approx 0.7071$) by carbonate rocks ($^{87}Sr/^{86}Sr \approx$ 0.7073–00709; Sr = 700–1000 ppm) (Civetta et al., 1991; Iannace, 1991) at Somma-Vesuvius has been quantitatively modelled by Pappalardo et al. (2004) and Piochi et al. (2006) who suggested that crustal contamination was a selective process involving thermal decomposition (decarbonation reactions) of the sedimentary wall rocks and exchange between magmas and fluids. Fulignati et al. (2004, 2005) also suggested similar conclusions on the basis of geochemical and mineralogical data collected on 79 and 1944 AD skarn ejecta. We recognize, however, that magma evolution was likely more complicated than as stated previously because no correlation has been found for δO^{18} and $^{87}Sr/^{86}Sr$ values, and because of the negative correlation between phenocryst abundance and values of $^{87}Sr/^{86}Sr$ (Figs. 5c and 11). Moreover, hornfels rhyolitic pumices characterized by $^{87}Sr/^{86}Sr$ higher than 0.711 and δO^{18} at around 15‰ have been found among ejecta in various pyroclastic deposits and have been interpreted as the result of the partial melting of the pelitic sediments during thermometamorphic event (Del Moro et al., 2001; Fulignati et al., 2005). This fact suggests the possible involvement of Miocene sediments in addition to carbonate during the evolution of magmas at the Somma-Vesuvius.

Fluid exchange between magmas and wall rocks could be more pervasive on magmas associated with high-explosive eruptions. Available data reveal relatively high values and a large range of δO^{18} for pumices from plinian and sub-plinian eruptions, and relatively low δO^{18} values and a smaller range for highly silica-undersaturated volcanic rocks from interplinian events (Figs. 5c, and 6a,b). The correlation for δO^{18} and chemical

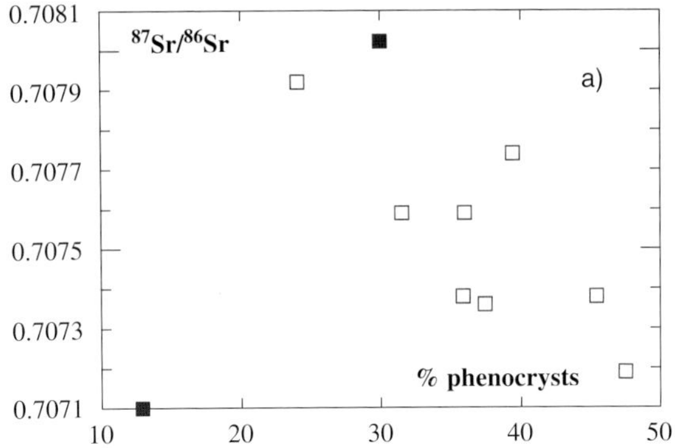

Figure 11. (a) $^{87}Sr/^{86}Sr$ versus phenocryst content in rocks from recent interplinian period of volcanism; (b) $^{87}Sr/^{86}Sr$ versus age of rocks. Symbols and source of data as in Figure 2.

differentiation indices (better defined for rocks from high-explosive eruptions), together with numerical considerations reported in Ayuso et al. (1998), data from Cortini et al. (2004) and the observed enrichment in some incompatible trace elements (La, Nb, Zr) of pumices from plinian eruptions (Fig. 2c), also support the effects of fluid exchange, rather than isotope fractionation determined by exsolution of gas from magma.

Magmas erupted during the post-1631 AD interplinian period are characterized by the decrease of the $^{87}Sr/^{86}Sr$ ratio with increasing phenocryst content down to typical values of clinopyroxenite (< 0.7071) (Del Moro et al., 2001). This relation can be attributed to (1) the entrapment of crystal mush generated during previous magma storage in the crust by rising magmas and/or (2) the accumulation/depletion of phenocrysts during magma movements through the crust towards the surface. In the first case, magmatic melts should be characterized by higher $^{87}Sr/^{86}Sr$ ratios. Otherwise, phenocrysts can be accumulated or be depleted in magma as a function of the ascent rate of magma towards the surface (see also Villemant et al., 1993). In particular, low ascent rate can result in crystal segregation and in longer time during which melt stay within wall rocks, thus producing rocks with lower crystal content and possibly higher crustal contamination. This second hypothesis is in agreement with evidence from Villemant et al. (1993) indicating that lavas derived from magmas experiencing volatile degassing generally contain lower crystal abundance than vesiculated fragments generated by gas overpressure. This idea is supported by evidence that magmas with the lowermost Sr isotope ratios erupted during the 1944 AD rose to the surface from 11–22 km depth (Marianelli et al., 1999). However, the repetitive and regular variation of $^{87}Sr/^{86}Sr$ values through time (Fig. 7) is consistent with the idea that residual magma or crystal mush remaining in the magmatic system after the end of the plinian (or sub-plinian) eruptive event, can be involved in subsequent eruptions (Civetta et al., 1991; Civetta and Santacroce, 1992; Cioni et al., 1995; Lima et al., 2003; Piochi et al., 2006).

$^{87}Sr/^{86}Sr$, δO^{18} and fluid inclusion data strongly suggest polybaric evolutionary processes of diverse parental magmas at Somma-Vesuvius. Evolutionary processes were dominated by crustal contamination and crystal entrapment, in addition to crystal fractionation and magma mixing. Evidence presented in this paper, in particular data shown in Figure 9, allows us to speculate that magmas with $^{87}Sr/^{86}Sr$ ratios of around 0.7071-3 and of 0.7074-5 derive from reservoirs probably located at different depths, i.e. > 12 km and at around 8–12 km, respectively. Magmas with higher Sr isotope compositions, for example those from Pompei and Avellino eruptions, evolved during storage in shallower magma chambers or, for example those from some of post-1631 AD interplinian eruptions, during the ascent through the conduit.

5. Conclusions

Available data in the literature furnish the possibility to preliminarily define the magma feeding system beneath the Somma-Vesuvius strato-volcano. It consists of three main levels of magma storage, the two deepest probably being long-lived reservoirs, and an uppermost crustal level that probably includes the volcanic conduit and hosted magmas during interplinian period of volcanism. The deeper level is located at depths exceeding 15 km and should furnish magma with $^{87}Sr/^{86}Sr$ ratios of < 0.7072 and δO^{18} < 8‰. The intermediate level occurs at around 8–12 km depth and supplies magmas with $^{87}Sr/^{86}Sr$ ratios between 0.7071 and 0.7074, and δO^{18} < 8‰ typically erupted both during interplinian (i.e. 1906 AD) and sub-plinian (472 AD, 1631 AD) events. The shallow level at around 5 km depth was the site of plinian magma chambers such as those of Pompei and Avellino eruptions. This type of magma feeding system fits with fluid and melt inclusions data (Belkin et al., 1985; Belkin and De Vivo, 1993; Marianelli et al., 1999; Cioni, 2000; Lima et al., 2003) indicating magma storage at 3.5–5 km, 8–10 km and > 12 km, with results of seismic (Zollo et al., 1996) and magnetotelluric (Di Maio et al., 1998) investigations indicating a discontinuity at 8–10 km depth, with seismic evidence of deeper magma storage extending up to 30 km depth (De Natale et al., 2001), and with the magnetized character of a narrow shallow crustal volume (Fedi et al., 1998). However, geophysical data do not indicate the occurrence of current magma storage at a depth of < 5 km, as vice versa is indicated by fluid and melt inclusion studies (Belkin et al., 1985; Belkin and De Vivo, 1993; Marianelli et al., 1999; Cioni, 2000; Lima et al., 2003).

Acknowledgements

The authors are thankful to A. Peccerillo for his constructive review, which helped to improve the final version of the manuscript. The paper has benefited from MIUR-PRIN funds to B. De Vivo (2003–2004). We also thank the Elsevier's Production Editor for editorial assistance.

References

Arnò, V., Principe, C., Rosi, M., Santacroce, R., Sbrana, A., Sheridan, M.F., 1987. Eruptive history. In: Santacroce, R. (Ed.), Somma-Vesuvius, Quaderni de "La Ricerca Scientifica", CNR, Italy, 251 pp.

Arrighi, S., Principe, C., Rosi, M., 2001. Violent strombolian and subplinian eruptions at Vesuvius during post-1631 activity. Bull. Volcanol. 63, 126–150.

Ayuso, R.A., De Vivo, B., Rolandi, G., Seal II, R.R., Paone, A., 1998. Geochemical and isotopic (Nd-Pb-Sr-O) variations bearing on the genesis of volcanic rocks from Vesuvius, Italy. J. Volcanol. Geotherm. Res. 82, 53–78.

Barberi, F., Leoni, L., 1980. Metamorphic carbonate ejecta from Vesuvius plinian eruptions: evidence of the occurrence of shallow magma chambers. Bull. Volcanol. 43, 107–120.

Beccaluva, L., Di Girolamo, P., Morra, V., Siena, F., 1990. Phlegraean Fields volcanism revisited: a critical re-examination of deep-eruptive systems and magma evolutionary processes. N. Jb. Geol. Paleont. Mh. H5, 257–271.

Belkin, H.E., De Vivo, B., 1993. Fluid inclusion studies of ejected nodules from plinian eruptions of Mt. Somma-Vesuvius. J. Volcanol. Geotherm. Res. 58, 98–100.

Belkin, H.E., De Vivo, B., Roedder, E., Cortini, M., 1985. Fluid inclusion geobarometry from ejected Mt. Somma-Vesuvius nodules. Am. Mineral. 70, 288–303.

Belkin, H.E., Kilburn, C.R.J., De Vivo, B., Trigila, R., 1993. Sampling and analytical chemistry of the recent Vesuvius activity (1631–1944). J. Volcanol. Geotherm. Res. 58, 273–290.

Bianco, F., Castellano, M., Milano, G., Vilardo, G., Ferrucci, F., Gresta, S., 1999. The seismic crises at Mt. Vesuvius during 1995 and 1996. Phys. Chem. Earth 24(111–112), 977–983.

Black, S., Macdonald, R., De Vivo, B., Kilburn, C.R.J., Rolandi, G., 1998. U-series disequilibria in young (AD 1944) Vesuvius rocks, preliminary implications for magma residence times and volatile addition. J. Volcanol. Geotherm. Res. 82, 97–111.

Brocchini, D., Principe, C., Castradori, D., Laurenzi, M.A., Gorla, L., 2001. Quaternary evolution of the southern sector of the Campanian Plain and early Somma-Vesuvius activity: insights from the Trecase 1 well. Miner. Petrol. 73, 67–91.

Caprarelli, G., Togashi, S., De Vivo, B., 1993. Preliminary Sr and Nd isotopic data for recent lavas from Vesuvius volcano. J. Volcanol. Geotherm. Res. 58, 377–381.

Chiarabba, C., Jovane, L., Di Stefano, R., 2005. A new view of Italian seismicity using 20 years of instrumental recordings. Tectonophysics 395, 251–268.

Chiodini, G., Marini, L., Russo, M., 2001. Geochemical evidence for the existence of high-temperature hydrothermal brines at Vesuvio volcano, Italy. Geochim. Cosmochim. Acta 65, 2129–2147.

Cioni, R., 2000. Volatile content and degassing processes in the AD 79 magma chamber at Vesuvius (Italy). Contrib. Mineral. Petrol. 140, 40–54.

Cioni, R., Civetta, L., Marianelli, P., Metrich, N., Santacroce, R., Sbrana, A., 1995. Compositional layering and syn-eruptive mixing of a periodically refilled shallow magma chamber: the AD 79 plinian eruption of Vesuvius. J. Petrol. 36, 739–776.

Cioni, R., Marianelli, P., Santacroce, R., 1998. Thermal and compositional evolution of the shallow magma chambers of Vesuvius, evidence from pyroxene phenocrysts and melt inclusions. J. Geophys. Res. 103(18), 277-18, 294.

Civetta, L., D'Antonio, M., de Lorenzo, S., Di Renzo, V., Gasparini, P., 2004. Thermal and geochemical constraints on the 'deep' magmatic structure of Mt. Vesuvius. J. Volcanol. Geotherm. Res. 133, 1–12.

Civetta, L., Galati, R., Santacroce, R., 1991. Magma mixing and convective compositional layering within the Vesuvius magma chamber. Bull. Volcanol. 53, 287–300.

Civetta, L., Santacroce, R., 1992. Steady state magma supply in the last 3400 years of Vesuvius activity. Acta Vulcanol. 2, 147–159.

Corrado, G., Rapolla, A., 1981. The gravity field of Italy: analysis of its spectral composition and delineation of a three-dimensional crustal model for central-southern Italy. Boll. Geof. Teor. Appl. 89, 17–29.

Cortini, M., Ayuso, R.A., De Vivo, B., Holden, P., Somma, R., 2004. Isotopic composition of Pb and Th in interplinian volcanics from Somma–Vesuvius volcano, Italy. Mineral. Petrol. 80(1–2), 83–96.

Cortini, M., Hermes, O.D., 1981. Sr isotopic evidence for a multi source origin of the potassic magmas in the Neapolitan area (South Italy). Contrib. Mineral. Petrol. 77, 47–55.

D'Antonio, M., Civetta, L., Orsi, G., Pappalardo, L., Piochi, M., Carandente, A., de Vita, S., Di Vito, M., Isaia, R., 1999. The present state of the magmatic system of the Campi Flegrei caldera based on a reconstruction of its behavior in the past 12 ka. J. Volcanol. Geotherm. Res. 91, 247–268.

D'Argenio, B., Pescatore, T., Scandone, P., 1973. Schema Geologico dell'Appennino Meridionale (Campania e Lucania). Proceedings of Moderne vedute della geologia dell'Appennino. Accademia Nazionale dei Lincei, 183, Roma, Italy.

De Astis, G., Pappalardo, L., Piochi, M., 2004. Procida Volcanic History: new insights in the evolution of the Phlegraean Volcanic District (Campania region, Italy). Bull. Volcanol. DOI 10.1007/s00445-004-0345-y, 66, 622–641.

Del Moro, A., Fulignati, P., Marianelli, P., Sbrana, A., 2001. Magma contamination by direct wall rock interaction: constraints from xenoliths from the walls of a carbonate-hosted magma chamber (Vesuvius 1944 eruption). J. Volcanol. Geotherm. Res. 112, 15–24.

De Natale, G., Troise, C., Pingue F., De Gori, P., Chiarabba, C., 2001. Structure and dynamics of the Somma-Vesuvius volcanic complex. Mineral. Petrol. 73(1–3), 5–22.

De Vivo, B., Ayuso, R.A., Belkin, H.E., Fedele, L., Lima, A., Rolandi, G., Somma, R., Webster, J.D., 2003. Chemistry, fluid/melt inclusions and isotopic data of lavas, tephra and nodules from >25 ka to 1944 AD of the Mt. Somma-Vesuvius volcanic activity. Mt. Somma-Vesuvius Geochemical Archive. Dipartimento di Geofisica e Vulcanologia, Università di Napoli Federico II, Open File Report 1-2003, 143 pp.

Di Maio, R., Mauriello, P., Patella, D., Petrillo, Z., Piscitelli, S., Siniscalchi, A., 1998. Electric and electromagnetic outline of the Mount Somma-Vesuvius structural setting. J. Volcanol. Geotherm. Res. 82(1–4), 219–238.

Federico, C., Aiuppa, A., Allard, P., Bellomo, S., Jean-Baptiste, P., Parello, F., Valenza, M., 2002. Magma-derived gas influx and water-rock interactions in the volcanic aquifer of Mt. Vesuvius, Italy. Geochim. Cosmochim. Acta 66, 963–981.

Fedi, M., Florio, G., Rapolla, A., 1998. 2.5D modelling of Somma-Vesuvius structure by aeromagnetic data. J. Volcanol. Geotherm. Res. 82(1–4), 239–247.

Ferrucci, F., Gaudiosi, G., Pino, N.A., Luongo, G., 1989. Seismic detection of a major Moho upheaval beneath the Campania volcanic area (Napoli, southern Italy). Geophys. Res. Lett. 16(11), 1317–1320.

Frondini, F., Chiodini, G., Caliro, S., Cardellini, S., Granieri, D., 2004. Diffuse CO_2 soil degassing at Vesuvio, Italy. Bull. Volcanol. 66(7), 642–651.

Fulignati, P., Gioncada, A., Sbrana, A., 1995. The magma chamber related hydrothermal system of Vesuvius, first mineralogical and fluid inclusion data on hydrothermalized subvolcanic and lavic samples from phreatomagmatic eruptions. Per. Mineral. 64, 185–187.

Fulignati, P., Marianelli, P., Métrich, N., Santacroce, R., Sbrana, A., 2004. Towards a reconstruction of the magmatic feeding system of the 1944 eruption of Mt Vesuvius. J. Volcanol. Geotherm. Res. 133, 13–22.

Fulignati, P., Marianelli, P., Sbrana, A., 1998. New insights on the thermometamorphic-metasomatic magma chamber shell of the 1944 eruption of Vesuvius. Acta Vulcanol. 10(1), 47–54.

Fulignati, P., Panichi, C., Sbrana, A., Caliro, S., Gioncada, A., Del Moro, A., 2005. Skarn formation at the walls of the 79 AD magma chamber of Vesuvius (Italy): mineralogical and isotopic constraints. N. Jb. Miner. Abh. 181/1, 53–66.

Gasperini, D., Blichert-Toft, J., Bosch, D., Del Moro, A., Macera, P., Albarede, F., 2002. Upwelling of deep mantle material through a plate window: evidence from the geochemistry of Italian basaltic volcanoes. J. Geophys. Res., 107, 2367–2376.

Gilg, H.A., Lima, A., Somma, R., Ayuso, R.A., Belkin, H.E., De Vivo, B., 1999. A fluid inclusion and isotope study of calc-silicate ejecta from Mt Somma-Vesuvius: evidence for interaction of high-temperature hypersaline fluids with the sedimentary basement. Proceedings of ECROFI XV, Potsdam, Germany. Terra Nostra 99(6), 118–120.

Gilg, H.A., Lima, A., Somma, R., Belkin, H.E., De Vivo, B., Ayuso, R.A., 2001. Isotope geochemistry and fluid inclusion study of skarns from Vesuvius. Mineral. Petrol. 73, 145–176.

Graham, D.W., Allard, P., Kilburn, C.R.J., Spera, F.J., Lupton, J.E., 1993. Helium isotopes in some historical lavas from Mount Vesuvius. J. Volcanol. Geotherm. Res. 58, 359–366.

Gupta, A.K., Fyfe, W.S., 2003. The Young Potassic Rocks. Uni. Book, New Daly, 370 pp.

Hermes, O.D. Cornel, W.C., 1981. Quenched crystal mush and associated magma compositions as indicated by intercumulus glasses from Mt. Vesuvius, Italy. J. Volcanol. Geotherm. Res. 9, 133–149.

Hofman, A.W., Jochum, K.P., Seufert, M., White, W.M., 1986. Nb and Pb in oceanic basalts: new constraints on mantle evolution. Earth Planet. Sci. Lett. 79, 33–45.

Iannace, A., 1991. Ambienti deposizionali e processi diagenetici in successione di piattaforma carbonatica del Trias Superiore nei Monti Lattari e Picentini (Salerno). Ph.D. thesis, Università "Federico II", Napoli, Italy, 221 pp.

Ippolito, F., D'Argenio, B., Pescatore, T., Scandone, P., 1975. Structural-stratigraphic units and tectonic framework of Southern Appennines. In: Squyres, C. (Ed.), Geology of Italy. Lybian Society of Earth Science, Libyan, Arab Republic, 11 pp.

Joron, J.L., Metrich, N., Rosi, M., Santocroce, R., Sbrana, A., 1987. Chemistry and petrography. In: Santacorce, R. (Ed.), Somma-Vesuvius. CNR Quad. Ric. Sci. 114, 105–174.

Landi, A., Bertagnini, A., Rosi, M., 1999. Chemical zoning and crystallization mechanisms in the magma chamber of the Pomici di Base plinian eruption of Somma-Vesuvius (Italy). Contrib. Mineral. Petrol. 135, 179–197.

Le Bas, M.J., Le Maitre, R.W., Streckeisen, A., Zanettin, B., 1986. A chemical classification of volcanic rocks based on the total alkali-silica diagram. J. Petrol. 27, 745–750.

Lima, A., Danyushevsky, L.V., De Vivo, B., Fedele, L., 2003. A model for the evolution of the Mt. Somma-Vesuvius magmatic system based on fluid and melt inclusion investigations. In: De Vivo, B., Bodnar, R.J. (Eds), Melt Inclusions in Volcanic Systems: Methods, Applications and Problems.Series: Development in Volcanology. Elsevier, Amsterdam, 272 pp.

Marianelli, P., Métrich, N., Sbrana, A., 1995. Shallow and deep reservoirs involved in magma supply of the 1944 eruption of Vesuvius. Bull. Volcanol. 61(1–2), 48–63.

Marianelli, P., Métrich, N., Sbrana, A., 1999. Shallow and deep reservoirs involved in magma supply of the 1944 eruption of Vesuvius. Bull. Volcanol. 61, 48–63.

Mastrolorenzo, G., Munno, R., Rolandi, G., 1993. Vesuvius 1906: a case study of a paroxysmal eruption and its relation to eruption cycles. J. Volcanol. Geotherm. Res. 58, 217–237.

Paone, A., 2005. Evidence of crustal contamination, sediment, and fluid components in the campanian volcanic rocks. J. Volcanol. Geotherm. Res. 138, 1–26.

Pappalardo, L., Civetta, L., D'Antonio, M., Deino, A.L., Di Vito, M.A., Orsi, G., Carandente, A., de Vita, S., Isaia, R., Piochi, M., 1999. Chemical and isotopical evolution of the Phlegraean magmatic system before the Campanian Ignimbrite (37 ka) and the Neapolitan Yellow Tuff (12 ka) eruptions. J. Volcanol. Geotherm. Res. 91, 141–166.

Pappalardo, L., Piochi, M., D'Antonio, M., Civetta, L., Petrini, R., 2002. Evidence for multi-stage magmatic evolution during the past 60 ka at Campi Flegrei (Italy) deduced from Sr, Nd and Pb isotope data. J. Petrol. 43(7), 1415–1434.

Pappalardo, L., Piochi, M., Mastrolorenzo, G., 2004. The 3800 yr BP–1944 AD magma plumbing system of Somma-Vesuvius: constraints on its behaviour and present satte through a review of isotope data. Ann. Geophys. 47(4), 1363–1375.

Pearce, J.A., Norry, M.J., 1979. Petrogenetic implications of Ti, Zr, Y, and Nb variations in volcanic rocks. Contrib. Mineral. Petrol. 69, 33–47.

Peccerillo, A., 2001. Geochemical similarities between the Vesuvius, Phlegraean Fields and Stromboli Volcanoes: petrogenetic, geodynamic and volcanological implications. Mineral. Petrol. 73, 93–105.

Peccerillo, A., 2003. Plio-Quaternary magmatism in Italy. Episodes 26, 222–226.

Peccerillo, A., Manetti, P., 1985. The potassic alkaline volcanism of central southern Italy: a review of the data relevant to petrogenesis and geodinamic significance. Trans. Geol. Soc. South Africa 88, 379–394.

Piochi, M., Ayuso, R.A., De Vivo, B., Somma, R., 2006. Crustal contamination and crystal entrapment during polybaric magma evolution at the Mt. Somma-Vesuvius volcano, Italy: geochemical and Sr isotope evidence. Lithos, 86, 303–329.

Piochi, M., Civetta, L., Orsi, G., 1999. Mingling in the magmatic system of Ischia (Italy) in the past 5 Ka. Mineral. Petrol. 66(4), 227 258.

Principe, C., Tanguy, J.C., Arrighi, S., Paiotti, A., Le Goff, M., Zoppi, U., 2004. Chronology of Vesuvius' activity from AD 79 to 1631 based on archeomagnetism of lavas and hisotrical sources. Bull. Volcanol. 66, 703–724.

Raia, F., Webster, J.D., De Vivo, B., 2000. Pre-eruptive volatile contents of Vesuvius magmas: constrains on eruptive history and behavior. I – the medieval and modern interplinian activities. Eur. J. Mineral. 12, 179–193.

Rittmann, A., 1933. Die geologisch bedingte Evolution und Differentiation des Somma-Vesuvs-magmas. Zs. Vulkanologie 15(1–2), 8–94.

Rolandi, G., Maraffi, S., Petrosino, P., Lirer, L., 1993. The Ottaviano eruption of Somma-Vesuvius (8000 y BP): a magmatic alternating fall and flow-forming eruption. J. Volcanol. Geotherm. Res. 58, 43–65.

Rolandi, G., Petrosino, P., Mc Geehin, J., 1998. The interplinian activity at Somma-Vesuvius in the last 3500 years. J. Volcanol. Geotherm. Res. 82, 19–52.

Rosi, M., Santacroce, R., 1983. The AD 472 "Pollena" eruption: volcanological and petrological data for this poorly-known, plinian-type event at Vesuvius. J. Volcanol. Geotherm. Res. 17, 249–271.

Rosi, M., Santacroce, S., Sheridan, M.F., 1987. Volcanic hazard. In: Santacroce, R. (Ed.), Somma-Vesuvius. Quaderni de "La Ricerca Scientifica", CNR, Italy, 251 pp.

Rudnick, R.L., Fountain, D.M., 1995. Nature and composition of the continental crust: a lower crustal perspective. Rev. Geophys. 33, 267–309.

Santacroce, R.A., Bertagnini, A., Civetta, L., Landi, P., Sbrana, A., 1993. Eruptive dynamics and petrogenetic processes in a very shallow magma reservoir: the 1906 eruption of Vesuvius. J. Petrol. 34, 383–425.

Savelli, C., 1967. The problem of rock assimilation by Somma-Vesuvius magmas, I. Compositions of Somma-Vesuvius lavas. Contrib. Mineral. Petrol. 16, 328–353.

Savelli, C., 1968. The problem of rock assimilation by Somma-Vesuvius magmas, II. Compositions of sedimentary rocks and carbonate ejecta from Vesuvius area. Contrib. Mineral. Petrol. 18, 43–64.

Scandone, R., Iannone, F., Mastrolorenzo, G., 1986. Stima dei parametri dinamici dell'eruzione del 1944 del Vesuvio. Boll. GNV 2, 487–512.

Somma, R., Ayuso, R.A., De Vivo, B., Rolandi, G., 2001. Major, trace element and isotope geochemistry (Sr-Nd-Pb) of interplinian magmas from Mt. Somma-Vesuvius (Southern Italy). Mineral. Petrol. 73, 121–143.

Taylor, S.R., Mclennan, S.M. (Eds), 1985. The Continental Crust: Its Composition and Evolution. Blackwell, Cambridge, 312 pp.

Trigila, R., De Benedetti, A., 1993. Petrogenesis of Vesuvius historical lavas constrained by Pearce element ratios analysis and experimental phase equilibria. J. Volcanol. Geotherm. Res. 58, 315–343.

Turi, B., Taylor, H.P., 1976. Oxygen isotope studies of potassic volcanic rocks of the Roman Province, Central Italy. Contrib. Mineral. Petrol. 55, 1–31.

Vilardo, G., Ventura, G., Girolamo, M., 1999. Factors controlling the seismicity of the Somma-Vesuvius volcanic complex. Volc. Seis. 20, 219–238.

Villemant, B., Trigila, R., De Vivo, B., 1993. Geochemistry of Vesuvius volcanics during 1631–1944 period. J. Volcanol. Geotherm. Res. 58, 291–313.

Vollmer, R., 1976. Rb–Sr and U–Th–Pb systematics of the alkaline rocks from Italy., Geochim. Cosmochim. Acta 40, 283–295.

Webster, J.D., De Vivo, B., Tappen, C., 2003. Volatiles, magmatic degassing and eruptions of Mt. Somma-Vesuvius: constraints from silicate melt inclusions, solubility experiments and modeling. In: De Vivo, B., Bodnar, R.J. (Eds), Melt Inclusions in Volcanic Systems. Methods, Applications and Problems. Series: Developments in Volcanology, Vol. 5, Elsevier, Amsterdam, pp. 207–226.

Webster, J.D., Raia, F., De Vivo, B., Rolandi, G., 2001. The behaviour of chlorine and sulfur during differentiation of the Mt. Somma-Vesuvius magmatic system. Mineral. Petrol. 73, 177–200.

Webster, J.D., Sintoni, M.F., De Vivo, B., 2005. The role of sulphur in promoting magmatic degassing and volcanic eruption at Mt. Somma-Vesuvius. In: De Vivo, B. (Ed.), Vesuvius and Ignimbrites of Campania Plain. Series Developments in Volcanology. Elsevier, Amsterdam.

Wilson, M. (Ed.), 1989. Igneous Petrogenesis. Unwin Hyman, London, 466 pp.

Zollo, A., Gasparini, P., Virieux, J., Le Meur, H., De Natale, G., Biella, G., Boschi, E., Capuano, P., De Franco, R., Dell'Aversana, P., De Matteis, R., Guerra, I., Iannaccone, G., Mirabile, L., Vilardo, G., 1996. Seismic evidence for a low-velocity zone in the upper crust beneath Mount Vesuvius. Science 274, 592–594.

Volcanism in the Campania Plain:
Vesuvius, Campi Flegrei and Ignimbrites
edited by B. De Vivo

Chapter 10

Input of deep-seated volatile-rich magmas and dynamics of violent strombolian eruptions at Vesuvius

A. Cecchetti[a], *P. Marianelli*[a,*], *N. Metrich*[b] *and A. Sbrana*[a]

[a]*Dipartimento di Scienze della Terra, Università di Pisa, Italy*
[b]*Laboratoire Pierre Süe, CNRS-CEA, Saclay, France*

Abstract

Melt inclusion data indicate that Vesuvius feeding system active after 1631 eruption consists of a shallow reservoir (P<100 MPa) and a vertically extended volume of crust (probably carbonate rocks) containing interconnected cracks filled by magma, at pressures >200 MPa. This work demonstrates that input of volatile-rich magma blobs causes the recent violent strombolian and subplinian eruptions at Vesuvius. Volatile-rich mafic magmas and associated exsolved gas bubbles rising from this deep storage system could trigger composite effusive-explosive eruptions and govern the transition from lava effusion to lava fountain phases. The results of this work highlight the role of magmatic volatiles and of the deep system in the explosive dynamics of the eruptions during this period of activity.

1. Introduction

After the subplinian 1631 eruption, Somma-Vesuvius volcano has experienced a three-century long period of open conduit semi-persistent activity (Santacroce, 1987). Volcanic activity varied from predominant lava effusions to strombolian and violent strombolian eruptions, with several complex eruptions having a mixed effusive-explosive character (Santacroce, 1987; Arrighi et al., 2001) and reaching anomalously high VEI (Volcanic explosivity index, Newall and Self, 1982). Recent works emphasize that the feeding system working in this period is a complex multistage crustal system (Marianelli et al., 1999; Lima et al., 2003; De Vivo et al., 2004; Marianelli et al., 2005). Generally, the working mode of the shallower parts of this system is relatively well known (Barberi et al., 1981; Belkin et al., 1993; Santacroce et al., 1993; Cioni et al., 1995; Marianelli et al., 1999; Cioni, 2000; Lima et al., 2003; Fulignati et al., 2004 and references therein). The objective of this paper is to shed light on the role of the deep portions of the feeding system, through melt inclusion (MI) studies, in the higher VEI explosive eruptions that specifically occurred in recent Vesuvius activity. MI are a very useful tool for clarifying the preeruptive P–T–X conditions of magmas and the processes affecting them, in order to understand the eruption styles with particular attention to degassing processes (Roggensack et al., 1997; Luhr, 2001; Metrich et al., 2001; Roggensack, 2001; Cervantes and Wallace, 2003; Webster et al., 2003).

*Corresponding author. *E-mail address*: marianelli@dst.unipi.it (P. Marianelli).

2. Analytical techniques

SEM-EDS microanalysis was carried out on minerals and glassy matrix using a Philips XL30 apparatus equipped with EDAX-DX4 (Dipartimento di Scienze della Terra, Pisa). Operating conditions were 20 kV voltage and ~0.1 nA beam current. A raster area of about 100 μm^2 was employed for glass analysis to reduce the light element loss. The accuracy of measurements was checked using certified minerals and glasses as reference standards (Marianelli and Sbrana, 1998).

Analyses of MI and host olivine compositions were carried out with a Cameca SX50 electron microprobe (Camparis, Paris). The major elements in MI were analyzed at 15 kV, with a counting time of 15–20 s, a beam current of 10 nA and a beam size of 10 μm; for minerals a beam current of 40 nA was used. S, Cl and P were determined at 15 kV with a beam current of 30 nA, a beam diameter of 15 μm and a counting time of 120 s. F was determined at 10 kV with a beam current of 60 nA, a beam diameter of 15 μm and a counting time of 125 s. Bulk rock compositions from selected eruptions were analyzed by XRF (spectrometer ARL 9400 XP, Dipartimento di Scienze della Terra, Pisa) using Claisse method (1957). Major elements concentrations were re-calculated using Lucas-Tooth and Price (1961) algorithm.

The dissolved H_2O and CO_2 contents of MI glasses were determined by infrared spectroscopy using a Nicolet Magna 560 spectrometer interfaced with a NicPlan microscope (Dipartimento di Scienze della Terra, Pisa). The quantitative procedures and band assignments described in Dixon et al. (1995) and Cervantes and Wallace (2003) were followed for this work. Quantitative measurements of dissolved total H_2O, CO_2, molecular H_2O and OH, were determined using the Beer–Lambert's law; $c=100\times AM/[\varepsilon\rho l]$, where A is the absorbance, M the molar mass (g·mol^{-1}), ε the molar absorptivity (L·mol^{-1}·cm^{-1}), ρ the glass density (g·cm^{-3}), and l the sample thickness (cm). The doubly polished sample thickness was measured using a petrographic microscope with a calibrated ocular. The precision of the thickness measurements varies from 2–3 μm depending on the proximity of a given inclusion to the edge of the wafer. Density was determined with a pycnometer at 2.64 ± 0.09 g cm^{-3} and 2.54 ± 0.11 g cm^{-3} on degassed K-tephritic and K-phonotephritic glass fragments, respectively. The water concentrations were calculated using the 3535 cm^{-1} absorption band for all the samples and the absorption coefficients for basalts (ε=67 L·mol^{-1}·cm^{-1}; Stolper, 1982). The baseline for the 3535 cm^{-1} band was assumed to be linear between 3800 and 2500 cm^{-1}. Using the major element compositions of MI as measured by electron microprobe, we calculate absorption coefficients (Dixon and Pan, 1995) for 1630 (ε=17 L·mol^{-1}·cm^{-1}, 1 σ=4), 4500 (ε=0.49 L·mol^{-1}·cm^{-1}, 1σ=0.02) and 5200 cm^{-1} bands (ε=0.64 L·mol^{-1}·cm^{-1}, 1σ=0.03). For CO_3^{2-}, the background obtained on a degassed K-tephritic glass was systematically subtracted and the absorbance measured at 1510 cm^{-1} after deconvolution taking into account the contribution of the H_2O molecular peak at 1630 cm^{-1}, and fitted as Gaussian peak. The molar absorptivity for carbonate was calculated at 379 L·mol^{-1}·cm^{-1} (1σ=8) for the tephritic compositions, using the ε equation $\varepsilon^{1525}=451-342[\mathrm{Na}/(\mathrm{Ca}+\mathrm{Na})]$, according to Dixon and Pan (1995). The precision of molecular water, hydroxyl group and carbonate group analyses based on multiple analyses of the same spot is estimated to be ~5%. Accuracies are limited by the uncertainties in the molar absorption coefficients and in the background correction procedures, and are estimated to be 10% for total water and carbonate, according to Dixon et al. (1995).

3. Results

3.1. Studied eruptions

The 1723, 1794, 1822, 1872, 1906 and 1944 eruptions were the most intense (VEI 2-3) over the 1637–1944 period of activity. These eruptions, characterized by mixed effusive-explosive dynamics always start with lava effusions followed by abrupt transitions to explosive phases: lava fountains, steady columns and occasionally phreatomagmatic activity. The explosive phases have a hawaiian/strombolian up to subplinian style (Arrighi et al., 2001) and are characterized by the deposition of widely dispersed lapilli and ash fallout layers.

The description of the main phases of the studied eruptions is hereafter summarized according to Santacroce (1987), Santacroce et al. (1993), Marianelli et al. (1999), and Arrighi et al. (2001).

- 1723: Small summit lava effusions followed by lava fountains (15 continuous and 3 pulsating, 106 h in total).
- 1794: Strong lava effusions from vents aligned along an ENE-WSW fracture located on the west side of the volcano. These are followed by lava fountains and by strong ash emission (13 days).
- 1822: Vigorous lava effusions from fractures located on higher slopes of the cone. These are followed by lava fountains (5 h), steady column phase (3 h) and strong ash emission (19 days).
- 1872: Lava flow from the slopes of the cone and successive lava fountains (three days for the main phase).
- 1906: Huge lava flows from the NNW-SSE fractures at the base of the cone followed by lava fountains (11 h), steady column phase (5 h) and strong ash emission (12 days).
- 1944: Lava flows from the summit crater followed by lava fountains (13 h in total), steady column phase and ash emission (seven days).

3.2. Bulk rock and groundmass compositions

The studied products consist of dark scoriae and coarse ashes collected from the widely dispersed lava fountain fallout blankets that cover the slopes and the plains surroundings Vesuvius. They are porphyritic with variable content in phenocrysts and ranges in composition from K-tephrites to K-phonotephrites (Table 1). Scoriae show variability in vesicularity and crystallinity of groundmass. The latter is glassy in vesicular scoria and crystalline with clinopyroxene (mainly Fs19-11, and rare Fs8-6), leucite, plagioclase (dominantly An76-44 and rare An83) and olivine (Fo78-68) microlites in less vesicular scoriae. The groundmass shows K-phonotephritic composition.

Systematic sampling of the explosive products emplaced during the most energetic lava fountains of these eruptions reveals the presence of Mg-rich crystals (olivine and diopside). Some of the olivines are still coated with K-tephritic glass, in which quench crystals of clinopyroxene Fs(9–10) drive the matrix glass compositions further towards K-phonotephrite (Fig. 1a, Table 1).

Table 1. Representative analyses of scoria bulk rocks, groundmasses and olivine coating glasses.

Eruption	Bulk rocks						Groundmasses			Olivine coating glasses		
	1794		1822		1872	1944						
SiO_2	49.37	49.04	48.99	48.97	48.70	47.90	47.94	48.12	46.64	48.15	48.15	47.20
TiO_2	0.90	0.94	0.91	0.93	1.03	0.94	1.18	1.29	1.5	1.06	1.06	1.25
Al_2O_3	12.46	13.37	12.89	15.32	16.30	14.29	18.73	19.55	17.38	16.63	16.63	14.54
FeO_{tot}	6.98	7.15	7.13	7.37	7.39	7.51	8.88	7.95	10.59	8.23	8.23	8.62
MnO	0.12	0.12	0.12	0.13	0.13	0.13	0.1	0.15	0.34	0.16	0.16	0.22
MgO	8.92	8.21	8.75	6.44	5.84	7.71	3.7	3.19	3.42	3.06	3.06	4.19
CaO	14.77	13.88	14.24	12.13	11.11	14.33	8.8	8.16	9.16	8.66	8.66	11.95
Na_2O	1.53	1.68	1.62	2.01	2.24	1.62	3.81	4.09	4.43	2.82	2.82	2.63
K_2O	4.33	4.96	4.72	6.01	6.56	4.90	5.65	6.24	5.06	7.54	7.54	5.42
P_2O_5	0.63	0.64	0.64	0.68	0.71	0.66	0.73	0.7	0.74	0.97	0.97	0.86
S	n.d.	n.d.	n.d.	n.d.	n.d.	n.d.	l.d	0.05	l.d.	0.09	0.09	0.17
Cl	n.d.	n.d.	n.d.	n.d.	n.d.	n.d.	0.48	0.52	0.73	0.54	0.54	0.53
Total	100.0	100.0	100.0	100.0	100.0	100.0	100.0	100.0	100.0	97.9	97.3	97.6
LOI	0.45	0.80	0.63	0.83	0.28	0.25						
CaO/Al_2O_3	1.19	1.04	1.10	0.79	0.68	1.00	0.47	0.42	0.53	0.52	0.52	0.82

Note: Bulk rocks major elements analyzed by XRF and re-calculated water-free to 100. Groundmass and glass analyses by EDS and WDS, respectively. n.d., not determined; ld, below detection limit.

3.3. *Mineral chemistry*

The studied samples show a primitive mineral assemblage (Mg-rich olivine, diopside and minor Cr-spinel) coexisting with an evolved paragenesis (leucite, salite, Fe-rich olivine, minor plagioclase and biotite). Olivine is present as both phenocrysts (Fo91-85) poorly but normally zoned with 0.2–1 Fo unit variation from the core to the rim, although some of them have thin Fe-rich rims (Fo78-73), and rare Fe-rich microphenocrysts (Fo78-56). Spinel (Cr#75-78) is observed as inclusion in olivine Fo90 only. Clinopyroxene texture is complex. Phenocrysts occur as unzoned diopside (Fs4-9), unzoned salite (Fs12-14) and salite showing oscillatory zoning (Fs11-15), in some cases having diopsidic cores (Fs4-9) surrounded by thick salitic rims (Fs11-15), whereas microlithes are mainly salite (up to Fs19). Leucite, having nearly stoichiometric composition , is common both as phenocrysts and as microlithes. Plagioclase is always present as microlithes and sporadically as phenocrysts or microphenocrysts (An91-72). Biotite is abundant only in 1822 deposits and rare in other eruptions (Table 2).

3.4. *Melt inclusions*

MI were analyzed in olivine, salite and leucite of the different eruptions. In this work only naturally glassy (unheated) MI were analyzed with the aim to avoid possible modifications in their initial dissolved volatile content.

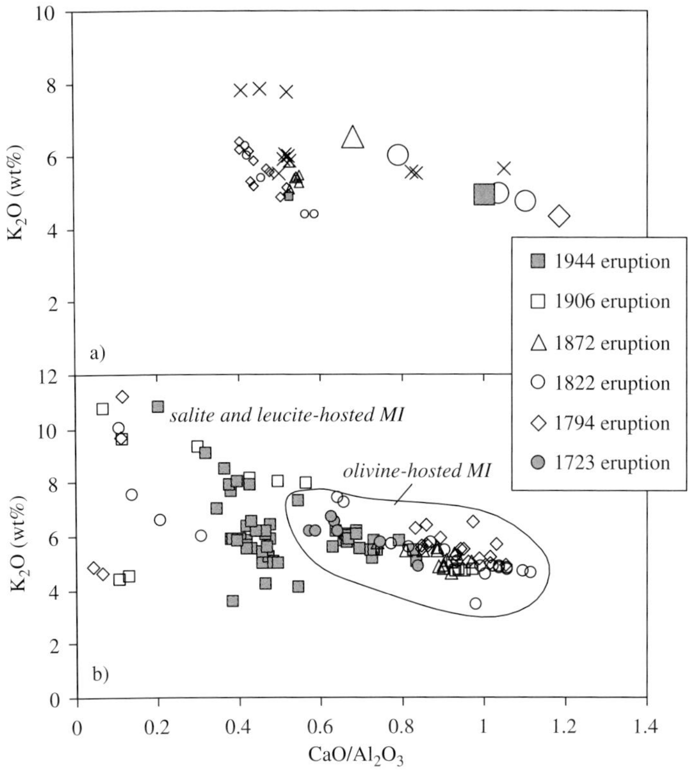

Figure 1. K_2O vs. CaO/Al_2O_3 diagrams. (a) Bulk rock compositions of studied scoriae (oversized symbols) and residual groundmasses (undersized symbols). Cross: glass coating olivine crystals. (b) Olivine-, pyroxene- and leucite-hosted MI. Data of olivine-hosted MI are from this work and Marianelli et al. (2005); data of leucite- and salite-hosted MI for 1794 and 1822 eruptions are from Vaggelli et al. (1993). *Note*: CaO/Al_2O_3 is here used as "differentiation index", as in olivine these elements behave as incompatible; therefore, this ratio in MI is independent of post-trapping evolution. Primitive melts evolve towards lower CaO and higher Al_2O_3 contents (olivine and Ca-rich diopsidic clinopyroxene crystallization results in CaO/Al_2O_3 ratio decreasing). In more evolved melts CaO/Al_2O_3 tend to increase due to crystallization of salite, leucite and plagioclase.

The olivine crystals display one or several MI, whose size varies from 50 to 200 μm, which are composed of brown glass and a bubble (Fig. 2A). As discussed by Danyushevsky et al. (2000, 2002), olivine-hosted MI can be affected by Fe-loss by post-trapping re-equilibration as a function of the cooling history. In this work after a careful inspection of MI and of compositional data, only compositions of MI not having suffered an extensive post-trapping diffusive re-equilibration have been discussed (Fig. 3). The analyses of olivine-hosted MI were corrected for the effects of post-trapping crystallization of olivine, on the walls of MI, by simulating reverse olivine fractionation using a software package of numerical modelling "Petrolog" (Danyushevsky et al., 2000; Danyushevsky, 2001). Olivine-melt equilibrium was calculated using the model of Ford et al. (1983); the Fe^{2+}/Fe^{3+} ratio in the melt is calculated following Borisov and Shapkin (1989) and assuming FO_2 close to the NNO buffer (Metrich and Clocchiatti, 1996). After the correction for post-trapping crystallization (<12%), olivine-hosted MI shows predominantly

Table 2. Representative chemical compositions of mineral phases.

	Clinopyroxene				*Olivine*			*Plagioclase*		*Biotite*	*Leucite*	*Spinel*
	Core	*Rim*	*Diopside*	*Salite*	*Core*	*Rim*						
SiO_2	50.70	45.77	50.68	46.30	40.59	40.84	36.40	47.39	44.56	39.49	55.06	0.09
TiO_2	0.54	1.64	0.71	1.43						0.27	0.09	0.89
Al_2O_3	3.65	8.77	3.54	7.97	0.44	0.15	0.17	32.93	35.52	16.83	23.34	10.49
FeO_{tot}	5.02	8.00	5.04	7.97	9.08	9.36	30.98	1.07	0.50	8.45	0.61	21.49
MnO	0.18	ld	0.11	0.12	0.16	0.19	0.76					0.80
MgO	15.20	11.84	15.17	12.24	49.14	49.06	30.96			20.97		12.82
CaO	24.31	23.76	24.42	23.65	0.41	0.30	0.73	16.13	18.45			
Na_2O	0.26	0.22	0.24	0.27				2.08	0.86	0.28	1.19	
K_2O								0.40	0.11	10.42	19.70	
NiO					0.18	0.10	ld					0.09
Cr_2O_3	0.12	ld	0.10	0.05								51.49
An								79.18	91.62			
Ab								18.48	7.73			
Or								2.34	0.65			
Fo					90.61	90.33	64.05					
Wo	49.23	51.12	49.37	50.42								
En	42.83	35.45	42.68	36.31								
Fs	7.93	13.43	7.95	13.26								
Cr/(Cr+Al)												0.76
K/(K+Na)											0.92	
Mg/(Mg+Fe)	0.84	0.73	0.84	0.73						0.74		

Note: Analyses of clinopyroxene are recalculated to six oxygens. Analyses of olivine are recalculated to four oxygens. Analyses of plagioclase are recalculated to eight oxygens. An (anortite), Ab (albite), Or (ortose), Fo (forsterite), Wo (wollastonite), En (enstatite), Fs (ferrosilite).
ld: below detection limit.

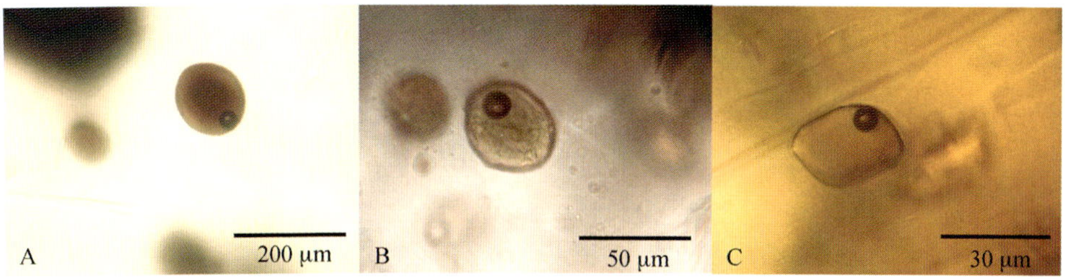

Figure 2. Transmitted light microphotographs of MI trapped in: (A) olivine, (B) leucite (C) salitic clinopyroxene.

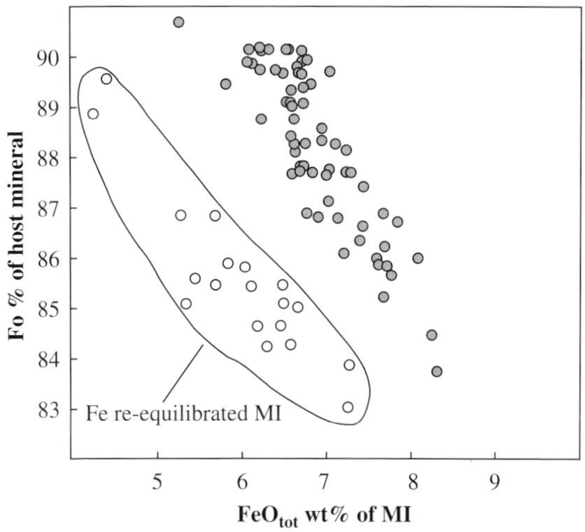

Figure 3. FeO_{tot} content of analyzed (not recalculated) MI plotted against the Fo content of host olivine. Grey symbols: well quenched, not affected by re-equilibration, MI from studied eruptions. Open symbols: MI affected by FeO-loss caused by a different (slow) cooling history of the hosting sample. For this reason these latter MI are not discussed further in this work.

K-tephritic compositions, however more evolved K-phonotephritic compositions are also present (Table 3). Inclusions in olivine (Fo_{90-87}) are representative of the most primary melts (CaO/Al_2O_3=1.1–0.84; K_2O=4.6–5.1 wt%, Fig. 1b and Fig. 4) of Vesuvius (Marianelli et al., 2005). The more evolved compositions are registered in 1723 and 1944 MI (Fig. 4). As a whole, these MI, although having a primitive composition (Mg#>60), display a very high dissolved volatile content (H_2O+CO_2+Cl+S+F ~5 wt%) as already reported by previous studies (Marianelli et al., 1995, 1999; Raia et al., 2000; Webster et al., 2001, 2003; Lima et al., 2003; Marianelli et al., 2005). The concentrations of H_2O (1.8–4.9 wt%) and CO_2 (1500–3500 ppm) are systematically high and variable (Fig. 5, Table 4) and are described in detail in Marianelli et al. (2005).

Salite and leucite phenocrysts contain primary rounded two-phase (glass+shrinkage bubble) MI (Fig. 2B,C). Since only natural unheated MI were analyzed, the influence of post-entrapment of the host phase within the inclusions on their chemistry has been evaluated. Consequences of this potential problem can be constrained through simple computations, following the procedure described by Webster et al. (2001). For example, we

Table 3. Analyses of MI in mineral phases from the studied eruptions.

Eruption	1723		1794		1822		1872		1906		1944		1906		1944	
SiO_2	46.86	48.03	45.62	48.10	46.90	49.41	46.01	45.78	47.84	48.37	47.82	46.77	51.27	49.98	47.34	50.07
TiO_2	1.64	1.6	1.12	1.01	1.37	1.00	1.08	1.13	1.21	0.96	1.22	1.22	1.26	1.07	1.20	1.03
Al_2O_3	17.53	16.73	13.51	12.52	15.27	12.48	14.28	13.21	13.00	12.63	14.92	14.70	17.77	17.37	18.66	18.78
FeO_{tot}	7.99	7.26	6.49	6.52	6.94	6.07	7.01	6.60	6.77	6.65	7.31	8.08	5.64	5.94	10.88	8.35
MnO	0.4	0.29	0.07	0.14	0.11	0.10	0.15	0.09	0.07	0.15	0.13	0.18	0.27	0.17	0.24	0.15
MgO	4.33	3.69	3.81	4.53	3.40	3.80	4.85	5.41	5.59	5.76	3.61	4.89	4.21	4.57	2.83	2.80
CaO	11.15	12.48	12.05	12.73	11.78	12.95	12.16	11.68	12.39	12.30	11.00	10.23	8.82	9.84	7.40	6.48
Na_2O	2.16	2.34	1.66	1.56	1.97	1.57	1.66	1.59	1.74	1.59	1.92	2.13	2.73	3.08	3.53	4.50
K_2O	6.78	6.3	6.04	5.28	5.86	5.03	5.39	5.30	4.76	4.73	5.83	5.48	8.03	7.97	5.81	7.02
P_2O_5	0.42	0.5	0.79	1.06	0.98	0.80	1.10	1.09	0.88	0.71	1.14	1.04	n.d.	n.d.	2.10	0.83
S	0.2	0.19	0.15	0.19	0.18	0.17	0.19	0.14	0.18	0.17	0.22	0.21	n.d.	n.d.	n.d.	n.d.
Cl	0.56	0.59	0.53	0.52	0.51	0.52	0.51	0.47	0.49	0.49	0.60	0.51	0.62	0.42	0.47	0.41
F	n.d.	n.d.	0.22	0.24	0.22	0.21	n.d.	n.d.	n.d.	n.d.	n.d.	n.d.	n.d.	n.d.	n.d.	n.d.
H_2O (wt%)	n.d.	n.d.	4.04	5.30	4.95	4.34	3.50	4.69	2.70	3.00	2.07	3.24	n.d.	n.d.	1.15	n.d.
CO_2 (ppm)	n.d.	n.d.	2965	2777	2200	2328	2505	2005	2063	1943	2042	2274	n.d.	n.d.	440	n.d.
Re-calculated composition																
SiO_2	46.52	47.26	44.63	47.18	46.05	48.04	45.52	45.29	47.09	47.64	46.90	46.23				
TiO_2	1.58	1.47	1.00	0.93	1.26	0.89	1.02	1.07	1.12	0.90	1.12	1.16				
Al_2O_3	16.91	15.32	12.10	11.49	13.97	11.08	13.52	12.56	12.06	11.81	13.67	13.94				
FeO_{tot}	8.23	7.83	7.03	6.95	7.56	6.72	7.28	6.79	7.05	6.90	7.94	8.38				
MnO	0.39	0.27	0.06	0.13	0.10	0.09	0.14	0.08	0.07	0.14	0.12	0.17				
MgO	5.66	7.10	7.77	7.68	6.63	8.19	6.83	7.23	8.37	8.22	6.74	6.76				

CaO	10.76	11.43	10.80	11.68	10.78	11.50	11.51	11.11	11.50	11.50	10.09	9.70				
Na_2O	2.08	2.14	1.48	1.43	1.81	1.39	1.57	1.52	1.61	1.49	1.76	2.02				
K_2O	6.54	5.77	5.41	4.85	5.36	4.47	5.11	5.03	4.42	4.43	5.34	5.20				
P_2O_5	0.41	0.46	0.71	0.97	0.90	0.72	1.04	1.04	0.82	0.67	1.04	0.98				
S	0.19	0.17	0.13	0.18	0.17	0.15	0.18	0.13	0.17	0.16	0.20	0.20				
Cl	0.54	0.54	0.48	0.48	0.46	0.47	0.48	0.44	0.46	0.45	0.55	0.48				
F	n.d.	n.d.	0.20	0.22	0.20	0.18	n.d.	n.d.	n.d.	n.d.	n. d.	n. d.				
H_2O (wt%)	n.d.	n.d.	3.62	4.88	4.54	3.86	3.32	4.47	2.51	2.81	1.90	3.07				
CO_2 (ppm)	n.d.	n.d.	2663	2555	2013	2072	2375	1909	1919	1821	1877	2161				
X_{Host}[b]	3	8	10	8	3	11	5	5	7	6	8	5				
CaO/Al_2O_3	0.64	0.75	0.89	1.02	0.77	1.04	0.85	0.88	0.95	0.97	0.74	0.70	0.50	0.57	0.40	0.35
Host mineral	Olivine	Olivine	Olivine	Olivine	Olivine	Olivine	Olivine	Olivine	Olivine	Olivine	Olivine	Olivine	Salite	Salite	Salite	Leucite
Fo/Fs[a]	85	88	90	89	87	90	88	89	90	90	86	86	10	9	13	
Saturation pressure (MPa)	n.d.	n.d.	348	398	373	325	308	360	225	236	201	276			60	

Note: Compositions of olivine-hosted MI were corrected for the effects of host mineral post-trapping crystallization (see text).
[a]Forsterite or Forrosilite content; [b]Fraction of host olivine which crystallized in MI; n.d., not determined.

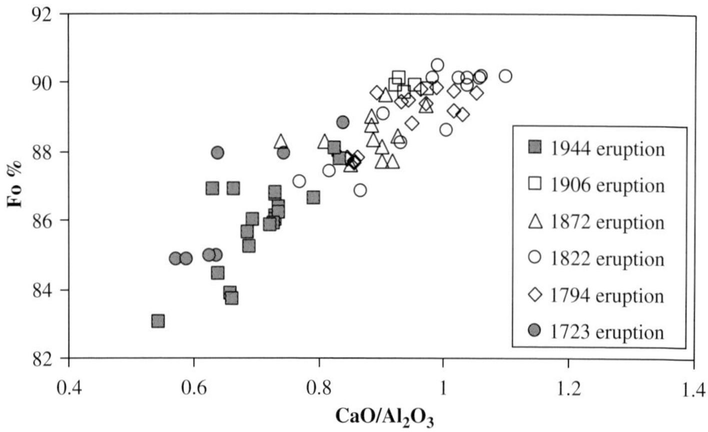

Figure 4. Composition of MI (CaO/Al_2O_3) vs. composition of host mineral (Fo content) for olivine-hosted MI. Data from this work and Marianelli et al. (2005).
Note: See note of Figure 1 caption.

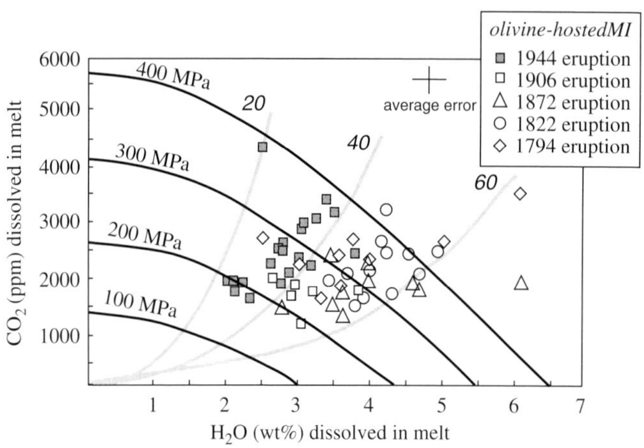

Figure 5. Dissolved H_2O and CO_2 contents in MI. Isobars and isopleths of constant vapour composition (20, 40, 60 mol% H_2O) are calculated using VolatileCalc (Newman and Lowernstern, 2002), for magmas containing 45 wt% SiO_2 (derived from the Π coefficient defined by Dixon, 1997). The curves of equilibrium between melt and constant gas composition represent the possible CO_2/H_2O ratio of the external gas that is required to generate the data arrays for the different samples (modified from Marianelli et al., 2005).

calculated that the range in CaO contents (~4–9 wt%) in salite-hosted MI would require 25 wt% post-trapping crystallization of host phase (containing ~24 wt% CaO), and should also affect the other melt components (25% relative). In contrast, the compositional trends of MI show covariations that differ from the relative 25% variability that could be a result of clinopyroxene crystallization. As a result, the variability of compositions of leucite and salite MI is not caused by the crystallization of the host phases on the inclusion walls. MI shows K-phonotephritic to K-tephriphonolitic (CaO/Al_2O_3 <0.55) compositions (Fig. 1b, Table 3), whose variability can be ascribed to the magma differentiation due to crystallization

Table 4. Representative of H_2O and CO_2 FTIR measurements for olivine-hosted MI.

Sample	Thickness (μm)	Abs H_2O (3535 cm^{-1})	H_2O_{tot} (wt%)	Abs CO_2 (1500–1400 cm^{-1})	$CO_{2\,tot}$ (ppm)
Eruption 1794					
vs98a 56 6	32	1.29	4.14	0.14	1877
vs98a 56 14	28	0.92	3.34	0.11	1772
vs98a 56 5	36	1.57	4.42	0.25	3040
vs98a 56 13	46	1.47	3.25	0.28	2634
vs98a 56 16	10	0.40	4.12	0.04	1741
vs98a 56 18	29	1.38	4.84	0.16	2431
Eruption 1822					
vs98a 52 8	43	1.32	3.17	0.22	2189
vs98a 52 2	43	1.72	4.06	0.32	3150
vs98a 52 3	18	0.88	4.92	0.10	2248
vs98a 52 9	24	0.90	3.89	0.12	2205
vs98a 52 1	53	1.74	3.33	0.21	1683
vs98a 52 5	37	1.52	4.22	0.16	1830
Eruption 1872					
vs0030 5	29	0.99	3.50	0.17	2505
vs0030 3	39	1.47	3.84	0.19	2092
vs0030 7	34	1.22	3.71	0.15	1939
vs0030 2	17	0.68	4.19	0.09	2253
vs0030 6	18	0.51	2.88	0.06	1507
vs0030 8	21	0.95	4.57	0.08	1739
Eruption 1944					
vs97109 3	25	0.29	1.14	0.06	1125
vs97109 5	35	0.75	2.19	0.18	2329
vs97109 6	26	0.80	3.14	0.12	2151
vs97109 7	50	1.06	2.15	0.12	1120
vs97109 8	32	0.91	2.87	0.19	2750
Eruption 1906					
vs1906 2	40	1.55	3.94	0.22	2523

mainly of clinopyroxene and leucite. The total volatile content (H_2O, CO_2, Cl, S) of such MI is low (<1.5 wt%) compared with that of olivine-hosted MI. This is in agreement with crystallization of leucite and salite from partially degassed melts, as hypothesized for melts residing in the 1944 shallow reservoir (Marianelli et al., 1999).

4. Discussion and conclusions

For each of the higher VEI (2–3) eruptions that occurred over the 1637–1944 period, the phenocrysts present in pyroclastic deposits can be ascribed to two paragenesis in equilibrium with different magmas: (a) leucite+salite+Fe-rich olivine±plagioclase± biotite and (b) Mg-rich olivine, diopside and Cr-spinel included in Mg-rich olivine. These two different mineral assemblages can be considered in equilibrium with K-tephritic and K-phonotephritic magmas, respectively, on the basis of the results of previous studies on the 1944 eruption (Marianelli et al., 1999). The coexistence of two parageneses in the same sample indicates that bulk rocks result from syneruptive mixing processes (Marianelli et al., 1995, 1999; Lima et al., 2003). In particular, MI studies on 1944 products have demonstrated that the two mineral assemblages are related to crystallization of magmas respectively in shallow reservoirs and in deep systems (Marianelli et al., 1999).

Assuming saturation conditions for trapped melts, the amount of volatiles (H_2O and CO_2) dissolved in MI provides an estimate of the minimum pressures of melt entrapment using solubility models (Dixon, 1997; Newman and Lowernstern, 2002). The calculated minimum pressure of saturation for olivine-hosted MI ranges between 200 and 400 MPa (Fig. 6), as reported by Marianelli et al. (2005). These are significantly higher than those estimated from K-tephriphonolitic volatile-poor MI from salite and leucite (10–60 MPa). The estimated saturation pressures would correspond to depths ≥8 km for MI in olivine and to depths <2 km for salite and leucite MI, assuming an average (volcanic, sedimentary silico-clastic and carbonate) rock density of 2600 kg m^{-3}.

These MI data testify, also for the 1723, 1794, 1822, 1872 and 1906 eruptions, the existence of a complex feeding system consisting of deep storage zones and a shallow reservoir, as already evidenced for 1944 eruption (Marianelli et al., 1999). The absence of relationship between the extent of melt differentiation (CaO/Al_2O_3 in olivine-hosted MI) and the trapping pressures, corroborate the hypothesis of melts arranged in a mush column. In our view, the variability of the MI composition and of their saturation pressures (Fig. 6) implies that the Vesuvius deep feeding system might have been formed by a vertically extended volume of crust-containing pockets or interconnected cracks filled by magma at depth >8 km (Marianelli et al., 2005). Olivines, formed at different depths and from different melts, were thus entrained and flushed upward by rising magmas. Such a process was also invoked by Danyushevsky et al. (2002) to explain the entrainment of

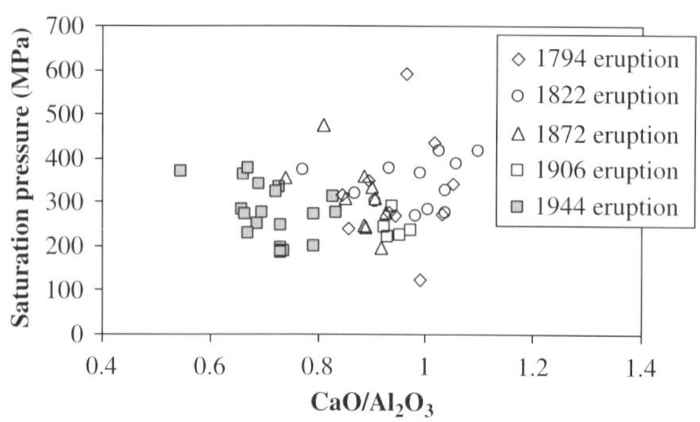

Figure 6. Composition of MI (CaO/Al_2O_3) vs. trapping (crystallization) pressure. Pressure estimates are from Marianelli et al. (2005).

olivine in magmas. A system consisting of a complex feeding column dominated by multiple mush zone environments (small magma chambers) with a large chamber at 12 km of depth beneath Vesuvius was suggested by Lima et al. (2003) and De Vivo et al. (2004) on the basis of fluid inclusions data for the post 79 AD eruptions.

The existence of a deep feeding system at Vesuvius, identified from MI study, is in agreement with seismic tomography data that highlighted the existence of a large (at least 400 km^2) low velocity layer (interpreted as an extended sill) at depth of about 8 km (Auger et al., 2001) and a deep magma root between 15 and 35 km as hypothesized from teleseismic data (De Natale et al., 2001). Interpretation of seismic data indicates that, at these depths beneath the Campanian area, tectonic structures separate different units piled up in the southern Appennine trust system (Mazzotti et al., 2000). The formation of deep magma reservoirs could be related to these tectonic discontinuities.

In addition, MI shows a large range of dissolved volatiles (H_2O=2.0–4.9 wt%; CO_2=1500–3500 ppm, Fig. 5) with an overall decrease in the H_2O/CO_2 ratio as the degree of differentiation increased particularly in the 1944 samples. The lowering of H_2O/CO_2 ratio in magmas might have enhanced the crystallization of the K-tephrite magma causing magma evolution towards K-phonotephritic compositions. However, the variability of the H_2O/CO_2 ratio is without any other evidence of change in the trace elements, Cl/F, S/Cl and Na_2O/K_2O ratios (Marianelli et al., 2005). These observations suggest that there was magma equilibration with a volatile phase having variable CO_2/H_2O ratio (Fig. 5). Carbonate-derived CO_2 through magma–wall rock interaction processes could contribute to modify the CO_2/H_2O ratio of the gas phase in equilibrium with the stored magmas. The dissolved H_2O lowering and CO_2 increasing, as observed in MI (Fig. 5), would be due to predominantly magmas rising from portions of the mush column in which magma–carbonates interaction was more effective. This phenomenon is clearly amplified in melts (i.e. MI) of the 1944 eruption (Fig. 5), although it also exists for the others. The interpretation of seismic data (Mazzotti et al., 2000) collected in Cilento area and extrapolated to Vesuvian area could support the hypothesis of magma ponding in carbonate series, since they indicate that crust beneath the volcano could be formed by a pile of carbonate units up to 20 km of depth (Southern Appennines duplex system).

The improved knowledge of the feeding system with the demonstration of the involvement of deep-seated, volatile-rich mafic magmas in the high VEI eruptions has strong implications on the observed eruption styles. In previous studies, the identification of phreatomagmatic phases in Vesuvius post-1631 activity has played an important role in the interpretation of dynamics of most explosive eruptions of this period of open conduit activity of Vesuvius (Arrighi et al. 2001 and references therein). In particular, the explosive phases of the eruptions of post-1631 period have been attributed to phreatomagmatic activity related to shallow aquifer(s) beneath the volcano (Bertagnini et al., 1991; Scandone et al., 1993; Raia et al., 2000). The deposits related to the transition from effusive to explosive phases in high-VEI eruptions record the occurrence of prolonged lava fountain phases with sometimes the establishment of sustained column. During such activity, the pressure in the conduit is higher than the hydrostatic pressure of the confining aquifer that prevents magma–water interaction (Barberi et al., 1988). Therefore, the explosiveness of these phases cannot be ascribed to phreatomagmatic processes.

Conversely, the present results highlight the role of magmatic volatiles and of the deep system in the explosive dynamics of the eruptions during this period of activity. In particular, the systematic occurrence of high-pressure juvenile crystals, grown from volatile-rich mafic magmas, in deposits related to the more intense phases of lava fountains is testified by the MI. It allows us to propose that rising CO_2–H_2O-rich magma blobs and related exsolved gas bubbles control the dynamics of the high VEI (2–3) eruptions and provoke the transition from passive effusions of degassed magmas residing in open conduit shallow reservoirs, towards intense lava fountains and sustained columns. Phreatomagmatic activity commonly develops in declining final phases of these eruptions, when pressure in conduits is dropping. The results of this work are interesting for the hazard mitigation at Vesuvius because of the role that the deep system has on the control of this type of explosive activity. VEI 2–3 eruptions are very common at Vesuvius in medieval and the recent period of activity and their impact on the Campanian area in Vesuvius surroundings has been very significant.

Acknowledgements

Constructive reviews by L. Danyushevsky and B. De Vivo have significantly improved the original manuscript. This research was partially supported by grants from the Gruppo Nazionale per la Vulcanologia-Istituto Nazionale di Geofisica e Vulcanologia (Italy) and Ministero Istruzione, Università e Ricerca (Italy).

References

Arrighi, S., Principe, C., Rosi M., 2001. Violent strombolian and subplinian eruptions at Vesuvius during post-1631 activity. Bull. Volcanol. 63, 126–150.

Auger, E., Gasparini, P., Virieux, J., Zollo, A., 2001. Seismic evidence of an extended magmatic sill under Mt. Vesuvius. Science 294, 1510–1512.

Barberi, F., Bizouard, H., Clocchiatti, R., Metrich, N., Santacroce, R., Sbrana, A., 1981. The Somma-Vesuvius magma chamber: a petrological and volcanological approach. Bull. Volcanol. 44, 295–315.

Barberi, F., Navarro, J.M., Rosi, M., Santacroce, R., Sbrana, A., 1988. Explosive interaction of magma with groundwater: insights from xenoliths and geothermal drilling. Rend. Soc. It. Min. Petrol. 43(4), 901–926.

Belkin, H.E., Kilburn, C.R.J., De Vivo, B., 1993. Sampling and major element chemistry of the recent (AD 1631–1944) Vesuvius activity. J. Volcanol. Geotherm. Res. 58, 273–290.

Bertagnini, A., Landi, P., Santacroce, R., Sbrana, A., 1991. The 1906 eruption of Vesuvius: from magmatic to phreatomagmatic activity through the flashing of a shallow depth hydrothermal system. Bull. Volcanol. 53, 517–532.

Borisov, A.A., Shapkin, A.I., 1989. New empirical equation of dependence of the Fe^{3+}/Fe^{2+} ratio in natural melts on their composition, oxygen fugacity and temperature. Geokhimiya 6, 892–897.

Cervantes, P., Wallace, P., 2003. Magma degassing and basaltic eruption styles; a case study of approximately 2000 year BP Xitle Volcano in central Mexico. J. Volcanol. Geotherm. Res. 120(3–4), 249–270.

Cioni, R., 2000. Volatile content and degassing processes in the AD 79 magma chamber at Vesuvius (Italy). Contrib. Mineral. Petrol. 140, 40–54.

Cioni, R., Civetta, L., Marianelli, P., Metrich, N., Santacroce, R., Sbrana, A., 1995. Compositional layering and syn-eruptive mixing of a periodically refilled shallow magma chamber: the AD 79 Plinian eruption of Vesuvius. J. Petrol. 36, 739–776.

Claisse, F., 1957. Accurate x-ray fluorescence analysis without internal standard. Norelco Rep. 4, 3-17, 17, 19, 95–106.

Danyushevsky, L.V., 2001. The effect of small amounts of H_2O on crystallisation of mid-ocean ridge and backarc basin magmas. J. Volcanol. Geotherm. Res. 110, 265–280.

Danyushevsky, L.V., Della Pasqua, F.N., Sokolov, S., 2000. Re-equilibration of melt inclusions trapped by magnesian olivine phenocrysts from subduction-related magmas; petrological implications. Contrib. Mineral. Petrol. 138(1), 68–83.

Danyushevsky, L.V., Sokolov, S., Falloon, T., 2002. Melt inclusions in olivine phenocrysts: using diffusive re-equilibration to determine the cooling history of a crystal, with implications for the origin of olivinephyric volcanic rocks. J. Petrol. 43, 1651–1671.

De Natale, G., Troise, C., Pingue, F., De Gori, P., Chiarabba, C., 2001. Structure and dynamics of the Somma-Vesuvius volcanic complex. Mineral. Petrol. 73, 5–22.

De Vivo, B., Fedele, L., Lima, A.M., 2004. Fluid and melt inclusions study from ejected Mt. Somma-Vesuvius nodules: new data to constrain evolution of the magmatic system. Workshop on Vesuvius and volcanism of Campania Plain, a cura di Benedetto De Vivo, Napoli 2004.

Dixon, J.E., 1997. Degassing of alkalic basalts. Am. Mineral. 82, 368–378.

Dixon, J.E., Pan, V., 1995. Determination of the molar absorptivity of dissolved carbonate in basanitic glass. Am. Mineral. 80, 1339–1342.

Dixon, J.E., Stolper, E.M., Holloway, J.R., 1995. An experimental study of water and carbon dioxide solubilities in mid-ocean ridge basaltic liquids. Part I: calibration and solubility models. J. Petrol. 36, 1607–1631.

Ford, C.E., Russel, D.G., Craven, J.A., Fisk, M.R., 1983. Olivine-liquid equilibria: temperature, pressure and composition dependence of the crystal/liquid cation partition coefficients for Mg, Fe^{2+}, Ca, Mn. J. Petrol. 24, 256–265.

Fulignati, P., Marianelli, P., Metrich, N., Santacroce, R., Sbrana, A., 2004. Towards a reconstruction of the magmatic feeding system of the 1944 eruption of Mt Vesuvius, J. Volcanol. Geotherm. Res. 133, 13–22.

Lima, A.M., Danyushevsky, L.V., De Vivo, B., Fedele, L., 2003. A model for the evolution of the Mt. Somma-Vesuvius magmatic system based on fluid and melt inclusion investigations. In: De Vivo, B., Bodnar, R.J. (Eds), Melt Inclusions in Volcanic Systems Methods, Applications and Problems.Series: Developments in Volcanology, Vol. 5, Elsevier, Amsterdam, pp. 227–249.

Lucas-Tooth, H.J., Price, B.J., 1961. A mathematical method for the investigation of interelements effects in x-ray fluorescence analysis. Metallurgia 64, 149–152.

Luhr, J.F., 2001. Glass inclusions and melt volatile contents at Paricutin Volcano, Mexico. Contrib. Mineral. Petrol. 142(3), 261–283.

Marianelli, P., Metrich, N., Santacroce, R., Sbrana, A., 1995. Mafic magma batches at Vesuvius: a glass inclusion approach to the modalities of feeding stratovolcanoes. Contrib. Mineral. Petrol. 120, 159–169.

Marianelli, P., Metrich, N., Sbrana, A., 1999. Shallow and deep reservoirs involved in magma supply of the 1944 eruption of Vesuvius. Bull. Volcanol. 61, 48–63.

Marianelli, P., Sbrana, A., 1998. Risultati di misure standard di minerali e di vetri naturali in microanalisi a dispersione di energia. Atti Soc. Tosc. Sci. Nat. 105, 57–63.

Marianelli, P., Sbrana, A., Metrich, N., Cecchetti, A., 2005. The deep feeding system of Vesuvius involved in recent violent strombolian eruptions. Geophys. Res. Lett. No. L02306, 32, 2.

Mazzotti, A., Stucchi, E., Fradelizio, G., Zanzi, L., Scandone, P., 2000. Seismic exploration in complex terrains: a processing experience in the Southern Apennines. Geophysics 65, 1402–1417.

Metrich, N., Bertagnini, A., Landi, P., Rosi, M., 2001. Crystallization driven by decompression and water loss at stromboli Volcano (Aeolian Islands, Italy). J. Petrol. 42(8), 1471–1490c.

Metrich, N., Clocchiatti, R., 1996. Sulfur abundance and its speciation in oxidized alkaline melts. Geochim. Cosmochim. Acta 60, 4151–4160.

Newhall, C.G., Self, S., 1982. The volcanic explosivity index (VEI): an estimate of explosive magnitude for historical volcanism. J. Geophys. Res. 87, 1231–1238.

Newman, S., Lowernstern, J.B., 2002. VolatileCalc: a silicate melt-H_2O-CO_2 solution model written in Visual Basic for excel. Comp. Geosci. 28, 597–604.

Raia, F., Webster, J.D., De Vivo, B., 2000. Pre-eruptive volatile contents of Vesuvius magmas: constraints on eruptive history and behavior. I. The medieval and modern interplinian activities. Eur. J. Mineral. 12(1), 179–193.

Roggensack, K., 2001. Unraveling the 1974 eruption of Fuego Volcano (Guatemala) with small crystals and their young melt inclusions. Geol. Boulder 29(10), 911–914.

Roggensack, K., Hervig, R.L., McKnight, S.B., Williams, S.N., 1997. Explosive basaltic volcanism from Cerro Negro Volcano; influence of volatiles on eruptive style. Science 277(5332), 1639–1642.

Santacroce, R., 1987. Somma-Vesuvius. Quad. Ric. Sci., Cons. Naz. delle Ric., 114, 230.

Santacroce, R., Bertagnini, A., Civetta, L., Landi, P., Sbrana, A., 1993. Eruptive dynamics and petrogenetic processes in a very shallow magma reservoir. J. Petrol. 34, 383–425.

Scandone, R., Giacomelli, L., Gasparini, P., 1993. Mount Vesuvius: 2000 years of volcanological observations. J. Volcanol. Geotherm. Res. 58(1–4), 5–25.

Vaggelli, G., De Vivo, B., Trigila, R., 1993. Silicate-melt inclusions in recent Vesuvius lavas (1631–1944): II. Analytical chemistry. J. Volcanol. Geotherm. Res. 58(1–4), 367–376.

Webster, J.D., De Vivo, B., Tappen, C., 2003. Volatiles, magmatic degassing and eruptions of Mt. Somma-Vesuvius; constraints from silicate melt inclusions, Cl and H_2O solubility experiments and modelling. In: De Vivo, B., Bodnar, R.J. (Eds), Melt Inclusions in Volcanic Systems Methods, applications and Problems. Developments in Volcanology, Vol. 5, Elsevier, Amsterdam, pp. 207–226.

Webster, J.D., Raia, F., De Vivo, B., Rolandi, G., 2001. The behavior of chlorine and sulfur during differentiation of the Mt. Somma-Vesuvius magmatic system. Mineral. Petrol. 73, 177–200.

Volcanism in the Campania Plain:
Vesuvius, Campi Flegrei and Ignimbrites
edited by B. De Vivo

Chapter 11

The role of sulfur in promoting magmatic degassing and volcanic eruption at Mt. Somma-Vesuvius

J.D. Webster[a,*], M.F. Sintoni[a,b] and B. De Vivo[b]

[a]Department of Earth and Planetary Sciences, AMNH, Central Park West at 79th St., New York, NY, USA
[b]Dipartimento di Scienze della Terra, Universitá di Napoli Federico II, via Mezzocannone 8, 80134 Napoli, Italy

Abstract

Past explosive and passive effusive eruptive activities of Mt. Somma-Vesuvius, Italy, are interpreted in terms of magma geochemistry and degassing processes through comparison of pre-eruptive volatile abundances, as determined through silicate melt compositions, and new experimental volatile solubility data for phonolite melt + H_2O + NaCl + KCl + $CaSO_4$ ± $CaCl_2$. These 200-MPa experiments reveal that small to modest levels of S in oxidized phonolitic melt have a substantial capacity to promote degassing by reducing Cl solubility in melt. This reduction in Cl solubility facilitates the exsolution of Cl- and S-enriched magmatic hydrosaline liquid with or without a coexisting vapor phase. Accounting for the elevated levels of S that occur in some Somma-Vesuvius melt inclusions means that S-enriched Somma-Vesuvius magmas should have exsolved a Cl- and S-charged magmatic hydrosaline liquid ± vapor, significantly earlier (i.e., at greater pressure and/or as a result of lesser bulk melt differentiation) than in the compositionally similar, S-poor Somma-Vesuvius magmas. With progressive magma evolution, after initial volatile phase exsolution, more compositionally evolved versions of the earliest primitive Cl- and S-rich volatile phases subsequently drove explosive plinian and sub-plinian eruptions.

1. Introduction and geologic background

Mt. Somma-Vesuvius is one of the most well-studied volcanoes on Earth. Its explosive plinian and sub-plinian eruptions are associated with magmas that differentiated from primitive, perhaps primary (Marianelli et al., 1995), potassium-enriched basaltic and tephritic through phonotephritic and tephriphonolitic to trachytic and phonolitic compositions (Belkin et al., 1998; Signorelli and Capaccioni, 1999; Signorelli et al., 1999). During the 30,000-year development of this composite volcano, fractional crystallization and magma mixing have, apparently, been the dominant mechanisms of magma evolution (Joron et al., 1987; Civetta and Santacroce, 1992; Belkin et al., 1993a,b; Trigila and De Benedetti, 1993; Ayuso et al., 1998; Belkin et al., 1998; Marianelli et al., 1999; Raia et al., 2000). Interestingly, magma evolution has played an important role in influencing the styles of eruptive behavior, because magma differentiation leading to explosive behaviors correlates with comparatively long repose periods and because the most explosive eruptions are associated with the most evolved melt compositions (Macedonio et al., 1990).

*Corresponding author. *E-mail address*: jdw@amnh.org (J.D. Webster).

Magmatic and volcanic processes at Mt. Somma-Vesuvius have been investigated through analyses of whole-rock samples and pumice separates (De Vivo et al., 2003). These samples have provided important constraints on magma evolution and as well as useful information on pre-eruptive magmatic abundances of some volatiles (Barberi et al., 1981; Belkin et al., 1985; Civetta and Santacroce, 1992; Belkin et al., 1993a,b; Santacroce et al., 1993; Trigila and De Benedetti, 1993; Villemant et al., 1993; Ayuso et al., 1998; Signorelli and Capaccioni, 1999; Signorelli et al., 1999). Investigations of silicate melt inclusions (MI) in magmatic phenocrysts, however, have provided improved knowledge on pre-eruptive magma chemistry, and in particular, crucial information on the abundances of volatiles and mobile fluxing components in magma. Recent MI investigations have, in fact, determined extreme enrichments of volatiles and associated mobile components, e.g., H_2O, Cl, S, F, P_2O_5, and CO_2, in some Mt. Somma-Vesuvius magmas (Vaggelli et al., 1993; Marianelli et al., 1995, 1999; Belkin et al., 1998; Cioni et al., 1998; Lima et al., 1999, 2003; Signorelli et al., 1999; Fulignati et al., 2000, 2001, 2004; Raia et al., 2000; Gilg et al., 2001; Webster et al., 2001).

Processes of magmatic degassing influence eruptive processes for most explosively erupting volcanoes. The exsolution and subsequent chemical evolution and physical expansion of one or more magmatic volatile phases, for instance, drive volcanic eruptions and play a fundamental role in controlling styles of eruption (De Vivo et al., 2005). Consequently, to better interpret magmatic and volcanic processes, with the goal of predicting future eruptions, it is crucially important to determine pre-eruptive volatile abundances and the depths (pressures) and timing of initial volatile phase(s) exsolution for samples representing prior eruptions. This allusion to timing of volatile exsolution refers to the extent of magma evolution at which a volatile phase first begins to exsolve. It is also essential to constrain how 'early' volatile phases change from primitive to more evolved compositions during progressive magma differentiation and/or ascent. Volatile phase exsolution begins as the emission of a CO_2-charged vapor phase in many magmas (Johnson et al., 1994), and some fractions of Somma-Vesuvius magmas show evidence of an important role for CO_2-H_2O in early if not initial degassing. However, other fractions of Somma-Vesuvius magmas indicate a significant role for Cl and S, consequently, we have focused herein on degassing processes for other primary magmatic volatiles, e.g., S and the dominant chloride species, in primitive Somma-Vesuvius 'fluids'.

For volatile-enriched magmas like those at Mt. Somma-Vesuvius, experimental constraints on the solubilities of the dominant volatiles (e.g., H_2O, CO_2, SO_2/H_2S, and the various Cl species) are necessary to interpret processes of volatile dissolution and exsolution. Most experiments have been limited, however, to binary volatile-bearing systems such as H_2O-CO_2, H_2O-S, and H_2O-Cl (Blank and Brooker, 1994; Carroll and Webster, 1994; Holloway and Blank, 1994; Dixon and Stolper, 1995; Dixon et al., 1995; O'Neil and Mavrogenes, 2002). New research has begun to reveal the solubility behavior of three volatiles (e.g., H_2O, S species, and Cl species) in melts of rhyodacitic (Botcharnikov et al., 2002, 2004a,b) and phonolitic (Webster, 2004a; Webster et al., 2004a,b) composition. The purpose of this report is to address and apply recent observations of the experimental work on H2O, S, and Cl dissolution in phonolitic melt to degassing of Mt. Somma-Vesuvius magmas. These preliminary results are intriguing because they show that the presence of geologically relevant abundances of S in alkaline magmas of intermediate silica contents have a major control on the solubility of Cl in melt, and hence, major implications for the depth and timing of volatile phase exsolution and for subsequent eruptive processes.

2. Results and discussion

For the purpose of this report, we consider magmatic volatile phases to include either a supercritical fluid; aqueous or aqueous-carbonic vapor; brine (which is equivalent to a water-bearing salt melt or a magmatic hydrosaline chloride liquid [herein abbreviated MHL]); or coexisting vapor plus MHL. This concern with nomenclature is not a useless exercise in semantics, because the type and number of volatile phases that exsolve have a significant effect on subsequent processes. Prior research involving S-poor melts, for example, has shown that passively erupted lava- and scoria-forming Mt. Somma-Vesuvius magmas were more likely to exsolve an MHL with or without vapor; whereas, the more H_2O-enriched, explosively erupting, pumice-forming magmas were more likely to exsolve vapor with or without MHL (Webster et al., 2003).

2.1. Experimental constraints

Hydrothermal experiments and empirical modeling clearly document a strong dependence of Cl solubility in silicate melt on the compositions of the melt and coexisting volatile phase(s) (Metrich and Rutherford, 1992; Carroll and Webster, 1994; Webster et al., 1999; Signorelli and Carroll, 2000; Webster and De Vivo, 2002). Chlorine solubility in phonolitic melt, for instance, increases quite strongly with decreasing (H_2O/Cl) in the system, and it also increases with melt CaO (Fig. 1a) (Webster et al., 2003) and MgO contents (Webster and De Vivo, 2002). A consequence of the observed relationship between Cl solubility and CaO and MgO in melt is that progressive evolution of the melt toward lower concentrations of alkaline earth metals, which is typical for most magmas, reduces Cl solubility and, hence, increases the tendency to exsolve a volatile phase or phases (Webster and De Vivo, 2002).

All experiments involved a natural pumiceous Somma-Vesuvius phonolite that erupted approximately 8000 years ago during the Ottaviano activities (i.e., sample S(9)2) which was misidentified as to its age in Webster et al. (2003)). Our experiments were conducted at 200 ± 10 MPa, temperatures of 924–1057°C, and at f_{O_2} ranging from +1.3 to + 3.3 log units above the NNO solid oxygen buffer. The temperatures and f_{O_2} of most S-enriched and Ca-enriched experiments, however, ranged from 924 to 960°C and +1.3 to +2.2 log units above NNO, respectively. In comparison, however, the range in intrinsic f_{O_2} of Somma-Vesuvius magmas is only poorly defined. Marini et al. (1998) calculated that f_{O_2} ranged from NN O+0.85 to NNO+1.2 for lavas (given estimated temperatures of 1100–1200°C) and NNO+1.2 to NNO+1.4 for explosively erupted pumices (with estimated temperatures of 800–850°C) with the assumption that the magmas involved behaved as open systems during degassing. The latter range overlaps with the lower part of the estimated f_{O_2} range for our experiments involving phonolitic melts. Note that sulfate is the dominant S species at all these oxygen fugacities (Carroll and Rutherford, 1988). It should also be noted that the temperatures used in our experiments exceed some estimated values for final, pre-eruption temperatures of Somma-Vesuvius phonolitic magmas. Constraints from prior experimental work are consistent with pre-eruptive magma temperatures as low as 750–800°C shortly before eruption (Scaillet and Pichavant, 2004), and thermobarometric work indicates temperatures as low as 780°C (Barberi et al., 1981; Joron et al., 1987). Nevertheless, during progressive magma

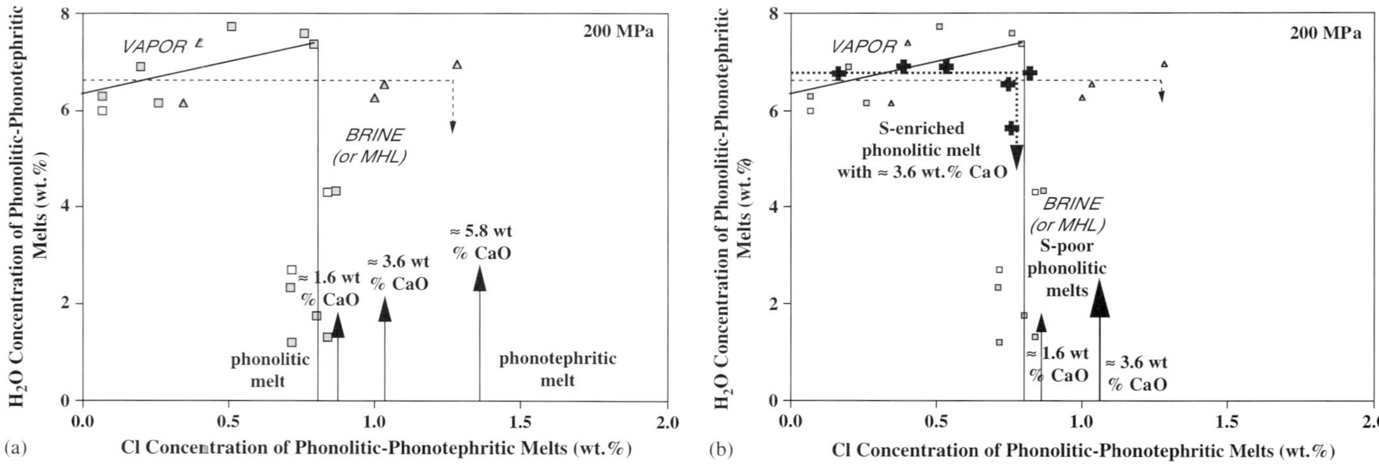

Figure 1. Plots showing experimentally determined solubilities of H_2O and Cl in vapor- and/or brine (magmatic hydrosaline liquid, MHL)-saturated silicate melts ranging from phonolitic to phonotephritic compositions at 200 MPa. Chemistry of melts from NaCl-KCl-H_2O runs (solid lines and squares), $CaCl_2$-NaCl-KCl-H_2O runs (triangles and dashed lines), and $CaSO_4$-NaCl-KCl-H_2O runs (filled crosses and dotted lines) are described in Table 1. Chlorine solubility varies strongly with (Cl/H_2O) ratio and CaO content of melts, whereas, H_2O solubility varies little (within analytical precision) as a function of varying CaO in melt. With increasing Cl in the system, the H_2O contents of melts remain relatively constant (or increase slightly) until melt becomes saturated in Cl under vapor-saturated conditions. Note that the increasing H_2O contents of NaCl-KCl-H_2O melts are due to increasing alkali contents of melt with increasing Cl abundances in the starting charges. With greater Cl abundances in the system (i.e., for MHL-saturated conditions), Cl contents of melt remain comparatively fixed while H_2O concentrations decrease. In (a) solid arrows (with progressively increasing lengths) denote modeled solubilities of Cl in MHL-saturated melts with 1.6, 3.6, and 5.8 wt.% CaO in melt, respectively. In (b) the influence of S on Cl solubility in melt is shown by the addition of 0.12 wt.% S (on average) to Cl- and S-charged MHL-saturated phonolitic melts containing 3.6 wt.% CaO as compared to the modeled solubility of Cl in S-poor phonolitic melt also containing 3.6 wt.% CaO. The former melts dissolve a maximum of 0.76 wt.% Cl; whereas, the latter dissolve 1.05 wt.% Cl. Thus, the presence of 0.12 wt.% S, on average, reduces Cl solubility by 28 relative %. Chlorine and S contents of glasses determined by electron microprobe, and H_2O concentrations determined by FTIR (filled symbols) and secondary ion mass spectrometry (open symbols) (methods described in Table 1).

evolution the phonolitic magmas would have passed through the higher range of temperatures used in our experiments. And, this is consistent with the observations of Cioni et al. (1998) that the upper portion of the Pompei magma chamber erupted white phonolitic pumice at temperatures 850–900°C and the upper realms of the Avellino magma chamber erupted phonolitic magma at 900–950°C. So, within the constraints on temperature and f_{O_2} that are currently available, our experiments are applicable to some if not many fractions of explosively erupted Somma-Vesuvius magmas. However, where appropriate, we address the potential influence of the small to modest differences in temperature and f_{O_2} between the magmas and our experiments in the discussion that follows.

Three sets of experiments involving the phonolite have been conducted (Table 1) (Webster et al., 2003, 2004a,b Webster, 2004a), and the results show that the addition of S leads to a dramatic reduction in Cl solubility. One set of experiments involved sample S(9)2 plus H_2O, NaCl, and KCl (Webster et al., 2003). The most Cl-enriched run-product glasses of these experiments contain an average of ≈1.6 wt.% CaO and 0.8 wt.% Cl (Fig. 1a), and represent phonolitic melt that was saturated in an MHL ± vapor at run conditions (Webster et al., 2003). To determine the influence of S on Cl and H_2O solubilities, a second set of experiments was conducted with sample S(9)2 plus $CaSO_4$, NaCl, KCl, and H_2O

Table 1. Starting Mt. Somma-Vesuvius pumice composition and average compositions of the most Cl-enriched run-product glasses representing magmatic hydrosaline liquid-saturated silicate melts.

Constituents (wt.%) (*n*)	Starting Phonolite S(9)2 (1)	NaCl-KCl Run glasses (6)	$CaCl_2$-NaCl-KCl Run glasses (3)	$CaSO_4$-NaCl-KCl Run glasses (5)
P_2O_5	0.02	0.02 ± 0.01	0.02 ± 0.01	0.02 ± 0.004
SiO_2	58.37	55.96 ± 1.76	56.80 ± 2.10	55.45 ± 1.72
S	0.01	0.01 ± 0.01	0.01 ± 0.01	0.12 ± 0.02
TiO_2	0.16	0.14 ± 0.02	0.19 ± 0.09	0.11 ± 0.03
Al_2O_3	19.74	19.86 ± 0.64	18.15 ± 1.0	19.17 ± 0.34
MgO	0.16	0.17 ± 0.04	0.17 ± 0.02	0.12 ± 0.02
CaO	1.80	1.57 ± 0.14	5.84 ± 1.48	3.6 ± 0.10
MnO	0.17	0.13 ± 0.03	0.10 ± 0.02	0.08 ± 0.01
FeO	1.86	1.64 ± 0.28	1.10 ± 0.19	1.46 ± 0.04
Na_2O	8.18	7.52 ± 0.33	5.16 ± 0.43	5.72 ± 0.03
K_2O	6.95	7.31 ± 0.57	5.28 ± 0.97	6.02 ± 0.20
F	0.72	0.65 ± 0.06	0.71 ± 0.13	0.45 ± 0.04
Cl	0.62	0.80 ± 0.06	1.26 ± 0.05	0.76 ± 0.03
Total (± 1σ)	98.78	95.80 ± 3.10	94.72 ± 3.87	93.23 ± 2.24
Max Cl solubility at 200 MPa	0.95	0.88	1.34	1.05

Note: Analyses by electron microprobe using methods of Webster et al. (2003); (*n*) is the number of samples. Experiments conducted at 200 MPa, temperatures of 924–1057°C, and f_{O_2} of NNO + 1.3 to NNO + 3.3 in Au and AuPd capsules with phonolite sample S(9)2 + H_2O + NaCl + KCl ± $CaCl_2$ ± $CaSO_4$ using methods of Webster et al. (2003, 2004a,b) and Webster (2004a). The most Cl-enriched and H_2O-poor run-product glasses were used to compute these averages. Maximum Cl solubility values were computed with model of Webster and De Vivo (2002) for magmatic hydrosaline liquid-saturated silicate melts with these bulk compositions at 200 MPa.

(Fig. 1b). The relative abundances of $CaSO_4$ added to the latter experiments were varied such that some of the experiments were saturated in $CaSO_4$ and some were not. The run-product glasses of the most Cl-enriched experiments (i.e., those representing MHL- ± vapor-saturated melts) contain, respectively, 0.12, 3.6, and 0.76 wt.% S, CaO, and Cl on average (Webster, 2004a; Webster et al., 2004a,b; Sintoni et al., manuscript in preparation). The Cl contents of these two sets of experiments cannot be compared directly, to constrain the influence of S on Cl solubility, because of the dramatic difference in their CaO concentrations. We have, however, determined the individual influence of CaO on Cl solubility through additional experiments and modeling.

A third set of experiments was conducted with S(9)2 plus $CaCl_2$, NaCl, KCl, and H_2O. The $CaCl_2$ was necessary to boost the CaO concentrations of the run-product glasses above that of the starting rock composition (i.e., to make the final melt composition comparatively more primitive). The glasses contain, respectively, 5.8, 1.26, and $\leq$ 0.01 wt.% CaO, Cl, and S on average. The enhanced Cl content of these glasses is primarily the result of the additional CaO in glass, which is consistent with observations of Webster and De Vivo (2002). Using the model of their study, we have computed the predicted maximum Cl concentrations of MHL-saturated silicate melts represented by the NaCl-KCl-H_2O glasses, the $CaCl_2$-NaCl-KCl-H_2O glasses, and the $CaSO_4$-NaCl-KCl-H_2O glasses (Table 1). The close agreement between the predicted Cl contents based on modeling and the experimentally determined Cl concentrations for the NaCl-KCl-H_2O glasses and $CaCl_2$-NaCl-KCl-H_2O glasses lends further support to the accuracy of the Cl solubility model. It is also clear that the predicted maximum Cl solubility of the $CaSO_4$-NaCl-KCl-H_2O glasses (based on modeling) is dramatically greater than the measured Cl abundances of the S-enriched run-product glasses. This difference in Cl contents reflects the influence of S on Cl dissolution in phonolitic melt, and is a consequence of the fact that the Cl solubility model does not, yet, account for the influence of S on Cl solubility. It should also be noted that within the precision of our analyses for H_2O in these glasses, the addition of 0.1–0.19 wt.% S has no detectable influence on H_2O solubility in phonolitic melt at 200 MPa (Fig. 1b).

The strong reduction in Cl solubility from $\approx$1.05 wt.% (as predicted) to 0.76 wt.% (as measured) represents a 28% relative change. This relationship may reflect the presence of elevated S concentrations in the melt and/or volatile phases (Webster, 2004a; Webster et al., 2004a,b). In particular, Cl and S may compete for similar 'sites' in the melt (e.g., for Ca and Fe perhaps), and as a result the presence of S reduces the maximum abundance of Cl that may dissolve in melt. If the reduction in Cl solubility is indeed a result of elevated S in the melt, then these initial volatile solubility experiments show that the addition of S to volatile phase-saturated phonolite melt results, on a molar basis, in one of the largest changes in Cl solubility yet to be recorded. Each additional mole of S^{6+} that dissolves in silicate melt (as SO_4^{2+}), reduces Cl solubility by 1–2 mol. This equals and may exceed the observed influences of Ti and P in diminishing Cl solubility in MHL-saturated silicate liquids (Webster and De Vivo, 2002).

Alternatively, the dissolution of elevated S concentrations in the associated volatile phase(s) may reduce the activity of Cl in the melt-volatile phase(s) system by simple dilution (assuming no strong positive deviations from ideal mixing in the volatile phase or phases which might serve to increase Cl activity) (Botcharnikov et al., 2004a,b). A reduction in the activity of Cl in the volatile phase(s) should lower Cl solubilities in the coexisting melt. Prior experiments on S-partitioning between volatile phases and silicate melts determined total

S concentrations (representing H_2S and SO_2) as high as several tens of wt.% in the associated volatile phase(s) (Scaillet et al., 1998; Keppler, 1999; Scaillet and Pichavant, 2003). These experiments involved haplogranitic and dacitic melts, and the abundances of S species in the run-product fluids were calculated by mass balance.

Interestingly, more recent experimental investigations of H_2O, S, and Cl solubilities in rhyodacitic melt and coexisting volatile phases (Botcharnikov et al., 2002, 2004a,b) exhibit results that are very similar to those determined for the Ottaviano phonolitic melt. In short, they observed that the addition of S to the rhyodacite experiments, conducted at 200 MPa, 850°C, and f_{O_2} of ≈NNO, reduced Cl solubility in the melt. In their experiments, however, the melts dissolved only ≤ 0.02 wt.% S (i.e., an order of magnitude less than that in our experiments). Consequently, the authors argued that it is unreasonable for such small quantities of S in melt to have such a dramatic influence on the Cl contents of the melt, and so the mixing properties of the H-, O-, Cl-, and S-dominated volatile phase(s) must exert a dominant control on the volatile solubilities of coexisting melt. We note the studies of Botcharnikov et al. (2002, 2004a,b) also report that increased Cl in rhyodacitic melts is associated with enhanced S solubility. Even though our results for alkaline potassic phonolitic melt are consistent with the studies on calc-alkaline rhyodacitic melt as regards the influence of S on Cl solubility, the experiments involving the former melts do not yet provide the data necessary to constrain the influence of Cl on S solubility.

Ongoing research with the Ottaviano phonolite is working to establish a robust method of determining the S concentrations of run-product fluids in $CaSO_4$-bearing experiments (Sintoni et al., manuscript in preparation). The lack of hard data on fluid compositions notwithstanding, we assume that the large partition coefficients (Scaillet et al., 1998; Keppler, 1999; Botcharnikov et al., 2002, 2004a,b; Scaillet and Pichavant, 2003; Wallace, 2003) for S (i.e., concentration of S in aqueous 'fluid'/concentration of S in silicate melt), for dacitic, haplogranitic, and rhyodacitic melts may also apply to molten phonolite. Thus, the volatile phase(s) of the phonolite experiments also probably contained similarly high S values (i.e., concentrations as high as 10 wt.%).

2.2. *Implications for magmatic degassing*

Volatile phase exsolution in magma occurs because of first boiling, second boiling, or a combination of both. First boiling is a result of: (1) the positive relationship between pressure and the solubility of most volatiles in silicate melts and (2) the reduction in confining pressure and the decreased volatile solubility in melt that result from magma ascent. Second boiling occurs because of the incompatible behavior of volatiles; the abundances of volatiles in residual melt increase as the temperature decreases and the extent of crystallization increases.

A poorly appreciated aspect of magmatic degassing is that once a volatile phase (or phases) exsolves from a silicate melt, the magma may remain in a state of volatile saturation depending on several factors. It is necessary that the system remains closed to volatile phase escape, pressure stays constant (or diminishes), the magma in question does not become more primitive in composition by mixing with less-evolved magma or by resorption of mafic minerals, or that the magma does not become less hydrous by resorption of anhydrous minerals (Holloway, 1976). A consequence of this is that even though many volcanic systems show evidence of pre-eruptive volatile phase exsolution, we currently have poor constraints on the timing of initial volatile degassing in erupting magmas. Magmas may, thus, be volatile phase saturated for a significant part of their history as they move through the shallow crust (Wallace, 2003).

Volatiles other than H_2O may cause 'early' exsolution of compositionally primitive volatile phases, and it follows that the compositions of magmatic volatile phases must evolve in concert with subsequent magma differentiation and ascent. It is well established that even though H_2O is the dominant and most abundant volatile in silicate magmas, other less-abundant volatiles can also exert a powerful control on magmatic degassing and resultant eruptive processes. Small amounts of CO_2 in felsic silicate melts, concentrations as low as several hundred to several thousand parts per million, for example, can greatly reduce the quantity of H_2O required for exsolution of a mixed CO_2- and H_2O-dominated volatile phase (Holloway, 1976; Holloway and Blank, 1994). And, 'early' exsolution of a CO_2-enriched volatile phase may represent a chief means of initial degassing in some basaltic magmas (Dixon and Stolper, 1995; Wallace and Anderson, 2000). Chlorine, in similarly small quantities, has been shown to have a comparable influence in promoting degassing processes in felsic to basaltic magmas (Webster et al., 1999; Webster and De Vivo, 2002; Webster, 2004b). In fact, Cl is fundamentally important to another form of second boiling in magmas. This degassing process is a result of the fact that the solubility of Cl is strongly dependent on melt composition. Chlorine solubility varies most effectively with Mg, Ca, Fe, and Al abundances of silicate melts (Webster et al., 1999; Webster and De Vivo, 2002), and consequently, a Cl-charged MHL with or without a coexisting vapor phase may exsolve directly from magma as Ca-, Mg-, Fe-, and Al-rich refractory minerals crystallize and the melt becomes increasingly depleted in these components. It is noteworthy in this regard that natural Somma-Vesuvius phonolitic melts also show modest variations in alkali contents and in their molar (Na/Na+K) ratio. For instance, the molar (Na/Na+K) of phonolitic glasses erupted during the Ottaviano, Avellino, and Pompeii (79 AD) activities varies from 0.48 to 0.68 (Rolandi et al., 1999; Signorelli et al., 1999; Signorelli and Carroll, 2000). These natural variations, however, have only a very small influence on Cl solubility in melt (and, hence, on subsequent degassing processes) as compared to the effects of Ca, Mg, and Fe given their much larger variabilities in these same melts. In summary, the solubility limit for Cl in melt decreases because of the change in melt composition (which promotes formation of an MHL) while the abundances of H_2O, CO_2, S, and Cl increase in the residual silicate melt. Thus, both processes work in concert to exsolve an MHL with or without vapor.

The fact that magmas may remain volatile saturated for extended portions of their history is fundamentally important because the evidence used to interpret volatile degassing (i.e., MI and experimental results), during the final stages of magmatism preceeding eruption, may have little bearing on the nature of magma and its volatile phase(s) well before eruption. The compositions of high temperature, magmatic gases, for instance, may not reflect the compositions of 'initial' magmatic volatile phases. Under closed-system conditions, for example, the presence of CO_2 and/or Cl in magma may lead to initial exsolution of a CO_2- and/or Cl-charged volatile phase, but with continued magma ascent and/or crystallization, the chemistry of this volatile phase must evolve toward a more H_2O-enriched composition simply because H_2O is the most abundant volatile. During the compositional evolution of the residual melt and volatile phases in magma, however, the relative abundances of CO_2:SO_2:H_2S:Cl^- species will also vary. The volume of the volatile phase will also increase with magma evolution. Thus, 'early' exsolution of a CO_2- or Cl-charged volatile phase or phases may actually initiate a sequence of processes that cause volatile-laden magmas to erupt explosively, even though the composition of the early volatile phase(s) may be quite different from that of the final 'vapor phase' in the volcanic crater. The expansion of a CO_2-bearing aqueous vapor will provide much more energy than that of an MHL to drive explosive volcanism, but the exsolution of an MHL is nonetheless significant in this regard because for some fractions of Mt. Somma-Vesuvius magma, at least, early MHL exsolution began the degassing processes, which ultimately caused violent eruptive behavior.

2.3. The role of sulfur in magmatic degassing at Mt. Somma-Vesuvius

The new experimental results for H_2O-, S-, and Cl-bearing Mt. Somma-Vesuvius phonolitic melts are important because they indicate that modest abundances of S in melt may also influence initial processes of degassing due to the strong relationship between the S content of volatile phase-saturated melt and Cl solubility. To better understand the magnitude of the effect of this relationship on volatile phase exsolution, it is useful to compare the Cl abundances of MI in Mt. Somma-Vesuvius rock samples with their S contents (Fig. 2). It is noteworthy that the MI in Somma samples (those $\geqslant$14,000 years in age) range from trachybasaltic to trachyandesitic compositions, and the younger Vesuvius MI erupted via passive-effusive non-plinian activities (also known as interplinian activities) and via explosive plinian/sub-plinian behaviors range from tephritic to tephriphonolitic compositions. Thus, it is inappropriate to compare, directly, the Cl abundances of the experimental phonolitic glasses with the MI because of the more primitive bulk compositions of the MI. Nevertheless, it is clear that MI in rocks erupted during the period of Somma activities contain lower S and Cl abundances, and that MI erupted during subsequent plinian/sub-plinian Vesuvian behaviors generally contain higher S contents which is consistent with prior observations (Webster et al., 2001). It is also clear that most Vesuvius MI erupted during passive-effusive, interplinian activities contain S abundances equivalent to those of our experiments, whereas many of the Vesuvius MI erupted via explosive activities contain significantly more S than that in the experimental glasses.

The presence of up to several thousand ppm S in the volatile phase-saturated melts, of the Mt. Somma-Vesuvius magmas, and of concomitantly elevated S in the coexisting volatile phase(s), should have had a dramatic influence on Cl solubility in the late-stage, evolved phonolitic melts. The effect of reducing Cl solubility for Cl-enriched magmas, like those at Somma-Vesuvius, is to enhance the likelihood of volatile phase exsolution (i.e., MHL exsolution). For example, the addition of 0.07–0.19 wt.% S to the phonolitic melts instigated a 28 relative% reduction in Cl solubility. In this context, however, we must reiterate that the Vesuvius MI are more primitive than the melt compositions employed in our experiments. Nevertheless, we can safely assume that the presence of these same levels of S in tephriphonolitic to phonotephritic melts should result in roughly similar reductions in Cl solubility given the results of Botcharnikov et al. (2004a,b) showing a strikingly similar influence of S on Cl solubility for calc-alkaline rhyodacitic melts. Note that the latter experiments were conducted at f_{O_2} near NNO, so the apparent influence of S on Cl solubility in silicate melts is felt over a significant range in oxygen fugacity. Consequently, the modest difference in f_{O_2} of our experiments relative to the estimated values for Somma-Vesuvius

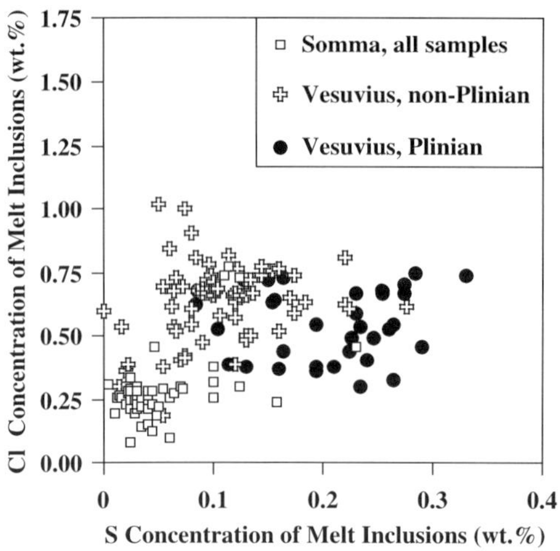

Figure 2. Plot of Cl versus S concentrations (wt.%) in melt inclusions from Mt. Somma (squares) (eruptive samples greater than 14,000 years in age), Mt. Vesuvius samples erupted under interplinian passive-effusive conditions <3800 years ago (crosses), and Mt. Vesuvius samples erupted under plinian/sub-plinian behaviors <3800 years ago (circles). Most melt inclusions from Vesuvius plinian/sub-plinian materials contain higher S contents and roughly similar Cl contents as melt inclusions in Vesuvius passive-effusive eruptive materials. See text for melt inclusion compositional data.

pumice-forming magmas (Marini et al., 1998) should not have modified the influence of S on Cl solubility to a significant extent. Thus, with all other factors being equal, the Vesuvius plinian/sub-plinian magmas, which contained generally higher S contents were comparatively more likely to exsolve a Cl- and S-charged MHL with or without vapor than the magmas erupted via passive-effusive behavior (and which contained roughly equivalent Cl contents) simply because of their elevated S abundances.

The significance of the relationship that links the abundance of S in melt to magmatic degassing can be expressed in another context that is perhaps more comprehensible. Water solubility in silicate melts varies strongly with pressure, so magma ascent increases the tendency to exsolve an aqueous vapor phase through first boiling. So, for a S-, CO_2-, and Cl-poor phonolitic magma in the shallow crust to experience a reduction in H_2O solubility of 28 relative% – during first boiling – it would have to undergo a pressure change of 200 to approximately 70 MPa (Webster et al., 2003). This pressure change is roughly equivalent to that which the magma would go through as it ascends from depths of 7–2.5 km, and it is consistent with the range in observed H_2O contents of Somma-Vesuvius MI and experimentally determined H_2O solubilities for this range in pressures (Marianelli et al., 1995, 1999; Belkin et al., 1998; Lima et al., 1999, 2003; Signorelli et al., 1999; Raia et al., 2000; Webster et al., 2001) and with experimental phase equilibrium constraints for phonolitic melts (Scaillet and Pichavant, 2004). The presence of a mere few thousand ppm S in phonolitic melt increases the tendency toward magmatic degassing that is equal to the drive for volatile exsolution experienced by a H_2O-dominated magma as it ascends through a distance of 4.5 km in the shallow crust.

Magma eruption dynamics (i.e., eruptive styles) are functions of the plumbing system that feeds magma to the volcanic edifice (Bower and Woods, 1998) and of the geochemistry of the magmas as they ascend toward the surface (Wallace and Anderson, 2000). Much of the prior research on Mt. Somma-Vesuvius has focused on its subsurface plumbing, in order to better understand past eruptive behaviors. Some investigations, for example, have attempted to constrain the depths at which magma ponds beneath the surface edifice, but there has not been strong agreement on the conclusions of the various studies (Belkin et al., 1985; Belkin and De Vivo, 1993; Marianelli et al., 1995, 1999; Gilg et al., 2001; Lima et al., 2003). Belkin et al. (1985), for instance, determined the densities of primary CO_2-bearing fluid inclusions in Somma-Vesuvius nodules and concluded that a small magma chamber lies at $\geqslant 3.5$ km depth and a larger chamber occurs at $\geqslant 12$ km depth from the surface. These conclusions are in accord with those of Lima et al. (2003) who concluded that the plumbing system likely resembles a complex feeder conduit composed of multiple 'mush zone' environments (small magma chambers) and includes a variety of local crystallization environments characterized by contrasting cooling rates and P–T conditions. In contrast, in their study of MI representing the plinian to sub-plinian Avellino, 79 AD, and 472 AD eruptions and the 1906 and 1944 interplinian eruptions, Marianelli et al. (1995) argued for a comparatively small chamber at shallow depths of 1–2 km, during open-system conditions, and larger and deeper magma chamber(s) at depths of 2–5 km during closed-system magma evolution. In a separate study of MI in 1944 eruptive products, Marianelli et al. (1999) considered that volatile-enriched magma derived from a deep chamber (i.e., at 11–22 km depth) entered the shallow chamber and subsequently mixed with magma already there. They also concluded that a CO_2- and H_2O-dominated volatile phase (or phases), which also contained S and Cl species, degassed comparatively early on, and this is consistent with observations of Signorelli et al. (1999) based on other evidence. The conclusions of Marianelli et al. (1999), however, are based on binary volatile solubility experiments that do not account for the influences of elevated S and Cl in melt.

Our experimental data provide new information bearing on the depths and timing of magmatic degassing at Mt. Somma-Vesuvius. Earlier research on Somma-Vesuvius MI, in combination with observations from volatile solubility experiments in the pseudo-system: phonolite-H_2O-Cl (Webster et al., 2003) concluded that vapor with or without MHL should have exsolved from compositionally evolved phonolitic magma when it had ascended to depths of $\leqslant 6$–7 km (i.e., via first boiling). Other work which focused on magma degassing via second boiling, accounted for the difference in bulk composition of the phonotephritic to tephriphonolitic MI and the more-evolved experimental phonolitic melts (Webster and De Vivo, 2002). This study determined that a typical Somma-Vesuvius phonotephritic melt would require approximately 40 wt.% crystallization to drive the residual melt to a phonolitic composition while simultaneously increasing the H_2O and Cl abundances of the residual melt to the point of MHL exsolution. This work is qualitatively consistent with long-standing observations that magma differentiation leading to explosive behavior at Mt. Somma-Vesuvius correlates with comparatively long repose periods and with the evolution of bulk melt compositions to phonolite (Macedonio et al., 1990). These studies, however, do not account for the effect of S on Cl solubility and degassing processes. For instance, Figure 3 shows the positive but small correlation between Cl solubility and pressure for MHL-saturated phonolitic and phonotephritic melts, and compares the effect of this 200–50 MPa pressure change on Cl solubility with the effect of increasing S in phonolitic melt (i.e., phonolitic melts containing 3.6 wt.% CaO) from 0.01 to 0.12 wt.%. Clearly, S-enriched melts like those at Somma-Vesuvius will exsolve a Cl- and S-charged MHL,

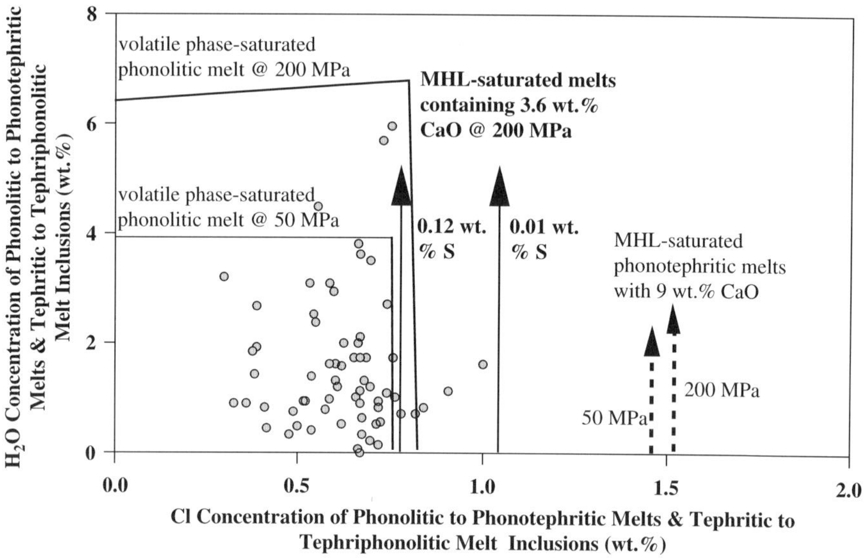

Figure 3. Plots showing experimentally determined solubilities of H_2O and Cl in vapor- and/or brine (magmatic hydrosaline liquid = MHL)-saturated silicate melts ranging from phonolitic to phonotephritic compositions at 200 and 50 MPa and tephritic to tephriphonolitic silicate melt inclusions (circles). Melts from phonolite-NaCl-KCl-H_2O runs (fine solid lines), phonolite-$CaCl_2$-NaCl-KCl-H_2O runs (bold solid arrow labeled 0.01 wt.% S), phonolite-$CaSO_4$-NaCl-KCl-H_2O runs (bold solid arrow labeled 0.12 wt.% S), and phonotephrite-NaCl-KCl-H_2O runs (dashed arrows). The influence of 0.12 wt.% S in silicate melt on Cl solubility (and the tendency for melt to exsolve a Cl- and S-charged MHL) far outweighs the effect of a pressure reduction of 200–50 MPa for phonolitic or phonotephritic melts on MHL exsolution.

with or without vapor, significantly earlier (i.e., at greater pressure and/or as a result of lesser bulk melt differentiation) than compositionally equivalent S-poor melts.

Recall that the temperatures used in our phonolite experiments exceed those estimated for Somma-Vesuvius phonolitic magma prior to eruption. The higher temperature range of our experiments (924–1057°C) may have influenced the solubility behaviors and the distribution of S and Cl between melt and the volatile phase(s). Sulfur solubility in silicate melts varies strongly with temperature (Carroll and Webster, 1994), and in fact Wallace (2003) has modeled the ability of temperature to modify the partitioning of S between vapor and felsic silicate melts. As temperature decreases from 1000–800°C at 220 MPa, the value of D_S (which is equivalent to the concentration of S in the volatile phase(s)/concentration of S in melt) increases in favor of the volatile phase by a factor of roughly 20 with oxygen fugacity held constant at NNO+1.5. Interestingly, the value of D_{Cl} (which is equivalent to the concentration of Cl in the volatile phase(s)/concentration of Cl in melt) increases in favor of the volatile phase by a factor of roughly 10 for Cl-enriched melts at 200 MPa. Thus, the S and Cl concentrations of the magmatic volatile phase(s) would have increased to values greater than those in our experiments as temperature decreased and the residual melt fractionated to evolved phonolitic compositions. To a first-order approximation, however, the increase in abundance of S in the volatile phase(s) would have been only twice that of Cl as a consequence of this difference in temperature.

In summary, there is evidence in support of 'early', perhaps initial, exsolution of primitive volatile phases from some fractions of Mt. Somma-Vesuvius magma as CO_2-charged 'fluids' (Marianelli et al., 1999); whereas other fractions of magma are consistent with 'early' exsolution of Cl- and S-charged MHL (with or without vapor) (Signorelli et al., 1999; Webster et al., 2003; Webster, 2004a; Webster et al., 2004a,b). In either case, these early volatile phase(s) set the stage for subsequent processes of volcanic eruption ranging from passive to explosive behaviors. Thus, future research must be directed at more precise determination of how the compositions and quantities of these primitive volatile phases progressively evolved with magma differentiation and ascent and how they ultimately drove ensuing eruptions.

The elevated CO_2 contents of some Somma-Vesuvius MI provide strong evidence of a significant role for this volatile in 'early' degassing and, perhaps, in subsequent eruptive processes. This report does not address the influence of CO_2 on degassing at Mt. Somma-Vesuvius, primarily because of the lack of relevant experimental data. Consequently, we are currently conducting the first preliminary experiments to determine, systematically, the influence of CO_2 on H_2O and Cl solubilities in Somma-Vesuvius phonolitic melts. We also note, in passing, that future experiments involving water, sulfur, and chlorine must be conducted at higher pressures to establish, quantitatively, the influence of S on the solubilities of Cl and H_2O for deeper and more primitive Somma-Vesuvius magmas.

Acknowledgements

We appreciate the thoughtful and detailed reviews of this manuscript by Michael Carroll and Bruno Scaillet. The research reported herein was partially supported by the NSF award EAR 0308866 to JDW and MIUR-PRIN 2003–2004 and PETIT-OSA projects to B. DeVivo.

References

Ayuso, R.A., De Vivo, B., Rolandi, G., Seal II, R.R., Paone, A., 1998. Geochemical and isotopic (Nb-Pb-Sr-O) variations bearing on the genesis of volcanic rocks from Vesuvius, Italy. J. Volcanol. Geotherm. Res. 82, 53–78.

Barberi, F., Bizouard, H., Clocchiatti, R., Métrich, N., Santacroce, R., Sbrana, A., 1981. The Somma-Vesuvius magma chamber: a petrological and volcanological approach. Bull. Volcanol. 44, 295–315.

Belkin, H.E., De Vivo, B., 1993. Fluid inclusion studies of ejected nodules from plinian eruptions of Mt. Somma-Vesuvius. In: De Vivo, B., Scandone, R., Trigila, R. (Eds), Vesusius. J. Volcanol. Geotherm. Res. 58, 89–100.

Belkin, H.E., De Vivo, B., Roedder, E., Cortini, M., 1985. Fluid inclusion geobarometry from ejected Mt. Somma-Vesuvius nodules. Am. Mineral. 70, 288–303.

Belkin, H.E., De Vivo, B., Török, K., Webster, J.D., 1998. Pre-eruptive volatile content, melt inclusion chemistry, and microthermometry of interplinian Vesuvius lavas (pre-1631 AD). J. Volcanol. Geotherm. Res. 82, 79–95.

Belkin, H.E., Kilburn, C.R.J., De Vivo, B., 1993a. Sampling and major element chemistry of the recent (AD 1631–1944) Vesuvius activity. J. Volcanol. Geotherm. Res. 58, 273–290.

Belkin, H.E., Kilburn, C.R.J., De Vivo, B., 1993b. Chemistry of the lavas and tephra from recent (AD 1631–1944) Vesuvius (Italy) volcanic activity. U.S. Geol. Surv. Open-File Report 93-399, 44pp.

Blank, J.G., Brooker, R.A., 1994. Experimental studies of carbon dioxide in silicate melts: solubility, speciation, and stable carbon isotope behavior. In: Carroll, M.R., Holloway, J.R. (Eds), Volatiles in Magmas. Rev. Mineral. 30, 157–186.

Botcharnikov, R.E., Holtz, F., Behrens, H., Sato, H., Koepke, J., 2002. Experimental study of Sulfur and Chlorine solubility in rhyodacite and andesite melts of Unzen Volcano, Japan. EMPG IX J. Conf. Abs. 7, p. 17.

Botcharnikov, R.E., Behrens, H., Holtz, F., Koepke, J., Sato, H., 2004a. Sulfur and chlorine solubility in Mt. Unzen rhyodacite melt at 850°C and 200 MPa. Chem. Geol. 213, 207–225.

Botcharnikov, R.E., Holtz, F., Behrens, H., Sato, H., Koepke, J., 2004b. Experimental study of sulfur and chlorine solubility in rhyodacite and andesite melts of Unzen volcano, Japan. J. Conf. Abs. 7, 17.

Bower, S.M., Woods, A.W., 1998. On the influence of magma chambers in controlling the evolution of explosive volcanic eruptions. J. Volcanol. Geotherm. Res. 86, 67–78.

Carroll, M.R., Rutherford, M.J., 1988. Sulfur speciation in hydrous experimental glasses of varying oxidation state: results from measured wavelength shifts of sulfur X-rays. Am. Mineral. 73, 845–849.

Carroll, M.R., Webster, J.D., 1994. Solubilities of sulfur, noble gases, nitrogen, chlorine, and fluorine in magmas. In: Carroll, M.R., Holloway, J.R. (Eds), Volatiles in Magmas, Rev. Mineral. 30, 231–279.

Cioni, R., Marianelli, P., Santacroce, R., 1998. Thermal and compositional evolution of the shallow magma chambers of Vesuvius: evidence from pyroxene phenocrysts and melt inclusions. J. Geophys. Res. 103(B8), 18277–18294.

Civetta, L., Santacroce, R., 1992. Steady state magma supply in the last 3400 years of Vesuvius activity. Acta. Vulcan. 2, 147–159.

De Vivo, B., Ayuso, R.A., Belkin, H.E., Fedele, L., Lima, A., Rolandi, G., Somma, R., Webster, J.D., 2003. Chemistry, fluid/melt inclusions and isotopic data of lavas, tephra and nodules from >25 ka to 1944 AD of the Mt. Somma-Vesuvius volcanic activity. Mt. Somma-Vesuvius Geochemical Archive. Dipartimento di Geofisica e Vulcanologia, Università di Napoli Federico II. Open File Report 1-2003, 143pp.

De Vivo, B., Lima, A., Webster, J.D., 2005. Volatiles in magmatic-volcanic systems. Elements 1, 19–24.

Dixon, J.E., Stolper, E.M., 1995. An experimental study of water and carbon dioxide solubilities in mid-ocean ridge basaltic liquids. Part II: applications to degassing. J. Petrol. 36, 1633–1646.

Dixon, J.E., Stolper, E.M., Holloway, J.R., 1995. An experimental study of water and carbon dioxide solubilities in mid-ocean ridge basaltic liquids. Part I: calibration and solubility models. J. Petrol. 36, 1607–1631.

Fulignati, P., Kamenetsky, V.S., Marianelli, P., Sbrana, A., Mernagh, T.P., 2001. Melt inclusion record of immiscibility between silicate, hydrosaline, and carbonate melts: applications to skarn genesis at Mount Vesuvius. Geology 29, 1043–1046.

Fulignati, P., Marianelli, P., Metrich, N., Santacroce, R., Sbrana, A., 2004. Towards a reconstruction of the magmatic feeding system of the 1944 eruption of Mt. Vesuvius. J. Volcanol. Geotherm. Res. 133, 13–22.

Fulignati, P., Marianelli, P., Santacroce, R., Sbrana, A., 2000. The skarn shell of the 1944 Vesuvius magma chamber. Genesis and P-T-X conditions from melt and fluid inclusion data. Eur. J. Mineral. 12, 1025–1039.

Gilg, H.A., Lima, A., Somma, R., Belkin, H.E., De Vivo, B., Ayuso, R.A., 2001. Isotope geochemistry and fluid inclusion study of skarns from Vesuvius. Mineral. Petrol. 73, 145–176.

Holloway, J.R., 1976. Fluids in the evolution of granitic magmas: consequences of finite CO_2 solubility. Geol. Soc. Am. Bull. 87, 1513–1518.

Holloway, J.R., Blank, J.G., 1994. Application of experimental results to C-O-H species in natural melts. In: Carroll, M.R., Holloway, J.R. (Eds), Volatiles in Magmas. Rev. Mineral. 30, 187–230.

Johnson, M.C., Anderson Jr., A.T., Rutherford, M.J., 1994. Pre-eruptive volatile contents of magmas. In: Carroll, M.R., Holloway, J.R. (Eds), Volatiles in Magmas. Rev. Mineral. 30, 281–330.

Joron, J.L., Métrich, N., Rosi, M., Santacroce, R., Sbrana, A., 1987. Somma-Vesuvius. III. Chemistry and petrography. In: Santacroce, R. (Ed.), "Quaderni de la ricerca scientifica", CNR Publ. 8, 105–174.

Keppler, H., 1999. Experimental evidence for the source of excess sulfur in explosive volcanic eruptions. Science 284, 1652–1654.

Lima, A., Belkin, H.E., Török, K., 1999. Primitive silicate-melt inclusions in compositionally diverse clinopyroxene phenocrysts from the medieval scorias from the Terzigno Formation: clue to understanding of Vesuvius magmatic processes. Mineral. Petrol. 65, 185–206.

Lima, A., Danyushevsky, L.V., De Vivo, B., Fedele, L., 2003. A model for the evolution of the Mt. Somma-Vesuvius magmatic system based on fluid and melt inclusion investigations. In: De Vivo, B., Bodnar, R.J. (Eds), Melt Inclusions in Volcanic Systems. Series: Developments in Volcanology, Vol. 5, Elsevier, Amsterdam, pp. 227–250.

Macedonio, G., Pareschi, M.T., Santacroce, R., 1990. Renewal of explosive activity at Vesuvius: models for the expected tephra fallout. J. Volcanol. Geotherm. Res. 40, 327–342.

Marianelli, P., Métrich, N., Santacroce, R., Sbrana, A., 1995. Mafic magma batches at Vesuvius: a glass inclusion approach to the modalities of feeding stratovolcanoes. Contrib. Mineral. Petrol. 120, 159–169.

Marianelli, P., Métrich, N., Sbrana, A., 1999. Shallow and deep reservoirs involved in magma supply of the 1944 eruption of Vesuvius. Bull. Volcanol. 61, 48–63.

Marini, L., Chiappini, V., Cioni, R., Cortecci, G., Dinelli, E., Principe, C., Ferrera, G., 1998. Effect of degassing on sulfur contents and $\partial^{34}S$ values in Somma-Vesuvius magmas. Bull. Volcanol. 60, 187–194.

Metrich, N., Rutherford, M.J., 1992. Experimental study of chlorine behavior in hydrous silicic melts. Geochim. Cosmochim. Acta 56, 607–616.

O'Neil, H.S.C., Mavrogenes, J.A., 2002. The sulfide capacity and the sulfur content at sulfide saturation of silicate melts at 1400°C and 1 bar. J. Petrol. 43, 1049–1087.

Raia, F., Webster, J.D., De Vivo, B., 2000. Pre-eruptive volatile contents of Vesuvius magmas: constraints on eruptive history and behavior. I – The medieval and modern interplinian activities. Eur. J. Mineral. 12, 179–193.

Rolandi, G., Maraffi, S., Petrosino, P., Lirer, L., 1999. The Ottaviano eruption of Somma-Vesuvio (8000 y BP): a magmatic alternating fall and flow-forming eruption. J. Volcanol. Geochem. Res. 58, 43–65.

Santacroce, R., Bertagnini, A., Civetta, L., Landi, P., Sbrana, A., 1993. Eruptive dynamics and petrogenetic processes in a very shallow magma chamber reservoir, the 1906 eruption of Vesuvius. J. Petrol. 34, 383–425.

Scaillet, B., Clemente, B., Evans, B., Pichavant, M., 1998. Redox control of sulfur degassing in silicic magmas. J. Geophys. Res. 103(B10), 23937–23949.

Scaillet, B., Pichavant, M., 2003. Experimental constraints on volatile abundances in arc magmas and their implications for degassing processes. In: Oppenheimer, C., Pyle, D.M., Barclay, J. (Eds), Volcanic Degassing. Geol. Soc. Spec. Pub., Vol. 213, pp. 23–52.

Scaillet, B., Pichavant, M., 2004. Crystallization conditions of Vesuvius phonolites. Geophys. Res. Abs. 6, p. 03764.

Signorelli, S., Capaccioni, B., 1999. Behavior of chlorine prior and during the 79 AD plinian eruption of Vesuvius (southern Italy) as inferred from the present distribution in glassy mesostases and whole-pumices. Lithos 46, 715–730.

Signorelli, S., Carroll, M.R., 2000. Solubility and fluid-melt partitioning of Cl in hydrous phonolitic melts. Geochim. Cosmochim. Acta 64, 2851–2862.

Signorelli, S., Vaggelli, G., Romano, C., 1999. Pre-eruptive volatile (H_2O, F, Cl, and S) contents of phonolitic magmas feeding the 3550-year old Avellino eruption from Vesuvius, southern Italy. J. Volcanol. Geotherm. Res. 93, 237–256.

Trigila, R., De Benedetti, A.A., 1993. Petrogenesis of Vesuvius historical lavas constrained by Pearce element ratios analysis and experimental phase equilibria. J. Volcanol. Geochem. Res. 58, 315–343.

Vaggelli, G., De Vivo, B., Trigila, R., 1993. Silicate melt inclusions in recent Vesuvius lavas (1631–1944). II. Analytical chemistry. J. Volcanol. Geotherm. Res. 58, 367–376.

Villemant, B., Trigila, R., De Vivo, B., 1993. Geochemistry of Vesuvius volcanics during 1631–1944 period. J. Volcanol. Geotherm. Res. 58, 291–311.

Wallace, P.J., 2003. From mantle to atmosphere: magma degassing, explosive eruptions, and volcanic volatile budgets. In: De Vivo, B., Bodnar, R.J. (Eds), Melt Inclusions in Volcanic Systems. Series: Developments in Volcanology, Vol. 5, Elsevier, Amsterdam, pp. 105–127.

Wallace, P.J., Anderson Jr., A.T., 2000. Volatiles in magmas. In: Sigurdsson, H. (Ed.), Encyclopedia of Volcanoes. Academic Press, New York, pp. 149–170.

Webster, J.D., 1992. Fluid melt interactions involving Cl-rich granites: experimental study from 2 to 8 kbar. Geochim. Cosmochim. Acta 56, 659–678.

Webster, J.D., 2004a. Magmatic volatiles and ore metal transport in hydrothermal systems. Penrose Conference Proceedings, Yellowstone National Park, September 2004, p. 53.

Webster, J.D., 2004b. The exsolution of magmatic hydrosaline melts. Chem. Geol. 210, 33–48.

Webster, J.D., De Vivo, B., 2002. Experimental and modeled solubilities of chlorine in aluminosilicate melts, consequences of magma evolution, and implications for exsolution of hydrous chloride melt at Mt. Somma-Vesuvius. Am. Mineral. 87, 1046–1061.

Webster, J.D., De Vivo, B., Tappen, C., 2003. Volatiles, magmatic degassing and eruptions of Mt. Somma-Vesuvius: constraints from silicate melt inclusions, solubility experiments and modeling. In: De Vivo, B., Bodnar, R.J. (Eds), Melt Inclusions in Volcanic Systems.Series: Developments in Volcanology, Vol. 5, Elsevier, Amsterdam, pp. 207–226.

Webster, J.D., Kinzler, R.J., Mathez, E.A., 1999. Chloride and water solubility in basalt and andesite liquids and implications for magmatic degassing. Geochim. Cosmochim. Acta 63, 729–738.

Webster, J.D., Raia, F., De Vivo, B., Rolandi, G., 2001. The behavior of chlorine and sulfur during differentiation of the Mt. Somma-Vesuvius magmatic system. Mineral. Petrol. 73, 177–200.

Webster, J.D., Sintoni, F., De Vivo, B., 2004a. Using magma geochemistry to interpret eruptive processes at Mt. Somma-Vesuvius and implications for forecasting eruptions. Workshop on Vesuvius and volcanism of the Campanian Plain, October 4–6, De Frede Editore Napoli (Abstracts), pp. 59–61.

Webster, J.D., Tappen, C., Sintoni, M.F., De Vivo, B., 2004b. Experimentally determined H_2O, S, and Cl solubilities in phonolite melt at 200 MPa: implications for volatile exsolution and eruption of Mt. Somma-Vesuvius magmas (abs.). EOS 85(17), V54B-03.

Volcanism in the Campania Plain:
Vesuvius, Campi Flegrei and Ignimbrites
edited by B. De Vivo

Chapter 12

Influence of hydrothermal processes on geochemical variations between 79 AD and 1944 AD Vesuvius eruptions

Annamaria Lima[a,*], *Benedetto De Vivo*[a], *Luca Fedele*[a] and *Maria Francesca Sintoni*[a,b]

[a]*University of Napoli "Federico II", Dipartimento di Scienze della Terra, 80134 Napoli, Italy*
[b]*Department of Earth and Planetary Sciences, American Museum of Natural History, New York, NY 10024-5192, USA*

Abstract

The authors discuss the compositional differences in homogenized silicate melt inclusions in olivine and clinopyroxene from cumulate nodules entrained in 79 AD plinian and 1944 AD interplinian eruptions. Such compositions are compared with those of melt inclusions in these same crystals in the volcanics and with bulk rock compositions in order to examine differences between the "closed system" represented by silicate melt inclusions and the "open system" represented by bulk rocks.

Homogenized melt inclusions (HMI) in olivine and clinopyroxene from cumulate nodules collected in 79 AD volcanics, show higher P_2O_5/K_2O, Na_2O/K_2O, Cl/K_2O and S/Cl values compared with HMI in olivine from nodules collected in 1944 AD volcanics. In addition, the marked geochemical differences as P_2O_5/K_2O, Na_2O/K_2O, Cl/K_2O and S/Cl variations between 79 AD and 1944 AD melt inclusions and 79 AD and 1944 AD bulk rocks have been interpreted as being strongly influenced by hydrothermal processes active before and during explosive eruptions. During the 1944 AD interplinian activity at Vesuvius, magmas continuously ascended through the system, undergoing simultaneous degassing and fractionation. The loosely cemented conduit within the volcano did not allow significant build up of volatile pressure, leading to frequent, non-violent interplinian eruptions. The 79 AD plinian explosive eruption occurred after a long repose time at closed-conduit condition. The conduit became closed as a result of the cooling of ascending magma at the end of a previous interplinian eruption.

1. Introduction

The present work is a summary of a more extended paper (Lima et al., in press). Its purpose is to highlight the more important aspects of a model proposed to explain geochemical variations in the Somma-Vesuvius eruptions as a function of eruptive mechanisms.

Mt. Somma-Vesuvius is a composite volcano situated on a NW-SE fault in the southernmost end of the Roman Province, south of the Campanian Plain. Its eruptive activity began at least 35 ka BP (Scandone et al., 1993), and according to Rolandi et al. (1998), the current Vesuvius cone was built by post-472 AD interplinian activity that followed the 472 AD subplinian eruption. Activity which can be attributed to the Mt. Somma system extends to as recent as 3.5 ka according to Webster et al. (2001).

---

*Corresponding author. *E-mail address:* anlima@unina.it (A. Lima).

Eruptive styles and cyclicity of the Mt. Somma-Vesuvius system have varied widely. Effusive lava flows and scoria eruptions were the most common type of volcanism during the last 3.5 ka, but they were periodically superseded by highly explosive plinian eruptions of pumice and ash, which also involved pyroclastic flows and surges.

Mt. Somma-Vesuvius products include basic, silica-undersaturated, potassic rocks, ranging from tephrite to phonolite and less-alkaline basaltic trachyandesites.

2. Previous studies

Santacroce et al. (1993), suggested that a single magma reservoir is located at a shallow depth within the limestone basement, whereas other authors, using isotopic and CO_2-rich fluid inclusion data, inferred the existence of two or more magma chambers (Cortini and Hermes, 1981; Belkin et al., 1985; Cortini et al., 1985; Belkin and De Vivo, 1993; Marianelli et al., 1999). High-resolution seismic tomography failed to resolve a shallow magma chamber, identifying instead a sharp transition to a low-velocity zone at >8 km, which may represent the top of a magma chamber (Zollo et al., 1998; Auger et al., 2001).

Lima et al. (2003) based on fluid (FI) and melt inclusions (MI) and on other petrologic data (De Vivo et al., 2003) suggest that the entire Somma-Vesuvius volcanic system resembles a complex feeding column which is dominated by multiple mush zone environments (small magma chambers), and thus includes a variety of local crystallization environments characterized by contrasting cooling rates and P–T conditions (Marsh, 1995). The shallower chamber with high aspect ratio occurs at a pressure of about 925–1000 bars. On the basis of petrography and mineral compositions, Lima et al. (2003) suggest that clinopyroxenes in cumulate nodules were formed at variable stages of fractionation of a single magmatic system. Higher concentrations of volatile elements in MI in more primitive phenocrysts imply that they crystallized at a higher pressure in a deeper magma chamber. The authors pointed out that the very tight major element compositional trends, formed by "basaltic" post-472 AD Vesuvius volcanic rocks, indicate that the composition of the erupting evolved melts has changed little since 472 AD, and that the magma chamber supplying post-472 AD interplinian eruptions is essentially in a steady-state condition.

Much of the information available on the Somma-Vesuvius hydrothermal system is derived from studies of skarn xenoliths. Skarns are Ca-Fe-Mg-Mn-rich silicate rocks formed by high-temperature metasomatic reactions mostly involving carbonate rocks (e.g. Einaudi et al., 1981). An aqueous fluid phase is always involved in such metasomatic reactions (e.g. Kwak, 1986; Meinert et al., 1997). At Somma-Vesuvius, skarn ejecta are characterized by a specific mineralogy that includes vesuvianite, wollastonite, anorthite, phlogopite, gehlenite, scapolite and clinopyroxene. Mineral zonation at the contact with the carbonate rocks is common in the plinian and subplinian eruption products. Skarn ejecta are potentially valuable sources of information to understand fluid evolution at magma chamber walls.

Gilg et al. (2001) calculated the pressure of skarn formation from the densities of CO_2 inclusions in wollastonite in skarn nodules. Results indicate trapping pressures from 660 to 1368 bars (assuming a formation temperature of ~1000°C and using Brown and Lamb, 1989, equation of state for CO_2). Gilg et al. (2001) also calculated the salinity of multiphase aqueous brine inclusions of 43–52 wt% NaCl eq., with total homogenization temperatures

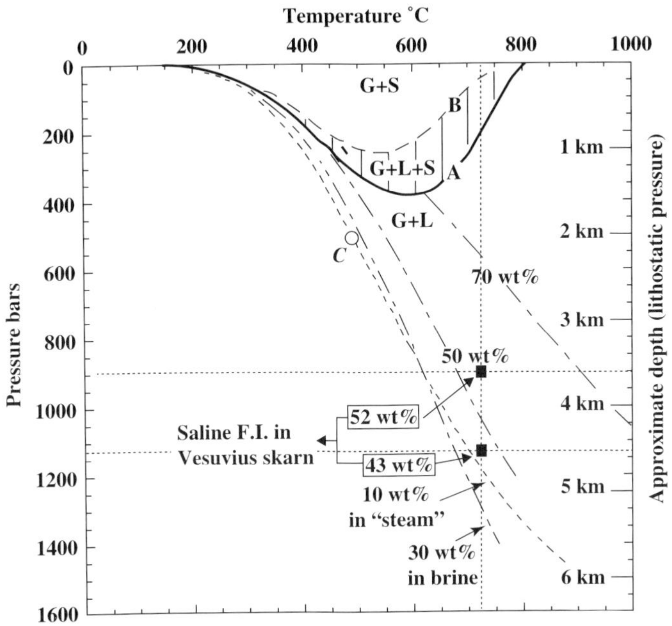

Figure 1. Temperature-depth diagram showing phase relations in the system NaCl-H_2O with depth corresponding to lithostatic conditions. Multiphase aqueous brine inclusions, yielding salinity between 43 and 53 wt% NaCl eq., found in Somma-Vesuvius skarn ejecta (Gilg et al., 2001) would be trapped at a pressure between 900 and 1100 bars, equivalent to about 3.6–4.5 km, assuming a temperature of 720°C. G, gas; L, liquid; S, solid; dot-dashed lines are contours of constant wt% NaCl dissolved in brine; short dashed line shows the boiling point curve for a 10 wt% NaCl solution at pressures and temperatures above its critical point C. Curve A shows the three-phase boundary G+L+S for the system NaCl-H_2O; curve B shows the three-phase boundary G+L+S for the system NaCl-KCl-H_2O, with Na/K in solution fixed by equilibrium with albite and K-feldspar at the indicated temperatures (after Fournier, 1987).

ranging between 720°C and 820°C (Fig. 1). Phase relations in the systems NaCl-H_2O and NaCl-KCl-H_2O provide a good, first approximation of how salinity is likely to vary in hydrothermal fluids (Cline and Bodnar, 1994) exsolved from crystallizing magmas, where fluid pressure (P_f) is controlled by lithostatic pressure (Fig. 1).

Gilg et al. (2001) deduce that there is no evidence for a convectively cooling hydrothermal system at the magma–carbonate wall rock interface at Somma-Vesuvius, based on lack of participation of externally derived fluids, such as meteoric waters or formational fluids. Chiodini et al. (2001) suggest that NaCl brines reside in the high-temperature reservoir beneath Vesuvius and influence the chemical composition of the gases discharged by the fumaroles at the bottom of the crater. During the present period of repose, geochemical evidence indicates that there was no input of fresh magma at shallow depths after the end of the last eruptive period (Chiodini et al., 2001; Lima et al., 2003).

3. Results

Compositional data of homogenized silicate melt inclusions (HMI) in olivine and clinopyroxene from cumulate nodules ejected by the 79 AD plinian and the 1944 AD interplinian eruptions have been compared with compositions of MI in these same crystals in the volcanics and with bulk rock compositions in order to examine differences between the "closed system" represented by HMI and the "open system" represented by bulk rocks.

New cumulate nodules have been sampled for the MI study. Sample DV506 has been collected at Terzigno in the 79 AD volcanic products and sample DV11 was collected on the slopes of Vesuvius, in the products of the 1944 activity. Variation diagrams for HMI in these nodule crystals are shown in Figure 2(A–D). Included for comparison are 1944 AD bulk rock compositions (Ayuso et al., 1998) and S3 cumulate nodule olivine MI data (Lima et al., 2003). Figure 2(E–H) for both 79 AD plinian and 1944 AD interplinian eruptions shows cumulate nodule MI average compositions, MI average compositions from volcanic crystals (Raia et al., 2000; Webster et al., 2001) and correlative average bulk rock compositions. Table 1 shows all average values and 1 sigma precision.

To plot 1944 AD interplinian volcanics shown in Figure 2(A–C), (Ayuso et al., 1998), compositions of the equilibrium olivine for each rock sample have been calculated using the olivine-melt equilibrium model of Ford et al. (1983), assuming an $Fe^{2+}e^{3+}$ value of 6 (just below the NNO buffer for Vesuvius compositions). MI compositions from phenocrysts in the volcanic rocks (Raia et al., 2000; Webster et al. 2001) have not been plotted in Figure 2(A–D) because host crystal Mg# are not available. Homogenized MI in olivine and clinopyroxene from cumulate nodule in 79 AD volcanics, show higher P_2O_5/K_2O, Na_2O/K_2O, Cl/K_2O and S/Cl values compared with HMI in olivine from nodules in 1944 AD volcanics (Fig. 2A–D). The latter also shows relatively constant Na_2O/K_2O values (~0.35, Fig. 2B,F).

Although nodule samples were collected in the same eruption where studied phenocrysts from volcanics were found, it is still important to determine when individual phenocrysts found in the ejecta formed or, alternatively, when MI were trapped in phenocrysts. We estimated that MI in olivine and clinopyroxene from nodules represent an earlier stage of magma evolution compared with MI trapped in phenocrysts from volcanics due to the fact that they have a less-evolved composition as showed in Figure 3.

Figure 2E and Table 1 show that the average P_2O_5/K_2O values for the 79 AD samples vary from 0.24 in cumulate-hosted MI to 0.17 (decrease of 29%) in MI from volcanic rock crystals with a strong decrease down to 0.02 (decrease of 88%) in bulk rock. The average P_2O_5/K_2O values for the 1944 AD interplinian vary from 0.18 in cumulate-hosted MI (Fig. 2E), to 0.12 in volcanics MI (decrease of 33%), whereas no variation between MI from volcanic rock crystals and bulk rock is observed. The trend of decreasing P_2O_5/K_2O ratio from MI in cumulate crystals to MI in volcanic rock crystals for both 79 AD and 1944 AD is a result of apatite crystallization that lowers the P concentration in the melt during magma evolution. Compared with bulk rock compositions, MI in 1944 AD volcanic rock crystals show no variation in P_2O_5/K_2O (there is no difference between closed- and open-system behavior). The large P_2O_5/K_2O depletion observed in 79 AD bulk rock could be explained by hydrothermal fluids that (as discussed in the following paragraphs) lowers the P in the magmatic system.

Average Na_2O/K_2O values vary in different ways (Fig. 2F). They decrease from 0.49 in cumulate-hosted MI to 0.37 (24%) in MI from volcanic rock crystals and then increase to 0.54 (31%) in bulk rock. The average Na_2O/K_2O values for 1944 AD (Fig. 2F) show only very weak variations from cumulate-hosted MI to MI from volcanic rock crystals to bulk rock composition (decrease of 3% and 6%, respectively). The 1944 AD MI and bulk rock

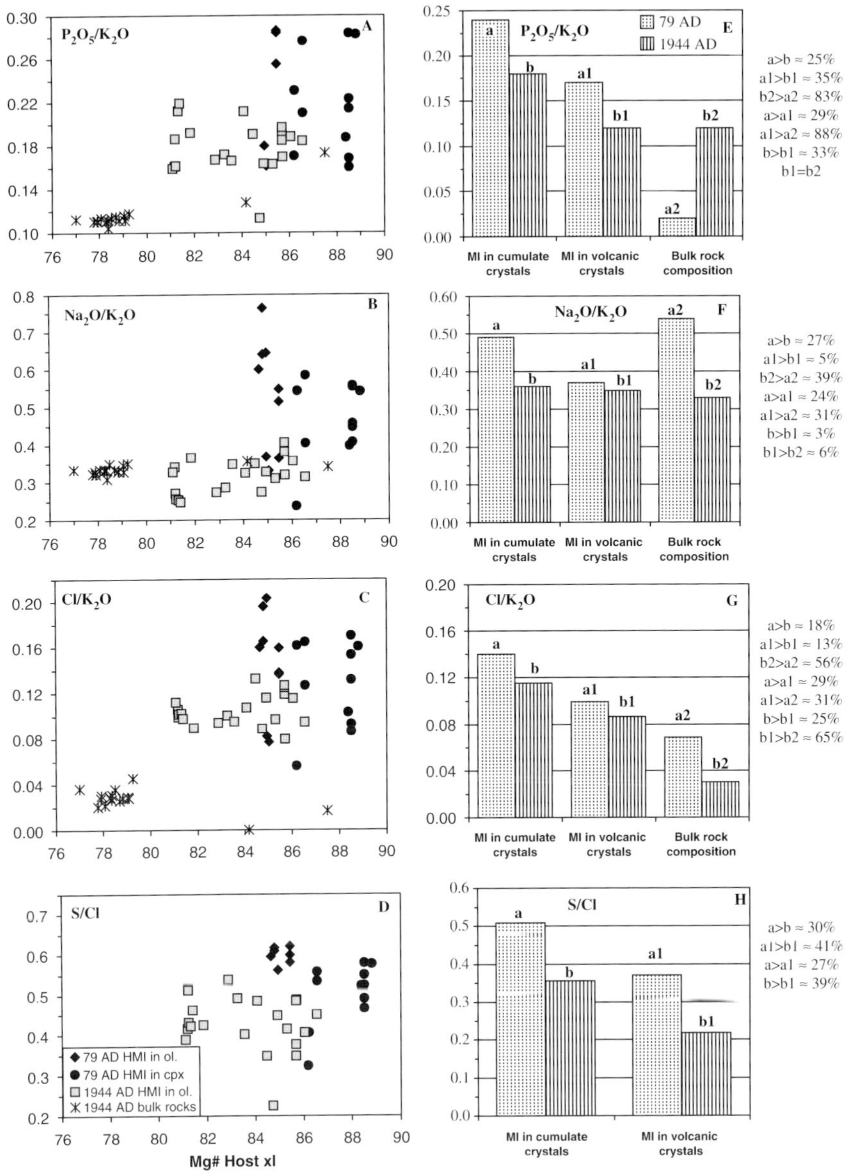

Figure 2. (A–D) Ratio of selected elements in homogenized MI in 79 AD and 1944 AD olivine and clinopyroxene from Somma-Vesuvius cumulate nodules, plotted against host mineral Mg#. In A–C the compositions of 1944 AD interplinian volcanics from Ayuso et al., 1998 are also shown for comparison. The compositions of the equilibrium olivine for each rock sample have been calculated using the olivine-melt equilibrium model of Ford et al., 1983, assuming an Fe^{2+}/Fe^{3+} value of 6. For Vesuvius compositions, this value corresponds to oxygen fugacity just below the NNO buffer. (E–H) Average compositions of 79 AD plinian cumulate nodules HMI, volcanic rock HMI and correlative bulk rock volcanic compositions; average compositions of 1944 AD cumulate nodules HMI, volcanic rock HMI and correlative bulk rock volcanic compositions. Volcanic rock crystal HMI compositions are from Webster et al. (2001) and Raia et al. (2000); volcanic bulk rock compositions are from Ayuso et al. (1998).

Table 1. P_2O_5/K_2O, Na_2O/K_2O, Cl/K_2O and S/Cl average values in cumulates olivine and clinopyroxene HMI, volcanic rock olivine and clinopyroxene HMI and average volcanic bulk rock compositions.

	Average value of	79 AD	1944 AD
P_2O_5/K_2O	MIs in nodule ol and cpx[a]	0.24 (0.06)	0.18 (0.01)
	MIs in volcanics ol and cpx[b]	0.17 (0.06)	0.12 (0.03)
	Volcanic compositions (b.r.)[c]	0.02 (0.01)	0.12 (0.02)
Na_2O/K_2O	MIs in nodule ol and cpx[a]	0.49 (0.12)	0.36 (0.03)
	MIs in volcanics ol and cpx[b]	0.37 (0.05)	0.35 (0.01)
	Volcanic compositions (b.r.)[c]	0.54 (0.07)	0.33 (0.01)
Cl/K_2O	MIs in nodule ol and cpx[a]	0.14 (0.04)	0.12 (0.02)
	MIs in volcanics ol and cpx[b]	0.10 (0.01)	0.09 (0.02)
	Volcanic compositions (b.r.)[c]	0.07 (0.005)	0.03 (0.01)
S/Cl	MIs in nodule ol and cpx[a]	0.52 (0.1)	0.36 (0.06)
	MIs in volcanics ol and cpx[b]	0.37 (0.1)	0.22 (0.05)
	Volcanic compositions (b.r.)[c]	<0.05	<0.05

ol, olivine; cpx, clinopyroxene.

Note: 1 sigma precision is given in parenthesis.
[a]This study data and Lima et al. (2003).
[b]Data from Webster et al. (2001) and Raia et al. (2000).
[c]Data from Ayuso et al. (1998).

compositions (Fig. 2F) show Na_2O/K_2O ratios that are relatively constant (averaging ~0.35). Sodium behaves as an incompatible element during magma fractionation. The increase of Na_2O/K_2O ratio in 79 AD bulk rock (~39% compared with 1944 AD bulk rock and ~31% compared with 79 AD MI in volcanic rocks) implies that during magma ascent there was Na enrichment in the magmatic system (as discussed in the following paragraphs).

The average Cl/K_2O values (Fig. 2G) for 79 AD decrease in a manner similar to cumulate-hosted MI to MI from volcanic rock crystals and to bulk rock (decrease of 29% and 31%, respectively). The average Cl/K_2O value for 1944 AD decreases from 0.12 to 0.09 (25%) from cumulate-hosted MI to MI from volcanic rock crystals (Fig. 2G). Bulk rocks show a strong Cl/K_2O decrease, up to 0.03 (65%). MI from the 79 AD and 1944 AD cumulate and volcanic crystals, and bulk rock compositions, show a progressive decrease in Cl/K_2O ratio. Somma-Vesuvius 79 AD and 1944 AD magmas were affected by Cl loss during their evolution. Bulk rock compositions from 1944 AD show a larger Cl/K_2O depletion compared with plinian 79 AD bulk rocks (~65% and ~31%, respectively, compared with correlative MI in volcanic rock crystals). This implies that removal of Cl was more efficient in pre-eruptive condition and/or during the 1944 AD interplinian eruption. On the other hand, Cl solubility is extremely sensitive to melt composition. Crystallization of Ca, Mg, Fe and Al-rich minerals dramatically reduces Cl solubility (Webster and De Vivo, 2002; Webster et al., 2003; Mathez and Webster, 2005), so the large difference in Cl/K_2O ratios in 79 AD and 1944 AD bulk rocks (56%) might reflect pre-eruptive magma composition and pressure condition.

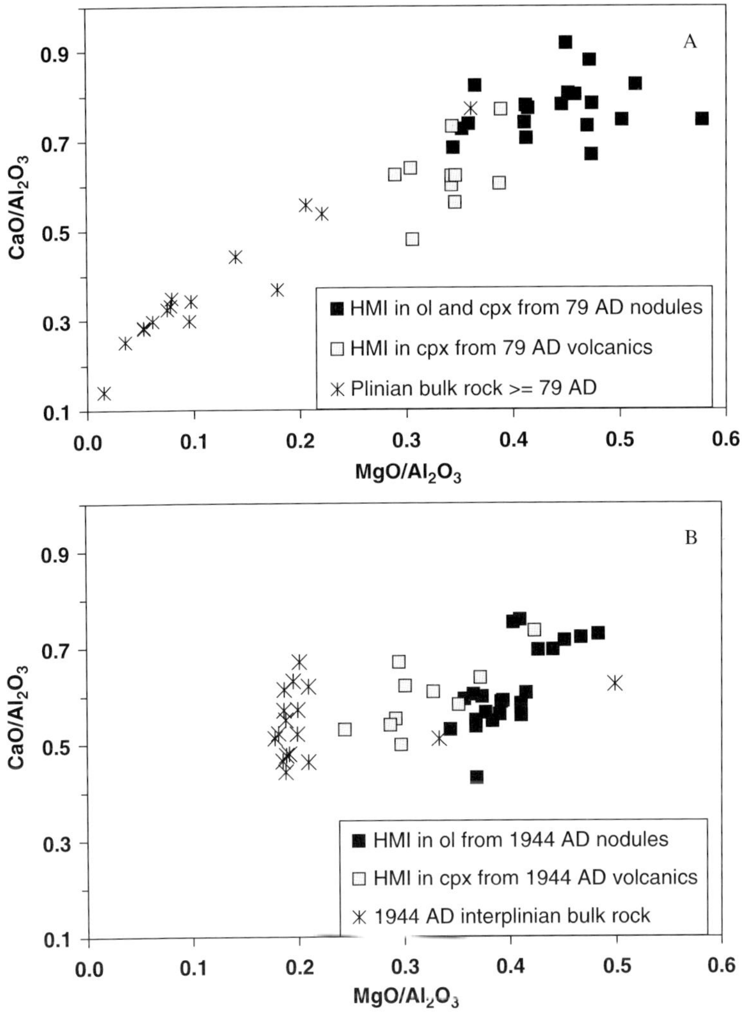

Figure 3. MgO/Al_2O_3 versus CaO/Al_2O_3 in HMI in olivine and clinopyroxene from cumulate nodules and from volcanic rocks, bulk rock compositions also are shown for 79 AD (A) and for 1944 AD (B) samples.

Average S/Cl values (Fig. 2H) can be compared only between cumulate and volcanic rocks MI data. There is a decrease from 0.52 to 0.37 (27%) for 79 AD and a decrease from 0.36 in cumulates MI to 0.22 (39%) in MI from volcanic rock crystals for 1944 AD (Fig. 2H). MI in olivine and clinopyroxene from cumulates and volcanic rocks show S/Cl depletion (~27% and ~39% for 79 AD and 1944 AD, respectively) for both plinian and interplinian events (Fig. 2H). These differences imply that S loss during early fractionation of the 1944 AD interplinian magma was more efficient. In addition, MI from 79 AD plinian crystals show an evolutionary trend with a higher S/Cl ratio compared with MI in 1944 AD interplinian crystals (Fig. 4). Experimental studies have determined that S solubility varies markedly with melt composition, especially with Fe content (Carroll and Webster, 1994). As a result,

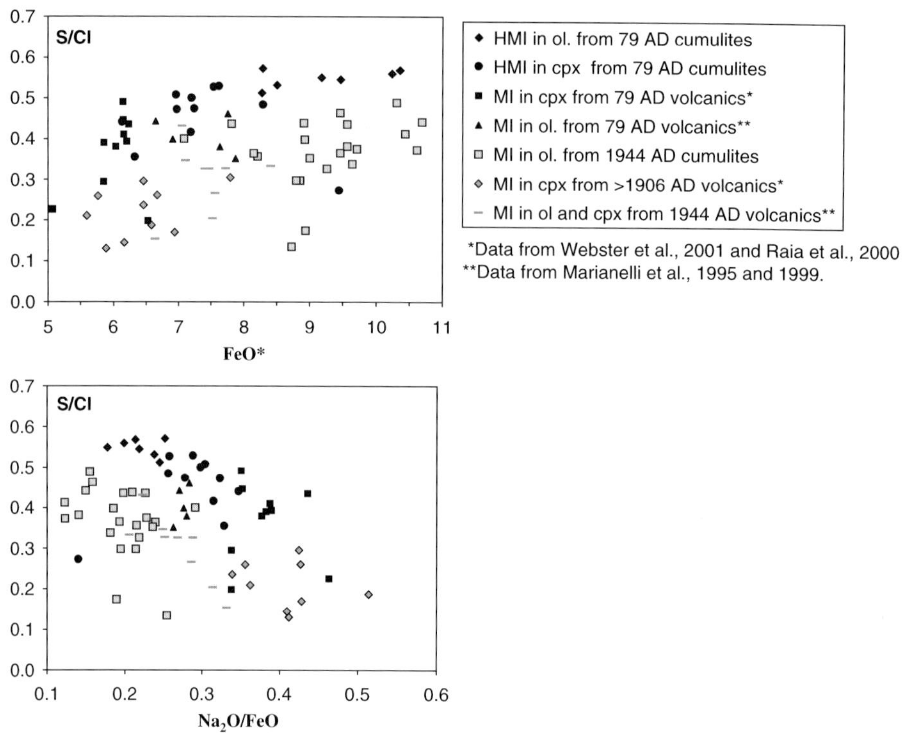

Figure 4. (A) S/Cl versus FeO* total iron and (B) S/Cl versus Na_2O/FeO^* in HMI in olivine and clinopyroxene from cumulate nodules and from volcanic rocks.

magmas may saturate and exsolve an S-rich aqueous volatile phase if the melts evolve to compositions characterized by lower S solubilities. Since the 79 AD plinian cumulate MI clearly display a higher S/Cl content compared with 1944 AD cumulate MI, we argue that HMI in 79 AD plinian nodules and in volcanic rock olivine and clinopyroxene have been trapped when the magma was in a state of volatile saturation under lithostatic pressure (close-conduit condition). This agrees with Lima et al. (2003) suggesting that crystallization at high pressure in a deep magma chamber is the cause of higher volatiles content in primitive phenocrysts MI, and with Webster et al. (2001), confirming that plinian and subplinian magma events have higher H_2O and S contents.

4. Evolution of the hydrothermal system at Somma-Vesuvius

The marked geochemical differences at Somma-Vesuvius between plinian (79 AD) and interplinian (1944 AD) products could represent the effects of a hydrothermal system present near the upper portion of the magma chamber.

During the 1944 AD activity, magmas continuously ascended through the system, undergoing simultaneous degassing and fractionation. This could explain part of the Cl/K_2O and S/Cl variations discussed in the previous paragraph. A loosely cemented conduit within the volcano would not allow significant build up of volatile pressure, leading to frequent, non-violent interplinian eruptions. This combination of ongoing magma supply and frequent eruptions in nearly steady-state conditions explains why there is little variation in the composition of the 1944 AD erupting melt (Fig. 2A–C). The uneven distribution of earlier formed phenocrysts in the ascending magmas explains compositional variations observed in the interplinian volcanic rocks (Lima et al., 2003).

The 79 AD plinian explosive eruption occurred after a long repose time at closed-conduit conditions. The conduit became closed as a result of the cooling of ascending magma at the end of a previous interplinian eruption. As described by Fournier (1999), closing of the system is enhanced by self-sealing processes, such as precipitation of newly formed hydrothermal minerals. Beneath Somma-Vesuvius (as modeled also for Campi Flegrei by De Vivo and Lima, this volume), near the upper portion of the shallow magma chamber, a carapace, surrounded by an impermeable rind, forms and acts as an interface between the brittle rocks above and the plastic rocks below. As shown in Figure 5A (modified from Fournier, 1999), in brittle rocks fluids circulate under hydrostatic pressure conditions. Here, the strain rate needed to cause shear failure of a preexisting open cracks is highly dependent on the coefficient of friction, on fluid pressure (P_f), and on the orientation of the fracture with respect to the stress field (Sibson, 1984). In the plastic rocks at lithostatic pressure conditions (where the least principal stress is the lithostatic load), brine and steam exsolve from crystallizing magma (as it is in the case of Vesuvius, Webster et al., 2001; Webster and De Vivo, 2002; Webster et al., 2003) and tend to accumulate in thin, overlapping horizontal lenses. In this region, the stress difference required to initiate plastic deformation is highly dependent on temperature, strain rate and rock type. When breaches of the self-sealed system occur (Fig. 5B), steam and brine are discharged into the brittle, hydrostatically pressured domain and faulting, brecciation, hydrothermal alteration and vein mineralization occur along with an explosive plinian eruption.

Several different mechanisms may play a role in triggering major breaches of the self-sealed zone (Fig. 5B). In 79 AD, an upward surge of new magma temporarily increased the local strain rate causing shear failure in a previously plastic material, allowing hypersaline brine and gas to be released quickly from the normally plastic region into the lower pressure and lower temperature brittle domain, where hydrothermal veins form as a result of decompression and cooling of the magmatic fluid. This mechanism produced at Somma-Vesuvius skarn-type mineralization including chlorides, sulfates and carbonates. In particular, fluoride-rich saline fluids led to the formation of accessory Ti-, Zr-, Th-, U- and REE-bearing minerals in vesuvianite skarns (Gilg et al., 2001; Fulignati et al., 2004). Based on the data from FI in the skarns (Fulignati et al., 2000; Gilg et al., 2001), the brittle-plastic transition beneath Somma-Vesuvius should occur at 900–1100 bars (3.6–4.5 km depth), at a temperature of about 720°C. Figure 5(A,B) shows an average indicated depth equivalent to 1000 bars pressure.

During plinian events, rapid expulsion of steam and brine into the overlying part of the system enhances hydraulic fracturing and brecciation. When fluid pressure (P_f) drops as a result of breaching the sealed zone, brines will boil producing a large volume of steam and also, if the pressure drops below about 300 bars, coexisting high-salinity brine or vapor plus salt.

At near-magmatic temperatures, decompression from lithostatic to hydrostatic Pf increases brines salinity and produces a lower-salinity vapor phase, shifting the system into the field of superheated steam plus solid salt at low pressure, with important implications

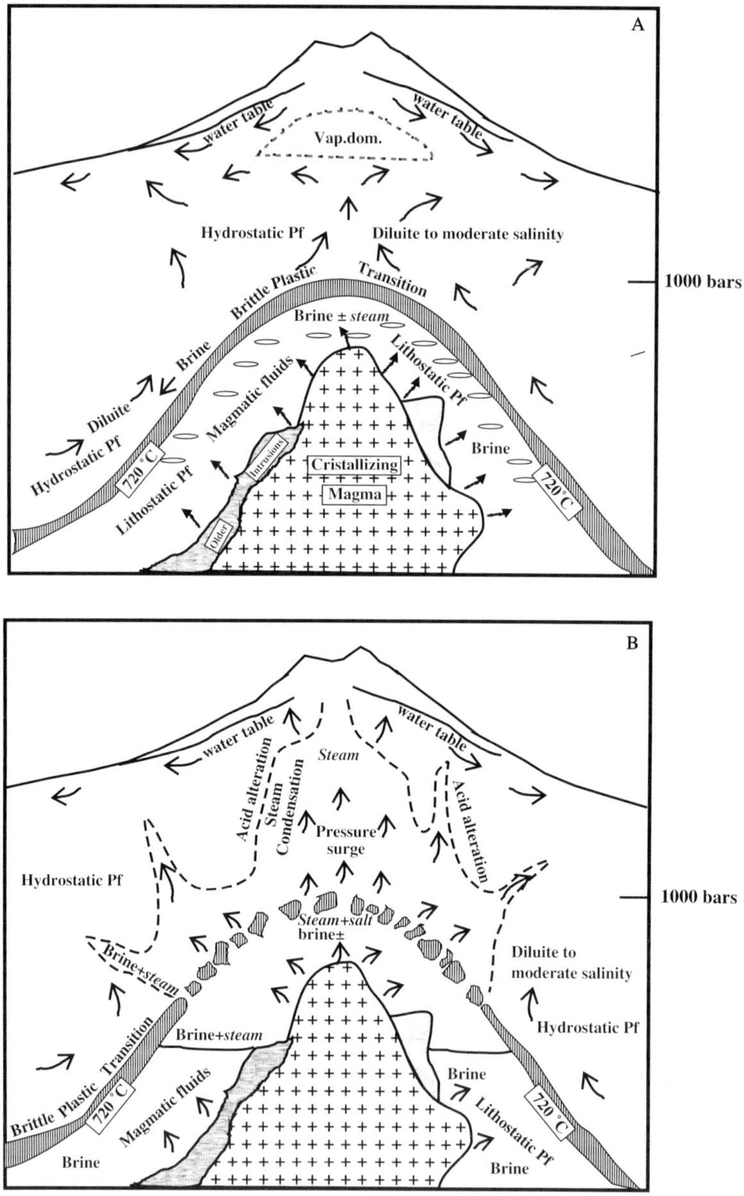

Figure 5. Schematic model (modified after Fournier, 1999) showing the transition from lithostatic to hydrostatic condition in the Somma-Vesuvius subvolcanic environment. (A) The brittle to plastic transition, consisting of a self-sealed zone, should occur at a pressure between 900 and 1100 bars (mean value is indicated) equivalent to about 3.6–4.5 km depth assuming a temperature of about 720°C. Temperature and fluid pressure (P_f) gradients across the interface are very steep. Dilute, dominantly meteoric water circulates at hydrostatic pressure in brittle rock, whereas highly saline brines of magmatic origin accumulate at lithostatic pressure in plastic rock. (B) An upward surge of magma temporarily increases the local strain rate to such a degree that previously plastic material undergoes shear failure in response to the stress difference. When the self-sealed zone is breached, steam and brine are discharged into the brittle, hydrostatically pressured domain. Faulting, brecciation, hydrothermal alteration and vein mineralization ensue along with explosive eruption.

on the amount of HCl that can be transported when (and if) the vapor phase escapes into overlying rock. HCl tends to partition into the vapor phase in equilibrium with boiling brine (Candela and Piccoli, 1995; William et al., 1995). Fournier and Thompson (1993) measured HCl concentration in the vapor phase finding it greater than predicted when NaCl precipitates (at a pressure below about 300 bars, Fig. 1); the reaction of NaCl with H_2O becomes important and hydrolysis reactions produce HCl and NaOH that remain preferentially in the melts (Veksler, 2004).

The above hydrothermal model could also explain the phosphorus depletion in 79 AD plinian bulk rocks and the loss of other metallic elements. In fact, phosphorus likely forms primary phosphate deposits depending on the physical characteristics of the vapor phase separating from boiling brines during decompression from lithostatic to hydrostatic Pf.

The escape of magmatic fluids from plastic rock at Vesuvius under normal circumstances and without a major perturbing event, is mostly represented by gas (Chiodini et al., 2001; Federico et al., 2002). Likely, during very long repose times, such as the one before the 79 AD plinian eruption, highly differentiated volcanic products were formed by in situ processes (Shinohara et al., 1995). A necessary condition for volatile transfer to be significant is melt supersaturation with respect to H_2O (Trial and Spera, 1990).

5. Conclusions

Compositions of HMI in olivine and clinopyroxene (from 79 AD and 1944 AD cumulate nodules and from 79 AD and 1944 AD volcanic rocks) and bulk rock compositions (from 79 AD and 1944 AD eruptions) have been compared to examine differences between the "closed system" represented by MI, and the "open system" represented by bulk rocks. P_2O_5/K_2O, Cl/K_2O and Na_2O/K_2O variations between 79 AD and 1944 AD MI and 79 AD and 1944 AD bulk rocks have been interpreted as being strongly influenced by hydrothermal processes active before and during explosive eruptions.

We propose a model where interplinian activity (e.g. 1944 AD) is the result of a steady state under the volcano, an "open-system" combination of ongoing magma supply and frequent eruptions which also limits the chemical variations of erupted products (Lima et al., 2003). On the other hand, plinian eruptions occur in "closed-system" conditions, created by cooling of magma at the end of previous interplinian eruptions when, most likely at a depth between 3.6 and 4.5 km beneath the Somma-Vesuvius, a carapace, surrounded by an impermeable rind, forms and acts as an interface between the brittle and plastic rocks (Fig. 5). Several different mechanisms may play a role in triggering major breaches of the self-sealed zone. An upward surge of new magma immediately before the 79 AD plinian eruption at Somma-Vesuvius temporarily increased the local strain rate to such a degree that previously plastic material underwent shear failure in response to a stress difference.

When a sealed zone is breached, fluid pressure (P_f) drops and brines boil producing a large volume of steam ± high-salinity brines or vapor plus salt. Hydrolysis reactions of NaCl with H_2O could explain the greater average Na_2O/K_2O content found in 79 AD bulk rocks at Vesuvius, while phosphorous enrichment could be explained by the formation of hydrothermal phosphate veins deposited as a result of decompression and cooling of the magmatic fluid.

Volatile-rich MI in olivine and clinopyroxene from 79 AD plinian nodules and from volcanic rock are trapped when the magma is in the condition of a "closed system" in a state of volatile saturation under lithostatic pressure.

Acknowledgements

The authors are very grateful to J.D. Webster for reviewing a preliminary version of the manuscript, and to R.J. Bodnar and L.V. Danyushevsky for their constructive review and helpful suggestions that improved the final version. The research benefited from funds from MIUR-PRIN grants in 2003 and 2004 to B. De Vivo.

References

Auger, E., Gasparini, P., Virieux, J., Zollo, A., 2001. Seismic evidence of an extended magmatic sill under Mt. Vesuvius. Science 294, 1510–1512.

Ayuso, R.A., De Vivo, B., Rolandi, G., Seal II, R.R., Paone, A., 1998. Geochemical and isotopic Nd-Pb-Sr-O variations bearing on the genesis of volcanic rocks from Vesuvius, Italy. In: Spera, F.J., De Vivo, B., Ayuso, R.A., Belkin, H.E. (Eds), Vesuvius Special Issue. J. Volcanol. Geotherm. Res. 82, 53–78.

Belkin, H.E., De Vivo, B., 1993. Fluid inclusion studies of ejected nodules from plinian eruptions of Mt. Somma-Vesuvius. In: De Vivo, B., Scandone, R., Trigila, R. (Eds), Vesuvius Special Issue. J. Volcanol. Geotherm. Res. 58, 89–100.

Belkin, H.E., De Vivo, B., Roedder, E., Cortini, M., 1985. Fluid inclusion geobarometry from ejected Mt. Somma-Vesuvius nodules. Am. Mineral. 70, 288–303.

Brown, P.E., Lamb, W.M., 1989. P–V–T properties of fluids in the system $H_2O \pm CO_2 \pm NaCl$: new graphical presentations and implications for fluid inclusion studies. Geochim. Cosmochim. Acta 53, 1209–1221.

Candela, P.A., Piccoli, P.M., 1995. Model ore-metal partitioning from melts into vapor and vapor/brine mixtures. Mineral. Assoc. Can. Short Course 23, 101–122.

Carroll, M.R., Webster, J.D., 1994. Solubilities of sulfur, noble gas, nitrogen, chlorine, and fluorine in magmas. Rev. Mineral. 30, 231–279.

Chiodini, G., Marini, L., Russo, M., 2001. Geochemical evidence for the existence of high-temperature hydrothermal brines at Vesuvio volcano, Italy. Geochim. Cosmoch. Acta 65, 2129–2147.

Cline, J.S., Bodnar, R.J., 1994. Direct evolution of brine from crystallizing silicic melt at the Questa, New Mexico, molybdenum deposit. Econ. Geol. 89, 1780–1802.

Cortini, M., Hermes, O.D., 1981. Sr isotopic evidence for a multi-source origin of the potassic magmas in the Neapolitan area S. Italy. Contrib. Mineral. Petrol. 77, 47–55.

Cortini, M., Lima, A., De Vivo, B., 1985. Trapping temperatures of melt inclusions from ejecta Vesuvian mafic xenoliths. J. Volcanol. Geotherm. Res. 26, 167–172.

De Vivo, B., Ayuso, R.A., Belkin, H.E., Fedele, L., Lima, A., Rolandi, G., Somma, R., Webster, J.D., 2003. Chemistry, fluid/melt inclusions and isotopic data of lava, tefra and nodules from >25 ka to 1944 AD of the Mt. Somma-Vesuvius volcanic activity. Mt. Somma-Vesuvius Geochemical Archive, Dipartimento di Geofisica e Vulcanologia, Università di Napoli Federico II, Open-File Report 1-2003, 143pp.

De Vivo, B., Lima, A., 2006. A hydrothermal model for ground movements (bradyseism) at Campi Flegrei, Italy. In: De Vivo, B. (Ed.), Volcanism in the Campania Plain: Vesuvius, Campi Flegrei and Ignimbrites. Series: Developments in Volcanology. Elsevier, Amsterdam, 9, 291–320.

Einaudi, M.T., Meinert, L.D., Newberry, R.J., 1981. Skarn deposits. Econ. Geol. 75th Anniv. Vol. 317–391.

Federico, C., Aiuppa, A., Allard, P., Bellomo, S., Jean-Baptiste, P., Parello, F., Valenza, M., 2002. Magma-derived gas influx and water-rock interactions in the volcanic aquifer of Mt. Vesuvius, Italy. Geochim. Cosmochim. Acta 66, 963–981.

Ford, C.E., Russell, D.G., Craven, J.A., Fisk, M.R., 1983. Distribution coefficients of Mg^{2+}, Fe^{2+}, Ca^{2+} and Mn^{2+} between olivine and melt. J. Petrol. 24, 256–265.

Fournier, R.O., 1987. Conceptual models of brine evolution in magmatic-hydrothermal system. U.S. Geological Survey Professional Paper 1350, 2, 1487–1506.

Fournier, R.O., 1999. Hydrothermal processes related to movement of fluid from plastic into brittle rock in the magmatic-epithermal environment. Econ. Geol. 94(8), 1193–1212.

Fournier, R.O., Thompson, J.M., 1993. Composition of steam in the system NaCl-KCl-H2O-quartz at 600°C. Geochim. Cosmochim. Acta 57, 4365–4375.

Fulignati, P., Marianelli, P., Santacroce, R., Sbrana, A., 2000. The skarn shell of the 1944 Vesuvius magma chamber. Genesis and P-T-X conditions from melt and fluid inclusion data. Eur. J. Mineral. 12, 1025–1039.

Fulignati, P., Marianelli, P., Santacroce, R., Sbrana, A., 2004. Probing the Vesuvius magma chamber-host rock interface through xenoliths. Geol. Mag. 141, 417–428.

Gilg, H.A., Lima, A., Somma, R., Belkin, H.E., De Vivo, B., Ayuso, R.A., 2001. Isotope geochemistry and fluid inclusion study of skarns from Vesuvius. Mineral. Petrol. 73, 145–176.

Kwak, T.A.P., 1986. Fluid inclusions in skarns carbonate replacement deposits. J. Metamorph. Petrol. 4, 363–384.

Lima, A., Danyushevsky, L.V., De Vivo, B., Fedele, L., 2003. A model for the evolution of the Mt. Somma-Vesuvius magmatic system based on fluid a melt inclusions investigations. In: De Vivo, B., Bodnar, R.J. (Eds), Melt Inclusions in Volcanic Systems: Methods, Applications and Problems. Series: Developments in Volcanology, Vol. 5. Elsevier, Amsterdam.

Lima, A., De Vivo, B., Fedele, L., Sintoni, M.F., in press. Geochemical variations between the 79 AD and 1944 AD Mt. Somma-Vesuvius volcanic products: constraints on the evolution of the hydrothermal system based on fluid and melt inclusions. Chem. Geol.

Marianelli, P., Métrich, N., Santacroce, R., Sbrana, A., 1995. Mafic magma batches at Vesuvius: a glass inclusion approach to the modalities of feeding stratovolcanoes. Cont. Mineral Petrol. 120, 159–169.

Marianelli, P., Metrich, N., Sbrana, A., 1999. Shallow and deep reservoirs involved in magma supply of the 1944 eruption of Vesuvius. Bull. Volcanol. 61, 48–63.

Marsh, B.D., 1995. Solidification fronts and magmatic evolution. Mineral. Mag. 60, 5–40.

Mathez, E.A., Webster, J.D., 2005. Partitioning behaviour of chlorine and fluorine in the system apatite-silicate melt-fluid. Geochim. Cosmochim. Acta 69, 1275–1286.

Meinert, L.D., Hefton, K.K., Mayers, D., Tasiran, I., 1997. Geology, zonation, and fluid evolution of the BigGossan Cu-Au skarn deposit, Ertsberg district, Irian Java. Econ. Geol. 92, 509–534.

Raia, F., Webster, J.D., De Vivo, B., 2000. Pre-eruptive volatile contents of Vesuvius magmas: constraints on eruptive history and behavior. I. The medieval and modern interplinian activities. Eur. J. Mineral. 12, 179–193.

Rolandi, G., Petrosino, P., McGeehin, J., 1998. The interplinian activity of Somma-Vesuvius in the last 3500 years. In: Spera, F.J., De Vivo, B., Ayuso, R.A., Belkin, H.E. (Eds), Vesuvius Special Issue. J. Volcanol. Geotherm. Res. 82, 19–52.

Santacroce, R., Bertagnini, A., Civetta, L., Landi, P., Sbrana, A., 1993. Eruptive dynamics and petrogenetic processes in a very shallow magma reservoir: the 1906 eruption of Vesuvius. J. Petrol. 34, 383–425.

Scandone, R., Giacomelli, L., Gasparini, P., 1993. Mount Vesuvius: 2000 years of volcanological observations. J. Volcanol. Geotherm. Res. 58, 5–25.

Shinohara, H., Kazahaya, K., Lowenstern, J.B., 1995. Volatile transport in a convecting magma column: implications for porphyry Mo mineralization. Geology 23, 1091–1094.

Sibson, R.H., 1984. Roughness at the base of the seismogenic zone: contributing factors. J. Geophys. Res. 89, 5791–5799.

Trial, A.F., Spera, F.J., 1990. Mechanisms for the generation of compositional heterogeneities in magma chamber. Geol. Soc. Am. Bull. 102, 353–367.

Veksler, I.V., 2004. Liquid immiscibility and its role at magmatic-hydrothermal transition: a summary of experimental studies. Geoch. Geol. 210, 7–31.

Webster, J.D., De Vivo, B., 2002. Experimental and modeled solubilities of chlorine in aluminosilicate melts. Consequences of magma evolution, and implications for magmatic brine exsolution at Mt. Somma-Vesuvius. Am. Mineral. 87, 1046–1061.

Webster, J.D., De Vivo, B., Tappen, C., 2003. Volatiles, magmatic degassing and eruptions of Mt. Somma-Vesuvius: constraints from silicate melt inclusions, solubility experiments and modeling. In: De Vivo, B., Bodnar, R.J. (Eds), Melt Inclusions in Volcanic Systems: Methods, Applications and Problems. Series: Developments in Volcanology, Vol. 5. Elsevier, Amsterdam.

Webster, J.D., Raia, F., De Vivo, B., Rolandi, G., 2001. The behavior of chlorine and sulfur during differentiation of the Mt. Somma-Vesuvius magmatic system. Mineral. Petrol. 73, 177–200.

William, T.J., Candela, P.A., Piccoli, P.M., 1995. The partitioning of copper between silicate melt and two-phases aqueous fluids: an experimental investigation at 1 kbar, 800°C and 0.5 kbar, 859°C. Contrib. Mineral. Petrol. 121, 388–399.

Zollo, A., Gasparini, P., Virieux, J., Biella, G., Boschi, E., Captano, P., de Franco, R., Dell'aversana, P., de Matteis, R., De Natale, G., Jannaccone, G., Guerra, I., Le Meur, H., Mirabile, L., 1998. An image of Mt. Vesuvius obtained by 2D seismic tomography. J. Volcanol. Geotherm. Res. 82, 161–173.

Volcanism in the Campania Plain:
Vesuvius, Campi Flegrei and Ignimbrites
edited by B. De Vivo

Chapter 13

Petrogenesis of the Campanian Ignimbrite: implications for crystal-melt separation and open-system processes from major and trace elements and Th isotopic data

Wendy A. Bohrson[a,*], *Frank J. Spera*[b], *Sarah J. Fowler*[c], *Harvey E. Belkin*[d], *Benedetto De Vivo*[e] *and Giuseppe Rolandi*[e]

[a]*Department of Geological Sciences, Central Washington University, Ellensburg, WA, 98926, USA*
[b]*Department of Earth Science and Institute for Crustal Studies, University of California, Santa Barbara, 93106, USA*
[c]*Department of Earth Sciences, University of California, Santa Barbara, 93106, USA*
[d]*956 National Center, U.S. Geological Survey, Reston, VA 20192, USA*
[e]*Dipartimento di Scienze della Terra, Università di Napoli Federico II, 80134 Napoli, Italy*

Abstract

The Campanian Ignimbrite is a large-volume trachytic to phonolitic ignimbrite that was deposited at ~39.3 ka and represents one of a number of highly explosive volcanic events that have occurred in the region near Naples, Italy. Thermodynamic modeling using the MELTS algorithm reveals that major element variations are dominated by crystal–liquid separation at 0.15 GPa. Initial dissolved H_2O content in the parental melt is ~3 wt.% and the magmatic system fugacity of oxygen was buffered along QFM+1. Significantly, MELTS results also indicate that the liquid line of descent is marked by a large change in the proportion of melt (from 0.46 to 0.09) at ~884°C, which leads to a discontinuity in melt composition (i.e., a compositional gap) and different thermodynamic and transport properties of melt and magma across the gap. Crystallization of alkali feldspar and plagioclase dominates the phase assemblage at this pseudo-invariant point temperature of ~884°C. Evaluation of the variations in the trace elements Zr, Nb, Th, U, Rb, Sm, and Sr using a mass balance equation that accounts for changing bulk mineral–melt partition coefficients as crystallization occurs indicates that crystal–liquid separation and open-system processes were important. Th isotope data yield an apparent isochron that is ~20 kyr younger than the age of the deposit, and age-corrected Th isotope data indicate that the magma body was an open system at the time of eruption. Because open-system behavior can profoundly change isotopic and elemental characteristics of a magma body, these Th results illustrate that it is critical to understand the contribution that open-system processes make to magmatic systems prior to assigning relevance to age or timescale information derived from such systems. Fluid–magma interaction has been proposed as a mechanism to change isotopic and elemental characteristics of magma bodies, but an evaluation of the mass and thermal constraints on such a process suggests large-scale interaction is unlikely. In the case of the magma body associated with the Campanian Ignimbrite, the most likely source of the open-system signatures is assimilation of partial melts of compositionally heterogeneous basement composed of cumulates and intrusive equivalents of volcanic activity that has characterized the Campanian region for over 300 kyr.

1. Introduction

The recognition that chemical and physical gradients preserved in ignimbrites yield insight into pre-eruptive magma dynamics (e.g., Hildreth, 1979; Smith, 1979) has profoundly

*Corresponding author. *E-mail address:* bohrson@geology.cwu.edu (W.A. Bohrson).

influenced the understanding of crustal magmatism. Studies during the last 25 years have revealed that, while the concept of reconstructing magma bodies from information about deposits is of fundamental value, significant challenges of interpretation are introduced by the dynamics of eruption (e.g., Spera, 1984; Wilson and Hildreth, 1997) as well as the panoply of mechanisms by which melt, exsolved volatile bubbles and crystals form, mix and unmix (e.g., Smith and Bailey, 1966; Hildreth, 1981; Worner and Schmincke, 1984a,b; Bacon and Druitt, 1988). Accurate reconstruction of the pre-eruptive structure of the magmatic plumbing system depends upon integration of field data, multi-scale petrological and geochemical studies of the full range of erupted products, and theoretical modeling. Also, a comprehensive understanding of the processes responsible for generating the physical and chemical gradients requires an approach that blends descriptive and quantitative constraints.

Although significant progress has been made toward a holistic understanding of how large-volume, intermediate to silicic composition crustal reservoirs form, two fundamental questions remain especially controversial: What processes quantitatively contribute to the major and trace element and isotopic heterogeneity of volcanic products of these large volume eruptions? And over what timescales do the associated magma bodies form and/or coalesce? The answers to these questions are different for different systems, but a detailed analysis that incorporates thermal, chemical, and mass constraints for each system is prerequisite for making generalized observations about the behavior of magmatic systems.

In this contribution, we highlight the impact that crystal–liquid separation has on melt compositional evolution and particularly focus on trace element and Th isotope evidence for open-system processes in the magma body associated with the Campanian Ignimbrite (CI), a large-volume pyroclastic deposit that outcrops in the vicinity of the densely populated city of Naples, Italy. This ~39.3 ka eruption is one of a number of explosive eruptions documented in the Campanian magmatic province, with the oldest dated at >300 ka (De Vivo et al., 2001). Thus, understanding the dynamics of the magma body associated with the CI, which is the largest identified eruption in the Mediterranean in the last 200 kyr (Barberi et al., 1978), will lend perspective to a much longer history of violent, destructive eruptions that have occurred over timescales of at least 3×10^5 years. Our interpretations, which build on previous contributions (e.g., Fisher et al., 1993; Orsi et al., 1996; Civetta et al., 1997; Signorelli et al., 1999; De Vivo et al., 2001; Pappalardo et al., 2002; Rolandi et al., 2003), utilize thermodynamic and quantitative mass balance modeling of major and trace element data and semi-quantitative limits on Th and Sr isotopes to evaluate the role of crystal-melt separation, magma-fluid interaction, and assimilation of wall rock on the geochemical evolution of the CI. We find that the major element trends in the CI are dominated by crystal-melt separation. Based on results of detailed MELTS calculations (Ghiorso and Sack, 1995), we provide strong evidence for a dramatic episode of multi-phase crystallization, in which, over a very small temperature interval centered around 884°C, the proportion of residual melt in the system changes from ~0.46 to 0.09. The behavior of trace elements in the system is quantitatively evaluated using the major element results. One of the significant conclusions that emerges is that dramatic changes in phase assemblage that occur during isobaric fractionation yield large variations in bulk mineral–melt partition coefficients (bulk K_{sm}) over very small temperature intervals. Quantitative assessment of trace element concentrations using a mass balance equation that properly accounts for changing bulk K_{sm} demonstrates the profound control solid-melt partitioning can have on trace element evolution as well as the challenges of interpretation (e.g., whether trace elements have been impacted by open-system processes or not) that emerge because, in some cases, published trace element mineral–melt partition coefficients (mineral–melt K_{sm}) vary significantly within a small compositional range. New Th isotope data, coupled with an evaluation of published Sr isotope data (Civetta et al., 1997), reveal evidence for open-system processes that most likely involve complex interaction between magma and intrusive equivalents of Campanian volcanics and/or cumulates formed in association with Campanian magmatism.

2. Previous work

2.1. Regional geology

The Campanian Plain is a region of southern Italy that encompasses the Phlegrean Fields (Campi Flegrei – burning fields), which is located west of Naples (Fig. 1a). The plain is structurally located within a graben formed in Mesozoic carbonate of the southern Apennine Mountains, which border the plain on the east and north. The graben formed during the Pliocene as a consequence of extension that occurred along the western margin of the Apennine chain, resulting in subsidence along the Tyrrhenian coast (Rosi and Sbrana, 1987; Scandone et al., 1991). The Phlegrean Fields are characterized by a long history of magmatism dating back to at least 300 ka, based on Ar/Ar ages of xenocrystic sanidine identified in pyroclastic material sampled in the Campanian Plain (De Vivo et al., 2001). This region of southern Italy is famous not only for eruptions in the Phlegrean Fields, including the most recent eruption of Monte Nuovo in 1538 AD, but also for the famed 79 AD eruption of Vesuvius that destroyed the villages of Pompeii and Herculaneum. Vesuvius most recently erupted in 1944 AD. The volcanology of the Naples area has been of interest for centuries, particularly because of the vivid descriptions of the 79 AD eruption of Vesuvius by Pliny the Younger. Continued scrutiny of the volcanology and petrology of this area is warranted by the potentially lethal combination of eruption from either the Phlegrean Fields or Vesuvius, coupled with the densely populated, highly urbanized areas in and around Naples; this combination presents significant challenges for volcanic hazard mitigation.

2.2. Volcanology and geochronology of the Campanian Ignimbrite

The Campanian Ignimbrite, a large-volume (~150 km^3, Civetta et al., 1997, to ~200 km^3 DRE, Rolandi et al., 2003) trachytic-phonolitic ignimbrite, was originally distributed over ~30,000 km^2 in and around Naples, Italy (Fisher et al., 1993). While previous ages for the CI have been reported at ~37 ka (e.g., Deino et al., 1992, 1994), more recently, the deposit has been precisely dated at 39.28±0.11 ka by incremental heating and total fusion $^{40}Ar/^{39}Ar$ geochronology (De Vivo et al., 2001). This age represents the weighted mean of results from 18 alkali feldspar separates derived from representative units within the CI from a variety of geographic locations. Based on the detailed work of De Vivo et al. (2001), the CI is composed of up to five physically distinct units. The gray tuff unit of the CI grades upward into a yellow tuff, the color of which reflects secondary mineralization by zeolites. In several sections throughout the Campanian Plain, the gray tuff is overlain by a lithic breccia that grades upward into either a weakly stratified yellow tuff and/or an incoherent pyroclastic

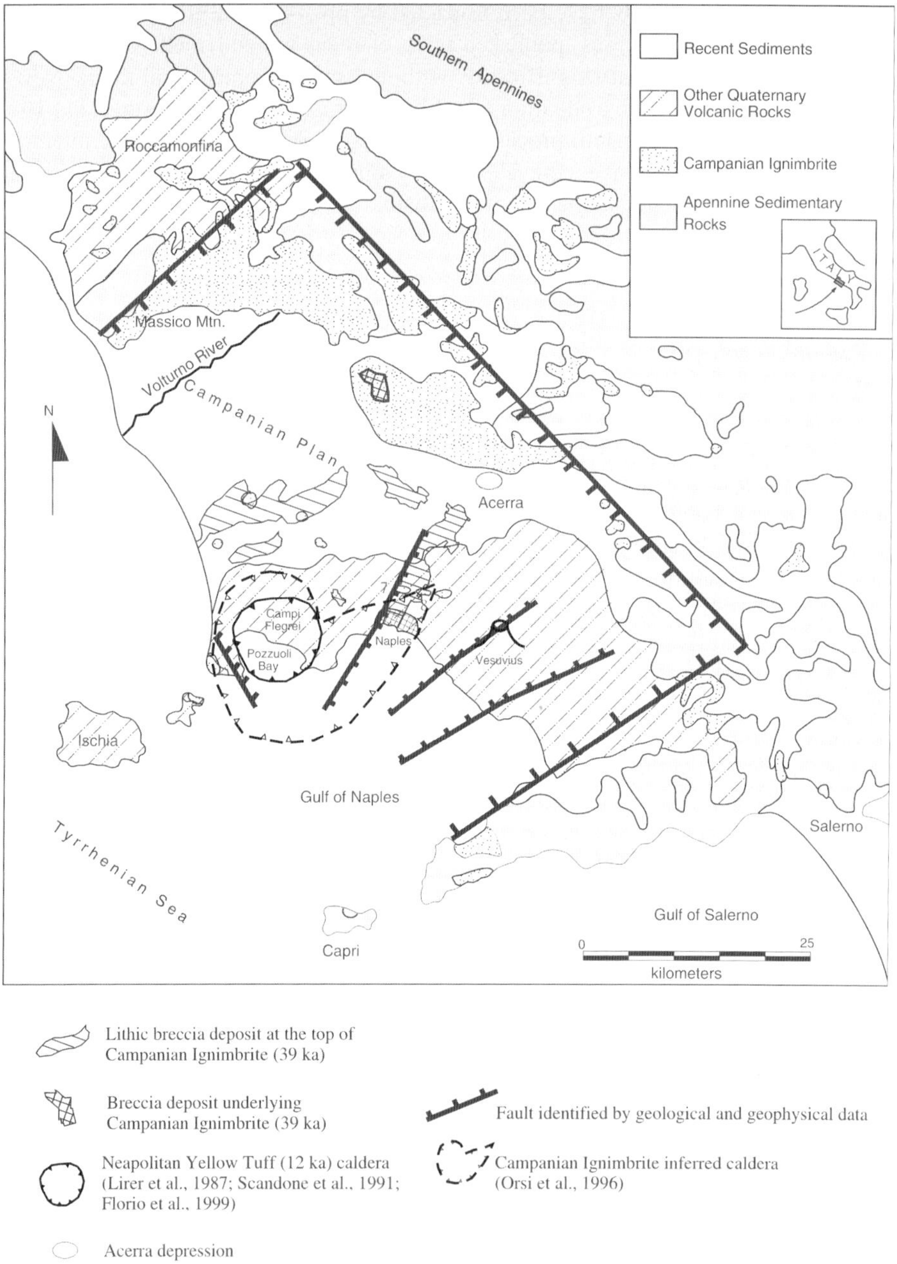

Figure 1. (a) Simplified map of the Campanian Plain (after De Vivo et al., 2001). (b) Locations of outcrops sampled for this study.

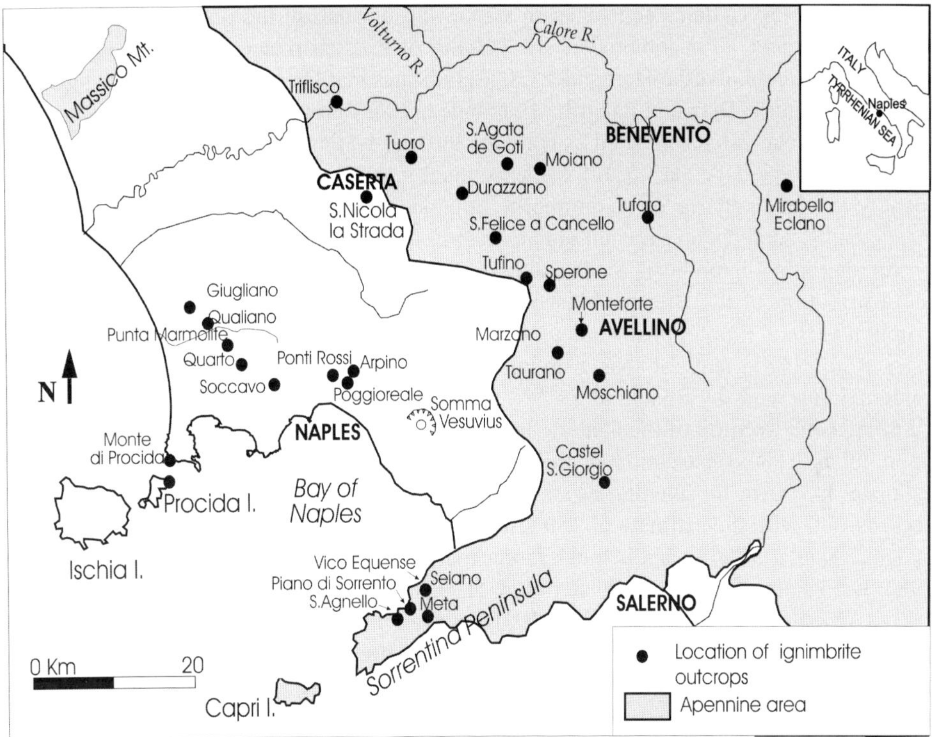

Figure 1. (*Continued*)

flow deposit. The lithic breccia is discontinuously exposed around the Campanian Plain, and in some locations shows evidence of proximal depositional characteristics. Significantly, outcrops with proximal facies characteristics are distributed in a belt that runs parallel to the Tyrrhenian Sea, from Naples north to Massico Mountain (De Vivo et al., 2001). A basal pumice unit has also been identified. Two samples of this unit are not included in the weighted mean age above; these samples yield ages of 40.16 ± 0.21 and 41.02 ± 0.26 ka. This unit may represent an event that is slightly older than that associated with the CI, or based on the behavior of the spectra, these ages more likely reflect incorporation of xenocrysts from lithic fragments (De Vivo et al., 2001).

The locations of the vent(s) of the CI are controversial, despite over a century of published inquiry on this topic (e.g., Scacchi, 1890). Barberi et al. (1978) noted that the NW-SE distribution of the ignimbrite is consistent with its eruption from a fissure associated with the Apennine front. Recent, more detailed work by De Vivo et al. (2001) and Rolandi et al. (2003) support this hypothesis; the characteristics and distribution of the lithic breccia and mapping of regions of maximum thickness of the CI suggest that the ignimbrite was fed from fissures, the locations of which are controlled by the local extensional tectonics associated with the evolution of the Apennine chain. Of interest to this hypothesis, a similar eruption mechanism has been proposed to explain the notable lack of caldera structures in the Sierra Madre Occidental (Mexico). Based on the size of this ignimbrite province, Swanson and McDowell (1984) estimate that 350 calderas equivalent in size to those found in the San Juan volcanic field would be required to accommodate the volume of erupted material, and yet, analysis by Aguirre-Díaz and Labarthe-Hernández (2003) indicates that fewer than 15 calderas have been identified. Aguirre-Díaz and Labarthe-Hernández (2003) propose that faults associated with Basin and Range extension acted as conduits through which large volumes of magma were explosively erupted. As part of the evidence cited in support of this hypothesis, the authors identify elongate patterns of co-ignimbrite lithic-lag breccia exposed adjacent to Basin and Range faults; such features are not unlike those in the Campanian region alluded to above.

An alternative hypothesis for the source of the CI is the Campi Flegrei caldera. Citing the distribution of the Breccia Museo, a lithic breccia interpreted to be associated with eruption of the CI, on opposite sides of a 12 km caldera, and evidence from drilling data, Rosi et al. (1983) suggest that eruption of the CI was associated with formation of a caldera. Later work by Rosi et al. (1996) used the distribution of proximal deposits of the CI to infer the location of the caldera, including specific locations for fault scarps formed upon collapse. Reconstruction of the size of the hypothesized caldera, coupled with an estimate of 700 m for average down drop, yields a collapse volume of 160 km^3, which accords well with the estimate of 150 km^3 for the CI by Civetta et al. (1997) and 200 km^3 by Rolandi et al. (2003), but is much larger than an earlier estimate of 80 km^3 (Rosi et al., 1983; De Natale et al., 1991). Fisher et al. (1993) report anisotropy of magnetic susceptibility (AMS) measurements for samples from the CI that suggest the pyroclastic current flowed radially outward from the Phlegrean Fields area. Although this contribution does not explicitly address the source of the CI, these data have been cited as support for a source within the Campi Flegrei area (e.g., Civetta et al., 1997). Further AMS work (Ort et al., 1999) suggest that eruptions of Piperno Tuff were fed by a central vent north of Pozzuoli (Fig. 1a), and tuffs that underlie the Breccia Museo may be related to eruptions from ring vents located on the northern and southern caldera margins. Rosi et al. (1996) note that all post-CI vents are located within the hypothesized caldera, but De Vivo et al. (2001) identify a unit located at Giugliano, 13 km outside of the identified caldera, with an age of 18.05±0.43 ka. Gravity and magnetic data provide supporting evidence for a ring structure in the Campi Flegrei area (Lirer et al., 1987; Scandone et al., 1991; Florio et al., 1999), but the associated eruptive event is still controversial; some attribute the structure to collapse associated with eruption of the Neapolitan Yellow Tuff (~12 ka, Lirer et al., 1987), whereas others attribute the caldera to eruption of the CI (Rosi et al., 1983). Another interpretation is that the depression is a nested caldera structure that formed in response to both eruptions (e.g., Orsi et al., 1996).

The Campanian Plain and Phlegrean Fields regions reflect a complex, integrated history of tectonic activity coupled with over 300 kyr of volcanism. Challenges presented by this complexity are exacerbated by exposure; many areas that may be critical to the interpretation of source are either covered by younger deposits and/or masked by intense urbanization. Resolution of the source controversy will require continued integrated volcanological, petrological, and geophysical studies.

2.3. Petrology and geochemistry of the Campanian Ignimbrite

The CI is a heterogeneous unit dominated by the compositional range phonolite to trachyte. Previously published ranges of major elements of single and composite pumice and glass from the ignimbrite include SiO_2 from 55.3 to 62.0 wt.% and MgO from 0.3 to 1.5 wt.% (Civetta et al., 1997; Pappalardo et al., 2002). A smaller number of samples of basal pumice, which is interpreted to represent Plinian fallout, has slightly higher SiO_2 concentrations, up to 62.4% (Signorelli et al., 1999). Matrix glass and glass inclusions from the basal pumice have SiO_2 and MgO (Signorelli et al., 1999) within the ranges cited above for pumice and glass from the ignimbrite. Average compositions of melt inclusions in clinopyroxenes from a subset of samples presented in this study, analyzed by Webster et al. (2003), extend the range of SiO_2 and MgO; those denoted as low-MgO melt inclusions have average SiO_2 and MgO of 57.8 and 2.9 wt.%, respectively, whereas high-MgO melt inclusions have 51.0 and 7.6 wt.%. Associated total alkalies for these melt inclusions extend the compositional range of melts of the CI event to basaltic trachyandesite (Fig. 2a). The highly alkaline nature of pumice and glass of the CI is further illustrated by K_2O and Na_2O concentrations of 5.2 to 9.9 and 2.0 to 6.6 wt.%, respectively, among pumice, glass inclusions, and matrix glass (Civetta et al., 1997; Signorelli et al., 1999; Pappalardo et al., 2002; Webster et al., 2003).

Trace elements display a range of concentrations within the CI and its associated fall-out deposit. For example, Sr varies from 17 to 777 ppm and Ba from 17 to 1111 ppm, and both show negative correlation with SiO_2. In contrast, elements such as Zr, Rb, and Sm, are positively correlated with SiO_2, and also display marked ranges in concentration (Civetta et al., 1997; Pappalardo et al., 2002).

Analysis of pumice, glass, and feldspar separates yield distinct Sr isotope values (Civetta et al., 1997; Pappalardo et al., 2002). $^{87}Sr/^{86}Sr$ of pumice ranges from 0.70728 to 0.70746, with the isotopic ratio of associated glass typically within $\pm$0.00002 of pumice. In contrast, most feldspar $^{87}Sr/^{86}Sr$ are less radiogenic than pumice and are characterized by a relatively narrow Sr isotope range from 0.70730 to 0.70734 (Civetta et al., 1997). Exceptions to this range include one feldspar that has $^{87}Sr/^{86}Sr$ of 0.70741 and is less radiogenic than its host pumice/glass, and one feldspar that has $^{87}Sr/^{86}Sr$ of 0.70748 and is more radiogenic than its associated pumice.

The phenocryst assemblage of the CI is dominated by sanidine and includes clinopyroxene, plagioclase, biotite, apatite, and spinel. Sanidine compositions range from Or_{58} to Or_{87}, and no zoning was detected within individual crystals (Civetta et al., 1997; Pappalardo et al., 2002). Plagioclase compositions vary widely, from An_{90} to An_{25}, and like sanidine, most crystals are compositionally homogeneous from core to rim (Civetta et al., 1997; Pappalardo et al., 2002). Clinopyroxene ranges from diopside to salite (Civetta et al., 1997; Webster et al., 2003), and detailed work by Webster et al. (2003) demonstrates that most clinopyroxene crystals are zoned, with higher MgO in the core. These authors note that the transition from higher to lower MgO is abrupt. Civetta et al. (1997) note that diopside is relatively rare in the clinopyroxene they analyzed, and the rims tend to be corroded.

Based on a variety of data, including those cited above, Civetta et al. (1997), Signorelli et al. (1999), Pappalardo et al. (2002), and Webster et al. (2003) propose models for the structure and petrogenetic evolution of the CI magma chamber. In this section, we briefly highlight ideas relevant to our study. Civetta et al. (1997) and Signorelli et al. (1999) hypothesize that the magma body was composed of two compositionally distinct layers of magma, separated by a compositional gap. Both contributions cite eruption dynamics and variable timing and extent of mixing between these two distinct layers to explain the geochemistry, petrology, and spatial and temporal distribution of compositionally distinct units. The upper layer, which is regarded as compositionally homogeneous and more

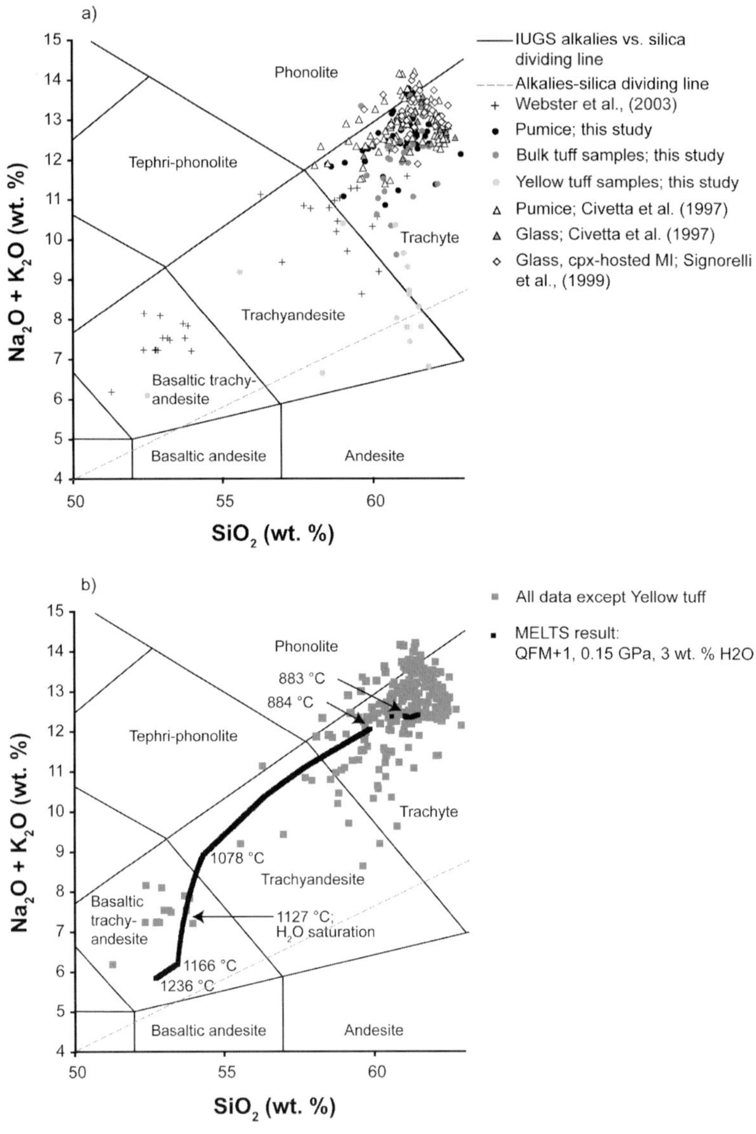

Figure 2. (a) Alkali–silica diagram using classification scheme of Le Maitre et al. (1989). Symbols are shown in legend. High-MgO melt inclusions of Webster et al. (2003) are dominantly basaltic trachyandesite and low-MgO melt inclusions are mostly trachyandesite to trachyte. Samples containing the clinopyroxene-hosted melt inclusions from the Webster et al. (2003) study are a subset of samples presented in this work. Most samples of the CI are trachyte to phonolite. Some samples, denoted as the yellow tuff unit, show depletions in alkalies. In addition to data from this study, also included are glass and pumice data from Civetta et al. (1997), and melt inclusion and glass data from Signorelli et al. (1999). (b) Alkalies–silica diagram with result of the best-fit MELTS model shown for comparison. Best-fit MELTS calculation was run in increments of 0.5°C, and therefore each black square represents a temperature decrease of 0.5°C. The MELTS curve is annotated with temperatures, including the MELTS-predicted compositional gap between 884 and 883°C. Note that MI data from Signorelli et al. (1999) and Webster et al. (2003) are *not* expected to lie on the predicted liquid line of descent due to post-entrapment crystallization. See text for discussion.

differentiated, evolved by crystal–liquid fractionation, as did the lower layer, which was modestly zoned. Sr isotope data are used to suggest that fluids interacted with the upper layer of the chamber, most likely after feldspar crystallized. Although the major element fractional crystallization model of Civetta et al. makes qualitative sense, there are some perplexing relationships among trace elements that hint at the need for a more detailed and realistic petrogenetic model. For example, an analysis using the reported initial and final melt concentrations for La, coupled with the observed partition coefficient of 0.3, indicates that removal of ~77% crystals is required; this amount of crystallization is much larger than the reported value of ~57%. The modeling results reported for Eu are also perplexing; the element's variation from 2.2 to 1.8 ppm in initial to final magma, respectively, cannot be described by a bulk partition coefficient of 0.9, as reported, because a bulk partition coefficient larger than 1 is required to accommodate the decrease in concentration as fractionation proceeds. The difference between the observed and calculated concentrations for Eu is reported to be 0, which implies a perfect match between the model and the observed data. Using the value of remaining melt calculated from the sum of the percentage of removed mineral phases (~57%), a bulk D of ~1.3 is required to accommodate the decrease in Eu from initial to final magma. An analysis of Ba, Sr, and Sc also yields bulk partition coefficient values that are different from those reported. These incongruities, taken together, suggest that fresh insight can be gained by a more comprehensive treatment of the phase equilibria coupled explicitly and self-consistently to trace element analysis.

Based on their analysis of matrix glass and glass inclusion data from pumice from the Plinian fall deposit, Signorelli et al. (1999) suggest the occurrence of pre-eruptive magma mixing, stimulated by input of trachybasalt into pre-existing magma. In this view, crystal fractionation of a variety of heterogeneous mixed magmas generated the evolved compositions. These authors also suggest that complexity in the distribution and compositional range of units within the fallout deposit precludes simple vertical stratification of layers in the magma body. Instead, they propose a vertically and laterally zoned body where compositional gradients are, in part, a function of distance from the source of mafic input. Thus, the degree of evolution increases as a function of distance from the source of mafic input. Unfortunately, it is impossible to test the validity of this model because the spatial distribution and eruptive history of the CI vent(s) are unlikely to ever be positively identified.

Webster et al. (2003) stress a role for magma mixing or mingling, and constraints provided by melt inclusion data indicate that mixing/mingling between more primitive and more evolved magma must have occurred shortly before eruption of the CI. Presence of MgO-rich melt inclusions in cores of diopsidic clinopyroxene led Webster et al. (2003) to suggest that these basaltic trachyandesite inclusions may represent primary mafic magma injected into the CI magmatic system; the compositional homogeneity of these inclusions also suggests that these primary magmas were likely derived from a common magmatic source(s). The compositional range of the low-MgO melt inclusions, from trachyandesite to trachyte, likely is a consequence of fractional crystallization, with or without mixing with primary magma.

Collectively, the models proposed by these authors emphasize roles for fractional crystallization, magma recharge and magma mixing/mingling, and contamination. Characterization of the chemical gradient(s) within the pre-eruptive CI magma body involves some (unspecified in detail) notion of vertical and/or lateral zonation. Despite these commonalities, central questions emerge from the earlier studies. These include quantitative analysis of the roles that crystal-melt separation, magma recharge and magma mixing/mingling, and crustal assimilation have in the evolution of the CI magma body, the relative timing of these processes as the magma body evolves, and the pre-eruptive structure of the magma plumbing system. The goal of our work is to quantify the aforementioned processes to the extent possible based on current descriptive knowledge of the products of the CI eruption. The current study is a preliminary examination of these issues; further details are presented elsewhere (Bohrson et al., in preparation; Fowler et al., submitted to Journal of Petrology).

3. Methods

One hundred and twenty-five samples of the CI were obtained from approximately 30 localities throughout the Campanian Plain (Fig. 1b). Based on the work of De Vivo et al. (2001), each sample was recovered from one of the five distinct units (gray tuff, yellow tuff, incoherent unit, lithic breccia, and basal pumice) of the CI *sensu lato*. Sample types include bulk pumice or bulk rock (which includes pumice plus matrix or matrix alone).

3.1. Analytical methods for major and trace element data

Major, minor, and trace elements were determined either by the U.S. Geological Survey Laboratories (Reston, VA and Denver, CO) or by Activation Laboratories (Ancaster, Ontario, CA). Before grinding, bulk tuff and pumice samples were examined and any alteration was removed; the sample was then washed in deionized water. Grinding and powdering were done with either mild steel or alumina disks. Major element oxides were determined in representative aliquots by WD-XRF after fusion with lithium metaborate/tetraborate. Trace elements were determined by three methods. (1) Inductively coupled plasma-mass spectrometry (ICP-MS) and inductively coupled plasma-atomic emission spectrometry (ICP-AES) were done on solutions using a lithium metaborate/tetraborate fusion procedure before sample dissolution with multiple acids; (2) pressed power pellets were exposed to the appropriate X-rays necessary to fluoresce the element of interest using EDXRF, and (3) standard instrumental neutron activation analysis (INAA) technique after irradiation in the USGS "TRIGA" reactor. FeO was determined by titration, total sulfur reported as SO_3 was determined by combustion/infrared spectroscopy, Cl by selective ion electrode or INAA, F by selective ion electrode, LOI and $H_2O+/-$ by gravimetric techniques, and CO_2 by infrared or colorimetric titration.

3.2. Calculation of parental melt composition and major element phase equilibria modeling

The motivation for applying detailed major element phase equilibria models is to have the ability to compare observed abundances with predictions based on closed-system crystal-melt separation. Differences between predictions and observations then shed light on issues including not only the role of crystal-melt separation, but also the role of assimilation of hydrothermally altered wall rock, lower crustal xenoliths, and/or meteoric fluids, the role of magma addition and mixing during recharge, as well the influence of mixing during high Reynolds number magma withdrawal. Below, the methods used to reconstruct the parental melt composition, and an overview of the phase equilibria model (i.e., the MELTS algorithm) are described.

3.2.1. Selection of parental melt composition

Basaltic rocks that may correspond to CI parental compositions have not been identified in the Campanian Plain region; however, Webster et al. (2003) identified glass inclusions entrapped within host clinopyroxene phenocrysts, which are hereafter referred to as Webster Melt Inclusions (WMI). WMI VE1#3 from the high-MgO MI group is *representative* of the least-evolved MI from this group and was therefore used as the basis for the parental melt composition. This sample was chosen, in part, because its incompatible trace element concentrations are lower than those of other high-MgO WMI from minimally deuterically altered CI units (i.e., gray tuff, lithic breccia, and incoherent tuff). In particular, the concentrations of incompatible trace elements such as Ce, Th, and U are generally lower in sample VE1#3 than in other high-MgO WMI. While we recognize that the high-MgO inclusions have a range of compositions, we note that using a different WMI sample as the basis for the parental melt composition does not alter the main conclusions of the petrologic interpretations presented in later sections of the paper.

As noted by Watson (1976), reaction between trapped melt inclusions and their host crystals is generally expected. Because the high-MgO WMI are found within clinopyroxene phenocrysts, the effects of post-entrapment crystallization on the composition of sample VE1#3 were accounted for by addition of clinopyroxene. Reconstruction of the major element composition of the parental melt (RPM) is based on the mass balance relation $y\,\overline{P}_i + (1-y)\,\overline{\mathrm{MI}}_i = \overline{\mathrm{RPM}}_i$, using the major element compositions of WMI sample VE1#3 ($\overline{\mathrm{MI}}_i$) and host pyroxene composition ($\overline{P}_i$) that corresponds to that of the WMI host crystal in sample VE1 (Webster et al., 2003; Table 1). The value of y, which is the extent of post-entrapment crystallization, was set to 0.2. We also used the method proposed by Kress and Ghiorso (2004) and found a similar result. Major element data for WMI VE1#3 and its clinopyroxene host, as well as the (hydrous) reconstructed parental CI melt (RPM) used as the initial condition in the MELTS isobaric simulations are given in Table 1. Also provided are the anhydrous reconstructed parental melt composition and the reconstructed concentrations for the trace elements, which are based on the measured abundances in MI VE1#3.

Table 1a. Major element data for melt inclusion and host clinopyroxene used in reconstruction of parental melt.

Oxide wt.%	SiO_2	TiO_2	Al_2O_3	FeO	MgO	MnO	CaO	K_2O	Na_2O	P_2O_5
VE1 Host cpx	52.49	0.37	2.13	3.96	16.48	0.12	24.28	0	0.17	0
VE1#3 MI	52.78	1.03	14.53	5.82	7.82	0.07	9.88	5.16	2.07	0.84

Table 1b. Reconstructed parental melt composition used in best-fit MELTS model.

Oxide wt.%	SiO_2	TiO_2	Al_2O_3	FeO	MgO	MnO	CaO	K_2O	Na_2O	P_2O_5	H_2O
Anhydrous	52.72	0.9	12.05	5.45	9.55	0.08	12.76	4.13	1.69	0.67	–
Hydrous	51.13	0.87	11.69	5.29	9.26	0.08	12.38	4.01	1.64	0.65	3

Trace element (ppm)	U	Th	Rb	Sr	Zr	Nb	Sm
	1.5	3.7	191	363	60	13.4	3

Note: Compositions used in MELTS modeling based on sample VE1#3 (Webster et al., 2003). Recalculated VE1#3 composition for X = 0.2. See text for discussion.

3.2.2. MELTS modeling

To calculate phase relations and major element variation diagrams, we used the MELTS algorithm (Ghiorso and Sack, 1995), which is a self-consistent, thermodynamic model of crystal-melt equilibria in which the system undergoes perfect separation of crystals from liquid. At specified pressure, temperature, and parental melt major element composition, the identity, composition, and proportion of phases in the multicomponent-multiphase system are computed in response to the extraction of enthalpy, both sensible and latent, from parental melt.

To carry out the MELTS calculations, constraints on the system's oxygen fugacity, pressure, and initial (dissolved) parental melt H_2O content are needed. In order to determine the best-fit parameters for the CI magmatic system, over 100 MELTS simulations were carried out over a grid of oxygen fugacity, pressure, and water content. The quality of the MELTS results was evaluated by comparison between the predicted liquid line of descent, phase compositions, and observed data. MELTS also predicts phase proportions, but we have not used these as a criterion because differential physical separation effects impact observed modal abundances. The density of fractionating phases varies widely, and both pre-eruptive and eruptive physical separation of crystals is expected and difficult to account for.

3.3. Trace element modeling

Based on the results of the phase equilibria calculations, it is possible to forward model the concentration of trace elements during the course of isobaric fractional crystallization. Deviations between calculated and observed compositions afford the possibility to better understand petrological processes controlling the magmatic evolution of the Campanian system. The starting point of a trace element assessment is the differential expression governing the concentration of a trace element in the melt as a function of the melt fraction during perfect crystal–liquid separation. For the moment, we ignore possible fractionation of trace elements into coexisting supercritical fluid. Although this assumption is implicitly made in many trace element studies, it may not always be tenable. The critical parameters governing the distribution of a trace element between solid-melt and supercritical fluid are the distribution ratio's K_{sm} and K_{sf} (solid-melt and solid-fluid partition coefficients, respectively), and the rate of change of the mass fraction of fluid with respect to the mass fraction of melt during fractional crystallization. In fact, the distribution of Rb, one of the trace elements used in this study, may be particularly sensitive to the existence of a fluid phase predicted to develop at $T \sim 1127°C$ during isobaric fractional crystallization at 0.15 GPa.

Ignoring the presence of a fluid phase, the well-known Rayleigh distillation mass balance relation in differential form is

$$\frac{dC_m}{df_m} = (K_{sm} - 1)\frac{C_m}{f_m} \tag{1}$$

which can be integrated

$$\int_{C_m^o}^{C_m} \frac{dC_m}{C_m} = \int_1^{f_m} \frac{(K_{sm} - 1)}{f_m} df_m \tag{2}$$

Because of the large differences in the instantaneous composition of solids being removed during fractional crystallization, the assumption that bulk solid-melt partition coefficient (K_{sm}) is constant is unacceptable. For example, K_{sm}^{Sr} as a function of fraction of melt ($f_m(T)$) calculated using mineral–melt partition coefficients (Table 2) and the proportion of phases returned from the MELTS simulation (Fig. 4) is shown in Figure 3. Across the crystallization interval, bulk K_{sm}^{Sr} varies by ~100 from the liquidus, where olivine first saturates, down to the near solidus, where plagioclase is present. As another example, bulk K_{sm} of Th varies by a factor of ~30 over the crystallization interval. Our approach is to determine *a priori*

Table 2a. Mineral–melt distribution coefficients used in trace element modeling for $T >$ feldspar-in (~884°C).

Element	Clinopyroxene	Spinel	Olivine	Apatite	References
Rb	0.10	0.15	0.01	0.40	1, 3, 20, 12
Sr	0.25	0[a]	0.02	1.3	8, –, 20, 21
Zr	0.12	0.71	0.06	0.64	7, 11, 19, 4
Nb	0.12	0.7	0.11	0.64[b]	8, 5, 3, –
Sm	0.75	0.01	0.02	4.5	17, 15, 2, 21
Th	0.03	0.10	0.02	1.6	8, 11, 20, 12
U	0.04	0.11	0.04	1.8	1, 11, 19, 10

[a]Distribution coefficient for Zr in spinel not reported in GERM compilation; estimated to be 0.
[b]Distribution coefficient for Nb in apatite not reported in GERM compilation; estimated using distribution coefficient for Zr.

Table 2b. Mineral–melt distribution coefficients for Zr used to calculate trends 1 and 2 for $T >$ feldspar-in (~884°C).

Element	Clinopyroxene	Spinel	Olivine	Apatite	References
Zr trend 1	0.12	0.71	0.06	0.1	7, 11, 19, –
Zr trend 2	0.12	0.71	0.06	0.64	7, 11, 19, 4

Table 2c. Mineral–melt distribution coefficients used in trace element modeling for $T <$ feldspar-in (~884°C).

Element	Alkali feldspar	Plagioclase	Biotite	Apatite	References
Rb	0.66	0.52	5.3	0.40	16, 20, 3, 12
Sr	3.9	2.8	0.31	2.1	16, 16, 3, 21
Zr	0.069	0.08	0.205	2.0	3, 20, 20, 12
Nb	0.15	0.57	9.5	2.0[a]	3, 3, 13, –
Sm	0.02	0.13	0.26	20.7	18, 10, 6, 12
Th	0.02	0.008	0.31	1.6	18, 10, 20, 10
U	0.04	0.014	0.08	1.8	18, 10, 20, 10

[a]Distribution coefficient for Nb in apatite not reported in GERM compilation; estimated using distribution coefficient for Zr.

Table 2d. Mineral–melt distribution coefficients for Zr used to calculate trends 1 and 2 for $T <$ feldspar-in (~884°C).

Element	Alkali feldspar	Plagioclase	Biotite	Apatite	References
Zr Trend 1	0.03	0.09	0.205	0.1	13, 20, 20, –
Zr Trend 2	0.36	0.36	0.59	2	9, 13, 3, 12

References: (1) Dostal et al. (1983); (2) Dunn and Sen (1994); (3) Ewart and Griffin (1994); (4) Fujimaki (1986); (5) Haskin et al. (1966); (6) Higuchi and Nagasawa (1969); (7) Keleman and Dunn, (1992); (8) Larsen (1979); (9) Leeman and Phelps (1981); (10) Luhr et al. (1984); (11) Mahood and Hildreth (1983); (12) Mahood and Stimac (1990); (13) Nash and Crecraft (1985); (14) Nagasawa and Schnetzler (1971); (15) Nielsen et al. (1992); (16) Philpotts and Schnetzler (1970); (17) Reid (1983); (18) Stix and Gorton (1990); (19) Villemant et al. (1981); (20) Villemant (1988); (21) Watson and Green (1981).

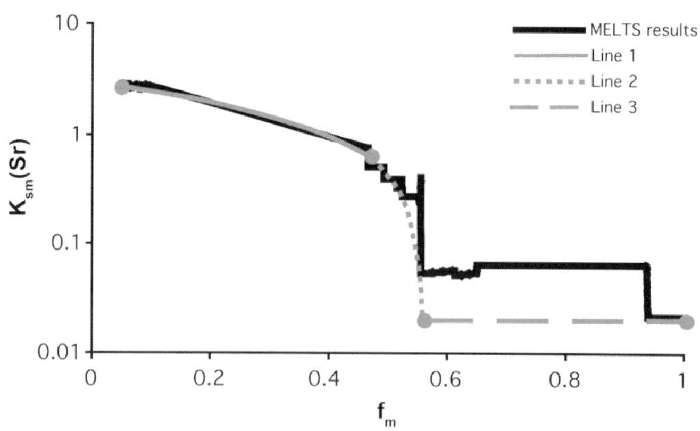

Figure 3. Fraction of residual melt, $f_m(T)$, vs. bulk partition coefficient for Sr, bulk K^{Sr}_{sm}. The heavy black line is the computed partition coefficient based on the solid assemblage predicted from MELTS and the partition coefficients listed in Table 2. The K^{Sr}_{sm} vs. $f_m(T)$ relationship is broken into three linear piecewise continuous segments as indicated by the line segments in gray in order to carry out the numerical integrations for the trace element evolution.

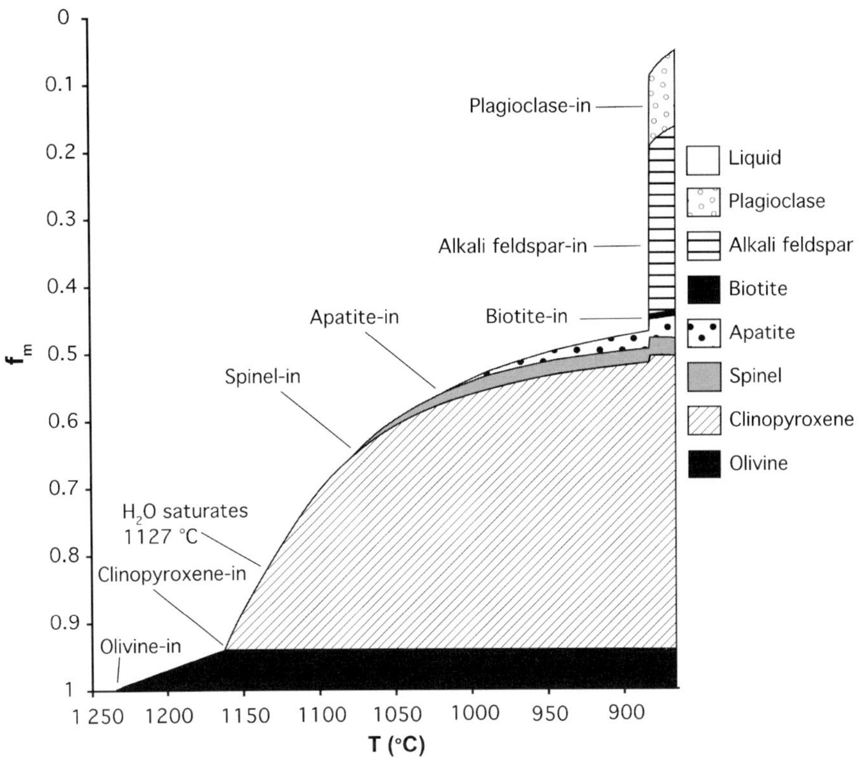

Figure 4. $f_m(T)$ vs. T (°C) for MELTS simulation of the reconstructed parental composition for the Campanian Ignimbrite. Fields represent cumulative proportion of mineral phases as saturation occurs. Note the distinct change in $f_m(T)$ at the point in which plagioclase and alkali feldspar become liquidus phases: $f_m(T)$ changes from ~0.46 to 0.09 at T ~ 884°C. See text for discussion.

from MELTS the variation of bulk K_{sm} for each trace element as a function of T or, equivalently, $f_m(T)$ and then to numerically integrate Equation (2) using a sequence of linear segmental parameterizations for K_{sm} as a function of $f_m(T)$ (Fig. 3). For a linear relationship between K_{sm} and $f_m(T)$, Equation (2) can be integrated analytically; in practice, it is faster and just as accurate to perform the integration numerically using a fourth-order Runge-Kutta scheme (e.g., Spera and Bohrson, 2001).

In summary, trace element concentrations were calculated from the liquidus temperature where $f_m(T) = 1$ down to $f_m(T) \sim 0.05$ using the numerically integrated form of Equation (2), taking explicit account of the variation in bulk K_{sm} for each trace element as a function of the fraction of remaining melt after removal of solids predicted from the MELTS simulation. The concentration of a trace element is then calculated using the initial concentration values ($C^o_{m,i}$) for the ith trace element listed in Table 1 and the numerically determined ratio C_m/C^o_m. As noted above, we neglect fractionation of an element into coexisting supercritical fluid phase in the results presented here, although for certain elements, specifically the alkalies (e.g., Rb; see Beswick, 1973) and for elements that strongly complex with halogens such as Cl, this assumption is suspect.

3.4. *Analytical methods for Th isotope data*

Th isotopic analyses of 12 samples of the CI were performed at the University of California, Los Angeles (UCLA), using a VG sector mass spectrometer equipped with a 30 cm electrostatic filter and a static collector array composed of a Faraday cup and an ion counting photomultiplier system. Thorium and uranium were run as metals. Abundance sensitivity on this instrument during data collection was <0.4 ppm at 1 amu. $^{232}Th/^{230}Th \pm 2\sigma$ obtained on a standard solution and international reference samples are (1) University of California, Santa Cruz Th standard: 170,587 ± 0.26% ($n = 35$), (cf. values of $17.03 \pm 9 \times 10^4$ and 17.05×10^4 reported in Reid, 1995); (2) AGV1 199,040±0.20% ($n = 5$); (3) JB1 332,302 ± 0.96% ($n = 5$) (cf. values of 20.06×10^4 and 33.54×10^4, respectively, reported in Reid, 1995). U and Th concentrations were measured by isotope dilution using a ^{229}Th- and ^{233}U-enriched spike. Elemental purification was achieved using anion exchange resin and 7N HNO_3.

4. Description of data

4.1. *Major and trace elements*

Representative major element data are presented in Table 3. Consistent with numerous previous studies on the CI, pumice and bulk/matrix samples analyzed in this study are trachytic to phonolitic (Fig. 2a). The typical phenocryst assemblage in pumice includes alkali feldspar, with lesser plagioclase and sparse to trace clinopyroxene, spinel, apatite, and biotite. Most samples are sparsely phyric; the maximum crystal content is ~10%, and thus, although our samples (and some of those of Civetta et al. (1997)) are pumice, major and trace element signatures are dominated by melt compositions. As a consequence of zeolitization, samples of the yellow tuff have compositions distinct from the pumice and bulk/matrix samples and will not be plotted or discussed further in this contribution. Portrayed in Figure 2a are data for the high-MgO and low-MgO melt inclusions of Webster et al. (2003), pumice and glass (by ICP) from Civetta et al. (1997) and matrix glass and clinopyroxene-hosted glass inclusions (by electron microprobe) from Signorelli et al. (1999). The Signorelli study focused on the Plinian phase of the eruption and therefore includes only data from the fall deposit (earliest part of the CI eruption).

Selected major oxide trends are illustrated in Figure 5a–d; all oxides are reported in wt.%. For most oxides, pumice and bulk samples are characterized by relatively coherent oxide–oxide trends. For example, SiO_2, Na_2O, and MnO vs. MgO form relatively tight negatively correlated arrays. CaO, P_2O_5, FeO, and K_2O are positively correlated arrays, although two pumice samples have distinctly lower K_2O compared to other samples with equivalent MgO. Fe_2O_3 and Al_2O_3 vs. MgO trends are more scattered.

A subset of the trace elements analyzed for these samples was chosen for detailed discussion in this study: Zr, Nb, Th, U, Rb, Sr, and Sm. Selected MgO (wt.%) vs. element trends are illustrated in Figure 6a–g; all trace elements are reported in ppm. These elements represent the range of element behavior in the CI. WMI trends for Zr, Nb, Th, and U show scattered changes in concentration at decreasing MgO. For these elements, a subset of the

Table 3. Selected major and trace element abundance data for samples of the Campanian Ignimbrite.

Sample	PontiR	IC-HB6a	ALT-1	SA-1b	SA-1a	AFGI-11	VE-1	AFGI-1	AFBP-1	AFBP-5	MP-1	ICB-9
type	Pumice	Pumice	Bulk	Bulk	Pumice	Pumice	Bulk	Pumice	Pumice	Pumice	Pumice	Pumice
SiO_2[a]	58.26	58.37	60.15	60.88	60.95	61.34	60.74	61.12	60.12	60.58	60.61	61.53
TiO_2	0.47	0.47	0.47	0.47	0.43	0.46	0.46	0.46	0.47	0.45	0.43	0.44
Al_2O_3	18.81	18.85	18.57	18.94	18.93	18.98	18.73	18.74	19.93	19.08	18.63	18.96
Fe_2O_3	1.89	2.22	2.32	2.75	2.54	1.71	2.49	2.80	1.11	0.88	1.80	3.08
FeO	2.44	2.12	1.44	1.01	1.08	1.89	1.05	0.79	2.30	2.52	1.39	0.23
MnO	0.10	0.10	0.16	0.17	0.12	0.19	0.20	0.21	0.24	0.23	0.23	0.20
MgO	1.44	1.38	0.92	0.79	0.79	0.68	0.62	0.59	0.46	0.42	0.40	0.38
CaO	4.09	4.02	3.23	2.64	2.46	2.17	2.25	2.19	1.79	1.71	1.75	1.76
Na_2O	2.66	2.61	4.44	4.23	4.40	4.63	5.20	5.32	5.39	5.54	6.39	5.78
K_2O	9.18	9.18	7.53	7.74	7.90	7.62	7.20	7.26	6.86	7.34	7.12	7.04
P_2O_5	0.27	0.25	0.18	0.16	0.15	0.09	0.12	0.09	0.04	0.05	0.09	0.08
SO_3	0.05	0.08	–	–	–	0.01	0.03	0.02	0.05	0.03	–	–
Cl	0.24	0.26	0.42	0.09	0.08	0.11	0.74	0.30	0.74	0.67	0.84	0.23
F	0.10	0.09	0.15	0.11	0.16	0.06	0.17	0.03	0.23	0.26	0.31	0.28
Nb[b]	21	20	64	70	55	87	93	101	112	114	114	104
Rb	225	230	315	340	305	361	380	362	356	406	420	375
Sm	6.4	6.3	11.6	12.7	9.6	9.85	13.8	14.8	15.6	15.1	14.8	14.1
Sr	760	800	355	265	270	202	150	128	101	44	22	20
Th	11.8	11.8	34.8	36.7	27.4	37.8	46.1	47.3	52.9	51.3	54.0	51.4
U	3.5	3.4	9.5	10.3	7.4	12.5	14.0	16.5	16.2	17.4	16.3	16.3
Zr	166	164	420	455	350	521	580	591	671	665	670	630

[a]Major elements in wt.%.
[b]Trace elements in ppm.

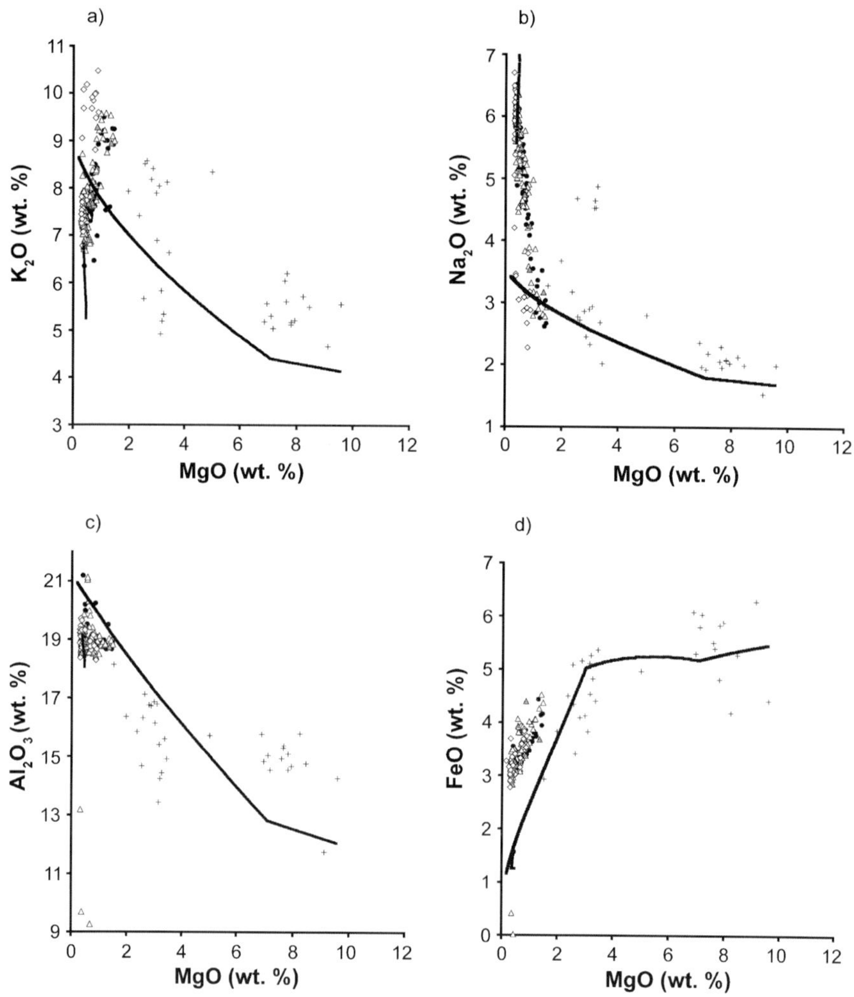

Figure 5. MgO (wt.%) vs. (a) K_2O (wt.%), (b), Na_2O (wt.%), (c) Al_2O_3 (wt.%), (d) FeO (wt.%). Symbols are the same as those in Figure 2. Best-fit MELTS model shown by black trends, and each black square represents a temperature decrease of 0.5°C. Gap in trends represents change in composition between 884 and 883°C.

low-MgO group is characterized by enriched trace element abundances (~2–4×) compared to the remaining low-MgO inclusions. These enriched melt inclusions have abundances similar to pumice with substantially less MgO. WMI Rb variations also show scattered changes with decreasing MgO, and the subset of low-MgO WMI noted above with more enriched abundances is absent. Sm has different behavior; concentrations among the WMI are generally similar to or higher than those of the pumice and glass/melt inclusions. For the elements discussed above, pumice and glass/melt show marked increases with decreasing MgO. Enrichment factors (defined as the ratio of element in the most MgO-rich pumice to least MgO-rich pumice) differ between the elements. Zr, Th, Nb, and U all have enrichment factors of ~6–8, whereas Rb and Sm have enrichment factors

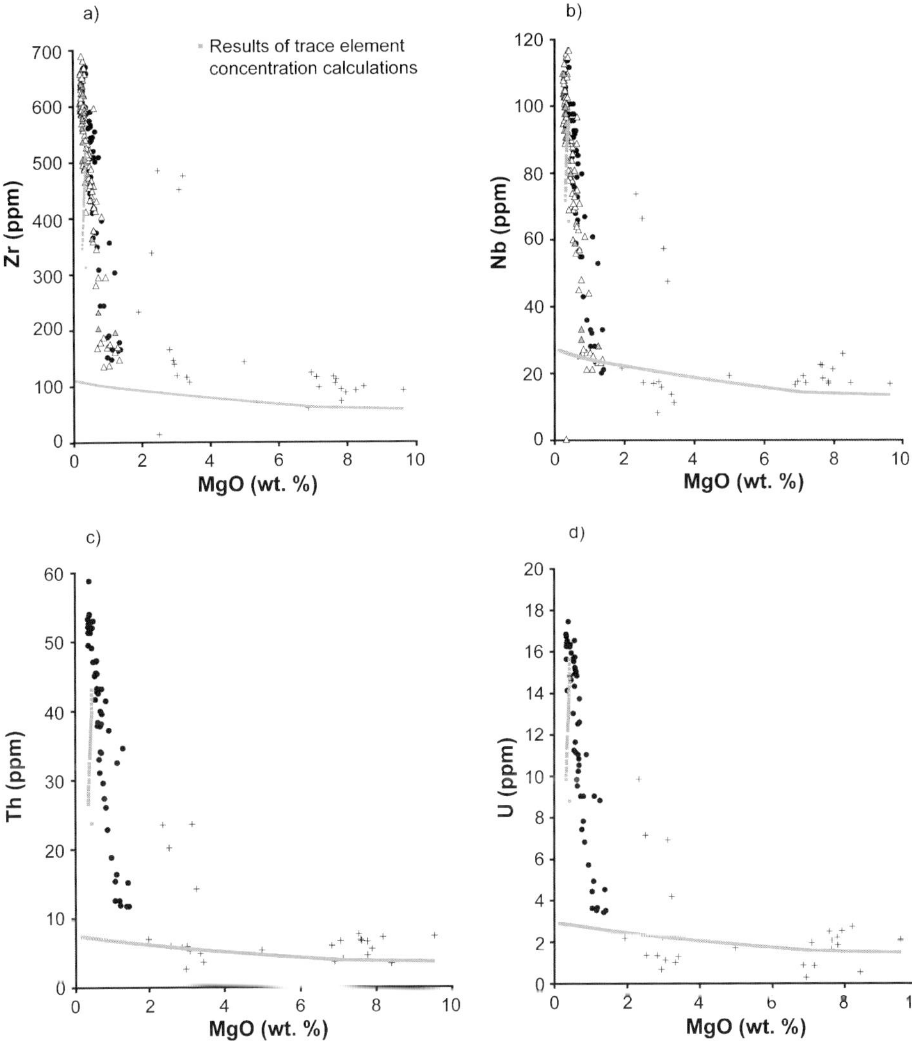

Figure 6. MgO (wt.%) vs. (a) Zr, (b) Nb, (c) Th, (d) U, (e) Rb, (f) Sm, (g) Sr. All trace elements in ppm. Symbols are the same as those in Figure 2. Best-fit MELTS model is shown by gray trends, and each gray square represents a temperature decrease of 0.5°C. Gap in trends represents change in composition between 884 and 883°C. MELTS trend after compositional gap not shown on (e), but arrow shows trajectory of trend at 884°C, which begins at 1255 ppm Rb.

of 3–4. With several exceptions, Sr in the high-MgO WMI group is relatively enriched, compared to the low-MgO group. The pumice and glass/melt inclusion data form a relatively tight array that decreases with decreasing MgO.

Element element trends for this subset of trace elements illustrate the systematics between U and Zr, Nb, Th, Rb, Sm, and Sr (Fig. 7a–f). For U vs. Zr, Nb, and Th, WMI form relatively tight clusters. Four or five inclusions plot at higher concentrations of U vs. trace element; these samples generally plot within the pumice and glass array, although

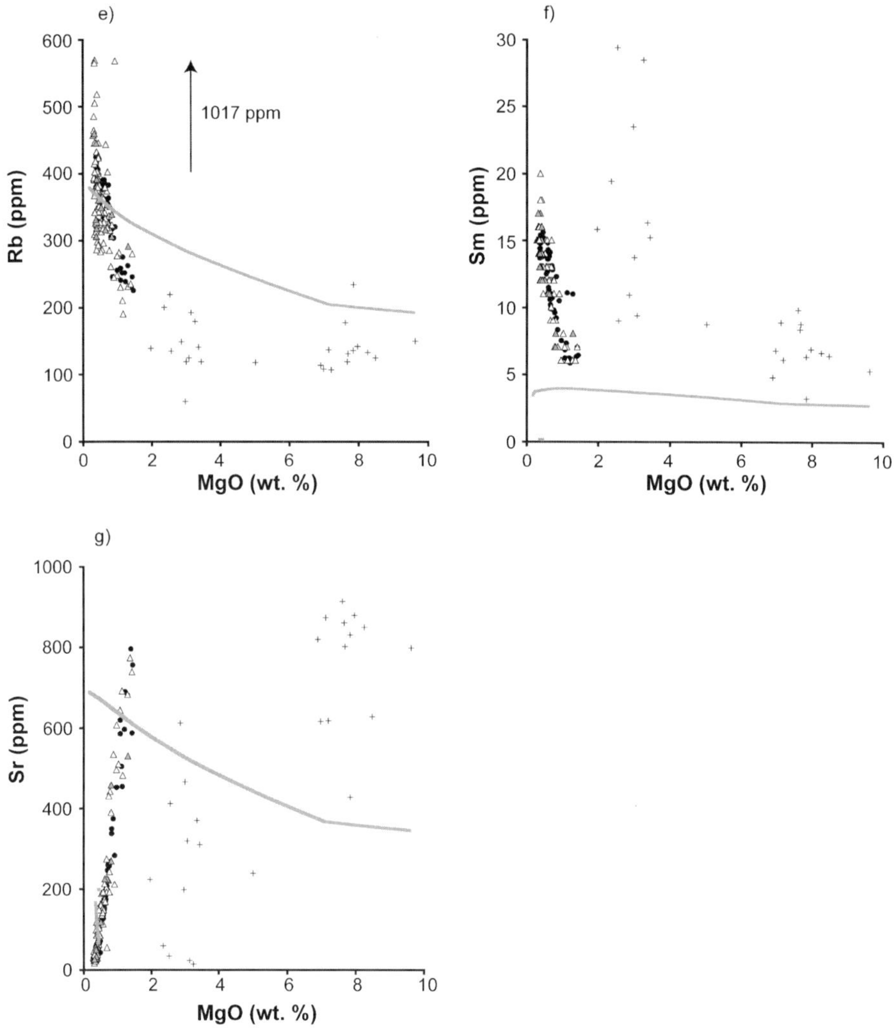

Figure 6. (Continued)

several melt inclusions plot above the primary U vs. Zr array. With few exceptions, pumice and glass samples form tight, positively correlated arrays. U vs. Rb has systematics similar to those described above: a tight cluster of WMI and a well correlated, positive array of pumice and glass. However, in contrast to the trends described above, the subset of WMI that plots at higher trace element concentrations fall slightly below the primary pumice and glass array. WMI exhibit variations of approximately a factor of 5 in Sm for a small range in U. Several samples, which plot at high U and Sm, are above the pumice and glass array, which is systematic and positively correlated. Like Sm, WMI have variable Sr over a small range of U (~factor of 5), and the high U melt inclusions plot below the pumice and glass array, which is somewhat scattered and broadly negatively correlated.

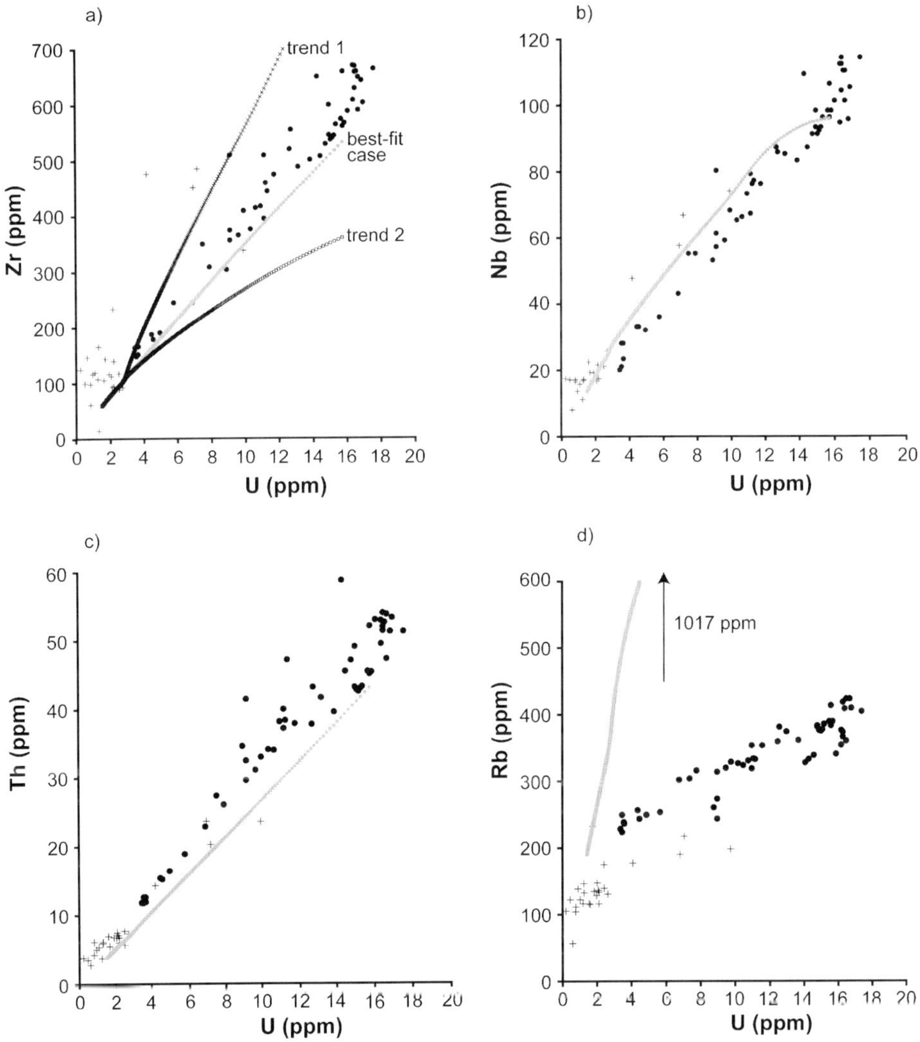

Figure 7. U (ppm) vs. (a) Zr, (b) Nb, (c) Th, (d) Rb, (e) Sm, (f) Sr. Symbols are the same as those in Figure 2. Best-fit MELTS model is shown by gray trends, and each gray square represents a temperature decrease of 0.5°C. In (a), trends 1 and 2 illustrate bracketing of observed trend based on use of variable mineral–melt partition coefficients, which are reported in Table 2. See text for discussion.

4.2. Th isotopes

Twelve samples of the Campanian Ignimbrite were analyzed for Th isotopes, and U and Th concentrations (Table 4). These samples are representative of the range of pumice and bulk rock compositions in our sample suite. They comprise eight samples of gray tuff (Sa-1a, Sa-1b, AFGI-1, AFGI-11, MP-1, ALT-1, VE-1, ICB-9), two samples of the lithic breccia (PONTIR, ICHB-6a), and two samples of the basal pumice (AFBP-1, AFBP-5).

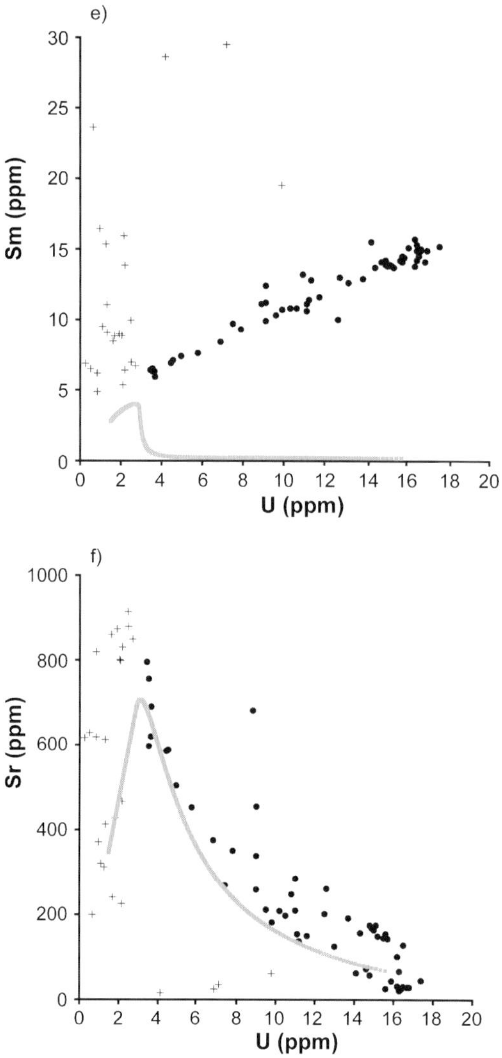

Figure 7. (Continued)

Measured thorium activity ratios, (^{230}Th)/(^{232}Th), range from 0.936 to 0.996, and (^{238}U)/(^{232}Th) range from 0.881 to 1.190 (Fig. 8a). Five samples are within 2.5% of secular equilibrium, five samples are enriched in ^{238}U (up to ~16% enrichment), and two samples are ^{230}Th-enriched (up to 8%). (^{230}Th)/(^{238}U) therefore vary from 0.837 to 1.08. Correlations with respect to (^{230}Th)/(^{232}Th) vs. (^{238}U)/(^{232}Th) and sample type (pumice vs. bulk; gray, basal pumice or lithic breccia) are lacking.

Measured Th isotope ratios were age-corrected using the mean ^{40}Ar/^{39}Ar age of 39.28±0.11 ka reported in De Vivo et al. (2001). It is important to note that 6 of the 12 samples analyzed for Th isotopes were included in the dating study, and thus, the age constraints on the these samples and the deposit as a whole are precisely documented.

Table 4. Selected major and trace element and Th isotopic data for samples of the Campanian Ignimbrite.

	Sample type	MgO (wt.%)	U (ppm)	Th (ppm)	Th/U	$^{232}Th/^{230}Th \times 10^{4a}$	$(^{238}U)/(^{232}Th)^b$	$(^{230}Th)/(^{232}Th)^b$	$(^{230}Th)/(^{238}U)^b$	$(^{230}Th)/(^{232}Th)_o^c$	$(^{230}Th)/(^{238}U)_o^c$
PONTIR	Pumice	1.44	3.62	11.22	3.10	19.39	0.979	0.958	0.979	0.950	0.970
ICHB6a	Pumice	1.38	3.63	11.10	3.06	19.86	0.993	0.936	0.942	0.911	0.917
ALT-1	Bulk	0.92	10.56	33.24	3.15	19.40	0.964	0.958	0.994	0.956	0.992
Sa-1b	Bulk	0.79	9.63	33.17	3.44	19.53	0.881	0.952	1.080	0.982	1.115
Sa-1a	Pumice	0.79	8.49	27.34	3.22	19.52	0.943	0.952	1.010	0.956	1.014
AFGI-11	Pumice	0.68	13.84	41.75	3.02	18.93	1.006	0.980	0.977	0.972	0.967
VE-1	Bulk	0.62	14.18	39.74	2.80	18.90	1.083	0.984	0.908	0.940	0.868
AFGI-1	Pumice	0.59	12.99	41.08	3.16	19.47	0.959	0.955	0.995	0.953	0.993
AFBP-1	Pumice	0.46	14.96	49.45	3.31	19.45	0.918	0.956	1.041	0.972	1.059
AFBP-5	Pumice	0.42	16.26	49.31	3.03	19.16	1.000	0.970	0.970	0.957	0.956
MP-1	Pumice	0.40	18.18	48.44	2.66	18.83	1.139	0.987	0.867	0.921	0.809
ICB-9	Pumice	0.38	18.13	46.24	2.55	18.65	1.190	0.996	0.837	0.912	0.767

[a] $^{232}Th/^{230}Th$ represent averages of at least two replicates except for samples PONTIR and ICHB6a, which are represented by only one analysis. Average in-run analytical precision for Th isotope analyses ranges from 0.23% to 0.49%. Typical reproducibility for Th isotope ratios is 1.5% or less. Typical reproducibility for U/Th is 2%.

[b] Activity ratios denoted by parentheses and calculated using the following decay constants: $\lambda_{230} = 4.9475 \times 10^{-11}\ yr^{-1}$; $\lambda_{230} = 9.1952 \times 10^{-6}\ yr^{-1}$; $\lambda_{230} = 1.55125 \times 10^{-10}\ yr^{-1}$ and using the weighted mean $^{232}Th/^{230}Th$.

[c] Age of 39.28 (± 0.11) ka, determined by incremental heating and total fusion $^{40}Ar/^{39}Ar$ geochronology (De Vivo et al., 2001) used for age correction on all samples.

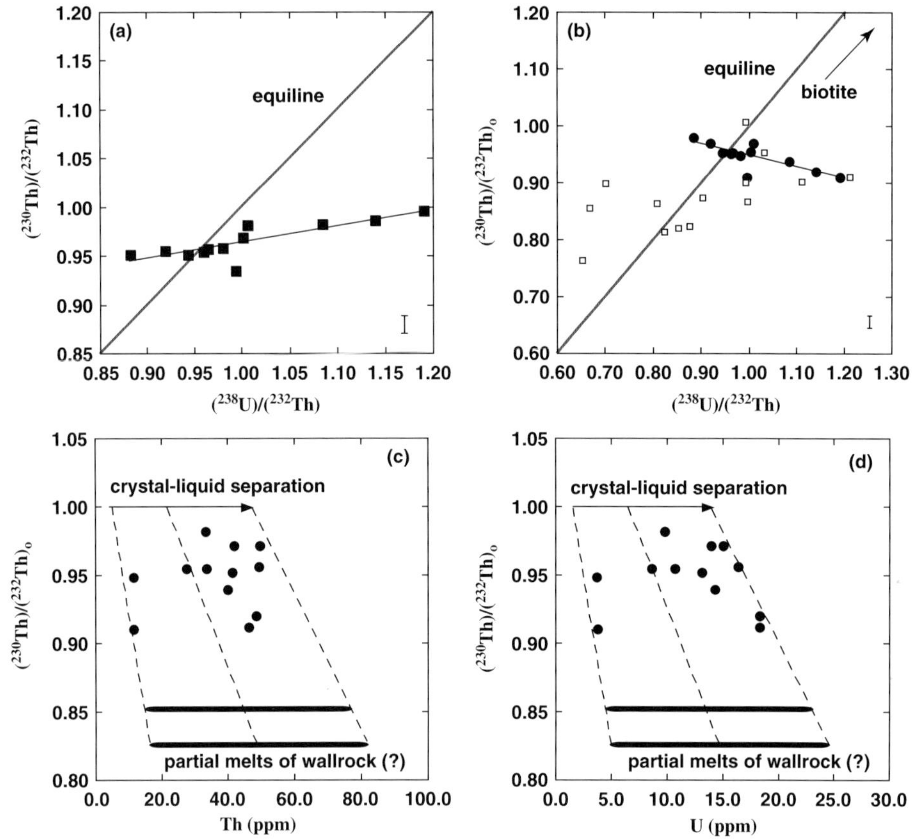

Figure 8. (a) $(^{238}U)/(^{232}Th)$ vs. measured $(^{230}Th)/(^{232}Th)$ for 12 samples of the CI. Analyses plot is a relatively well-correlated linear array (R = 0.82), which gives an apparent age of 19.6±4.7 ka. Note that this is ~20 kyr younger than the 39.28±0.11 ka age of the deposit (De Vivo et al., 2001). (b) $(^{238}U)/(^{232}Th)$ vs. $(^{230}Th)/(^{232}Th)_o$. Samples, which were all age-corrected to 39.28±0.11 ka, plot on a relatively well-correlated linear array (R = 0.77) characterized by a negative slope. (c) Th (ppm) vs. $(^{230}Th)/(^{232}Th)_o$. (d) U (ppm) vs. $(^{230}Th)/(^{232}Th)_o$. In both (c) and (d), note lack of simple correlation between isotope ratio and concentration. Also shown are arrows to qualitatively illustrate effects of crystal–liquid separation, and possible mixing lines between fractionating melt and partial melts of heterogeneous wall rock characterized by relatively low $(^{230}Th)/(^{232}Th)_o$. See text for discussion of qualitative open-system model.

Initial Th activity ratios, $(^{230}Th)/(^{232}Th)_o$, vary from 0.911 to 0.982, and $(^{230}Th)/(^{238}U)_o$ range from 0.767 to 1.115 (Fig. 8b). Three samples are within 1.5% of secular equilibrium, seven samples are ^{238}U-enriched by up to ~23%, and two samples are ^{230}Th-enriched by up to ~12%. The data plot as a relatively coherent linear, with a regression coefficient of 0.77.

Figure 8c,d illustrates variations between $(^{230}Th)/(^{232}Th)_o$ and Th (ppm) and U (ppm). Neither plot exhibits systematic relationships between activity ratio and concentration. The two samples with the lowest U and Th concentrations, PONTIR and ICHB-6a, identified as lithic breccias, have $(^{230}Th)/(^{232}Th)_o$ that range from ~0.91 to 0.95. Samples with higher abundances of Th and U exhibit a slightly greater range of $(^{230}Th)/(^{232}Th)_o$ from ~0.91 to 0.98.

5. Discussion

5.1. Magma chamber processes

Ignimbrites have been interpreted as deposits that represent rapid evacuation of a magma chamber (e.g., Smith, 1979; Hildreth, 1981), and thus, the primary compositional and isotopic heterogeneity observed in the deposit reflects the range of magmas that were in the magma chamber just prior to evacuation. Such magmas form by processes that occur as melt forms and as melt + crystals + vapor ascend through and are stored in the crust. For our data set, comparison of the observed data with the MELTS model reveals that crystal–liquid separation played an important if not dominant role in developing the observed major element compositional heterogeneity. Detailed reconstructions of selected trace element trends highlights the critical control that phase assemblage and mineral–mineral partition coefficients (K_{sm}) have on melt trace element abundances. Imprecise knowledge of K_{sm}, in some cases, makes it difficult to ascertain the contribution from crystal–liquid separation, compared to other processes such as crustal assimilation, but several elements provide strong evidence for open-system processes. Th disequilibria data and published Sr isotope data (Civetta et al., 1997) support this suggestion by providing definitive evidence of open-system processes in the Campanian magma plumbing system.

5.1.1. Crystal–liquid equilibria: major element discussion

The results of isobaric fractional crystallization starting from RPM composition listed in Table 1 are summarized in Figures 2b, 4, and 5a–d. Best-fit MELTS parameters include pressure = 0.15 GPa, crystallization along the QFM+1 oxygen buffer, and initial dissolved water content of the RPM of 3 wt.%, which is within the range of water contents measured on melt inclusions reported by Webster et al. (2003). The temperature interval used to compute the liquid line of descent is 0.5°C. Figure 2b shows the calculated trend (anhydrous basis) on an alkali–silica diagram. Upon fractional crystallization of parental trachyandesite, the melt evolves to higher total alkali (K_2O plus Na_2O) and higher silica. The most striking feature of this diagram is the presence of a compositional gap beginning at $T \approx 884$°C. This temperature defines a pseudo-invariant point in thermodynamic space that behaves practically like a eutectic point in a simple binary component system. That is, as heat (enthalpy) is isobarically extracted from the system, the fraction of melt changes from ~0.46 to ~0.09 essentially isothermally ($T \approx 884$°C), and melt simultaneously saturates in alkali feldspar, plagioclase, and biotite. At this temperature, the melt composition and properties change rather abruptly. The difference in composition of the two distinct liquids across the gap is ~2 wt.% for Na_2O and Al_2O_3, ~1.5 wt.% for K_2O, 1 wt.% for SiO_2 and CaO (Fig. 5), and ~0.5 wt.% for H_2O (Table 5). In fact, across the gap, both the composition of the melt and the bulk magma (melt plus supercritical fluid) change rather significantly.

Figure 4 shows the abundance of the crystalline phases removed as a function of $f_m(T)$. In addition to the rapid change in fraction of melt at the pseudo-invariant temperature, the other important features of Figure 4 are the following: (1) olivine is the liquidus phase; (2)

Table 5. Characteristics predicted by best-fit MELTS model of melts at pseudo-invariant point.

f_m	T (°C)	SiO_2	TiO_2	Al_2O_3	FeO	MgO	MnO	CaO	K_2O	Na_2O	P_2O_5	H_2O
0.46	884	57.11	0.24	19.97	1.10	0.18	0.15	4.37	8.24	3.26	0.76	4.59
0.09	883	57.82	0.36	18.13	1.19	0.35	0.79	3.53	6.45	5.28	0.85	5.22

melt saturates in H_2O (discrete supercritical fluid develops) at ~1127°C; and (3) there is a 100°C temperature interval dominated by clinopyroxene fractionation. The order of solids removal is olivine, clinopyroxene, spinel, apatite, and then simultaneous removal of alkali feldspar, plagioclase, and biotite at the pseudo-invariant point temperature, 884°C.

In Figure 5a–d, MgO variation diagrams are shown for K_2O, Na_2O, Al_2O_3, and FeO. In each case, the MELTS predictions are in reasonable agreement with glass and pumice data, suggesting that crystal fractionation is the mechanism that dominates evolution of the major elements in the CI. The offset between the calculated and observed trend on the FeO–MgO diagram (Fig. 5d) is most likely attributed to the imposition of a fixed oxygen buffer curve of QFM+1.

In summary, a reconstruction of the CI parental melt based on glass inclusions entrapped in clinopyroxene phenocrysts has been used as an initial condition to develop the liquid line of descent, assuming perfect closed-system fractionation at fixed pressure (0.15 GPa or ~5 km depth) along the QFM+1 oxygen buffer. When one considers that the effects of assimilation are not accounted for, comparison between predicted and observed data is reasonable and allows a “reference” state to be defined, upon which a trace element analysis can be carried out.

5.1.2. Crystal–liquid equilibria: trace element discussion

The goal of numerically modeling trace element trends generated by crystal–liquid equilibria is to quantitatively assess the degree of enrichment or depletion possible through separation of melt and solids. In this section, we integrate results of the best-fit MELTS model with the numerical formulation described above to reconstruct trace element abundances during crystal fractionation allowing for the change in bulk K_{sm}.

Results of the best-fit trace element models are illustrated in Figures 6 and 7. Before interpreting these trends in the context of the open- vs. closed-system question, it is useful to review the “robustness” of such model trends. A minimum of three types of input informs the final trace element results: the choice of parental composition, the phase proportions returned from the MELTS model, and the choice of mineral–melt K_{sm}. Several observations are relevant to the interpretations provided below. Variations in the parental composition will obviously lead to variations in trace element concentrations. The main impact changing the parental composition will have (all other modeling parameters being the same) to translate up or down the parts of the trend that are shown from ~10 wt.% MgO to ~2 wt.% MgO (Fig. 6). The trajectory of the trends at lower MgO is primarily controlled by the bulk K_{sm}, particularly because the proportion of solids removed at this stage is so high. Thus, although some fine-tuning of the parental composition might improve the major element–trace element and/or trace element–trace element trends, given the range evident from the high-MgO melt inclusions of Webster et al. (2003), it is unlikely that reasonable variations in the parental composition will drastically change the interpretations of trace element behavior. As noted above, the best-fit MELTS model is the result of approximately 100 simulations, which differ in detail but fundamentally return the same results: olivine and clinopyroxene at high temperatures (liquidus and subliquidus), followed by onset of apatite and spinel crystallization. At ~884°C, a large mass of crystals, dominated by alkali feldspar and plagioclase, form. Crystallization at this pseudo-invariant point is a feature of all of the MELTS results, as is onset of apatite formation. Thus, although the details of the MELTS model might change with further examination, the best-fit model, which is subject to the assumptions relevant to describing a magmatic system using the MELTS approach, are robust. The final variable in the trace element modeling, which is the one that is perhaps the most problematic in producing robust results, is the choice of mineral–melt K_{sm}. In this study, mineral–melt K_{sm} were chosen based on assessment of the range of values available in the Geochemical Earth Reference Model (GERM) Partition Coefficient compilation (http://earthref.org/) (All values in the GERM compilation are referenced, and references relevant to this study are included in Table 2). Inspection of the compilation for a particular phase and element, even within a restricted compositional range (e.g., silicic or basaltic), in most cases, reveals large differences in mineral–melt K_{sm} (e.g., Rb in basaltic plagioclase ranges from 0.016 to 0.3; Sr basaltic clinopyroxene ranges from 0.04 to 0.449; Th in apatite ranges from 1.6 to 41). In many cases, when applying partition coefficients to model a particular suite of rocks, it is difficult to develop independent criteria that allow a rational choice to be made among the values. The approach we adopted was to apply different K_{sm} from the compilation until the observed trends were suitably bracketed, and then final K_{sm} were chosen and the best-fit case was calculated. These best-fit cases are summarized in Figure 7. An example of bracketing is shown in Figure 7a (U vs. Zr). One model trend (trend 1) plots above the observed trend, whereas the other model trend (trend 2) plots below. The critical point about these model trends is they have very different implications for understanding trace element variations. In the case of trend 2, the implication is that crystal–liquid separation cannot produce enrichments that are sufficient to describe the data. This implies the system was open, and therefore another process, such as crustal assimilation, was important. However, permissible changes in K_{sm} (based on available data) yield a trend (trend 1) that plots above the observed data. Trend 1 implies that crystal–liquid separation can produce concentrations that are more enriched than those observed. In this case, elements may have been scavenged from the magma, and such scavenging might be caused by transport of the element into a vapor phase, in the case of a vapor-saturated magma. In several cases (e.g., Th, Sm, Rb), application of the most extreme mineral–melt K_{sm} listed in the compilation that were relevant to the broad magma compositional range failed to yield model trends that reproduced or bracketed the data. In these cases, the model case shown is that which employs the most extreme, allowable K_{sm}.

Figures 7a, b and f illustrate results for U vs. Zr, Nb, and Sr. For these elements, observed and model trends show similar systematics. This suggests, to a first order, that the observed enrichments may be attributed to crystal–liquid separation. U, Zr, and Nb all behave incompatibly, increasing in abundance with decreasing MgO (Fig. 6a,b,d). At ~1.5 wt.% MgO, the marked increase in observed and model abundances of these elements is interpreted to result from the large change in crystallinity associated with invariant point crystallization at ~884°C. Because $f_m(T)$ changes dramatically over a small temperature interval, the consequences on trace element abundances are profound. Sr is initially incompatible (i.e., increases in abundance with decreasing MgO, Fig. 6g), and then, for $T <$ 884°C, Sr becomes compatible because plagioclase and alkali feldspar crystallize (Fig. 7f).

Figures 7c and e illustrate model vs. observed results from U vs. Th and Sm. These elements provide an interesting window into the behavior of elements typically considered incompatible. In the case in which apatite is an important crystallizing phase, because the K_{sm} for Th and Sm can be very high (e.g., order 2–90; e.g., Nagasawa, 1970; Mahood and Stimac, 1990), these elements will be very efficiently removed from the melt at ≈884°C. In the case of Th, the best-fit trace element trend uses the apatite–melt K_{sm} that is least compatible of those reported (e.g., 1.6 of the range 1.6 to 41; Mahood and Stimac, 1990; Bea et al., 1994). The U-Th trend is positively correlated, indicating that Th is enriched during crystal–liquid separation, but the model trend is not as Th-enriched as the observed trend. Even with an apatite–melt Th partition coefficient of 1 (results not shown), Th is still not as enriched in the model as observed. Assuming the MELTS model is robust with respect to phase proportions, this result implies either that Th is not compatible in apatite, a result not supported by the partition coefficient database, or that Th may be impacted by open-system processes. (Because the MELTS and trace element modeling assume perfect fractional crystallization, it is possible that apatite did not effectively separate from its host melt. However, this explanation seems unlikely because P_2O_5 decreases with decreasing MgO, and Th increases as P_2O_5 decreases, suggesting apatite was fractionated.) The case of Sm is more extreme because, compared to Th, it has much higher apatite–melt K_{sm}. The best-fit case illustrated in Figures 6f and 7e employed the lowest mineral–melt K_{sm} in the GERM compilation (Table 2), and yet, the model trend grossly underestimates the observed Sm abundance; in fact, the model predicts that the melt will have almost no Sm. Although the detailed modeling is not yet complete, the LREE and other MREE in the CI behave similarly to Sm. Because these elements tend to be compatible (to varying degrees) in apatite, equivalent model systematics may emerge. Again, the model vs. observed results indicate either that the Sm apatite–melt partition coefficient is highly inaccurate, or that most of the Sm is related to some open-system contribution.

Figure 7d shows the U vs. Rb results. The illustrated model trend employs mineral–melt K_{sm} from the GERM compilation that are the most compatible for each phase; despite this, the model Rb trend grossly overestimates the observed abundances of Rb. This suggests that the GERM K_{sm} are not relevant, that there is some yet unidentified trace phase that removed Rb, and/or that a separate process, such as interaction with a vapor, has removed Rb from the melt. The critical parameters governing the distribution of a trace element between solid, melt, and supercritical fluid are the solid-melt (K_{sm}) and fluid-melt (K_{fm}) partition coefficients and the rate of change of the mass fraction of fluid (f_f) with respect to $f_m(T)$ during isobaric fractional crystallization. If the bulk partition coefficient for Rb between melt and vapor exceeds ~3, as it may based on laboratory experiments on sanidine and biotite (Beswick, 1973), then the discrepancy between the calculated trend assuming Rb is totally insoluble in the supercritical fluid and the observed trend can be rationalized. The role of melt-vapor fractionation is easily estimated. For insoluble behavior with $K_{fm}^{Rb} = 0$ and $K_{sm}^{Rb} = 0$ the situation is identical to standard solid-melt Rayleigh fractionation and enrichment in residual melt is expected $C_m^{Rb}/C_{m,o}^{Rb} = \sim 2$. If however $K_{fm} = 5$ (i.e., Rb is decidedly soluble in fluid relative to melt), then the melt will be depleted in Rb such that $C_m^{Rb}/C_{m,o}^{Rb} = \sim 0.1$. If this hypothesis of supercritical fluid fractionation is valid for Rb, one expects other geochemically similar elements will exhibit similar behavior. The study by Bohrson et al. (in preparation) addresses this issue further.

A final comment about all of the model trends is relevant to understanding magma chamber processes. The trace element mass balance model presented here represents perfect fractional crystallization and evaluates compositional changes in very small increments of fraction of melt. Pumice likely represents some integration of parcels of melt with distinct histories, and thus, some of the discrepancy between model and observed trends may be the result of magma and eruption dynamics that mix discrete masses of distinct-composition melt. An interesting test of this hypothesis would be to collect small spatial resolution data of glass in the pumice. Signorelli et al. (1999) represents just such a study because their glass data is the result of microprobe analyses, and thus, are small-spatial resolution samplings of the major element chemistry of the Campanian melt. In some cases, the MELTS model reproduces parts of the glass data trends. No trace element data are provided in the Signorelli et al. (1999) study, but the prediction would be that the *in situ* trace element data would fall close to or on the model trends, thus implying that the pumice analyses reflect mixing of melts that occurs in the chamber and/or upon eruption.

In summary, results of numerical trace element modeling indicate that Sm and Th are over-enriched compared to model values, suggesting these elements have been affected by processes other than simple equilibrium crystal–liquid separation. General trends for U vs. Nb, Zr, and Sr indicate that the abundances of these elements may be dominantly controlled by crystal-melt separation, but the range of K_{sm} available in the literature does not preclude some contribution from other sources (i.e., open system). Finally, the model predicts Rb should be more abundant in the Campanian samples, indicating that Rb may have been scavenged by vapor that was in equilibrium with melt.

5.1.3. Open-system processes: Th and Sr isotopes

Radiogenic isotopes are effective in identifying and quantifying open-system processes because isotopes of the same element are not fractionated by (high-temperature) equilibrium crystal–liquid separation. Thus, for the closed-system case, $^{87}Sr/^{86}Sr$ and $(^{230}Th)/(^{232}Th)$ of melts will be the same as those of newly formed crystals. Likewise, in the case of a zoned magma body, all of the compositionally distinct masses of magma will share a common initial isotope ratio. The closed-system description assumes that the timescale of crystallization is short compared to the half-life of the parent isotope, and thus no (or insignificant) radiogenic in-growth occurs during the process. In the case of Rb, which has a half-life of ~4.9×10^{10} yr, this assumption is robust for magmatic processes, except in rare cases in which Rb/Sr is extremely high. The half-life of ^{230}Th is relatively short (~7.5×10^{4} yr), and thus Th disequilibria have been studied in an attempt to understand crystal and magma residence, and fractionation times.

$(^{238}U)/(^{232}Th)$ vs. $(^{230}Th)/(^{232}Th)$ is illustrated in Figure 8a. The striking observation related to this figure is the relatively coherent linear array that the data define. Linear regression yields a correlation coefficient (R) of 0.82, and an apparent isochron age of 19.6±4.7 ka. Such a result is problematic because the deposit (including 6 of the 12 samples analyzed for Th isotopes) has been precisely dated at 39.28±0.11 ka (De Vivo et al., 2001). Age-corrected Th data are plotted in Figure 8b. *The data clearly illustrate that the melts represented by pumice of the CI were not in Th isotopic equilibrium upon eruption.* The negative slope of the array also suggests that the system was open with respect to Th isotopes. The slope of the array is important because a positively correlated array might be interpreted to reflect a magma residence time; in contrast, a negative array (in the absence of an open system) would indicate timescales less than 0.

One of the interesting aspects of the Th isotope data is the relatively well-correlated trend in $(^{238}U)/(^{232}Th)$ vs. $(^{230}Th)/(^{232}Th)_o$. Despite this, simple correlations between age-corrected Th isotope ratio and concentration of U or Th are lacking (Fig. 8c,d), as are correlations with other elements/oxides (not shown; e.g., MgO, Na_2O, Zr, Nb). A good example of the lack of simple systematics is the observation that samples with very similar, high Th concentrations (AFBP-1, AFBP-5, MP-1, and ICB-9) have $(^{230}Th)/(^{232}Th)_o$ from ~0.91 to 0.97. Similarly, samples with low Th concentrations (PONTIR and ICHB-6a) have $(^{230}Th)/(^{232}Th)_o$ of ~0.95 and 0.91, respectively. These observations collectively preclude simple two-component mixing as the explanation for the Th isotopic signatures.

The interpretation of the open-system history of Th in the magma body is further complicated by the following observations: (1) there are no systematic correlations between elements or oxides and the degree of disequilibrium, $(^{230}Th)/(^{238}U)_o$. This, taken together with the range of U, Th. vs. $(^{230}Th)/(^{232}Th)_o$ noted above, means it is difficult to reliably identify the least contaminated sample or establish the systematics of U-Th fractionation during differentiation. (2) Examination of the $(^{238}U)/(^{232}Th)$ of the high-MgO WMI indicates they are heterogeneous (0.45 to 1.36) and have a range that encompasses and exceeds those of the CI samples. If this range represents potential fractionation that has affected less-differentiated melts of the Campanian system, then it is difficult to systematically define characteristics of U-Th fractionation in more differentiated melts. That is, it is unclear if the range $(^{238}U)/(^{232}Th)$ in the pumice was inherited from heterogeneous parental magmas, was due to differentiation processes, or both. (3) Although the number of samples analyzed for Th is somewhat restricted, there does not appear to be any correlation between $(^{230}Th)/(^{238}U)_o$ or $(^{230}Th)/(^{238}U)_o$ and sample locality.

Sr isotope data for Campanian samples published by Civetta et al. (1997) include acid-leached data for pumice, glass, and feldspar. Figure 9 illustrates the systematics of $^{87}Sr/^{86}Sr$ vs. Sr concentration (by ICP) in the pumice. Note that no Sr concentration data are available for the glass and feldspar data, so Figure 9 is intended to illustrate pumice Sr isotope-concentration systematics and provide a visual comparison of $^{87}Sr/^{86}Sr$ for pumice, glass, and feldspar from the same sample. Observations and interpretations relevant to this discussion from Civetta et al. (1997) include (1) samples were acid-leached, and residue and leachate (with only a few exceptions) yielded similar $^{87}Sr/^{86}Sr$; thus the authors conclude it is unlikely that the range in pumice $^{87}Sr/^{86}Sr$ is the result of post-eruptive alteration. (2) Pumice $^{87}Sr/^{86}Sr$ are broadly correlated with Sr concentration; the least Sr-rich samples are those with the most radiogenic $^{87}Sr/^{86}Sr$. (3) Feldspars exhibit isotopic disequilibrium with pumice. For most feldspar, $^{87}Sr/^{86}Sr$ range is small, and feldspar signatures are less radiogenic than pumice or glass. Exceptions include two samples that have relatively radiogenic $^{87}Sr/^{86}Sr$; these are associated with pumice that has relatively low Sr concentration (Fig. 9).

The hypothesis that was favored by Civetta et al. (1997) to explain the disequilibria and the change in $^{87}Sr/^{86}Sr$ among the pumice (including the systematic relationship with Sr concentration) was contamination after feldspar formation. They call on a model that potentially involves interaction between hydrothermal fluids and magma prior to eruption.

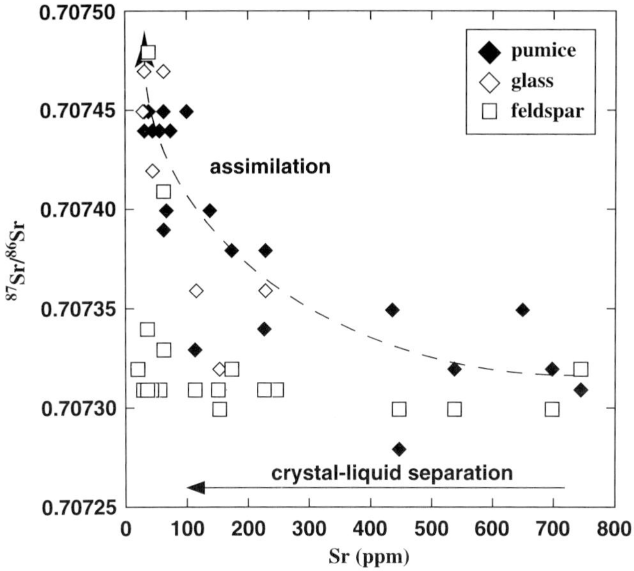

Figure 9. Sr (ppm) vs. $^{87}Sr/^{86}Sr$ of pumice, glass, and feldspar data of Civetta et al. (1997). Symbols shown in legend. Note disequilibrium between feldspar, pumice and glass and the general trend defined by decreasing Sr (ppm) and increasing $^{87}Sr/^{86}Sr$. Arrow shows qualitative effect of assimilation of wall rock with relatively radiogenic Sr isotope signature.

The authors indicate that selective assimilation of 20% of a seawater-like fluid can explain the range of $^{87}Sr/^{86}Sr$ in the pumice. Citing the same data set, Pappalardo et al. (2002) note that this interaction could occur on the timescales of seconds. These authors reference Palacz and Wolff (1989), which provides (among other types of data) Sr isotopes on the Granadilla Pumice, Tenerife. Feldspar $^{87}Sr/^{86}Sr$ range from 0.70314–0.70318, and vigorous acid-leached pumice range from 0.7032–0.7052. Thus, like the Campanian data set, there is Sr isotope disequilibria between feldspar and acid-leached glass. The model of Palacz and Wolff posits vesiculating magma and hydrothermal fluids (with high $^{87}Sr/^{86}Sr$) interact just prior to or during eruption. Limited diffusion of Sr occurs into the bubble walls before quenching. Palacz and Wolff (1989) suggest that diffusion at magmatic temperatures is fast enough to impart radiogenic signatures on the timescale of eruption.

Tracer diffusion of Sr at 900°C in CI melt containing several wt.% dissolved H_2O is estimated at $\sim 10^{-13}$ m^2/s (Spera, 2000). The characteristic timescale for Sr diffusive transport, $\tau_{Sr} \sim \delta^2/D$, where δ is a diffusion length scale, τ_{Sr} is ~10 s for micron scale. Although this timescale is small, its significance with respect to isotopic exchange of Sr during eruption is not clear. The concentration of Sr in CI melt (>20 ppm) exceeds the Sr concentration in seawater (~9 ppm), so diffusion of seawater Sr into melts is not suggested. If concentrated brine interacts with the magma, the chemical potential gradient might possibly allow for Sr transport into the melt. But there is a huge energy requirement for fluid heating, and once subsolidus temperatures are reached, Sr diffusion slows dramatically. For a diffusion event of one-day duration at 900°C, the diffusion scale is 0.1 mm – this seems rather small.

Magma–fluid interaction has been provided as an explanation for open-system signatures in Phlegrean Fields as well as other locations (e.g., Hildreth et al., 1984; Clocchiatti et al., 1988; Villemant, 1988; Michaud, 1995; Villemant et al., 1996). Several observations, some of which are noted in papers referenced above, are relevant to this hypothesis. First, any proposal that invokes addition of fluid to a magma body must grapple with the fluid conditions therein. It is unlikely that fluid will flow into a fluid-saturated magma because of lack of a H_2O chemical potential gradient (e.g., Taylor, 1974, 1977, 1980, 1986). Second, fluids need to come to total or partial thermal equilibrium with magma at temperatures of ~900–1100°C and must therefore be heated considerably from shallow meteoric values of ~100–200°C. For fluids at crustal temperatures, such heating requires enthalpy delivered from the magma body. A simple calculation of energy balance places reasonable limits on the mass of fluid that can be involved in such an interaction. For a single mass of magma (i.e., no recharge), the maximum energy available to heat crustal fluid is the sum of the sensible and latent heat. For a magma similar in composition and properties to the Campanian ($T_{liquidus}$ = 1236°C; $T_{solidus}$ = 866°C, ΔH = 396 kJ kg^{-1}, C_{pm} = 1484 J kg^{-1} K^{-1}) and heat capacity of supercritical H_2O of 4 kJ kg^{-1} K^{-1}, the maximum fluid to magma ratio is ~0.3. We note that this is an extreme case because the calculation assumes all of the thermal potential of the magma body is used for heating of initial cool fluid and the chamber must crystallize fully, thus rendering an eruption impossible. However, if we neglect these extreme limitations for purposes of illustration, a third constraint, one of mass balance, can be constructed for magma–fluid interaction. In this case, we use the Sr isotope characteristics of the Campanian samples of Civetta et al. (1997). Using a nominal feldspar $^{87}Sr/^{86}Sr$ value of 0.70731 as an initial value and the range of Sr concentration and $^{87}Sr/^{86}Sr$ from the pumice, the required fluid to magma ratios to invoke the observed isotopic shift range from ~0.3 to ~0.9. To decrease the high water to magma ratio to ~0.3 (the approximate limit from thermal balance), the concentration of Sr in a hydrothermal fluid would have to be three times that in seawater, assuming seawater-like $^{87}Sr/^{86}Sr$ of ~0.709.

Although it is unlikely that the Th isotopes characteristics of the Campanian samples are due to magma–fluid interaction, particularly because the best-fit MELTS model indicates the Campanian magmatic system was vapor-saturated in the compositional range represented by the ignimbrite samples included in this study, it is instructive to examine the mass balance arguments because the concentration of Th in seawater is orders of magnitude less than that found in typical igneous rocks. Seawater has a variable but high to extremely high (^{230}Th)/(^{232}Th) (4-300, Roy-Barman et al., 1996). Despite this, large water-to-magma ratios are typically required to produce isotopic changes in magmas because of the very low (pg/g) concentration of Th in typical seawater. For example, for an average (^{230}Th)/(^{232}Th) of 100 and a Th concentration of ~0.008 ppb (the highest reported for either seawater or mid-ocean ridge hydrothermal fluids (Chen et al., 1986; Chen, 1987) water-to-magma ratios required to produce a (^{230}Th)/(^{232}Th) range from ~0.91 to 0.98, assuming the magma has ~11 ppm Th, are on the order of tens to hundreds. Clearly such interactions are not possible based on thermal constraints noted above, and even if such interaction could occur, these high water-to-magma ratios would predict seawater-like Sr isotopes (~0.709) in CI pumice, which is not observed.

In summary, hypotheses that invoke fluid–magma interaction must be evaluated in the context of chemical potential gradients for fluid, and allowable thermal (energy) and mass balances. Because some magmas, at least for part of their evolution, are fluid-saturated (indeed the model CI parental magma volatile saturates at ~1127°C), it is unlikely that fluid will enter the magma in the first place. Even if this constraint is relaxed, the high heat capacity of fluid requires a great deal of energy to be extracted from the magma body to heat it from the ambient crustal to magmatic temperatures. Such demanding thermal requirements imply high degrees of crystallization, which, because of the physical changes that accompany crystal formation (e.g., increase in viscosity), may impact the eruptability of the body. Finally, fluids are typically characterized by low concentrations of elements compared to most rocks. Simple mass balance calculations like those above can therefore provide limitations on hypotheses that invoke fluids to explain chemical and isotopic variations in magmas. In the case of the CI, we conclude that it is highly unlikely that the open-system signatures are due to direct interaction between magma and fluid.

The explanation for the Th and Sr isotopic heterogeneity may lie in interaction between magma and wall rock, which might include intrusive parts of the long-lived Campanian magmatic system. Several lines of evidence bear on such an open-system model. Characterization of U-Th disequilibria associated with products (tephra, lava, and cumulate nodule) of the 1994 eruption of Vesuvius (Black et al., 1998) demonstrate that whole-rocks and mineral phases (leucite, biotite, pyroxene, and a magnetic separate) are dominated by (^{230}Th)/(^{232}Th) similar to or lower than the CI samples (Fig. 8b). (^{238}U)/(^{232}Th) ranges from Th- to U-enriched, and some samples show relatively extreme behavior (e.g., tephra biotite = 5.1 (not shown); cumulate magnetic separate = 0.65). The cumulate is interpreted as being composed of phases that are cogenetic with the 1944 magma or that represent an earlier stage of magmatism on Vesuvius. Beneath Vesuvius, it is likely that there are intrusive equivalents of the 1944 erupted products, because typical intrusive to extrusive relationships (Crisp, 1984) predict that large masses of magma remain in the crust relative to those erupted. Thus, if cumulates and intrusive equivalents present in the magma storage-transport system of Vesuvius undergo *in situ* decay, wall rock that is isotopically heterogeneous and in some cases, characterized by extreme (^{230}Th)/(^{232}Th) and (^{238}U)/(^{232}Th), may be present.

Although we recognize the character of the wall rock beneath the Campanian region might be different from that beneath Vesuvius, the similarity in (^{230}Th)/(^{232}Th) of the 1944 Vesuvius samples and our Campanian samples lends support to the possibility of similar basement beneath the Campanian region. Fedele et al. (this issue) provide some constraints on the nature of cumulate material beneath the Campanian region. They describe trachytic xenoliths from the Breccia Museo (a unit erupted in the Campanian Plain) that are interpreted to represent crystal accumulations on the associated magma chamber walls and floor. In addition to feldspar and clinopyroxene (among other phases), these xenoliths are characterized by a spectrum of accessory minerals such as U-bearing thorite, and U- and REE-bearing phosphates. Many of these accessory minerals have typical hydrothermal textures and have been interpreted to reflect interaction between solids and hydrothermal fluids. As a consequence, whole-rock trace element data of these nodules show concentrations of elements such as U, Th, Zr, and REE that are, in some cases, similar to those of the pumice in the CI (e.g., Zr abundances in nodules range from 285 to 658; Th ranges from 20 to 48). Using the whole-rock nodule U and Th data, (^{238}U)/(^{232}Th) ranges from ~0.4 to 1.3, which encompasses and exceeds (^{238}U)/(^{232}Th) of the CI samples. Although no

(^{230}Th)/(^{232}Th) are available for these samples, if the range is similar to those in this study and the Vesuvius study, then aging of such material will provide basement that is isotopically heterogeneous. In addition, the nodules may provide a source of trace elements and if extensive hydrothermal interaction has affected basement rocks, then their Sr isotope signatures would be displaced toward the value of modern seawater, ~0.709.

A reasonable hypothesis regarding the open-system signatures of the CI system is that magmas interacted with heterogeneous wall rock composed of cumulates and intrusive equivalents of the volcanic products exposed in the Campanian region. Sr isotope disequilibria between feldspar and pumice/glass suggest that assimilation occurred largely during feldspar growth, particularly because of the two feldspar samples that have relatively high ^{87}Sr/^{86}Sr; assuming these signatures are primary, these feldspars would not have such elevated signatures if assimilation occurred strictly after feldspar growth. Figures 8c,d may provide additional evidence that assimilation occurred as the magma body was undergoing crystal–liquid separation. These figures illustrate that, despite the relatively systematic relationship shown on the age-corrected isochron diagram (Fig. 8b), there is no simple correlation between Th, U concentration and isotope ratio. This precludes two-component mixing, and our attempts to model such data arrays with energy-constrained assimilation-fractional crystallization (EC-AFC, Spera and Bohrson, 2001; Bohrson and Spera, 2001) failed to reproduce the isotope-concentration relationships. The EC-AFC formulation does not model zoning in magma bodies, and therefore, the failure of the model to reproduce the observed trends may be tied to the compositional heterogeneity that is produced as the magma body evolves. The Th, U vs. isotope array may reflect assimilation in a magma body that is undergoing compositional zonation. Thus, as feldspar (and other phases) form, the residual liquid becomes compositionally zoned. At the same time, heterogeneous basement that has been heated to its solidus or above can partially melt and be incorporated into this evolving magma body (Fig. 8c,d). Thus, magmas with distinct Th and U concentrations interact with partial melts of heterogeneous basement to yield this compositionally complex suite of samples.

The idea that assimilation is occurring as the magma body is zoning may also be supported by the Sr concentration-isotope data of Civetta et al. (1997). Examination of Figure 9 (and other element vs. ^{87}Sr/^{86}Sr trends not shown) shows that there are pumice samples that range in Sr concentration from ~650 to 100 ppm. ^{87}Sr/^{86}Sr of these vary from 0.70733 to 0.70735, which is likely to be close to reproducibility. Thus, this implies that magmas with very different Sr concentrations have similar ^{87}Sr/^{86}Sr. Likewise, pumice with ~100 ppm Sr range in ^{87}Sr/^{86}Sr from 0.70733 to 0.70745. We do note the general trend of increasing ^{87}Sr/^{86}Sr with decreasing Sr concentration; such a trend is consistent with general mass balance constraints for assimilation affecting low-Sr magmas. However, the ranges cited above highlight potential complexity of such a process. Like U, Th vs. (^{230}Th)/(^{232}Th), there seem to be complex relationships between isotopic and elemental signatures. This may provide support for the idea that assimilation occurred as magma chamber zoning and crystal growth occurred.

The hypothesis that partial melts of cumulates and intrusive equivalents of magmas that have erupted on the Campanian Plain represent a potential assimilant may also be consistent with the trace element data. Trace element numerical model results provide poor fits for Sm and Th, which strongly implicates open-system processes. Some open-system contribution to Nb, Zr, and U are permissible, based on the modeling results. Partial melting of rocks similar to the nodules studied by Fedele et al. (this issue) as well as intrusive material hypothesized to be present in the magmatic storage-transport system would contribute trace elements to the evolving magma body.

Detailed quantification of the proposed model is precluded for several reasons. First, the highly heterogeneous nature of the basement, coupled with the uncertainty in the uncontaminated magma composition, introduces challenges in defining the end-members of the assimilation-fractional crystallization process. The trace element analysis illustrates how sensitive the closed-system trace element trends are to choice of mineral–melt partition coefficients; because of this, it is difficult to identify the contribution that open-system processes makes to trace element mass balance. Finally, the mass, species (trace element and isotope), and energy conservation equations for a system that is zoning as it evolves have not yet been formulated. Previous work (Spera and Bohrson, 2001, 2002, 2004; Bohrson and Spera, 2001, 2003) has demonstrated that energy-constrained assimilation-fractional crystallization can lead to non-monotonic element–isotope trends, and thus, full understanding of the systematics of the CI data may require formulation of conservation equations that examine the effects of open-system processes on a zoned magma body.

6. Summary of model, relationship to previous work, and open questions

Based on the results of MELTS simulations, crystal–liquid separation played a critical role in the evolution of the Campanian magma body. Of particular interest is the occurrence of a compositional gap, which is a consequence of a nearly isothermal crystallization at ~884°C. Several authors (e.g., Civetta et al., 1997; Pappalardo et al., 2002) have identified a compositional gap in the CI that was used in support of the idea that the magma chamber included two compositionally distinct layers that mixed during eruption. We suggest that the compositional gap is a consequence of the process of crystal–liquid separation, as predicted by the phase equilibria results discussed earlier.

The crystallization event at ~884°C caused drastic changes in the state of the magmatic system: ~40% crystallization took place over a small temperature interval, which led to a decrease in the proportion of residual melt from ~0.46 to 0.09. Most oxide and element trajectories show distinct changes in response to this event. MELTS results indicate that, as crystallization proceeds, the phase assemblage evolves, which leads to drastic changes in bulk solid-melt partition coefficients. Numerical modeling, based on a mass balance equation that accommodates changes in bulk partition coefficients, yielded some model trends that effectively reproduced observed trends. In other cases, model trends provide evidence of processes other than closed-system crystallization. As a result of the MELTS modeling and implementation of a mass balance equation that correctly accommodates changes in bulk partition coefficients, the critical need for better constraints on mineral–melt partition coefficients is highlighted. Several examples shown here illustrate that, because of the permissible range of mineral–melt partition coefficients, it is difficult to constrain whether open-system processes have impacted the behavior of particular trace elements. This limitation seriously hampers attempts to quantify the evolution of the magmatic system.

$(^{230}Th)/(^{232}Th)$ vs. $(^{238}U)/(^{232}Th)$ yields an apparent isochron of 19.6±4.7 ka, which is approximately 20 ka younger than the eruption age of the deposit. The regression coefficient for this array is 0.87, which indicates good correlation between (^{230}Th)/(^{232}Th) vs. (^{238}U)/(^{232}Th). Because precise and independent age evidence is available, it is clear that this linear array does not have significance with respect to the eruption age. *Because Th, as well as other isotope systems, have been used to provide information about timescales associated with formation of magma bodies, our results indicate that the effects that open-system magmatic processes have on the species (isotope) balance must be assessed prior to assigning relevance to timescale information derived from these systems.* This result also points to the value of acquiring independent age or timescale information, and thus highlights the critical need for studies that integrate constraints from a range of perspectives.

Age-corrected Th isotope data suggest the magma body was open with respect to Th, and published Sr data (Civetta et al., 1997) are consistent with this. Fluids have been hypothesized to play a role in generating the open-system signature in the CI (Civetta et al., 1997; Pappalardo et al., 2002), but thermal and mass balance analysis suggests that assimilation of basement rock is more likely. Th isotopic data for Vesuvius (Black et al., 1998), and U-Th data for cumulate nodules from the Breccia Museo (Fedele et al., this issue) suggest that isotopically heterogeneous basement may reside beneath the Campanian Plain. Assimilation of partial melts of such basement may contribute isotopic and elemental heterogeneity to the CI magmas. Accessory phases in the nodules from the Breccia Museo (Fedele et al., this issue) suggest that hydrothermal fluids may have affected these rocks, and thus it is possible that the assimilant bears a hydrothermal imprint. Oxygen isotopes should reveal more about this, as would Nd isotopes; in the case of oxygen, some deviation from magmatic values would be expected. However, because the abundance of Nd is so low in typical seawater-like fluids (and other hydrothermal fluids), Nd isotopes may be dominated by a magmatic signature.

Finally, constraints on the timing of assimilation were suggested by Civetta et al. (1997) based on Sr isotope disequilibria. They hypothesized that assimilation occurred after feldspar formation. Phase equilibria modeling however indicates that both alkali feldspar and plagioclase precipitate continuously from 884°C to the solidus. Thus, if fluids interacted with melt at supersolidus temperatures, feldspar should record the contamination event. Reevaluation of those data, coupled with complex trends in Th isotope–U, Th concentration space, and Sr isotope–Sr concentration space introduces the possibility that assimilation occurred as the magma body was zoning. Thus, crystal growth, zoning in the melt and assimilation may have occurred simultaneously. Additional work on mineral phases would better elucidate the complex question of timing.

Better understanding of the behavior of magma chambers undergoing compositional zoning, crystallization, assimilation, eruption, and even recharge hinge, in part, on conservation models that accommodate mass, species, and energy constraints. These types of models, coupled with characterization of volcanic products at a range of scales, hold great promise for improving our ability to develop comprehensive images of crustal magmatic systems.

Acknowledgments

The authors would like to thank Dr. Mark Ghiorso for assistance with the MELTS modeling, Dr. Frank Ramos for assistance with collection of the Th isotopic data, and Dr. Ray

Macdonald and Dr. A. Peccerillo for helpful reviews. The National Science Foundation (to FJS and WAB) and the University of California Presidential Postdoctoral Program (to WAB) provided support for this work.

References

Aguirre-Díaz, G., Labarthe-Hernández, G., 2003. Fissure ignimbrites: fissure-source origin for voluminous ignimbrites of the Sierra Madre Occidental and its relationship with basin and range faulting. Geology 31, 773–776.

Bacon, C.R., Druitt, T.H., 1988. Compositional evolution of the zoned calc-alkaline magma chamber of Mount Mazama, Crater Lake, Oregon. Contrib. Mineral. Petrol. 98(2), 224–256.

Bea, F., Pereira, M.D., Stroh, A., 1994. Mineral/leucosome trace-element partitioning in a peraluminous migmatite (a lase ablation-ICP-MS study). Chem. Geol. 117, 291–312.

Barberi, F., Innocenti, F., Lirer, L., Munno, R., Pescatore, T., Santacroce, R., 1978. The Campanian Ignimbrite: a major prehistoric eruption in the Neapolitan area (Italy). Bull. Volcanol. 41, 10–31.

Beswick, A.E., 1973. An experimental study of alkali metal distributions in feldspars and micas. Geochim. Cosmochim. Acta 37, 183–208.

Black, S., Macdonald, R., De Vivo, B., Kilburn, C.R.J., Rolandi, G., 1998. U-series disequilibria in young (A.D. 1944) Vesuvius rocks: preliminary implications for magma residence times and volatile addition. J. Volcanol. Geotherm. Res. 82, 97–111.

Bohrson, W.A., Spera, F.J., 2001. Energy-constrained open-system magmatic processes II: application of energy-constrained assimilation – fractional crystallization (EC-AFC) model to magmatic systems. J. Petrol. 42(5), 1019–1041.

Bohrson, W.A., Spera, F.J., 2003. Energy-constrained open-system magmatic processes IV: geochemical, thermal and mass consequences of energy-constrained recharge, assimilation and fractional crystallization (EC-RAFC). Geochem. Geophys. Geosys. (G3) doi: 10.1029/2002GC00316.

Chen, J.H., 1987. U, Th, and Pb isotopes in hot springs on the Juan de Fuca ridge. J. Geophys. Res. 92, 11411–11415.

Chen, J.H., Edwards, R.L., Wasserburg, G.J., 1986. ^{238}U, ^{234}U, and ^{232}Th in seawater. Earth Planet. Sci. Lett. 80, 241–251.

Civetta, L., Fisher, R.V., Heiken, G., Ort, M., Orsi, G., Pappalardo, L., 1997. Geochemical zoning, mingling, eruptive dynamics and depositional processes – the Campanian Ignimbrite, Campi Flegrei caldera, Italy. J. Volcanol. Geotherm. Res. 75(3–4), 183–219.

Clocchiatti, R., Joron, J.-L., Treuil, M., 1988. The role of selective contamination in the evolution of recent historic lavas of Mt. Etna. J. Volcanol. Geotherm. Res. 34, 241–249.

Crisp, J.A., 1984. Rates of magma emplacement and volcanic output. J. Volcanol. Geotherm. Res. 20, 177–221.

Deino, A., Curtis, G., Rosi, M., 1992. $^{40}Ar/^{39}Ar$ dating of the Campanian Ignimbrite, Campanian Region, Italy. Int. Geol. Congress, Kyoto, Japan, 3, 633.

Deino, A., Southon, I., Terrasi, F., Campajola, L., Orsi, G., 1994. ^{14}C and $^{40}Ar/^{39}Ar$ dating of the Campanian Ignimbrite, Phlegrean Fields, Italy. Abstract ICOG, 3, 633.

De Natale, G., Pingue, F., Allard, P., Zollo, A., 1991. Geophysical and geochemical modeling of the 1982–1984 unrest phenomena at Campi Flegrei caldera (southern Italy). J. Volcanol. Geotherm. Res. 48, 199–222.

De Vivo, B., Rolandi, G., Gans, P.B., Calvert, A., Bohrson, W.A., Spera, F.J., Belkin, H.E., 2001. New constraints on the pyroclastic eruptive history of the Campanian volcanic Plain (Italy). Mineral. Petrol. 73, 3–31.

Dostal, J., Dupuy, C., Carron, J.P., Dekerneizon, M.L., Maury, R.C., 1983. Partition-coefficients of trace-elements – application to volcanic-rocks of St-Vincent, West-Indies. Geochim. Cosmochim. Acta 47, 525–533.

Dunn, T., Sen, C., 1994. Mineral/matrix partition-coefficients for ortho-pyroxene, plagioclase, and olivine in Basaltic to Andesitic systems – a combined analytical and experimental-study. Geochim. Cosmochim. Acta 58, 717–733.

Ewart, A., Griffin, W.L., 1994. Application of proton-microprobe data to trace-element partitioning in volcanic-rocks. Chem. Geol. 117, 251–284.

Fedele, L. et al., this issue. Magmatic-hydrothermal fluid interaction and mineralization in alkali-syenite nodules from the Breccia Museo pyroclastic deposit, Naples, Italy.

Fisher, R.V., Orsi, G., Ort, M., Heiken, G., 1993. Mobility of a large-volume pyroclastic flow – emplacement of the Campanian ignimbrite, Italy. J. Volcanol. Geotherm. Res. 56(3), 205–220.

Florio, G., Fedi, M., Cella, F., Rapolla, A., 1999. The Campanian Plain and Phlegrean Fields: structural setting from potential field data. J. Volcanol. Geotherm. Res. 91, 361–379.

Fujimaki, H., 1986. Partition-coefficients of Hf, Zr, and REE between zircon, apatite, and liquid. Contrib. Mineral. Petrol. 94, 42–45.

Ghiorso, M.S., Sack, R.O., 1995. Chemical mass transfer in magmatic processes IV. A revised and internally consistent thermodynamic model for the interpolation and extrapolation of liquid-solid equilibria in magmatic systems at elevated temperatures and pressures. Contrib. Mineral. Petrol. 119, 197–212.

Haskin, L.A., Frey, F.A., Schmitt, R.A., Smith, R.H., 1966. Meteoritic, solar and terrestrial rare-earth distributions. Phys. Chem. Earth 7, 167–321.

Higuchi, H., Nagasawa, H., 1969. Partition of trace elements between rock-forming minerals and the host volcanic rocks. Earth Planet. Sci. Lett. 7, 281–287.

Hildreth, W., 1979. The Bishop Tuff: evidence for the origin of compositional zonation in silicic magma chambers. Geol. Soc. Am. Special Paper, 180, 43–75.

Hildreth, W., 1981. Gradients in silicic magma chambers: implications for lithospheric magmatism. J. Geophys. Res. 86(B11), 10153–10192.

Hildreth, W., Christiansen, R.L., O'Neil, J.R., 1984. Catastrophic isotopic modification of rhyolitic magma at times of caldera subsidence, Yellowstone Plateau volcanic field. J. Geophys. Res. 89(B10), 8339–8369.

Keleman, P.B., Dunn, J.T., 1992. Depletion of Nb relative to other highly incompatible elements by melt/rock reaction in the upper mantle. EOS 73, 656–657.

Kress, V.C., Ghiorso, M.S., 2004. Thermodynamic modeling of post-entrapment crystallization in igneous phase. J. Volcanol. Geotherm. Res. 137, 247–260.

Larsen, L.M., 1979. Distribution of REE and other trace-elements between phenocrysts and peralkaline undersaturated magmas, exemplified by rocks from the Gardar Igneous Province, South Greenland. Lithos. 12, 303–315.

Leeman, W., Phelps, D., 1981. Partitioning of rare earths and other trace elements between sanidine and coexisting volcanic glass. J. Geophys. Res. 86, 10193–10199.

Le Maitre, R.W., Bateman, P., Dudek, A., Keller, J., Lameyre, J., Le Bas, M.J., Sabine, P.A., Schmid, R., Sorensen, H., Streckeisen, A., Woolley, A.R., Zanettin, B., 1989. A Classification of Igneous Rocks and Glossary of Terms: Recommendations of the International Union of Geological Sciences Subcommission on the Systematics of Igneous Rocks. Blackwell Scientific Publications, Oxford, UK.

Lirer, L., Luongo, G., Scandone, R., 1987. On the volcanological evolution of Campi Flegrei. EOS Trans. Am. Geophys. Union 68, 226–234.

Luhr, J.F., Carmichael, I.S.E., Varekamp, J.C., 1984. The 1982 eruptions of El Chichon volcano, Chiapas, Mexico: mineralogy and petrology of the anhydrite-bearing pumices. J. Volcanol. Geotherm. Res. 23, 69–108.

Mahood, G.A., Hildreth, E.W., 1983. Large partition coefficients for trace elements in high-silica rhyolites. Geochim. Cosmochim. Acta 47, 11–30.

Mahood, G.A., Stimac, J.A., 1990. Trace-element partitioning in pantellerites and trachytes. Geochim. Cosmochim. Acta 54, 2257–2276.

Michaud, V., 1995. Crustal xenoliths in recent hawaiites from Mount Etna, Italy: evidence for alkali exchanges during magma-wall rock interaction. Chem. Geol. 122(1–4), 21–42.

Nash, W.P., Crecraft, H.R., 1985. Partition coefficients for trace elements in silicic magmas. Geochim. Cosmochim. Acta 49, 2309–2322.

Nagasawa, H., 1970. Rare earth concentrations in zircon and apatite and their host dacite and granites. Earth Planet. Sci. Lett. 9, 359–364.

Nagasawa, H., Schnetzler, C.C., 1971. Partitioning of rare earth, alkali, and alkaline earth elements between phenocrysts and acidic igneous magmas. Geochim. Cosmochim. Acta 35, 953–968.

Nielsen, R.L., Gallahan, W.E., Newberger, F., 1992. Experimentally determined mineral-melt partition coefficients for Sc, Y and REE for olivine, orthopyroxene, pigeonite, magnetite and ilmenite. Contrib. Mineral. Petrol. 110, 488–499.

Orsi, G., De Vita, S., Di Vito, M., 1996. The restless, resurgent Campi Flegrei nested caldera (Italy): constraints on its evolution and configuration. J. Volcanol. Geotherm. Res. 74(3–4), 179–214.

Ort, M.H., Rosi, M., Anderson, C.D., 1999. Correlation of deposits and vent locations of the proximal Campanian Ignimbrite deposits, Campi Flegrei, Italy, based on natural remnant magnetization and anisotropy of magnetic susceptibility characteristics. J. Volcanol. Geotherm. Res. 91(2–4), 167–178.

Palacz, Z.A., Wolff, J.A., 1989. Strontium, neodymium and lead isotope characteristics of the Granadilla Pumice, Tenerife: a study of the causes of strontium isotope disequilibrium in felsic pyroclastic deposits. In: A.D. Saunders (Ed.), Magmatism in the Ocean Basins. Geological Society Special Publication, 42, pp. 147–159.

Pappalardo, L., DiVitro, M., Orsi, G., Carandente, A., Fisher, R.V., Civetta, L., deVita, S., 2002. Timing of magma extraction during the Campanian Ignimbrite eruption (Campi Flegrei Caldera). J. Volcanol. Geotherm. Res. 114(3–4), 479–497.

Philpotts, J.A., Schnetzler, C.C., 1970. Phenocryst-matrix partition coefficients for K, Rb, Sr and Ba, with applications to anorthosite and basalt genesis. Geochim. Cosmochim. Acta 34, 307–322.

Reid, F., 1983. Origin of the rhyolitic rocks of the Taupo Volcanic Zone, New Zealand. J. Volcanol. Geotherm. Res. 15, 315–338.

Reid, M.R., 1995. Processes of mantle enrichment and magmatic differentiation in the eastern Snake River plain. Th isotope evidence. Earth Planet. Sci. Lett. 131, 239–254.

Rolandi, G., Belluci, F., Heizler, M.T., Belkin, H.E., De Vivo, B., 2003. Tectonic controls on the genesis of Ignimbrites from the Campanian Volcanic Zone, southern Italy. Mineral. Petrol. 79, 3–31.

Rosi, M., Sbrana, A. (Eds), 1987. Phlegrean Fields, 114. Consiglio Nazionale Delle Ricerche, Rome, 175 pp.

Rosi, M., Sbrana, A., Principe, C., 1983. The Phlegrean Fields: structural evolution, volcanic history, and eruptive mechanisms. J. Volcanol. Geotherm. Res. 17, 273–288.

Rosi, M., Vezzoli, L., Aleotti, P., Censi, M.D., 1996. Interaction between caldera collapse and eruptive dynamics during the Campanian Ignimbrite eruption, Phlegrean Fields, Italy. Bull. Volcanol. 57, 541–554.

Roy-Barman, M., Chen, J.H., Wasserburg, G.J., 1996. ^{230}Th-^{232}Th systematics in the central Pacific Ocean: the sources and the fates of thorium. Earth Planet. Sci. Lett. 139, 351–363.

Scacchi, A., 1890. La regione vulcanica fluorifera della Campanian (II Edition). Mem. Regio. Comm. Geol. It., 4.

Scandone, R., Bellucci, F., Lirer, L., Rolandi, G., 1991. The structure of the Campanian Plain and the activity of the Neapolitan volcanoes (Italy). J. Volcanol. Geotherm. Res. 48, 1–31.

Signorelli, S., Rosi, M., Vaggelli, G., Francalanci, L., 1999. Origin of magmas feeding the Plinian phase of the Campanian Ignimbrite eruption, Phlegrean Fields (Italy): constraints based on matrix-glass and glass-inclusion compositions. J. Volcanol. Geotherm. Res. 91(2–4), 199–220.

Smith, R.L., 1979. Ash-flow magmatism. Geol. Soc. Am. Special Paper 180, 5–27.

Smith, R.L., Bailey, R.A., 1966. The Bandelier Tuff – a study of ash-flow eruption cycles from zoned magma chambers. Bull. Volcanol. 29, 83–104.

Spera, F.J., 1984. Some numerical experiments on the withdrawal of magma from crustal reservoirs. J. Geophys. Res. 89(B10), 8222–8236.

Spera, F.J., 2000. Physical properties of magmas. In: H. Sigurdsson (Ed.), Encyclopedia of Volcanoes. Academic Press, San Francisco, pp. 171–190.

Spera, F.J., Bohrson, W.A., 2001. Energy-constrained open-system magmatic processes I: general model and energy-constrained assimilation and fractional crystallization (EC-AFC) formulation. J. Petrol. 42(5), 999–1018.

Spera, F.J., Bohrson, W.A., 2002. Energy-constrained open-system magmatic processes III: energy-constrained recharge, assimilation and fractional crystallization (EC-RAFC). Geochem. Geophys. Geosyst. (G3) doi: 10.1029/2002GC00315.

Spera, F.J., Bohrson, W.A., 2004. Open-system magma chamber evolution: an energy-constrained geochemical model incorporating the effects of concurrent eruption, recharge, variable assimilation and fractional crystallization (EC'RAχFC). J. Petrol. 45(12), 2459–2480.

Stix, J., Gorton, M.P., 1990. Variations in trace-element partition-coefficients in Sanidine in the Cerro Toledo Rhyolite, Jemez Mountains, New-Mexico – effects of composition, temperature, and volatiles. Geochim. Cosmochim. Acta 54, 2697–2708.

Swanson, E.R., McDowell, F.W., 1984. Calderas of the Sierra Madre Occidental volcanic field western Mexico. J. Geophys. Res. 89, 8787–8799.

Taylor, H.P., Jr., 1974. The application of oxygen and hydrogen isotope studies to problems of hydrothermal alteration and ore deposition. Econ. Geol. 69, 843–883.

Taylor, H.P., Jr., 1977. Water/rock interactions and the origin of H_2O in granitic batholiths. J. Geol. Soc. Lond. 133, 509–558.

Taylor, H.P., Jr., 1980. The effects of assimilation of country rock by magmas on $^{18}O/^{16}O$ and $^{87}Sr/^{86}Sr$ systematics in igneous rocks. Earth Planet. Sci. Lett. 47, 243–254.

Taylor, H.P., Jr., 1986. Igneous rocks: II. Isotopic case studies of CircumPacific magmatism. In: Valley, J.W., Taylor, J.H.P, O'Neil, J.R. (Eds), Reviews in Mineralogy, Vol. 16, Stable Isotopes in High Temperature Geological Processes, Mineral Society of America, pp. 273–316.

Villemant, B., 1988. Trace element evolution in the Phlegrean Fields (Central Italy): fractional crystallization and selective enrichment. Contrib. Mineral. Petrol. 98, 169–183.

Villemant, B., Boudon, G., Komorowski, J.-C., 1996. U-series disequilibria in arc magmas induced by water-magma interaction. Earth Planet. Sci. Lett. 140, 259–267.

Villemant, B., Jaffrezic, H., Joron, J.L., Treuil, M., 1981. Distribution coefficients of major and trace-elements – fractional crystallization in the alkali basalt series of Chaine-Des-Puys (Massif Central, France). Geochim. Cosmochim. Acta 45, 1997–2016.

Watson, E.B., 1976. Glass inclusions as samples of early magmatic liquid: determinative method and application to a South Atlantic basalt. J. Volcanol. Geotherm. Res. 1, 73–84.

Watson, E.B., Green, T.H., 1981. Apatite/liquid partition coefficients for the rare earth elements and strontium. Earth Planet. Sci. Lett. 56, 405–421.

Webster, J.D., Raia, R., Tappen, C., De Vivo, B., 2003. Pre-eruptive geochemistry of the ignimbrite-forming magmas of the Campanian Volcanic Zone, Southern Italy, determined from silicate melt inclusions. Mineral. Petrol. 79, 99–125.

Wilson, C.J.N., Hildreth, W., 1997. The Bishop Tuff: new insights from eruptive stratigraphy. J. Geol. 105(4), 407–439.

Worner, G., Schmincke, H.U., 1984a. Mineralogical and chemical zonation of the Laacher See tephra sequence (east Eifel, W. Germany). J. Petrol. 25(4), 805–835.

Worner, G., Schmincke, H.U., 1984b. Petrogenesis of the zoned Laacher See tephra. J. Petrol. 25(4), 836–851.

Volcanism in the Campania Plain:
Vesuvius, Campi Flegrei and Ignimbrites
edited by B. De Vivo

Chapter 14

A hydrothermal model for ground movements (bradyseism) at Campi Flegrei, Italy

B. De Vivo* and A. Lima

Dipartimento di Scienze della Terra, Universita di Napoli Federico II, Via Mezzocannone 8 – 80134 Napoli, Italy

Abstract

Ground movements (bradyseism) at Campi Flegrei, Italy, have been explained by a classical model that involves the intrusion of new magma to shallow depth, or by models which emphasize both the magmatic and aquifer effects. The authors describe a model for the ground deformations that involves only hydrothermal fluids, of magmatic or meteoric/marine origin, with no direct involvement of the magma, other than as a heat source. They explain the bradyseism at Campi Flegrei by a hydrothermal model, using the porphyry systems (Henley and McNabb, 1978; Burnham, 1979; Fournier, 1999) as an analogue of the Campi Flegrei system. In this view, Campi Flegrei might very well represent a modern analogue of a mineralized porphyry system, as has been demonstrated for White Island, New Zealand (Rapien et al., 2003). The authors used fluid and melt inclusion data from Campi Flegrei and other volcanoes of the Neapolitan area (Vesuvius, Ponza and Ventotene) to demonstrate the linkage with porphyry systems. Fluid inclusions in all the above volcanic systems show clear evidence of various stages of silicate melt/hydrosaline melt/aqueous fluid/CO_2 immiscibility during the magmatic evolution and its transition from magmatic to hydrothermal stage, comparable to the plastic, lithostatic domain in porphyry systems. In contrast, convectively driven fluids are found only in the volcaniclastic sediments of the Campi Flegrei caldera (in the geothermal wells of San Vito and Mofete fields), and are representative of the brittle, hydrostatic domain. The coexistence of liquid-dominated and vapor-dominated inclusions in the same fluid inclusion assemblage is strong evidence of boiling conditions during inclusion trapping, whereas fluid inclusions with daughter crystals trapped in samples from deeper, hotter levels indicate a high concentration of solute (brines), as confirmed by drilling.

The scenario suggested by fluid inclusion data indicates that the Campi Flegrei system receives an influx of saline water (magmatic + seawater), localized in aquifers at depths of ~2.5–3 km. The fluids are heated by the underlying crystallizing magma and remain under lithostatic pressure for long periods. The pressure in the upper, apical part of the magma chamber increases as water exsolves from the magma and causes uplift of the overlying rocks (positive bradyseism). When the system ruptures, due to the increasing pressure, the regime changes from lithostatic to hydrostatic, resulting in boiling, hydraulic fracturing, volcanic tremors and finally pressure release leading to deflation of the ground. Afterward, the system begins to seal again due to the precipitation of newly formed minerals and a new phase of positive bradyseism will occur only after several years when the system "reloads" under new lithostatic pressure conditions. In this scenario, a hydrothermal eruption can still occur, but only if the fluids pass from lithostatic to hydrostatic pressure when the overlying rocks have a thickness <500 m. If this happens, the hydrothermal eruption could trigger a magmatic eruption, as was probably the case of the Monte Nuovo eruption in 1538 AD.

*Corresponding author. *E-mail address*: bdevivo@unina.it (B. De Vivo).

1. Introduction

The phenomenon of slow, vertical, ground movements in the Campi Flegrei area, Italy, has been known since Roman times. The ground movement is named bradyseism from the Greek words meaning, literally, "slow movement".

The most recent intense deformation in Campi Flegrei occurred in 1538, in 1970–1972 and in 1982–1984. The ground deformation of 1538 AD culminated with the eruption of Monte Nuovo, whereas no eruptions occurred in 1970–1972 and 1982–1984. Probably many other bradyseismic events occurred in the Campi Flegrei before 1538 leaving no historical reference to their occurrence. In the last 2000 years at Campi Flegrei, there has been an eruption only in 1538 AD.

Secular deformations and intense unrests are typical manifestations of activity at calderas, and a distinctive feature of such deformation episodes is that they are generally not followed by eruptions (Dzurisin and Newhall, 1984).

Two different mechanisms have been proposed for Campi Flegrei bradyseism. The first explains the phenomenon with the classical model of intrusion of new magma at shallow depth (Corrado et al., 1976; Barberi et al., 1984). The second one ascribes (Oliveri del Castillo and Quagliariello, 1969) the ground movements to the effect of heating and expansion of ground water (Casertano et al., 1976). In recent years, different researchers have presented models, emphasizing both the magmatic and aquifer effects, to explain the observed features (Bianchi et al., 1987; Gaeta et al., 1998; Bonafede, 1991; De Natale et al., 1991; Dvorak and Berrino, 1991; De Natale and Pingue, 1993; De Natale et al., 1997; Troise et al., 1997; Scandone et al., this volume). In particular, De Natale et al. (2001) developed a mechanical fluid-dynamic model to explain the main features of Campi Flegrei deformation. The model involves two processes: the first one is related to the elastic response of the shallow crust to increasing pressure within a shallow magma chamber; the second involves the fluid dynamics of shallow aquifers in response to increasing pressure and/or temperature at depth.

We extend the contribution of Oliveri del Castillo and Quagliariello (1969) by describing an active role for the deformations at Campi Flegrei to only hydrothermal fluids, both of magmatic and/or meteoric/seawater origin, and not to the magma. In our model, the magma plays only the role of a "furnace" to heat the system. To support our view we use data obtained from fluid inclusions from geothermal boreholes at Campi Flegrei (De Vivo et al., 1989), as well as data from fluid and melt inclusions in the nearby sub-volcanic systems of the Pontine Islands (De Vivo et al., 1995; Belkin et al., 1996) and Vesuvius (Lima et al., 2003, this volume).

Our model is based on the model of the porphyry systems (Henley and McNabb, 1978; Burnham, 1979; Fournier, 1999, and references therein) as an analogue of the Campi Flegrei sub-volcanic system. In other words, the Campi Flegrei represents a modern analogue of mineralized systems associated with former magmatic systems (Beane and Titley, 1981; Beane, 1982; Beane and Bodnar, 1995; Roedder and Bodnar, 1997). White Island (New Zealand) is an example of an active magmatic system that represents an embryonic copper porphyry system that has not reached the productive stage of copper mineralization (Rapien et al., 2003). The Burnham model is used extensively and accepted by many ore deposits experts. In this model, the active role played by hydrothermal fluids in the transport and deposition of mineralization is amply demonstrated. The occurrence of boiling fluids and hydrothermal explosions is well documented as well, where the magmas located at depths between 5 and 7 km act as a sub-volcanic heating engine of the system.

2. History of Campi Flegrei bradyseism and models proposed to explain the deformation

Initial studies concerning the slow movements at Campi Flegrei come from the observations of sea level markers on the archaeological site previously called Serapis Temple (but really a marketplace, Macellum) in Pozzuoli. Boreholes left by a marine mollusk (*Lithodomus lithophagus*) have been found on the columns of this monument. These bores (and the shells left in them) record ancient relative sea level changes. This interesting movement attracted the attention of many other researchers (Breislak, 1792; Forbes, 1829; Niccolini, 1839, 1845; Babbage, 1847; Lyell, 1872; Gunther, 1903), and Parascandola (1947), who reconstructed the history of secular ground movements at Campi Flegrei using the boreholes. Dvorak and Mastrolorenzo (1991) updated the studies of Parascandola by reconstructing the history of vertical movements at Campi Flegrei (Fig. 1). It has to be stressed that the above studies document the ground movements since Roman times but no data are available on the phenomenon before Roman times. We have a record of ground deformation at Campi Flegrei only for the past 2000 years, which in terms of geological phenomena represents a very short time span.

Historically, the first documented episode of fast uplift at Campi Flegrei is the one associated with the 1538 Monte Nuovo eruption (the ground rose about 7 m before the eruption). Bradyseism was again active in 1970–1972, when uplift of 1 m was registered; this episode was followed by subsidence of 30 cm. It has to be stressed that in the period 1970–1972, the uplift began to be recorded well after it had begun. The total uplift was probably on the order of 1.7 m and began in 1969. The uncertainty in our knowledge of the

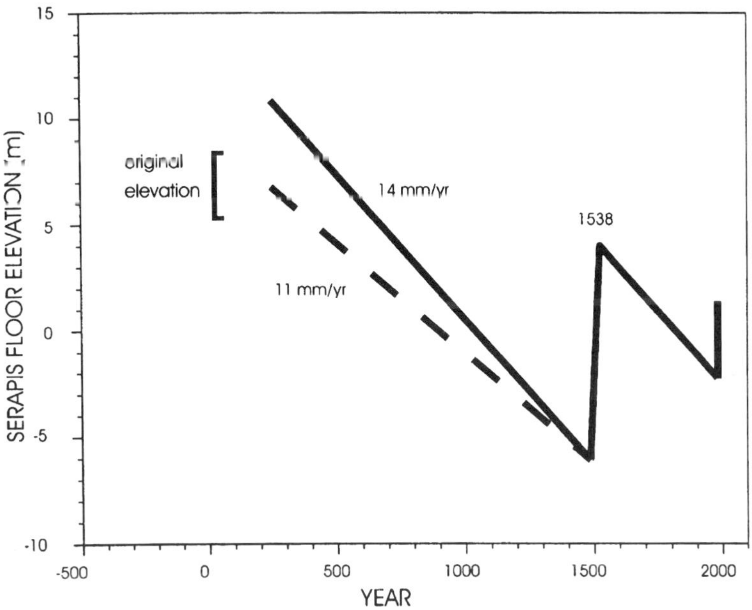

Figure 1. Vertical movements at Serapis temple, Pozzuoli (from Dvorak and Mastrolorenzo, 1991).

uplift that occurred only 33 years ago emphasizes that our knowledge about this phenomenon in pre-historic times is minimal.

A new episode of uplift of 1.8 m occurred from 1982 to 1984, then in 1985 was followed by a slow subsidence that continues today (Fig. 2). One of these episodes caused the evacuation of about 30,000 people from the city of Pozzuoli, and the building of a new town at Monte Ruscello, about 3 km away from the most active caldera center, but still inside the Campi Flegrei caldera! The areal distribution of ground movement at Campi Flegrei, with its circular symmetry, is shown in Figure 3.

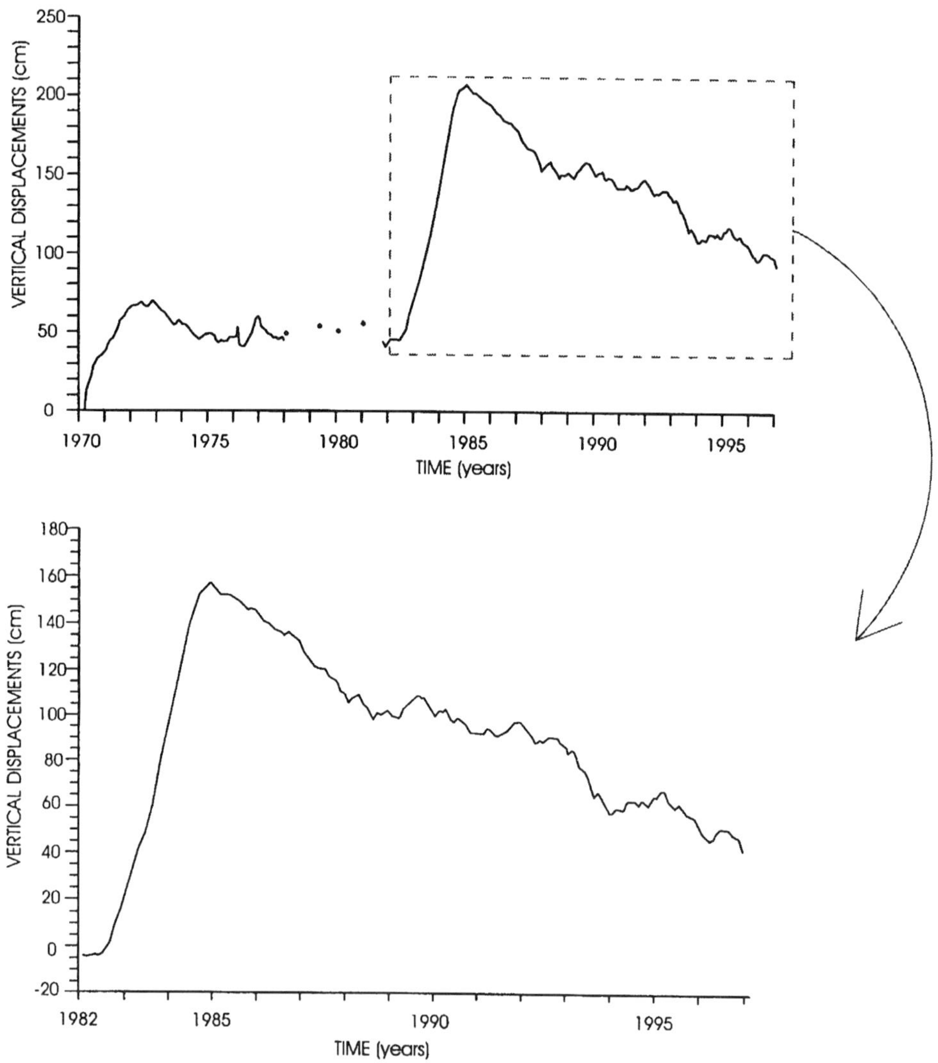

Figure 2. Vertical ground displacement at Pozzuoli harbour in the 1970–1996 period, as recorded by tide gauge (continuous line) and levelling data (closed circle) (from De Natale et al., 2001).

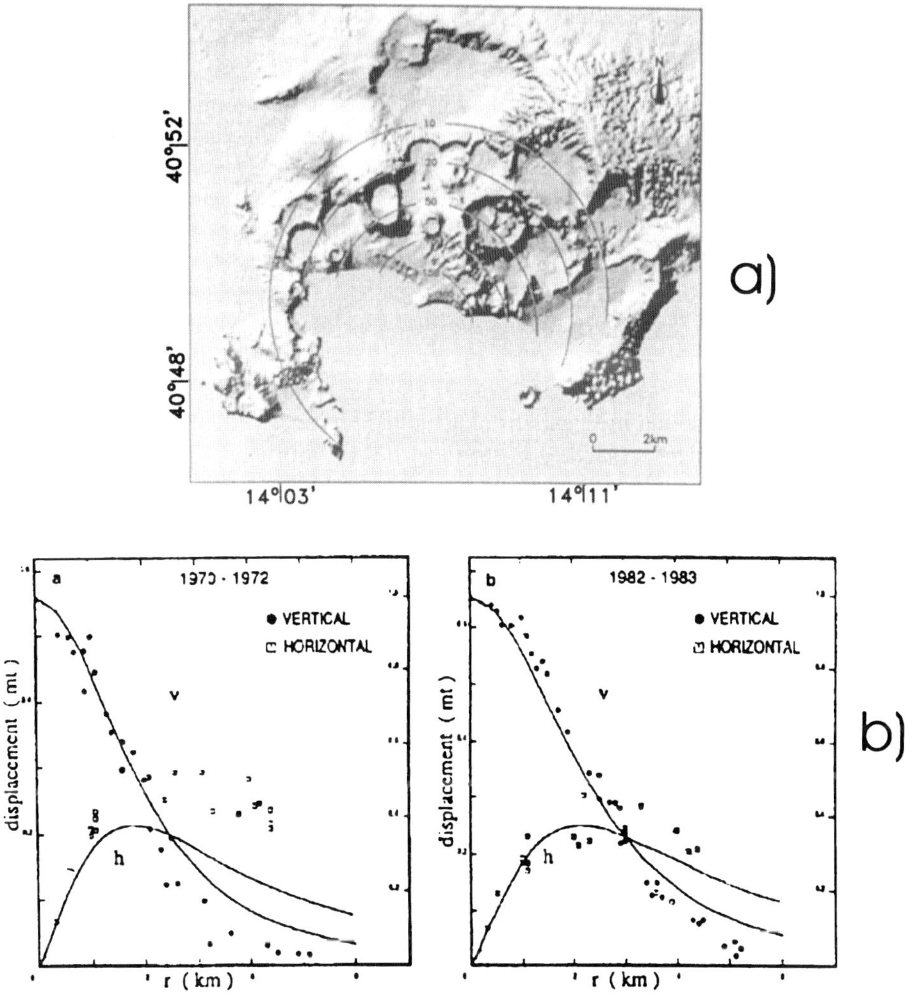

Figure 3. (a) Contours of vertical elevation at Campi Flegrei area in the period 1982–1985. (b) Ground deformation data as function of the distance from Pozzuoli, measured in the period 1970–1972 and 1980–1983; the fit with the best Mogi model is also shown (depth = 2.5 km for 1970–1972 and depth = 3 km for 1982–1983) (from De Natale et al., 2001).

Various models have been proposed to explain the deformation at Campi Flegrei. One purely mechanical model – mostly popular in the 1970s – attribute the unrest episodes to the intrusion of new magma at shallow depth (Corrado et al., 1976; Berrino et al., 1984; Bonafede et al., 1986; Bianchi et al., 1987). An alternative model, which has gained favor in recent years, explains the unrests mainly as a result of heating and expansion of aquifers (Oliveri del Castillo and Quagliariello, 1969; Casertano et al., 1976). Other researchers have published papers that explain the unrest mainly as a result of fluid-dynamic processes in the shallow geothermal system (Bonafede, 1990, 1991; De Natale, 1991; De Natale et al., 2001; Trasatti et al., 2005). Cortini et al. (1991) and Cortini and Barton (1993) suggested that the bradyseism is governed by the internal dynamics of the Campi Flegrei volcanic system whose physicochemical details are unknown. Analysis of the ground elevation at Pozzuoli, performed using nonlinear dynamics, showed that the Campi Flegrei system underwent a phase transition during the evolution from the sinking stage to that of uplift. They also suggested that the dynamics of Campi Flegrei, which appears fairly predictable on a timescale of a few days, could be driven by convection in the magma chamber. Cubellis et al. (2002) explain the different eruptive phases of Campi Flegrei in terms of chaotic convective cells operating inside the magma chamber, where the time evolution is hypothesized to be governed by three non-linear Lorenz equations (Lorenz, 1963). De Natale et al. (2001) make a comprehensive review of all pros and cons of the different models referenced above.

Seismicity at Campi Flegrei occurs only during unrest episodes (Corrado et al., 1976; De Natale et al., 1995). During 1982–1984 more than 15,000 earthquakes occurred, ranging from 0.4 to 4.2 in magnitude (De Natale and Zollo, 1986). De Natale et al. (1995) show that earthquake locations and mechanisms indicate the presence of faults that are associated with the inner caldera collapse structure and dip inward. Gravimetric and seismic methods (Berrino et al., 1984, 1992; AGIP, 1987; Berrino and Gasparini, 1995) show a marked Bouguer minimum centered at Campi Flegrei caldera (Fig. 4). Other seismic studies (Aster and Meyer, 1988; Ferrucci et al., 1992) define the 3D shallow velocity structure of the area and the location of the top part of the shallow magma chamber (Fig. 5). Ground deformations and seismicity are associated with the presence of intense fumarolic and hydrothermal activity, concentrated in the crater of Solfatara, where CO_2 and H_2O fluxes are particularly intense and probably represent magmatic degassing (Chiodini et al., 2001).

The above models for Campi Flegrei – and particularly the ones that propose a cause–effect relationship between ground uplift and the occurrence of an eruption – should be compatible with similar observations from other recently active calderas in the world (e.g., Rabaul, Mc Kee et al., 1989; Long Valley, Hill, 1984; Yellowstone, Dzurisin and Yamashita, 1987; see De Natale et al., 2001). However, studies of similar systems worldwide

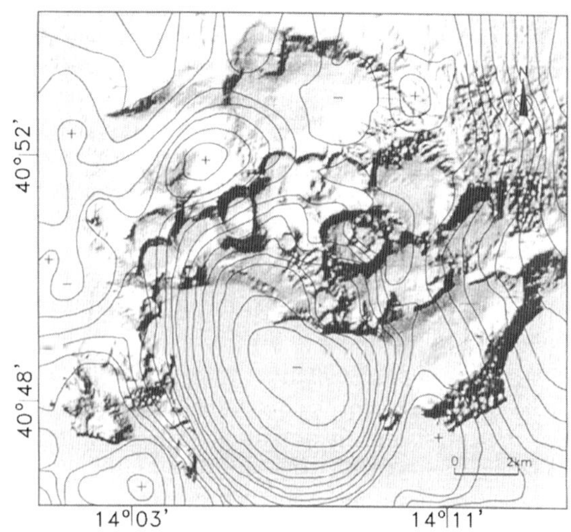

Figure 4. Bouguer anomaly contours at Campi Flegrei (from Scandone et al., 1991).

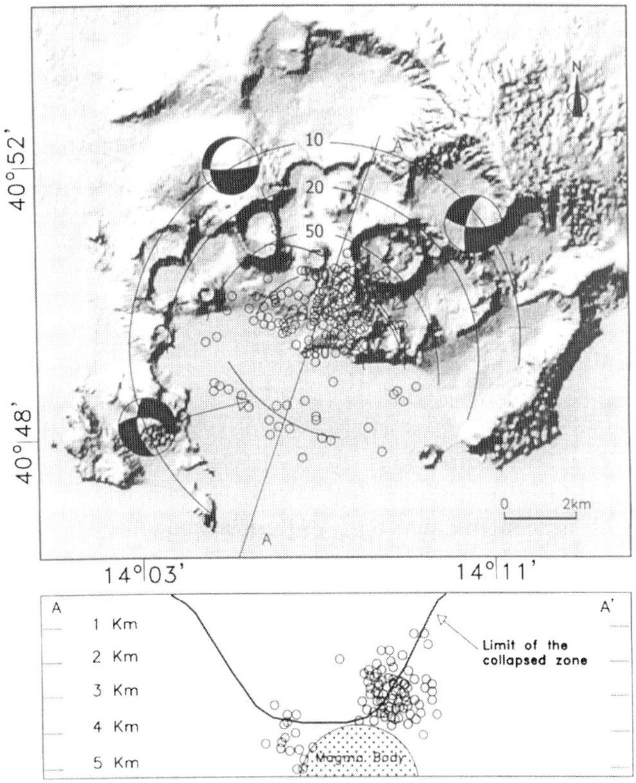

Figure 5. Map of various geophysical observations at Campi Flegrei. The contours of vertical elevation (in meters) are shown together with earthquake hypocenters; the projection of the collapsed zone as modelled from gravity anomalies is superimposed on the depth section of hypocenters. Composite focal mechanisms computed for three different seismic zones are also shown. Also shown is the location of magma chamber as inferred by Ferrucci et al. (1992) (from De Natale et al., 2001).

indicate that deformation episodes at calderas are generally not followed by eruptions (Dzurizin and Newhall, 1984). The eruption is an exception rather than the norm.

With these observations in mind, we describe the Campi Flegrei volcanic system and its vertical movements as the modern analogue of a mineralized porphyry system (Henley and McNabb, 1978; Burnham, 1979; Fournier, 1999; and references therein).

3. Magma and hydrothermal fluids

The role of hydrothermal fluids in the formation of Cu and Mo mineralization in magmatic systems (known as porphyry Cu and Mo) has been recognized since the 1960s (Burnham, 1967, 1979). In the following pages, we summarize briefly some relevant parts of this model from papers of Burnham (1979) and Fournier (1999).

According to the Burnham (1979) model, the first stage in the formation of a porphyry copper system involves the intrusion of water-undersaturated granodioritic magma.

Initially, the system is open and any volatiles generated during solidification of the melt escape from the system. Eventually, an impermeable rock rind develops at the top of the magma chamber, isolating the underlying magma from the overlying rocks. The magma becomes a closed system and only conductive loss of heat to the wall rocks is permitted. As the crystallization front migrates downward, the melt becomes saturated in water and an H_2O-saturated carapace composed of crystals+melt+fluid develops at the top of the magma chamber and below the crystalline rind (Fig. 6a).

The H_2O-saturated carapace migrates downward into the magma chamber as crystallization proceeds and plays a critical role in the development of porphyry Cu–Mo systems. The overlying crystalline rind serves as a barrier to the migration of volatiles, both outward to the wall rocks and inward from the wall rock. The H_2O-saturated carapace, especially in its upper parts, is the site of accumulation of an aqueous fluid phase as the system cools and evolves (Fig. 6b). The process of resurgent boiling (second boiling) is a natural consequence of cooling a melt saturated with respect to H_2O or one or more crystalline phases. The overall reaction, H_2O-saturated melt $\Rightarrow$ crystals + "vapor", takes place with the evolution of heat which is essentially the heat of crystallization (the heat of vaporization of H_2O from the melt is negligible, according to Burnham and Davis, 1974).

The exsolution of water from crystallizing hydrous melts produce high internal overpressures in the magma chamber. In isolated small pockets in crystallizing H_2O-saturated magma, values of ΔP_{in} (internal overpresure) $\geq$ 5 Kb are possible. Overpressures on the order of 400 bars are expected in igneous rocks deforming viscoelastically. An overpressure of this magnitude is enough to cause brittle failure of the overlying rocks (Burnham, 1972; Koide and Bhattacharji, 1975). The fractures are concentrated in and above the apical parts

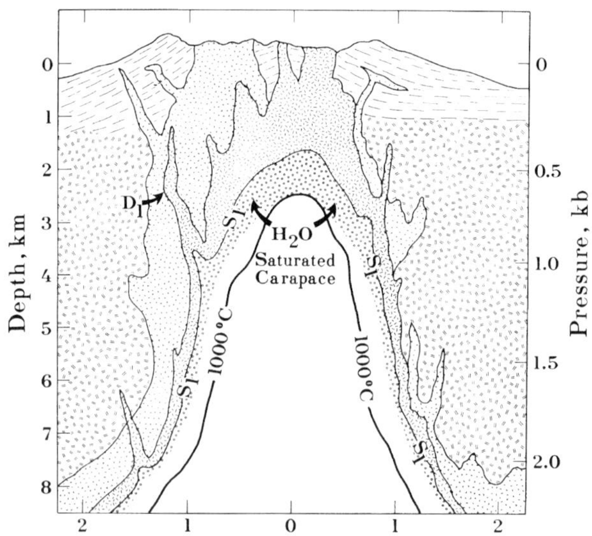

Figure 6a. Schematic cross section through a hypothetical granodiorite porphyry stock and associated dike (D_1). S_1 represents the H_2O-saturated solidus at this arbitrarily chosen initial stage in the development of a porphyry copper system and the circle pattern represents the zone of H_2O-saturated magma (H_2O-saturated carapace) (from Burnham, 1979).

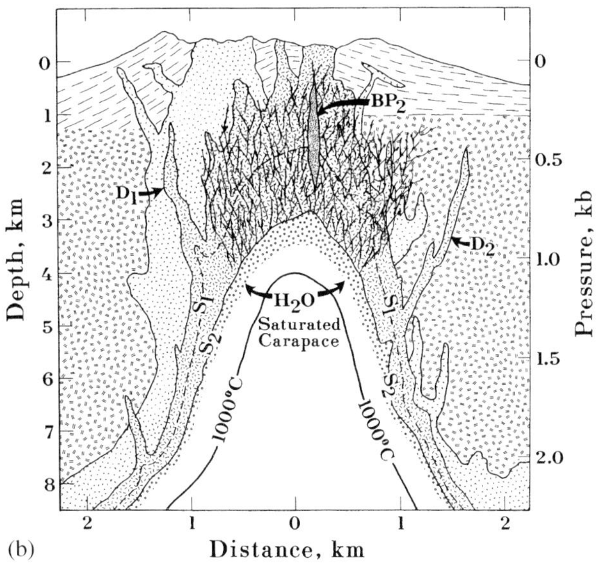

(b)

Figure 6b. Schematic cross section as in Figure 6a, except at a later (second) stage of solidification. BP_2 and D_2 schematically represent a breccia pipe and dike that formed as a result of wall rock failure between stages 1 and 2. Chaotic line pattern represents extensive fracture system that also developed during this period of activity and retreat of the H_2O-saturated carapace (from Burnham, 1979).

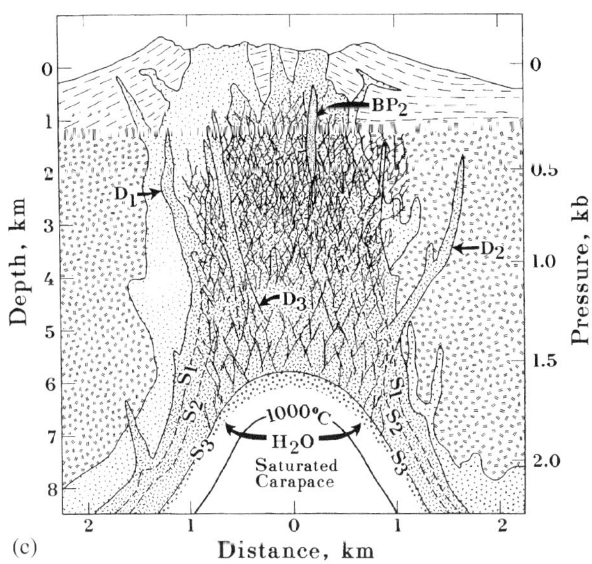

(c)

Figure 6c. Schematic cross section as in Figure 6a,b, except at a stage of waning magmatic activity in the development of a porphyry copper–molybdenum system (from Burnham, 1979).

of the stock (Fig. 6b), and tend to be steep, but their orientation depends upon the regional stress field.

An additional overpressure, referred to as "telluric pressure" that arises from differences in density between magma and wall rocks is superimposed on ΔP_{in} under quiescent conditions (Burnham, 1979). The "telluric pressure" is dependent on the height of the magma column above the depth of isostatic compensation, that is, the depth at which the wallrocks yield by plastic deformation under lithostatic pressure. It is this "telluric overpressure" that represents the driving force for intrusion as indicated in Figure 6a and which prevents a reduction in magma pressure that would have resulted from reduction in volume by crystallization.

The maximum mechanical energy ($P_t \Delta V_r$) released in the reaction H_2O-saturated melt $\Rightarrow$ crystals + "vapor" is enormous (Burnham, 1979), although it is only 1% of the total thermal energy content of the magma. At the depths indicated in Figures 6a–c, the mechanical energy released from the H_2O-saturated carapace presumably is expended mainly in fracturing a much larger volume of rocks. In the earlier stage of fracturing, the enclosing impermeable rocks are stretched laterally and may not be breached completely. Fluids penetrate this myriad of fractures and extend them outward and upward by hydraulic fracturing. This action results in lowering the fluid pressure in each fracture to below lithostatic pressure, except near the top of the fracture. With time, the H_2O-saturated carapace retreats to progressively deeper levels in the stock.

If major fractures breach the overlying rocks, breccia dikes and pipes are very likely to form (see BP_2 in Fig. 6b). If the breach occurs in the thinner, lateral flanks of the carapace, more normal dike intrusions result (D_2 in Fig. 6a,b). The mechanics of breccia pipe formation are hence visualized as primarily due to internal overpressure in the carapace (and not to contraction on cooling). As pressure decreases, the heat is lost to the wall rocks, and the magma dike devolatilizes and is quenched. In response to this process, more magma rises into the system until internal pressures are restored to near previous values.

At this stage, the magma system is restored to the same state as it was prior to the fracturing, except for the fact that the H_2O-saturated carapace is shifted downward in the magma chamber. In addition, the myriad of narrow fractures outside the solidus boundary promotes loss of H_2O and heat to the fracture system until the fractures become healed by precipitation of new minerals (mainly quartz). Further cooling of the magma leads to reactivation of the same process that operated before. The end result is a chimney-like fracture system (Fig. 6c) that serves to channel ore-bearing fluids and heat from the underlying magma system to higher levels in the stock. The above model of Burnham (1967, 1979) describes the mechanism that controls the deposition of ore (mostly Cu and Mo) in the classical "porphyry systems" (Lowell and Guilbert, 1970; Roedder, 1971; Wallace, 1974; Gustafson and Hunt, 1975; Henley and McNabb, 1978).

The transition from magmatic to epithermal conditions in a shallow sub-volcanic environment, such as in the case of Campi Flegrei, is shown in Figure 7. Dilute waters (dominantly meteoric) circulate through brittle rocks (Fig. 7a) at hydrostatic pressure at temperatures $<370°C$. The transition from brittle to plastic behavior occurs across the impermeable rocks between the underlying H_2O-saturated magma and the overlying highly fractured rocks at temperatures of 400–800°C. Within the lithostatic regime, hypersaline, magmatic brine and "steam" accumulate in relatively thin, horizontal, lenses or network of fractures that have limited vertical interconnectivity (when the least principal stress is the lithostatic load). When

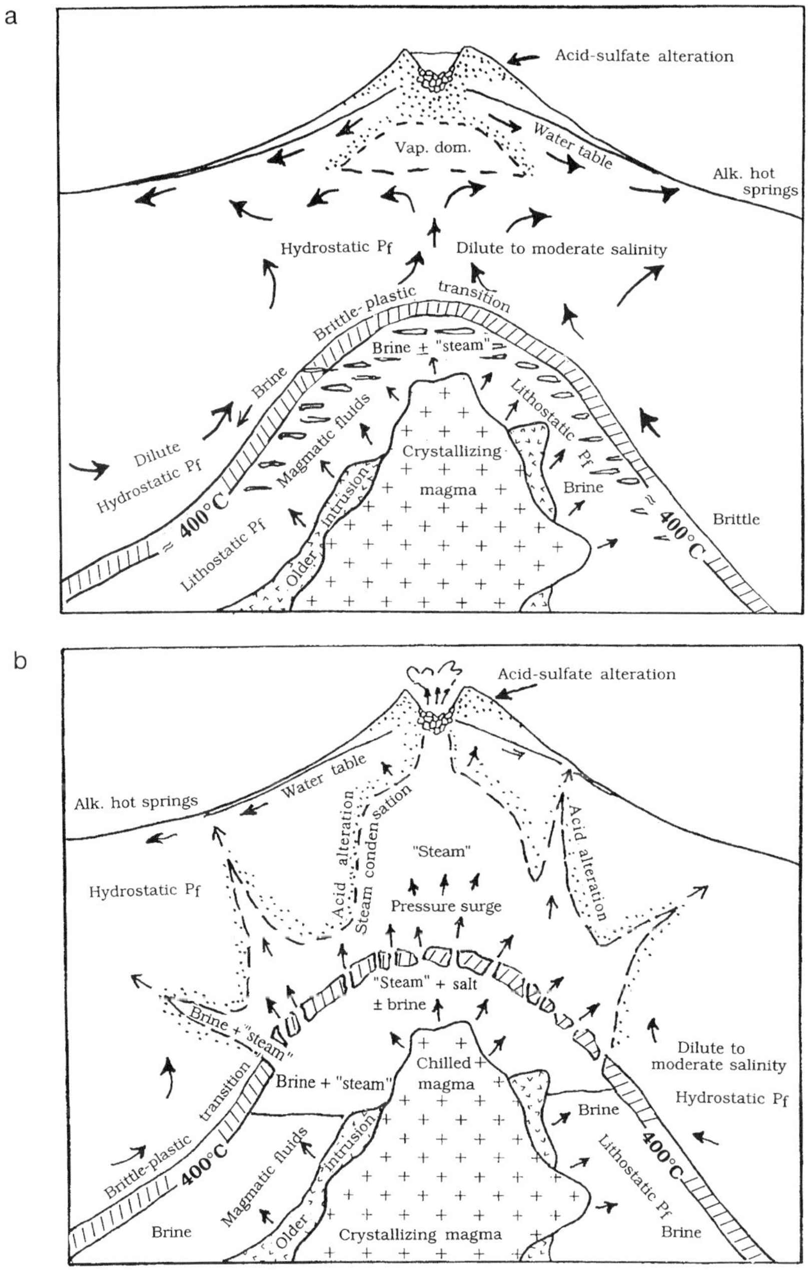

Figure 7. Schematic model of the transition from magmatic to epithermal conditions in a subvolcanic environment where the tops of intruded plutons are at depths in the range 1–3 km. (a) The brittle to plastic transition occurs at about 370–400°C and dilute, dominantly meteoric waters circulate at hydrostatic pressure in brittle rock, while highly saline, dominantly magmatic fluid at lithostatic pressure accumulates in plastic rocks. (b) Episodic and temporary breaching of a normally self-sealed zone allows magmatic fluid to escape into the overlying hydrothermal system (from Fournier, 1999).

magmatic bodies have been repeatedly intruded to relatively shallow depths, a large volume of 400–800°C plastic rocks develops, and the brittle–plastic interface can be as shallow as 2 km. In the SV1 well at Campi Flegrei, temperatures of 400°C have been found at ≈3000 m (De Vivo et al., 1989). Thus, sub-volcanic magmatic-hydrothermal systems are subdivided into a plastic, lithostatic domain, and an overlying hydrostatically pressured hydrothermal domain in which rocks deform brittlely. Several processes may play a role in triggering major breaches in the self-sealed, impermeable region that separates the plastic (magmatic) and brittle (hydrothermal) domains. One process is continued degassing of crystallizing magma and accumulation of the evolved fluid in the H_2O-saturated carapace beneath the impermeable barrier until the latter becomes stretched sufficiently to rupture by tensile failure (Northon and Cathles, 1973; Philipps, 1973; Burnham, 1979, 1985). A variation on this process is exsolution of volatiles at depth and transport to the top of the chamber by convection of the magma (Shinoara et al., 1995).

Another mechanism for breaching the self-sealed zone is upward injection of a new pulse of magma from depth. This would increase the strain rate within the overlying rocks to such a degree that the brittle to plastic transition temporarily migrates to a deeper and hotter enviroment. Fluids in the initially plastic rock would be at lithostatic pressure, and the change from plastic to brittle behavior with increasing strain rate would result in breaching of the self-sealed zone by shear failure in response to a small stress difference. A new pulse of magma would add volatiles upon crystallization and would heat previously existing brines trapped in horizontal lenses, inducing boiling. The resulting rapid expansion of the fluids would cause an additional increase in the strain rate, possibly leading to failure of the overlying rocks.

Large seismic events that affect permeability and rates of flow within an overlying hydrostatically pressured hydrothermal system may also lead to breaching of the self-sealed zone. In fact, the increase in flow rates may result in extraction of heat at the base of the circulating system more rapidly than heat can be supplied by conduction through plastic rocks from below. This leads to a progressive downward decrease in temperature and simultaneous expansion of the permeable region as a result of volumetric contraction and tensile cracking (Lister, 1974; Northon and Knapp, 1977; Northon and Knight, 1977; Carrigan, 1986).

Whichever mechanism produces the breaches in the self-sealed system, faulting, brecciation, hydrothermal alteration and ore deposition occur during this active process. This domain is characterized by hypersaline brines coexisting with "steam". For example, fluids present during the formation of "porphyry systems" are very hot (>400°C, and as high as 725°C), very saline (>30% to 60% salts), and consist of two phases, liquid and vapor (Roedder, 1984; Heinrich et al., 1992; Cline and Bodnar, 1994; Bodnar, 1995).

4. Fluid and melt inclusion data from Campi Flegrei and other Neapolitan sub-volcanic systems

The composition of residual liquids evolves during magmatic differentiation and these changes are recorded by fluid inclusions trapped in minerals (Roedder, 1984). Both magmatic and hydrothermal phenomena can thus be studied from observations of melt and fluid inclusions in igneous and hydrothermal minerals that crystallized at different depths and at various stages in the evolution of a magmatic system (De Vivo et al., 2005).

The observed composition of melt inclusions is determined by the composition of the original trapped melt and by the action of any later alteration processes, especially those due to hydrothermal activity. Hydrothermal changes frequently mask the original melt composition, particularly when the host igneous rock is introduced into the water table environment (Touret and Frezzotti, 1993; Student and Bodnar, 2004). Melt inclusion studies have thus concentrated mostly on the evolution of aqueous fluids in sub-volcanic systems in both ancient (porphyry ore deposits) and modern geothermal systems. Indeed, it was recognized from such studies that the majority of hydrothermal ore deposits in volcanic rocks or their subjacent plutonic suites were formed within geothermal systems similar to those active today (Henley and Ellis, 1983). At the same time, alteration processes have restricted the potential of fluid inclusion studies to follow the evolution of a melt during magma solidification from supraliquid temperatures (Roedder and Coombs, 1967; Frezzotti, 1992). Fortunately, however, melt evolution can be investigated unambiguously when magma immiscibility occurs in fluid inclusions (Roedder, 1992).

"Granitoid" xenoliths entrained in volcanic units from Campi Flegrei (and from subvolcanic systems of other volcanoes in the Neapolitan area, including Vesuvius and the Pontine Islands (Ponza and Ventotene) provide samples of the deep environment. The source of such "granitoids" is the plastic, lithostatic, domain as discussed in the models of Burnham (1979) and Fournier (1999), where the fluids (melts + hydrosaline melts) trapped as inclusions represent magmatic fluids evolved from a subjacent crystallizing magma. In contrast, inclusions that trap vapor plus brines represent the plastic-brittle transition zone. In the Campi Flegrei system, information obtained from studies of fluid inclusions from geothermal boreholes (De Vivo et al., 1989) furnish data on the hydrostatic domain, which is characterized by dilute to moderate salinity fluids.

Geothermal exploration in the Campi Flegrei started in the years 1939 to 1954; this program was resumed in 1978 by a joint venture between the national utilities AGIP and ENEL and the Italian Geodynamic Project of CNR (Rosi and Sbrana, 1987) (Fig. 8). Several wells were drilled to depths up to 3 km (Carella and Guglielminetti, 1983). At shallow depths partially hydrothermally altered volcanic, volcaniclastic and sedimentary rocks are found. At greater depth, their thermometamorphic equivalents are encountered. The deep wells indicate the presence of a saline water-dominated geothermal field with multiple reservoirs. A fluid inclusion study (De Vivo et al., 1989) of cores from Mofete (MF1, MF2, MF5) and San Vito (SV1, SV3) geothermal wells indicates that the hydrothermal minerals were precipitated from aqueous fluids (containing ± CO_2) that were moderately saline (3–4 wt% NaCl equiv.) to hypersaline (up to 49 wt% NaCl equiv.), and were boiling. Three types of primary fluid inclusions were found in authigenic K-feldspar, quartz, calcite and epidote: (A) two-phase (liquid + vapor), liquid-rich inclusions with a range of salinity; (B) two-phase (liquid + vapor), vapor-rich inclusions with low salinity; and (C) three-phase (liquid + vapor + crystals – NaCl), liquid-rich inclusions with hypersaline fluids (Fig. 9).

Data from selected core samples reveal a general decrease in porosity and increase in bulk density with increasing depth and temperature. Hydrothermal minerals commonly fill fractures and pore spaces and define a zonation pattern, similar in all five wells studied, in response to increasing depth (pressure) and temperature. A greenschist facies assemblage, defined by albite + actinolite, gives way to amphibolite facies, defined by plagioclase (andesine) + horneblende, in the San Vito well at about 380°C. The fluid inclusion salinity values

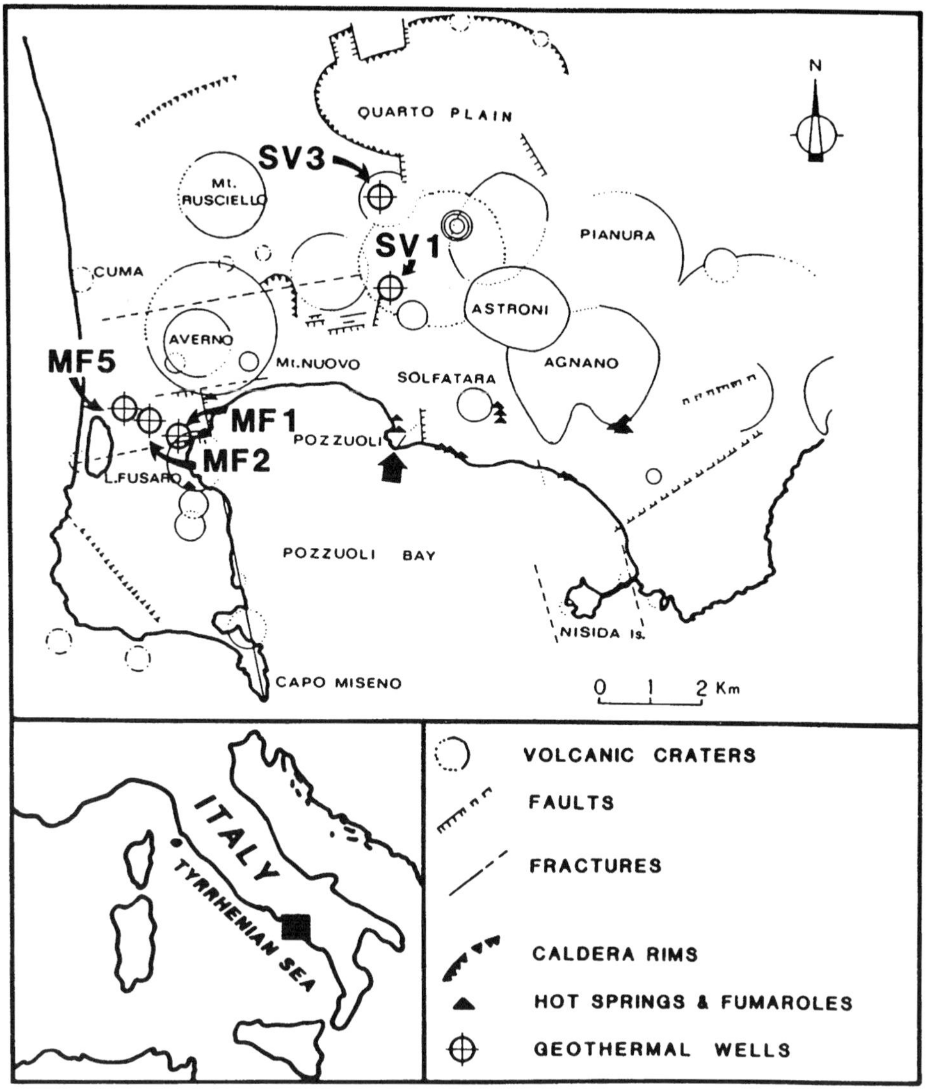

Figure 8. Volcano-tectonic schematic map of the Campi Flegrei volcanic complex showing the location of the studied geothermal wells. MF = Mofete, SV = San Vito. The arrow points to Pozzuoli, the center of uplift during the 1982–1984 bradyseismic crisis (from Barberi et al., 1984).

mimic the saline and hypersaline values found by drilling. Fluid inclusion (V/L) homogenization temperatures increase with depth and generally correspond to the extrapolated down-hole temperature. However, fluid inclusion data for MF5 and mineral assemblage for SV3, indicate a fossil, higher-temperature regime compared with the present one. The $^{87}Sr/^{86}Sr$ ratio in SV3 cores shows an approach to equilibrium with a fluid similar to modern seawater (De Vivo et al., 1989).

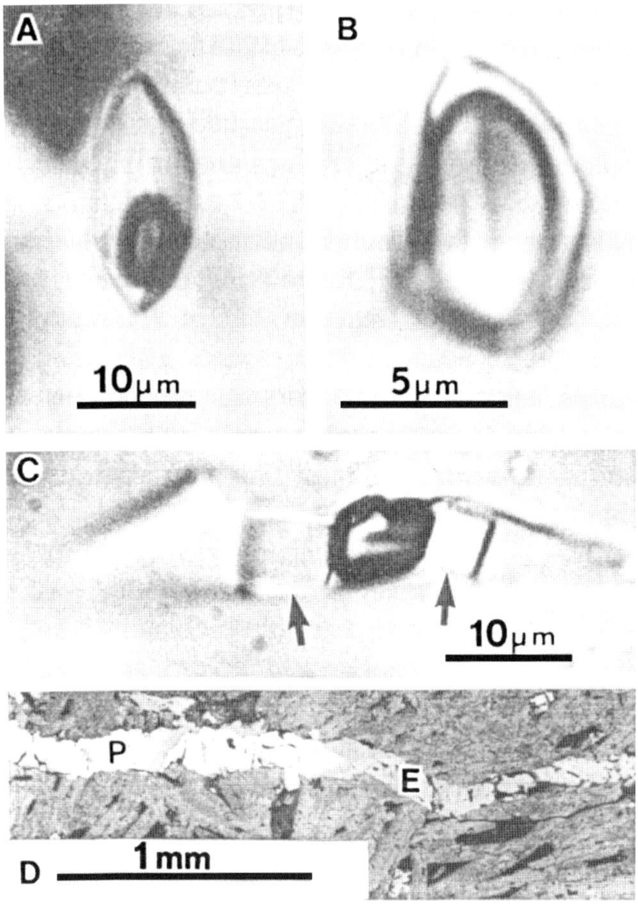

Figure 9. (A) Type A primary inclusion in calcite from San Vito 3-1948, Th = 290°C and 4 wt% NaCl equivalent. (B) Type B primary inclusion in quartz from San Vito 1-2676, Th = 321°C. (C) Type C inclusion from Mofete 5-2610 in quartz. Arrows point to halite cubes (daughter crystals). TmNaCl = 380°C, Th (V+L) = 390°C, salinity = ~44 wt% NaCl equivalent (assumes vapor present NaCl+H_2O solution). (D) Typical filled fracture in Mofete 2-1824. The fracture oriented parallel to the scale bar is filled with pyrite (P) and epidote (E) (from De Vivo et al., 1989).

It has to be stressed that the common coexistence of type (A) and type (B) inclusions (with similar homogenization temperature) in samples from some of the studied cores suggests that the fluids were trapped while they were in a boiling condition and that a major component of the trapped fluids is CO_2. On the other hand, boiling conditions are also suggested by the association of calcite and adularia (Browne and Ellis, 1970; Simmons and Christensen, 1994) in the hydrothermal alteration mineralogy of different core samples.

Boiling, as explained by Henley and Mc Nabb, 1978, Burnham (1979) and Fournier (1999), is a common mechanism associated with ore deposition in the porphyry systems. A fluid in the lithostatic pressure domain (such as fluids in the plastic domain surrounding a crystallizing magma) may undergo boiling and effervescence during a pressure release caused by hydraulic/seismic fracturing. This process may permit the passage of fluids from the lithostatic to the hydrostatic domain. This process also results in the deposition of minerals in microcracks, fractures and breccias (Fig. 9) to re-seal the system and return the fluids to the lithostatic domain. The high salinity in the fluids trapped in newly formed minerals in the hydrostatic domain has been attributed to concentration processes affecting seawater-derived fluids. Fournier (1999) indicates that a brine is unlikely to form from a moderately saline solution (with seawater composition) by purely upward adiabatic decompression, especially when the heat source is relatively "shallow" and the fluids are heated to above 400–450°C, which are characteristics of the Campi Flegrei system.

The Campi Flegrei hydrothermal system represents a modern analogue of an ancient mineralized system and may have the potential for significant ore formation at depth, as has been suggested at White Island, New Zealand (Rapien et al., 2003). Although no major mineralization has yet been found at Campi Flegrei, galena, sphalerite, pyrrhotite, pyrite, arsenopyrite and hematite are found in minor amounts in fractures. In fact, the Campi Flegrei hydrothermal system has two characteristics, which would favor the formation of ore deposition. First, is the recognition of boiling. Boiling of a hydrothermal solution, especially one carrying significant amounts of CO_2, can be an effective way to deposit ore minerals (Cunningham, 1978, 1985). When the solution boils, the dissolved gases are strongly partitioned into the vapor phase. This can increase the pH and destabilize various metal complexes (sulfur species), and cause their precipitation. Second, brine stratification is present in the form of reservoirs with different salinities (Guglielminetti, 1986) (Table 1), and fluid mixing across the brine interface is a significant mechanism for ore mineral precipitation (Williams and McKibben, 1987; McKibben et al., 1988). Evidence that fluid inclusions in the Campi Flegrei record a transition from magmatic to hydrothermal conditions is found by Fedele et al. (this volume), who conducted fluid inclusion studies and SEM-EDS and electron microprobe analysis on late-stage minerals such as apatite, Zr-bearing minerals (zircon and baddaleyite), pyroclore group minerals, thorite and phosphate. Complex daughter mineral assemblages found in multiphase fluid inclusion of the xenoliths are evidence of high solute entrapment. Abundances of chlorides, sulfides and, to a lesser extent, sulfates and carbonates, suggest that the fluid inclusions trapped a hypersaline/sulfur-rich fluid (possibly with minor CO_2) which likely exsolved from a crystallizing magma. Microthermometry on secondary hypersaline fluid inclusions suggests two possible scenarios for fluid trapping: (1) circulation of non-boiling, high-temperature (up to 525°C), high salinity fluids which were trapped under decreasing P–T conditions; (2) circulation of boiling, hypersaline fluids, trapped at low pressure and temperatures up to 300°C. All data suggest that the xenoliths record the effects of a hydrothermal phase, possibly associated with a transition from a magma-dominated to a fluid-dominated stage at the margins of a magma chamber. The presence of base metals and tungsten minerals in the xenoliths suggests a potential for mineralization which is similar to that observed in the alkaline volcanic systems of Pontine Islands and Mt. Somma-Vesuvius, where fluids escaping from the upper part of a sub-volcanic magma chamber and trapped in the plastic, lithostatic domain have also been found in feldspathoid-bearing syenite xenoliths.

At Ventotene, De Vivo et al. (1995) found gabbroic cumulate and alkali syenite xenoliths. The gabbroic cumulates contain only silicate melt inclusions ± vapor bubble ± droplets of an opaque phase and rarely some CO_2. The alkali syenite xenoliths have three types of fluid inclusions: (1) single-phase vapor and silicate melt inclusions;

Table 1. Well chemistry data (in ppm) from Campi Flegrei (from De Vivo et al., 1989).

Mofete field, chemistry of water separated at atmospheric pressure (Guglielminetti, 1996)

	Shallow reservoir		Intermediate reservoir	Deep reservoir
	Mofete 1 550–896 m	Mofete 1 1273–1606 m	Mofete 2 1275–1989 m	Mofete 5 2310–2699 m
Na	14320	20860	10600	85160
K	1760	1880	2467	43380
Ca	792	2124	1005	53950
B	178	183	295	231
Sr	49	58	30	1310
As	13	17	22	nd
Li	36	46	28	480
Mn	10	28	52	5510
Fe	1	3	1	9450
SiO_2	568	690	938	210
Cl	25304	37800	21169	313850
HCO_3	116	77	85	TR
SO_4	72	7	12	TR
TDS	42860	65509	37880	515902
pH	7.5	6.5	6.0	4.5

Mofete field, water chemistry calculated at reservoir conditions (Guglielminetti and Tore, 1985; Guglielminetti, 1986) — San Vito 1, composition of fluid sampled during purge tests (Bruni et al., 1985)

	Shallow reservoir		Intermediate reservoir	Sample	A	B
	Mofete 1 550 896 m	Mofete 1 1273–1606 m	Mofete 2 1275–1989 m			
Na	10025	12589	5090	Na	11750	6280
K	1230	2342	1180	K	8000	4025
Ca	555	1281	480	Ca	3290	1980
B	125	110	140	Mg	1120	540
Sr	34	41	14	Li	47	26
As	9	11	11	F	5	5
Li	25	28	13	SiO2	369	246
Mn	7	17	25	Cl	37755	20024
Fe	1	2	1	HCO3	nd	24
Mg	nd	5	0.61	TDS	62336	33150
Ba	nd	2.8	0.49	pH	3.2	4.4

(*Continued*)

Table 1. (*Continued*)

	Mofete field, water chemistry calculated at reservoir conditions (Guglielminetti and Tore, 1985; Guglielminetti, 1986)			San Vito 1, composition of fluid sampled during purge tests (Bruni et al., 1985)		
	Shallow reservoir		Intermediate reservoir	Sample	A	B
	Mofete 1 550–896 m	Mofete 1 1273–1606 m	Mofete 2 1275–1989 m			
Cr	nd	0.07	0.14			
Cu	nd	0.01	<0.03			
Pb	nd	0.36	<0.29			
Zn	nd	0.12	<0.02			
SiO_2	398	417	450			
Cl	17710	22810	10200			
HCO_3	81	46	41			
SO4	50	4	6			
TDS	30000	39500	18200			
CO2*	nd	17642	31980			
H_2S*	nd	236	1117			

*flashable gas, maximum content.
nd, not determined; TDS, total dissolved solids; TR, trace.

(2) two-phase silicate melt + salt, silicate melt + CO_2 (V), aqueous (L + V), and silicate melt + vapor inclusions; (3) three-phase and multiphase inclusions: CO_2 (L) + CO_2 (V) + H_2O; silicate melt + saline melt + H_2O ± birefringent or opaque-trapped minerals; H_2O + salt + silicate glass ± birefringent trapped minerals. The characteristics of the fluid phases trapped in xenoliths indicate that they evolved in a pressure regime varying from lithostatic to hydrostatic during the crystallization and cooling history of the host intrusive rocks. The gabbro crystallized at $T \sim$ 900–1000°C, around 1 to 1.4 kb, whereas the alkali syenite crystallized at $T \sim$ 600–700°C, between ~200 and 400 bars; the transition rock crystallized between 800°C and 1000°C (Fig. 10). The Ventotene xenoliths show clear evidence of various stages of silicate melt – hydrous saline melt – aqueous fluid – CO_2 fluid immiscibility (Roedder, 1992, and references therein) during the magmatic evolution and its transition from magmatic to hydrothermal stage. Fluids from the hydrostatic, brittle domain are recorded in the alkali syenite only by very subordinate secondary H_2O inclusions, which were trapped in the later stage of the hydrothermal process at much lower temperature (130–290°C), but at pressures relatively close to those of alkali syenite crystallization. The trapping of dilute to moderate salinity fluids in the hydrostatic domain occurred after breaching of the impermeable rocks separating the plastic from the brittle zones (see Fig. 7). At Ponza Island, Belkin et al. (1996) found evidence similar to that found in Ventotene in feldspathoid-bearing syenite xenoliths entrained in trachyte. Fluid inclusions show evidence of heterogeneous immiscible phases during crystallization, represented by silicate melt + CO_2 + H_2O and

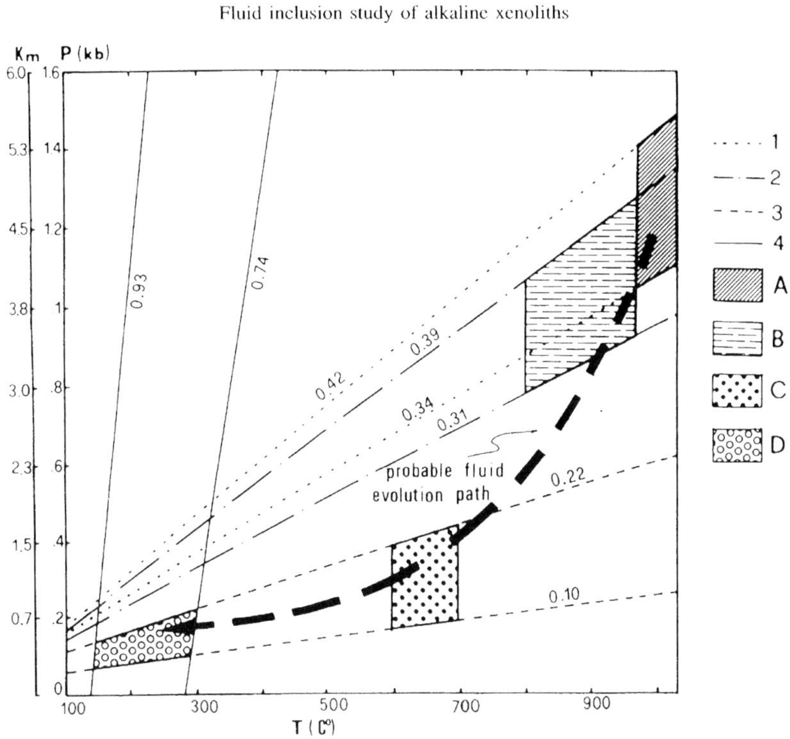

Figure 10. P–T plot for fluid inclusions in Ventotene xenoliths. Isochores for CO_2 and H_2O–NaCl are from Brown and Lamb (1989). (1) CO_2 isochores from VT-2A (gabbro); (2) CO_2 isochores from VT-4 (transition rock); (3) CO_2 isochores from VT-5B (alkali syenite); (4) H_2O–NaCl isochores from secondary inclusions (VT-4; VT-5). Field A, B and C represent the P–T conditions of crystallization of gabbro (A), transition rocks (B) and alkali syenite (C); field D represents the P–T conditions of late-stage aqueous secondary inclusion trapping (from De Vivo et al., 1995).

silicate melt/hypersaline/sulfur-rich aqueous inclusions. In contrast to Ventotene, however, no evidence for immiscibility between silicate melt and hydrosaline melt was found in Ponza. The potassium feldspar contains primary $H_2O + CO_2$ inclusions with small amounts of silicate melt and hypersaline/sulfur-rich aqueous inclusions. In the latter, small amounts of silicate melt are sometimes present. Silicate melt inclusions are more commonly observed coexisting with hypersaline/sulfur-rich aqueous inclusions along healed fractures. The evolution of the fluid phases under sub-solidus conditions is represented by secondary aqueous + CO_2 inclusions. These inclusions indicate moderate- to high-temperature hydrothermal fluids (between 359°C and 424°C) with low-to-moderate salinity (between 2.9 and 8.5 wt% NaCl equiv.). As for Ventotene, the xenoliths record the magmatic/hydrothermal transition, most probably in the upper part of the magma chamber. The peripheral zone, also in Ponza, has become enriched in various incompatible elements, including Th, U, REE, Zr, Y and others, as reported for other potassium-enriched magmatic systems (Tait, 1988; Turbeville, 1992; Belkin et al., 1994; Federico et al., 1994; Renzulli et al., 1995).

At Mt. Somma-Vesuvius, a mineralogical, fluid inclusion and isotope study (Fulignati et al., 2001; Gilg et al., 2001) carried out on skarn-bearing xenoliths found evidence of immiscibility in good agreement with the findings from the Pontine Islands (De Vivo et al., 1995; Belkin et al., 1996; Kamenetsky et al., 2003). In particular four major types of fluid inclusions were trapped in various minerals (wollastonite, vesuvianite, gehlenite, clinopyroxene and calcite: (a) primary silicate melt inclusions; (b) $CO_2 \pm H_2S$-rich vapor inclusions; (c) multiphase aqueous brine inclusions, with sylvite and halite daughter minerals; (d) complex chloride–carbonate–sulfate–fluoride–silicate-bearing saline melt inclusions, indicating immiscibility between silicate melt–aqueous chloride-rich liquid-carbonate/sulfate melt. The study also indicates that there is no evidence – as for Ponza and Ventotene – for a convectively cooling hydrothermal system at the magma-carbonate wall rock interface, indicating a lack of participation of externally derived fluids, such as meteoric waters or formational fluids. Salt-rich aqueous fluids must have separated from a silicate melt (magma) during cooling, infiltrated the carbonate rocks, and reacted with them to form skarns.

The saline melt inclusions contain high concentrations of Cl, carbonate, sulfate and sometimes F. These saline fluids might be related to the extremely Cl- and SO_2-rich melts of the Plinian to sub-Plinian eruptions (Webster et al., 2001; Lima et al., this volume). As found also in Ponza and Campi Flegrei, the fluids played an important role in the transport of Ti, Zr, Th, U and REE from the magma to the wall rocks.

The Pb, Nd, C and O isotope compositions of skarn and the presence of silicate melt inclusion-bearing wollastonite nodules indicate assimilation of carbonate wall rock by the alkaline magma at moderate depth (<5 km) which promoted exsolution of CO_2-rich vapor and complex saline melts from a locally contaminated magma.

5. A model of the bradyseism at Campi Flegrei

Fluid and melt inclusion studies indicate that the Campi Flegrei system is subdivided into a shallower permeable zone, where convectively driven fluids (encountered in the San Vito and Mofete wells) circulate, and a deep plastic, lithostatic domain dominated by magmatic fluids. The occurrence of two distinct fluid regimes is thought to be related to the presence of a large caldera filled by volcaniclastic sediments. The fluids have mostly either meteoric or seawater origin, with a magmatic component. This difference, i.e., the lack of a caldera filled with sediments above the volcanic plumbing systems of the other Neapolitan volcanic centers, might explain the occurrence of the bradyseism in the Campi Flegrei, as opposed to the other nearby volcanic areas where such ground movement has never been observed.

The drilling of MF1, MF2 and MF5 penetrated three reservoirs (Carella and Guglielminetti, 1983). A shallow reservoir was found in MF1 with a temperature of 250°C and a calculated reservoir fluid composition between 30,000 and 40,000 ppm TDS (total dissolved solids). A more dilute, intermediate reservoir was observed in MF2 with a temperature of 337°C and calculated fluid composition of ~18,000 ppm TDS. Well MF5 briefly produced hypersaline fluids from 2700 m (>500,000 ppm TDS, corresponding to 150,000 ppm in the reservoir) with a bottom hole temperature of 347°C (Table 1). The measured pCO_2 in MF1 at 1606 m (295°C) was 26 bars and in MF2 at 1989 m (345°C) was 51 bars (AGIP internal report), and is in agreement with the Carella and Guglielminetti (1983) estimate of the CO_2 content of the Mofete field of between 1 and 5 mol%. The data from San Vito field indicate fluid compositions that resemble compositions reported from the shallow and intermediate reservoirs of the Mofete field.

The coexistence of liquid-dominated and vapor-dominated inclusions indicates that the system, at some stage, contained a two-phase fluid. This coexistence would reflect a causal relationship between boiling and mineral deposition with the resulting inclusion trapping. The inclusion salinities (Figs. 11 and 12) reflect the chemical characteristics of the fluids found by drilling (Table 1). The variation of salinity with depth and location of MF1, MF2 and MF5, agrees with the fluid systematics found by Carella and Guglielminetti (1983), who stress the existence of multiple reservoirs, separated by impermeable layers. The shallow reservoir found at MF2 produced moderately saline fluids whereas the deep reservoir of MF5 contains a hypersaline fluid. Zones of low transmissivity prevent a major inflow of groundwater from penetrating and diluting the deep reservoirs. Hydrothermal activity in the Campi Flegrei is related to the onset of volcanic activity in the area (~50,000 ybp).

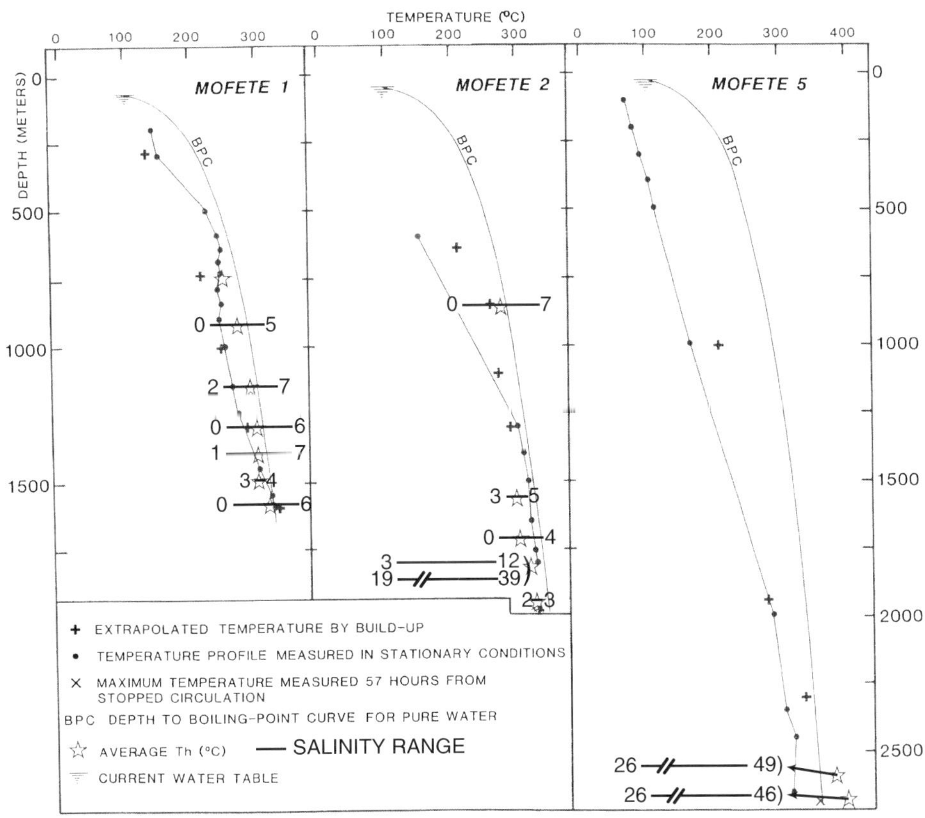

Figure 11. A comparison of the Mofete wells (1, 2 and 5) temperature profile with the average fluid inclusion homogenization temperature and range of salinity (wt% NaCl equivalent). The depth to the boiling-point curve for pure water is shown calculated to the current water table (from De Vivo et al., 1989).

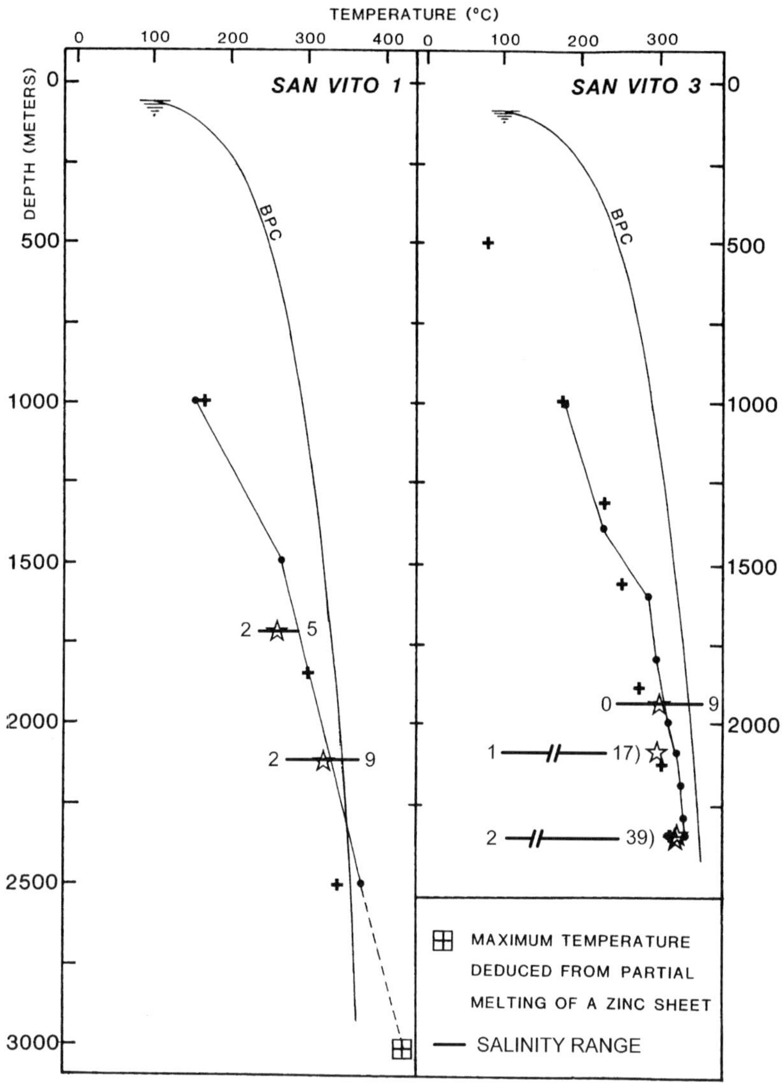

Figure 12. A comparison of the San Vito wells (1, 2 and 5) temperature profile with the average fluid inclusion homogenization temperature and range of salinity (wt% NaCl equivalent). The depth to the boiling-point curve for pure water is shown calculated to the current water table (from De Vivo et al., 1989).

This is confirmed by geochronology of zircons from alkali syenite nodules, which gives an age of ~50,000 ybp (Fedele et al., this volume). On the other hand, the homogenization temperatures of fluid inclusions found in hydrothermal minerals from the wells (except for MF5) correlate with the current thermal regime and the general characteristics of the fields (Figs. 11 and 12). This suggests that the host minerals are relatively recent and/or that the fields have remained stable during the later stages of Campi Flegrei evolution. Drilling data suggests that deep MF5 fluids have cooled subsequent to inclusions trapping.

In summary, primary fluid inclusions trapped in various authigenic minerals yield homogenization temperatures and salinities that mimic the temperature increase with depth and the salinities of the sampled reservoirs. The coeval nature of liquid-rich and vapor-rich inclusions is strong evidence of boiling conditions during inclusion trapping. Daughter crystals (halite) in core samples from deeper, hotter, levels indicate a high concentration of solute, as confirmed by drilling. This means that the above fluids are trapped in the brittle zone (see Fig. 7), in a hydrostatic pressure domain, immediately above the brittle-plastic transition zone. Conversely, the fluid inclusions that record immiscibility (in alkali syenite xenoliths) are trapped in the lithostatic domain and contain mostly fluids of magmatic origin.

The overall scenario of fluid inclusions and isotope characteristics is compatible with a Campi Flegrei geothermal field with a deep influx, from magmatic + seawater sources, of saline water which remains localized in aquifers at a depth of ~2.5–3 km. These aquifers are mostly under lithostatic pressure, being heated by an underlying, crystallizing magma body that is not moving upwards. The magma, in other words, does not increase the pressure on the overlying load of volcanoclastic sediments, but it furnishes to the overlying system heat and fluids, which escape from the apical portion of the magma chamber as shown in the Burnham model (1979).

The brittle–ductile transition zone is not fixed in space, but migrates up as the system is recharged with magma and migrates downward as the system cools. This is predicted in the Burnham model (Burnham et al., 1979) and documented in porphyry-mineralized systems (Beane and Titley, 1981; Beane, 1982; Beane and Bodnar, 1995; Roedder and Bodnar, 1997). The vertical and lateral extension of the hydrothermal system is relatively limited because of the presence of the overlying impermeable rocks.

In our model of Campi Flegrei, in addition to the inner, ~5–7 km deep, impermeable layer separating the plastic, lithostatic region from the brittle (hydrostatic) domain, there is a second, outer, shallower, impermeable layer. This is at a depth of ~2–2.5 km in the permeable volcaniclastic rocks, and its impermeability is due to the precipitation of newly formed minerals (as a consequence of the boiling process), which heal all fractures and faults that extend toward the surface. The continuous supply of heat and inflow of magmatic fluids creates an overpressure primarily against the inner, deeper, impermeable layer (transition from plastic to brittle domain), and subsequently also against the outer impermeable layer separating the more dilute surficial from the deep (at 2700 m) hypersaline geothermal aquifers (Carella and Guglielminetti, 1983). These internal overpressures, concentrated in the apical part of a crystallizing magma, can reach extremely elevated values ($\Delta P_{in} \geq 5$ kb). The amount of this overpressure, caused only by H_2O/CO_2-saturated fluids, would be enough to explain the positive bradyseism, which occurs episodically at Campi Flegrei.

When ΔP_{in} is high enough to breach the deeper impermeable layer and also the more surficial impermeable layer within the brittle domain, then hydrofracturing occurs, followed by intense boiling and consequent hydrothermal circulation. This is the moment when volcanic tremors occur and when the positive bradyseism is at its peak, as shown by geophysical experiments (Leet and Malone, 1990; Leet, 1991), which demonstrate that intense boiling in sub-volcanic environments can be responsible for volcanic tremor. In other words, in order to explain the seismic tremors, it is not necessary to invoke the presence of a magma, which rises toward the surface. At this point, the fluids that were in a lithostatic pressure regime pass into the hydrostatic domain, and the intense boiling causes the deposition of newly formed hydrothermal minerals; the mineral deposition, in turn, self-seals the system and leads to the restoration of the lithostatic domain.

Immediately after breaching of the impermeable layer, which separates the plastic from the brittle zone, deflation begins and lasts until the lithostatic domain is restored by the deposition of newly formed minerals (self-sealing process). In order to start a new episode of positive bradyseism it is necessary that the fluids escaping from the underlying crystallizing magma and/or fluids of external source (meteoric influx) "reload" the system, generating an overpressure in the apical portion of the magma body/chamber.

Considering that the hypocenters of the earthquakes registered during the last bradyseismic event of 1982–1984 were located at ~4 km depth (which should correspond to the location of the zone above the impermeable layer separating the plastic from the hydrostatic domains), we assume that a magma chamber, from which the magmatic fluids and the heat derive, should be located at a depth of ~5–7 km.

Our model, which is supported by the ideas of Cortini et al. (1991) and Cortini and Barton (1993), explains both the inflation and the deflation of the ground surface at Campi Flegrei, whereas a model which suggests that positive bradyseism is caused by the intrusion and rise of new magma toward the surface would explain only the positive event (inflation), but not the negative event (deflation). The crystallizing magma acts only as the "furnace" of the system, heating up and, at the same time, supplying fluids to underground, confined hydrothermal systems. Because of the depths at which these processes occur, we consider an eruption a very improbable event, because the ΔP_{in} is not enough to breach completely 5–7 km of overlying load. The case would be very different if a new batch of magma were intruded and rise toward the surface in Campi Flegrei. But, so far, no clear geophysical evidence of such occurrence has been reported. In our model an eruption could occur, as was the case of Monte Nuovo in 1538 AD. This would happen only if the process described above occurred with a load of volcanic rock/sediment <500 m. In this case, the high-pressured fluids passing from lithostatic to hydrostatic pressure domain could give rise to a hydrothermal eruption which in turn could trigger a magmatic eruption.

6. Conclusions

To explain the bradyseism at Campi Flegrei, we propose a model that emphasizes the role of hydrothermal fluids as opposed to the classical model of intrusion of new magma at shallow depths (Barberi et al., 1984; Corrado et al., 1976), suggesting that ground deformations could be generated by conductive heating of the hydrothermal fluids overlying the magmatic chamber. The authors describe a hydrothermal model based on the model of porphyry systems (Henley and McNabb, 1978; Burnham, 1979; Fournier, 1999). Campi Flegrei might represent a modern analogue of an ancient mineralized porphyry system (Beane and Titley, 1981; Beane and Bodnar, 1995; Roedder and Bodnar, 1997; Sasada, 2000), as has been suggested for White Island, New Zealand (Rapien et al., 2003). Deep, aqueous fluids, heated by an underlying crystallizing magma, would remain under lithostatic pressure for a long period. The heat and fluids supplied by the magma would determine the overpressure in the upper, apical, part of the magma chamber through accumulation of volatile phases confined by impermeable rocks, which cause uplift of the overlying rocks (positive bradyseism). A crisis occurs when the conditions change from lithostatic to hydrostatic pressure, with consequent boiling (De Vivo et al., 1989), hydraulic fracturing, volcanic tremor and then pressure release. At this point, the area experiences a sudden change from maximum inflation, followed by pressure release to beginning of subsidence (deflation of the ground). Afterward, the system, saturated with boiling fluids, begins to seal again due to the precipitation of newly deposited minerals. The beginning of a new positive bradyseism phase will occur only after several years when the system "reloads" under new lithostatic pressure conditions. This model is capable of explaining both inflation and deflation of ground, whereas a model which is restricted to the magma rising process as responsible for the bradyseism, would explain only the inflation. In the above scenario, a hydrothermal eruption can still occur, but only if the fluids pass from lithostatic to hydrostatic pressure when the overlying rocks have a thickness <500 m. If this happens, a hydrothermal eruption could trigger a magmatic eruption, which was probably the case of the Monte Nuovo eruption in 1538 AD.

Acknowledgments

The authors are grateful to the thorough reviews of Massimo Cortini (University of Napoli Federico II, Italy), Charles G. Cunningham (U. S. Geological Survey, Reston, Va, USA), Giuseppe De Natale (INGV, Napoli, Italy) and Robert J. Bodnar (Virginia Tech, Blacksburg, VA, USA), which have helped to greatly improve the final manuscript.

References

AGIP, 1987. Geologia e geofisica del sistema geotermico dei Campi Flegrea. Int. Report, p. 17.

Aster, R.C., Meyer, R.P., 1988. Three-dimensional velocity structure and hypocenter distribution in the Campi Flegrei caldera, Italy. Tectonophysics 149, 195–218.

Babbage, C., 1847. Observation on the Temple of Serapis at Pozzuoli near Naples. R. and J.E. Taylor, London, p. 35.

Barberi, F., Corrado, G., Innocenti, F., Luongo, G., 1984. Phlegraean Fields 1982 – 1984: brief chronicle of a volcano emergency in a densely populated area. Bull. Volcanol. 47(2), 175–185.

Beane, R.E., 1982. Hydrothermal alteration in silicate rocks. In: Titley, S.R. (Ed.), Advances in geology of the porphyry copper deposits of southwestern North America. University of Arizona Press, Tucson, pp. 117–137.

Beane, R.E., Bodnar, R.J., 1995. Hydrothermal fluids and hydrothermal alteration in porphyry copper deposits. Arizona Geol. Soc. Digest 20, 83–93.

Beane, R.E., Titley, S.R., 1981. Porphyry copper deposits. Part II. Hydrothermal alteration and mineralization. Econ. Geol. 75th Anniversary Vol. pp. 235–263.

Belkin, H.E., Cavarretta, G., Tecce, F., 1994. Petrogenesis of leucite-bearing nodules from Colli Albani volcano ejecta, Latium, Italy. Abstracts Int. Mineral. Assoc., 16th Gener. Meeting, 4–9 September, Pisa, Italy, p. 35.

Belkin, H.E., De Vivo, B., Lima A., Torok, K., 1996. Magmatic (silicate/sulfur-rich/CO_2) immiscibility and zirconium and rare-earth element enrichment from alkaline magma chamber margins: evidence from Ponza island, Pontine Archipelago. Eur. J. Mineral. 8, 1401–1420.

Berrino, G., Corrado, G., Luongo, G., Toro, B., 1984. Ground deformation and gravity changes accompanying the 1982 Pozzuoli uplift. Bull. Volcanol. 47, 187–200.

Berrino, G., Gasparini P., 1995. Ground deformation and caldera unrest. Cahiers du Centre Europeen de Geodynamique et de Sismologie 8, 41–55.

Berrino, G., Rymer, H., Brown, G.C., Corrado, G., 1992. Gravity-height correlation for unrest at calderas. J. Volcanol. Geotherm. Res. 53, 11–26.

Bianchi, R., Cordini, A., Federico, C., Giberti, G., Lanciano, P., Pozzi, J.P., Sartoris, G., Scandone R., 1987. Modelling of surface ground deformation in volcanic areas: the 1970–1972 and 1982–1984 crises of Campi Flegrei, Italy. J. Geophys. Res. 92 (B13), 14139–14150.

Bodnar, R.J., 1995. Fluid inclusion evidence for a magmatic source for metals in porphyry copper deposits. Miner. Assoc. Canada Short Course 23, 139–152.

Bonafede, M., 1990. Axi-symmetric deformation of a thermo-poro-elastic half-space: inflation of a magma chamber. Geophys. Int. 103(2), 289–299.

Bonafede, M., 1991. Hot fluid migration: an efficient source of ground deformation. Application to the 1982–85 crisis at Campi Flegrei, Italy. In: Luongo, G., Scandone, R. (Eds), Campi Flegrei. J. Volcanol. Geotherm. Res. 48, 187–198.

Bonafede, M., Dragoni, M., Quareni, F., 1986. Displacement and stress fields produced by a centre of dilatation and by a pressure source in visco-elastic half-space: application to the study of ground deformation and seismic activity at Campi Flegrei, Italy. Geophys. J. R. Astr. Soc. 87, 455–485.

Breislak, S., 1792. Essai mineralogiques sur le Solfatare de Pouzzole. Part 3, Observations sur l'exterieur du crater de la Solfatare. Giaccio, Naples, pp. 170–177.

Brown, P.E., Lamb, W.M., 1989. P-V-T properties of fluids in the system $H_2O \pm CO_2 \pm NaCl$: new graphical presentations and implications for fluid inclusion studies. Geochim. Cosmochim. Acta 53, 1209–1221.

Browne, P.R.L., Ellis, A.J., 1970. The Oaki-Broadlands hydrothermal area, New Zealand: mineralogy and related geochemistry. Am. J. Sci. 269, 97–131.

Bruni, P., Chelini, W., Sbrana, A., Verdiani, G., 1985. Deep exploration of the S. Vito area (Pozzuoli-NA) – Well S. Vito 1. In: Strub, A.S., Ungemach, P. (Eds), European Geothermal Update. Proc. Third Int. Seminar, pp. 390–406.

Burnham, C.W., 1967. Hydrothermal fluids at the magmatic stage. In: Barnes, H.L. (Ed.), Geochemistry of hydrothermal ore deposits. Holt, Rinehart and Winston, New York, pp. 34–76.

Burnham, C.W., 1972. The energy of explosive volcanic eruptions. Earth Mineral. Sci., The Pennsylvania State University, 41, 69–70.

Burnham, C.W., 1979. Magmas and hydrothermal fluids. In: Barnes, H.L. (Ed.), Geochemistry of hydrothermal ore deposits. Wiley, New York, pp. 71–136.

Burnham, C.W., 1985. Energy release in subvolcanic environments: implications for breccia formation. Econ. Geol. 80, 1515–1522.

Burnham, C.W., Davis, N.F., 1974. The role of H_2O in silicate melts: II. Thermodynamics and phase relations in the system $NaAlSi_3O_8$-H_2O to 10 kilobars, 700 to 1000°C. Am. J. Sci. 274, 902–940.

Carella, R., Guglielminetti, M., 1983. Multiple reservoirs in the Mofete fields, Naples, Italy. Workshop on Geothermal Reservoir Engineering, Stanford, 12 pp.

Carrigan, C.R., 1986. A two-phase hydrothermal cooling model for shallow intrusions. J. Volcanol. Geotherm. Res. 28, 175–192.

Casertano, L., Oliveri del Castillo, A., Quagliariello, M.T., 1976, Hydrodynamics and geodynamics in the Phlegraean Fields area of Italy, Nature 264, 154–161.

Chiodini, G., Marini L., Russo, M., 2001. Geochemical evidence for the existence of high-temperature hydrothermal brines at Vesuvius volcano, Italy. Geochim. Cosmochim. Acta 65(13), 2129–2147.

Cline J.S., Bodnar, R.J., 1994. Direct evolution of brine from a crystallizing silicic melt at the Questa, New Mexico, molybdenum deposit. Econ. Geol. 89, 1780–1802.

Corrado, G., Guerra, I., Lo Bascio, A., Luongo, G., Rampoldi, F., 1976. Inflation and microearthquake activity of Phlegraean Fields, Italy. Bull. Volcanol. 40(3), 169–188.

Cortini, M., Barton, C.C., 1993. Non linear forecasting analysis of inflation-deflation patterns of an active caldera (Campi Flegrei, Italy). Geology 21, 239–242.

Cortini, M., Cilento, L., Rullo, A., 1991. Vertical ground movements in the Campi Flegrei caldera as a chaotic dynamic phenomenon. J. Volcanol. Geotherm. Res. 48, 103–114.

Cubellis, E., Di Donna, G., Luongo, G., Mazzarella, A., 2002. Simulatine the mechanism of magmatic processes in the Campi Flegrei area (southern Italy) by the Lorenz equations. J. Volcanol. Geotherm. Res. 115, 339–349.

Cunningham, C.G., 1978. Pressure gradients and boiling as mechanisms for localizing ore in porphyry systems. J. Res. Geol. Survey 6, 745–754.

Cunningham, C.G., 1985. Characteristics of boiling-water-table and carbon dioxide models for epithermal gold deposition. In: Tooker, E.W. (Ed.), Geologic characteristics of sediment- and volcanic-hosted disseminated gold deposits – Search for an occurrence model. U. S. Geol. Survey Bull. 1646, 43–46.

De Natale, G., Petrazzuoli, S.M., Pingue, F., 1997. The effect of collapse structures on ground deformations in calderas. Geophys. Res. Lett. 24(13), 1555–1558.

De Natale, G., Pingue, F., 1993. Ground deformations in collapsed caldera structures. J. Volcanol. Geotherm. Res. 57, 19–38.

De Natale, G., Pingue, F., Allard, P., Zollo, A., 1991. Geophysical and geochemical modeling of the Campi Flegrei caldera. In: Luongo, G., Scandone, R. (Eds), Campi Flegrei. J. Volcanol. Geotherm. Res. 48, 199–222.

De Natale, G., Troise, C., Pingue, F., 2001. A mechanical fluid-dynamical model for ground movements at Campi Flegrei caldera. J. Geodyn. 32, 487–517.

De Natale, G., Zollo, A., 1986. Statistical analysis and clustering features of the Phlegraean Fields earthquake sequence (May 1984). Bull. Seism. Soc. Amer. 76(3), 801–814.

De Natale, G., Zollo, A., Ferraro, A., Virieux, J., 1995. Accurate fault mechanism determinations for a 1984 earthquake swarm at Campi Flegrei caldera (Italy) during an unrest episode: implications for volcanological research. J. Geophys. Res. 100 (B12), 24167–24185.

De Vivo, B., Barbieri, M., Belkin, H.E., Chelini, W., Lattanzi, P., Lima, A., Tolomeo, L., 1989. The Campi Flegrei (Italy) geothermal system: a fluid inclusion study of the Mofete and San Vito fields. J. Volcanol. Geotherm. Res. 36, 303–326.

De Vivo, B., Lima, A., Webster, J.D., 2005. Volatiles in magmatic-volcanic systems. Elements 1, 19–24.

De Vivo, B., Torok, K., Ayuso, R.A., Lima, A., Lirer, L., 1995. Fluid inclusion evidence for magmatic/saline/CO_2 immiscibility and geochemistry of alkaline xenoliths from Ventotene island, Italy. Geochim. Cosmochim. Acta 59(14), 2941–2953.

Dvorak, J.J., Berrino, G., 1991. Recent ground movement and seismicity activity in Campi Flegrei, Southern Italy: episodic growth of a resurgent dome. J. Geophys. Res. 96(B2), 2309–2323.

Dvorak, J.J., Mastrolorenzo, G., 1991. The mechanism of recent vertical crustal movements in Campi Flegrei caldera, Southern Italy. Geol. Soc. Amer. Spec. Pap., 263.

Dzurisin, D., Newhall, C.G., 1984. Recent ground deformation and seismicity at Long Valley (California), Yellowstone (Wyoming), the Phlegrean Fields (Italy) and Rabaul (Papua, new Guinea). In: Hill, D.P., Bailey, R.A., Ryall, A.S. (Eds), Proceedings of Workshop XIX; Active tectonic and magmatic processes beneath Long Valley Caldera, eastern California. Open-File Report – U.S. Geological Survey, pp. 784–829.

Dzurisin, D., Yamashita, K.M., 1987. Vertical surface displacements at Yellowstone caldera, Wyoming, 1976–1986. J. Geophys. Res. 92(B13), 13753–13766.

Fedele, L., Tarzia, M., Belkin, H.E., De Vivo, B., Lima, A., Lowenstern, J.B. (this volume). Magmatic-hydrothermal fluid interaction and mineralization in alkali-syenite nodules from the Breccia Museo pyroclastics deposit, Naples, Italy.

Federico, M., Peccerillo, A., Barbieri, M., Wu, T.W., 1994. Mineralogical and geochemical study of granular xenoliths from the Alban Hills volcano, Central Italy: bearing on evolutionary processes in potassic magma chambers. Contrib. Mineral. Petrol. 115, 384–401.

Ferrucci, F., Hirn, A., Virieux, J., De Natale, G., Mirabile, L., 1992. P-SV conversions at a shallow boundary beneath Campi Flegrei caldera (Naples, Italy): evidence for the magma chamber. J. Geophys. Res. 97(B11), 15351–15359.

Forbes, J.D., 1829. Physical notice in the Bay of Naples, Number 5. On the temple of Juppiter Serapis at Pozzuoli and the phenomena which it exhibits. Edin. J. Sci., New Series 1, 260–286.

Fournier, R.O., 1999. Hydrothermal processes related to movement of fluid from plastic into brittle rock in the magmatic-epithermal environment. Econ. Geol. 94(8), 1193–1211.

Frezzotti, M.L., 1992. Magmatic immiscibility and fluid phase evolution in the Mount Genis granite (southeastern Sardinia, Italy). Geochim. Cosmochim. Acta 56, 21–33.

Fulignati, P., Kamenetsky, V.S., Marianelli, P., Sbrana, A., Mernagh, T.P., 2001. Melt inclusion record of immiscibility between silicate, hydrosaline, and carbonate melts: applications to skarn genesis at Mount Vesuvius. Geology 29, 1043–1046.

Gaeta, F.S., De Natale, G., Peluso, F., Mastrolorenzo, G., Castagnolo, D., Troise, C., Pingue, F., Mita, D.G., Rossano, S., 1998. Genesis and evolution of unrest episodes at Campi Flegrei caldera: The role of thermal fluid dynamical processes in the geothermal system, J. Geophys. Res. 103 (B9), 921–933.

Gilg, H.A., Lima, A., Somma, R., Belkin, H.E., De Vivo, B., Ayuso, R.A., 2001. Isotope geochemistry and fluid inclusion study of skarns from Vesuvius. Mineral. Petrol. 73, 145–176.

Guglielminetti, M., 1986. Mofete geothermal field. Geothermics 15, 781–790.

Guglielminetti, M., Tore, G., 1985. Pit tests in the wells Mofete 1 and 2. In: Strub, A.S., Ungemach, P. (Eds), European Geothermal Update. Proc. Third Int. Seminar, pp. 473–483.

Gunther, R.T., 1903. The submerged Greek and Roman foreshore near Naples. Archaeologia 58, 62.

Gustafson, L.B., Hunt, J.P., 1975. The porphyry copper deposit at El Salvador, Chile. Econ. Geol. 70, 857–912.

Heinrich, C.A., Ryan, C.G., Mernagh, T.P., Eadington, P.J., 1992. Segregation of ore metals between magmatic brine and vapor – a fluid inclusion study using PIXE microanalysis. Econ. Geol. 87, 1566–1583.

Henley, R.W., Ellis, A.J., 1983. Geothermal systems ancient and modern: a geothermal review. Earth Sci. Rev. 19, 1–50.

Henley, R.W., McNabb, A., 1978. Magmatic vapor plumes and ground water interaction in porphyry copper emplacement. Econ. Geol. 73(1), 1–20.

Hill, D.P., 1984. Monitoring unrest in a large silicic caldera. The Long Valley-Inyo craters volcanic complex in the East-central California. Bull. Volcanol. 47(2), 371–395.

Kamenetsky, V.S., De Vivo, B., Naumov, V.B., Kamenestky, M.B., Mernagh, T.P., Van Achterbergh, E., Ryan, C.G., Davidson, P., 2003. Magmatic inclusions in the search for natural silicate-melt immiscibility: methodology and examples. In: De Vivo, B., Bodnar, R.J. (Eds), Melt inclusions in volcanic systems. Methods, Applications and Problems. Series "Developments in Volcanology 5", Elsevier, Amsterdam, pp. 65–82.

Koide, H., Bhattacharji, S., 1975. Formation of fractures around magmatic intrusions and their role in ore localization. Econ. Geol. 70, 781–799.

Leet, R.C., 1991. Investigation of hydrothermal boiling and steam quenching as possible sources of volcanic tremor and geothermal ground noise. PhD thesis. University of Washington, Seattle, WA, p. 282.

Leet, R.C., Malone, S.D., 1990. Boiling in hydrothermal systems may cause volcanic tremor. AGU 1990 Fall Meeting. EOS, Transactions 71(43), 1675.

Lima, A., Danyushevsky, L.V., De Vivo, B., Fedele, L., 2003. A model for the evolution of the Mt. Somma-Vesuvius magmatic system based on fluid and melt inclusion investigations. In: De Vivo, B., Bodnar, R.J. (Eds), Melt inclusions in volcanic systems. Methods, Applications and Problems. Series "Developments in Volcanology 5", Elsevier, Amsterdam, pp. 227–249.

Lima, A., De Vivo, B., Sintoni, M.F., An hydrothermal model to explain geochemical variations between 79 A.D. and 1944 A.D. Vesuvius eruptions.

Lister, C.R.B., 1974. On the penetration of water into hot rock. Geophys. J. Astron. Soc. 39, 465–509.

Lyell, C., 1872. Principles of geology. London, pp. 164–179.

Lorenz, E.N., 1963. Deterministic non-periodic flow. J. Atmos. Sci. 20, 130–141.

Lowell, J.D., Guilbert, J.M., 1970. Lateral and vertical alteration-mineralization zoning in porphyry ore deposits. Econ. Geol. 65, 373–408.

McKee, C., Mori, J., Talai, B., 1989. Microgravity changes and ground deformation at Rabaul caldera, 1973–1985. In: Latter, J.H. (Ed.), IAVCEI Proc. in Volcanol. 1, Volcanic Hazards, Springer, Wien, pp. 389–428.

McKibben, M.A., Williams, A.E., Okubo, S., 1988. Metamorphosed Plio-Pleistocene evaporates and origins of hypersaline brines in the Salton Sea geothermal system, California: fluid inclusion evidence. Geochim. Cosmochim. Acta 52, 1047–1056.

Niccolini, A., 1839. Tavola cronologica-metrica delle varie altezze tracciate dal mare fra la costa di Amalfi ed il promontorio di Gaeta nel corso di diaciannove secoli. Napoli, Flautina, pp. 11–52.

Niccolini, A., 1845. Descrizione della gran terna puteolana volgarmente detta Tempio di Serapide. Napoli, Stamperia Reale.

Northon, D.L., Cathles, L.M., 1973. Breccia pipe products of exsolved vapour from magmas. Econ. Geol. 68, 540–546.

Northon, D.L., Knapp, R., 1977. Transport phenomena in hydrothermal systems: nature of porosity. Am. J. Sci. 277, 913–936.

Northon, D.L., Knight, J.E., 1977. Transport phenomena in hydrothermal systems: cooling plutons. Am. J. Sci. 277, 937–981.

Oliveri del Castillo, A., Quagliariello, M.T., 1969. Sulla genesi del bradisismo flegreo. Atti Associazione Geofisica Italian, 18th Congress, Napoli, pp. 557–594.

Parascandola, A., 1947. I fenomeni bradisismici del Serapeo di Pozzuoli, Napoli, Privately published, pp. 156.

Philipps, W.L., 1973. Mechanical effects of retrograde boiling and its probable importance in the formation of some prophyry ore deposits. Inst. of Mining and Metallurgy Trans., Sect. B 82, B90–B98.

Rapien, M.H., Bodnar, R.J., Simmons, S.F., Szabo, C.S., Wood, C.P., Sutton, S.R., 2003. Melt inclusion study of the embryonic porphyry copper system at White Island, New Zealand. Econ. Geol., Spec. Publ. 10, 41–59.

Renzulli, A., Upton, B.G.J., Nappi, G., 1995. Magma chamber processes preceding the Pitigliano Formation eruption (Latera volcanic complex, central Italy): evidence from cognate plutonic clasts. Acta Vulcanologica 7(1), 55–74.

Roedder, E., 1971. Fluid inclusion studies on the porphyry-type ore deposits at Bingham, Utah, Butte, Montana, and Climax, Colorado. Econ. Geol. 66, 98–120.

Roedder, E., 1984. Fluid inclusions. Reviews in Mineralogy, Mineral. Soc. Am., Vol. 12, pp. 644.

Roedder, E., 1992. Fluid inclusion evidence for immiscibility in magmatic differentiation. Geochim. Cosmochim. Acta 56, 5–20.

Roedder, E., Bodnar, R.J., 1997. Fluid inclusion studies of hydrothermal ore deposits. In: Barnes, H.L. (Ed.), Geochemistry of hydrothermal ore deposits, 3rd ed. Wiley, New York, pp. 657–698.

Roedder, E., Coombs, D.S., 1967. Immiscibility in granitic melts, indicated by fluid inclusions in ejected granitic blocks of Ascension island. J. Petrol. 8, 417–451.

Rosi, M., Sbrana, A. (Eds), 1987. Phlegraean Fields. CNR, Quaderni de "La Ricerca Scientifica", Monog. 114(9), 175.

Sasada, M., 2000. Igneous-related active geothermal system versus porphyry copper hydrothermal system. Proceeding World Geothermal Congress 2000, Kyushu – Tohoku, Japan, May 28 – June 10, pp. 1691–1693.

Scandone, R., Bellucci, F., Lirer, L., Rolandi, G., 1991. The structure of the Campanian Plain and the activity of the Neapolitan volcanoes (Italy). J. Volcanol. Geotherm. Res. 48, 1–31.

Scandone, R., Giacomelli, L., Speranza, F.F. The volcanological history of the volcanoes of Naples: a review (this volume).

Shinoara, H., Kazahaya, K., Lowenstern, J.B., 1995. Volatile transport in a convecting magma column: implications for porphyry Mo mineralization. Geology 23, 1091–1094.

Simmons, S.F., Christensen, B.W., 1994. Origins of calcite in a boiling geothermal system. Amer. J. Sci. 294(3), 361–400.

Student, J.J., Bodnar, R.J., 2004. Silicate melt inclusions in porphyry copper deposits: Identification and homogenization behavior. Can. Mineral. 42(5), 1583–1599.

Tait, S.R., 1988. Samples from the crystallizing boundary layer of a zoned magma chamber. Contrib. Mineral. Petrol. 100, 470–483.

Touret, J.L.R., Frezzotti, M.L., 1993. Magmatic remnants in granites. Bull. Soc. Geol. Franc., II 229–242.

Trasatti, E., Giunchi, G., Bonafede, M., 2005. Structural and rheological constraints on source depth and overpressure estimate at the Campi Flegrea caldera, Italy. J. Volcanol. Geotherm. Res. 144, 105–118.

Troise, C., De Natale, G., Pingue, F., Zollo, A., 1997. A model for earthquake generation during unrest crises at Campi Flegrei and Rabaul calderas. Geophys. Res. Lett. 24(13), 1575–1578.

Turbeville, B.N., 1992. Relationships between chamber margin accumulates and pore liquids: evidence from arrested in situ processes in ejecta, Latera caldera, Italy. Contrib. Mineral. Petrol. 110, 429–441.

Wallace, S.R., 1974. The Henderson ore body: elements of discovery, reflections. Am. Inst. Min. Engr. Trans. 256, 216–227.

Webster, J.D., Raia, F., De Vivo, B., Rolandi, G., 2001. The behaviour of chlorine and sulfur during differentiation of the Mt. Somma Vesuvius magmatic system. Mineral. Petrol. 73, 177–200.

Williams, A.E., McKibben, M.A., 1987. A brine interface in the Salton Sea geothermal system, California: mechanism for active ore formation. Geol. Soc. Am., Abstr. prog. 19, 890.

Author Index

Subject Index